PHASE IN OPTICS

SERIES IN CONTEMPORARY CHEMICAL PHYSICS

Editor-in-Chief: M. W. Evans *(York University, Toronto)*

Associate Editors: S Jeffers *(York University, Toronto)*
D Leporini *(University of Pisa, Italy)*
J Moscicki *(Japiellonian University, Poland)*
L Pozhar *(The Ukrainian Academy of Sciences)*
S Roy *(The Indian Statistical Institute)*

World Scientific Series in Contemporary Chemical Physics – Vol. 15

PHASE IN OPTICS

Vlasta Peřinová, Antonín Lukš and Jan Peřina

Palacký University, Olomouc, Czech Republic

World Scientific
Singapore • New Jersey • London • Hong Kong

Published by

World Scientific Publishing Co. Pte. Ltd.

P O Box 128, Farrer Road, Singapore 912805

USA office: Suite 1B, 1060 Main Street, River Edge, NJ 07661

UK office: 57 Shelton Street, Covent Garden, London WC2H 9HE

Library of Congress Cataloging-in-Publication Data
Peřinová, Vlasta, 1943–
    Phase in optics / Vlasta Peřinová, Antonín Lukš, Jan Peřina.
      p.   cm. -- (World Scientific series in contemporary chemical
physics ; vol. 15)
    ISBN 981023208X
    1. Quantum optics.   2. Optical phase conjugation.   3. Phase space
(Statistical physics)   I. Lukš, Antonín, 1944– . II. Peřina,
Jan, 1936– . III. Title. IV. Series.
QC446.2.P465   1998
535--dc21

                                     98-28356
                                       CIP

British Library Cataloguing-in-Publication Data
A catalogue record for this book is available from the British Library.

Copyright © 1998 by World Scientific Publishing Co. Pte. Ltd.

All rights reserved. This book, or parts thereof, may not be reproduced in any form or by any means, electronic or mechanical, including photocopying, recording or any information storage and retrieval system now known or to be invented, without written permission from the Publisher.

For photocopying of material in this volume, please pay a copying fee through the Copyright Clearance Center, Inc., 222 Rosewood Drive, Danvers, MA 01923, USA. In this case permission to photocopy is not required from the publisher.

This book is printed on acid-free paper.

Printed in Singapore by Uto-Print

Q C
446
.2
P465
1998
PHYS

Preface

The knowledge about the wave nature of particles, as a seed of particle optics, was one of the sources of quantum theory and introduced the concept of phase factors. At high speed optics improved our understanding and use of the wave nature of light in the classical regime.

But quantum field theory as a second-quantized theory has been plagued with problems of promotion of the classical phase to the quantum one since the beginning. In this theory the quantum phase could be expected only where the harmonic oscillator would reappear.

When quantum optics came into being, the radiation modes began to be studied as quantum harmonic oscillators and the hopes started to come true. But not without corrections.

The purpose of this book is to discuss these questions. To pursue this goal we pay great attention to the optical phase in classical regime, and also present information about particle optics and about the geometric (Berry and Pancharatnam) phases. We expound the coherent state technique, which has been useful for founding quantum optics. The most important part of this book discusses typical problems of the phase of a quantum harmonic oscillator and thus those of an optical mode. We take into account the contemporary literature concerning the simultaneous measurement of the position-like and momentum-like field quadratures and the reconstruction of a quantum state.

This book may be considered a continuation of previous monographs by one of the authors on Coherence of Light (Van Nostrand Reinhold, London 1972, second edition D. Reidel, Dordrecht 1985) and on Quantum Statistics of Linear and Nonlinear Optical Phenomena (D. Reidel, Dordrecht 1984; second edition Kluwer, Dordrecht 1991), and on Quantum Optics and Fundamentals of Physics (Kluwer, Dordrecht 1994) co-authored by Z. Hradil and B. Jurčo, which also cover the prerequisites for reading this book.

The book will be useful to research workers in general optics, quantum optics and electronics, optoelectronics, and nonlinear optics, as well as to students of physics, optics, optoelectronics, photonics, and optical engineering.

The text of the book has been worked out as follows: Chapters 2, 3, and 5 have been written by J. Peřina except sections 3.1, 3.9, 3.10, 5.3, and 5.4. All the other

v

text is by V. Peřinová and A. Lukš.

We would like to thank Mgr. R. Böhm for carefully typing a part of the manuscript. We are obliged to Dr. Jan Peřina Jr. and Ing. J. Křepelka for careful preparation of figures. We express our gratitude to Professors Emil Wolf and Malvin Carl Teich for their kind help and deep interest in this work. V. P. and A. L. acknowledge the Grant Agency of Czech Republic for supporting the investigation of quantum optical interpretation of phase of the light wave and of the use of quantum phase in the analysis of optical processes and optics measurements by grant 202/93/0011. J. P. would like to thank the Grant Agency of Czech Republic and Czech Ministry of Education for supports of this research by grants 202/96/0421 and VS96028, respectively.

Olomouc, December 1997   Vlasta Peřinová
                                                  Antonín Lukš
                                                  Jan Peřina

Contents

Chapter 1

Introduction

Phase plays the central role in classical optics and is crucial for explanation of basic classical optical phenomena, such as interference and diffraction. However, it does not only play the role of fundamental optical quantity, but has also practical meaning in nontraditional imaging techniques, such as optical holography and phase conjugation. Therefore methods of measurement of phase and its retrieval methods are important directions in classical optics. The most efficient methods of the phase measurement are homodyne and heterodyne detections.

Classical picture of the phase resting on the assumption of strong optical fields, well produced by means of lasers, is drastically changed when low-intensity optical beams are considered. In this case quantum statistical properties of such beams are substantial for their complete description and phase-space methods of quantum statistical physics can be employed. Especially, the orderings of field operators have important physical meanings and the corresponding phase-space descriptions can be based on the operator generating functions, variously ordered quantum characteristic functions, resolutions of the identity, and the corresponding quasidistributions. In this connection the resolutions of the identity in terms of the coherent states have been developed by R. J. Glauber and other authors and generalizations have included the earlier Wigner function related to the symmetrical ordering of field operators. Nonclassical features of light beams, such as squeezing of vacuum fluctuations and sub-Poissonian photon statistics, in relation to quantum noise of optical beams, quantum oscillations in photocount distributions, collapses and revivals of photon number and its variance, etc., can be obtained using these methods, including radiation in interaction with matter. Also losses can be included in the framework of irreversible quantum dynamics and generation and decay of quantum macroscopic superposition states can be considered.

The difficulties with the definition of the quantum phase of extremely weak optical fields exhibit multiplicity of quantum phase concepts. The canonical quantum phase consists in a definition of quantum phase distribution, but it has also motivated many proofs of non-existence of an Hermitian phase operator. Adopting a "conservative" point of view, one may say that the quantum phase related to the antinormal ordering of field operators is confined to another definition of quantum phase distribution. In general, quantum phase concepts related to various orderings of field operators can be defined which are subjected to the same confinement. From an opposite "non-conservative" point of view any quantum phase concept is associated with an Hermitian phase operator, which is still less "well behaved". D. T. Pegg and S. M. Barnett have demonstrated which parameter should be introduced in order that one obtains a sequence of well-behaved Hermitian phase operators and regains the canonical phase distribution on letting the parameter go to infinity. Operational quantum phase concepts each emerge from an analysis of experimental scheme serving for quantum phase measurement.

All these concepts are useful in interpretation of phase-dependent measurements at quantum levels. In particular, they are appropriate for quantum interferometry which besides provides new theoretical and experimental information about photon paths and the visibility of interference fringes, about their possible restoration (optical eraser), about two-photon interference phenomena, wave–particle behaviour of a single photon, about nonlocal behaviour of quantum systems and Einstein–Rosen–Podolsky correlations and about applications of these terms to quantum cryptography, quantum communications, and quantum computing. Important applications are in quantum measurements, especially in the field of nondemolition measurements enabling to perform in principle arbitrarily accurate quantum measurement of an observable provided that quantum noise arising in the measurement is related to the conjugate observable. In this way a kind of quantum noise engineering can be founded capable to manipulate the quantum noise. Interesting ways of measuring quantum phase are provided by nonlinear optical processes, namely optical parametric processes (second-harmonic and second-subharmonic generations, parametric generation and amplification, frequency conversion), Kerr effect, four-wave mixing, phase conjugation, Raman scattering, etc. These nonlinear processes can also be used in nonlinear optical couplers composed of nonlinear waveguides, which represent effective elements for generation and transmission of nonclassical light. All these quantum features can be observed not only with photons, but also with electrons, neutrons, atoms and ions, and recently with Bose–Einstein condensates of atoms. For instance, atom interferometry provides a new tool for very precise measurements of a sensitivity surpassing that of the optical interferometers by several orders.

Many of these results have been dealt with in recent books by Vogel and Welsch (1994), Walls and Milburn (1994), Peřina, Hradil and Jurčo (1994), Mandel and Wolf (1995), and Leonhardt (1997), in review articles by Lynch (1995), Tanaś, Miranowicz and Gantsog (1996), and in topical reviews by Carruthers and Nieto (1968) and Pegg and Barnett (1997).

In chapter 2 we illustrate the role of classical phase in explanation of basic classical optical phenomena and point out its importance for the holographic and phase conjugation methods. We also provide some knowledge about phase retrieval procedures.

In chapter 3 we touch on the operator-valued measure and emphasize the importance of a positive operator-valued measure for quantum theory of measurement. We proceed with a brief review of the coherent state technique, including the coherent states of a harmonic oscillator and various resolutions of the identity, in other words, particular operator-valued measures. We introduce squeezed coherent and atomic coherent states. We mention the application of these methods to analysis of nonclassical features of light beams. We treat the irreversible quantum dynamics of radiation modes decaying into reservoirs or of those interacting with matter. We provide a picture of the theory of continuous photon-number measurement. In this chapter we cover in part the topics of quantum macroscopic superposition states and postpone the exposition of their quantum phase properties into the next chapter.

In chapter 4 we illustrate the meaning of the terms action–angle variables in classical mechanics and in the "old" quantum theory using the Bohr–Sommerfeld quantization condition. We concentrate ourselves, above all, on the case of harmonic oscillator and that of plane rotator. We expound some knowledge on geometric phase, comprising both its discovery in the adiabatic approximation of the evolution of a quantum mechanical system (Berry's phase) and in classical optics (Pancharatnam's phase). After an exposition of the quantum phase problem we contemplate the statistical treatment of angle or phase, which, of course, can be based on any solution of this problem. Then we concentrate on the idea of two optical phase operators which is inherent to the Susskind–Glogower formalism. Orderings of exponential phase operators lead to "phase" distributions inside the unit circle (disk) except the antinormal ordering that yields the expected phase distribution on the unit circle. We treat both the solution of the quantum phase problem using enlargement of the original Hilbert space of the optical mode and the solution of the quantum phase problem using restriction of the original Hilbert space. We introduce definitions of various optical phase representations and a derivation of "new" Wigner function for number and phase with respect to each solution.

In the case of two-mode light field we sense novel conditions for a quantum phase-difference definition. We devote a section to special states with extremal number–phase properties. We treat the quantum phase from phase-space quasidistributions (those of the complex amplitude or of two canonically conjugate quadratures). We expound here phase properties of real states of an optical field and pay attention to single-mode and two-mode superposition states.

We deal with the eight-port scheme for simultaneous detection of canonically conjugate quadratures and with how it is related to the quantum phase from a quasidistribution of the complex amplitude. We expound thoroughly the problem of simultaneous measurement of canonically conjugate quadratures and mention various detection techniques of the statistical operator (or state) of an optical mode.

In chapter 5 we discuss quantum interferometry, quantum nondemolition measurements, and expound the knowledge from the quantum estimation theory, returning to the quantum interferometry. We gather results on phase dependences of quantum dynamics and on time dependences of the quantum phase statistics of nonlinear optical processes. We characterize atom interferometry as an extension of optical interferometry.

Chapter 2

Phase in classical and nonlinear optics

In this chapter we expound the optical phase from the classical point of view. In the classical region there are no ambiguities in the definition of the phase, which represents a crucial notion of classical optics. The most efficient use of this notion is demonstrated in classical linear optics by phenomena of interference and diffraction of light and by holographical method of imaging, which is a fundamental optical method. The method of phase conjugation, which is no less fundamental, involves the phase in a nonlinear optical process. All these phenomena will be discussed in this chapter together with methods of phase retrieval applicable when the optical phase cannot be obtained by means of a direct measurement.

2.1 Interference of light

The necessary property of light beams for observation of interference phenomena is coherence of optical beams, representing from the point of view of quantum theory a boson cooperative phenomenon. In order to demonstrate the interference of light as closely related to phase properties we assume partially coherent light beams, which provide reduced visibility of interference pattern. For more complete description of coherence we refer the reader to books [Peřina (1985), Saleh and Teich (1991), Mandel and Wolf (1995)], where classical and quantum approaches to optical coherence and interference have been developed and their relation has been expounded.

The electromagnetic field is a real vectorial field. Considering for simplicity only one scalar component of the field at the space point $\mathbf{x}$ and at time t denoted as $V(\mathbf{x}, t)$, we can decompose this field in terms of the spectral components $\tilde{V}(\mathbf{x}, \nu)$ with the use of the Fourier integral

$$V(\mathbf{x}, t) = \int_{-\infty}^{\infty} \tilde{V}(\mathbf{x}, \nu) \exp(-i2\pi\nu t) \, d\nu, \qquad (2.1.1)$$

5

where the spectral function $\tilde{V}(\mathbf{x}, \nu)$ is defined by the inverse Fourier transformation. Similarly as for monochromatic waves it is convenient to replace the real functions $\cos(2\pi\nu t)$ and $\sin(2\pi\nu t)$ by the complex function $\exp(-i2\pi\nu t)$ for the sake of simplified calculations, we can introduce such a complex representation for polychromatic optical fields by putting

$$V(\mathbf{x}, t) = \int_0^\infty \tilde{V}(\mathbf{x}, \nu) \exp(-i2\pi\nu t)\, d\nu, \qquad (2.1.2)$$

which leads to the description of real electromagnetic field provided that the cross-symmetry condition $\tilde{V}^*(\mathbf{x}, \nu) = \tilde{V}(\mathbf{x}, -\nu)$ holds, expressing the spectral components for negative frequencies in terms of the spectral components for positive frequencies. Even if this convention represents rather a simplification of calculations in classical optics, it is fundamental in quantum optics, enabling us to distinguish between annihilation and creation of a photon.

A basic experiment to demonstrate the interference of two classical waves is the Young two-slit experiment, as illustrated in figure 2.1. Placing a quadratic detector, such as the eye, photographic plate or photodetector at the point $Q(\mathbf{x})$ of the screen, we can write for the spectral intensity detected at this point

$$I(\mathbf{x}, \nu) = \langle |\tilde{V}(\mathbf{x}, \nu)|^2 \rangle, \qquad (2.1.3)$$

where the brackets mean an ensemble average over all realizations of the spectral field and

$$\tilde{V}(\mathbf{x}, \nu) = a(\mathbf{x}, \mathbf{x}_1, \nu)\tilde{V}(\mathbf{x}_1, \nu) + a(\mathbf{x}, \mathbf{x}_2, \nu)\tilde{V}(\mathbf{x}_2, \nu), \qquad (2.1.4)$$

where $a(\mathbf{x}, \mathbf{x}_j, \nu) \equiv a_j(\nu)$, $j = 1, 2$, represent general spectral propagation factors respecting changes in propagation direction and in intensities. Thus we arrive at the two-beam spectral interference law

$$\begin{aligned} I(\mathbf{x}, \nu) &= I_1(\mathbf{x}, \nu) + I_2(\mathbf{x}, \nu) \\ &\quad + 2\sqrt{I_1(\mathbf{x}, \nu)I_2(\mathbf{x}, \nu)}\, |g(\mathbf{x}_1, \mathbf{x}_2, \nu)| \cos[\psi(\mathbf{x}_1, \mathbf{x}_2, \nu)], \end{aligned} \qquad (2.1.5)$$

where

$$I_j(\mathbf{x}, \nu) = |a_j(\nu)|^2 G(\mathbf{x}_j, \mathbf{x}_j, \nu), \quad j = 1, 2, \qquad (2.1.6)$$

are the spectral intensities contributed at the point $Q(\mathbf{x})$ from the pinholes $P_j(\mathbf{x}_j)$, $j = 1, 2$, $\psi(\mathbf{x}_1, \mathbf{x}_2, \nu)$ is the phase of $g(\mathbf{x}_1, \mathbf{x}_2, \nu)$, the spectral degree of coherence is defined as

$$g(\mathbf{x}_1, \mathbf{x}_2, \nu) \equiv g_{12}(\nu) = \frac{G(\mathbf{x}_1, \mathbf{x}_2, \nu)}{\sqrt{G(\mathbf{x}_1, \mathbf{x}_1, \nu)G(\mathbf{x}_2, \mathbf{x}_2, \nu)}}, \qquad (2.1.7)$$

and the spectral correlation function is defined as

$$G(\mathbf{x}_1, \mathbf{x}_2, \nu) = \langle \tilde{V}^*(\mathbf{x}_1, \nu)\tilde{V}(\mathbf{x}_2, \nu) \rangle. \qquad (2.1.8)$$

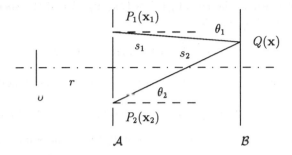

Figure 2.1: Scheme of the Young two-slit experiment for demonstration of partial coherence; σ is an extended source, $\mathcal{A}$ is a screen with two small pinholes $P_1(\mathbf{x}_1)$ and $P_2(\mathbf{x}_2)$, and $\mathcal{B}$ is a screen of observation with a typical point $Q(\mathbf{x})$; r, s_1, s_2 are the corresponding distances and θ_1, θ_2 inclination angles.

The quantities $G(\mathbf{x}_j, \mathbf{x}_j, \nu)$, $j = 1, 2$, are the spectral intensities at the slits $P_j(\mathbf{x}_j)$. The visibility of interference pattern is then determined by

$$C(\mathbf{x}, \nu) = \frac{I_{\max}(\mathbf{x}, \nu) - I_{\min}(\mathbf{x}, \nu)}{I_{\max}(\mathbf{x}, \nu) + I_{\min}(\mathbf{x}, \nu)}$$

$$= \frac{2}{\sqrt{\frac{I_1(\mathbf{x}, \nu)}{I_2(\mathbf{x}, \nu)}} + \sqrt{\frac{I_2(\mathbf{x}, \nu)}{I_1(\mathbf{x}, \nu)}}} |g(\mathbf{x}_1, \mathbf{x}_2, \nu)| \qquad (2.1.9)$$

and the spectral degree of coherence fulfills the following inequality

$$0 \leq |g(\mathbf{x}_1, \mathbf{x}_2, \nu)| \leq 1. \qquad (2.1.10)$$

If $|g| = 0$ the beams are spectrally incoherent, if $|g| = 1$ they are spectrally coherent exhibiting the maximum visibility, and if $0 < |g| < 1$ the beams are partially coherent in the spectral region.

Such cases of interference in the spectral space can be observed if the time of observation τ is much larger than the coherence time $\tau_c = 1/\Delta\nu$, where $\Delta\nu$ is the spectral halfwidth of light. On the other hand, if $\tau \ll \tau_c$ the complementary interference phenomenon can be observed in real space–time. Let us assume that there is a time delay $\tau = (s_1 - s_2)/c$, c being the velocity of light, between both the beams so that the spectral correlation function $G(\mathbf{x}_1, \mathbf{x}_2, \nu)$ is replaced by $G(\mathbf{x}_1, \mathbf{x}_2, \nu) \exp(-i2\pi\nu\tau)$ and that the propagators are approximately independent of frequency as in the quasi-monochromatic approximation, where the frequency halfwidth $\Delta\nu$ is much less than the mean frequency ν_0, so that they can be taken at the mean frequency. Then integrating over ν in (2.1.5) and using the Wiener–Khinchin theorem, which relates the function of mutual coherence $\Gamma_{12}(\tau) \equiv \Gamma(\mathbf{x}_1, \mathbf{x}_2, \tau)$ and the spectral correlation function $G_{12}(\nu) \equiv G(\mathbf{x}_1, \mathbf{x}_2, \nu)$,

$$\Gamma_{12}(\tau) = \int_0^\infty G_{12}(\nu) \exp(-i2\pi\nu\tau)\, d\nu, \qquad (2.1.11)$$

we arrive at the interference law for two partially coherent quasimonochromatic beams

$$I(\mathbf{x}) = I_1(\mathbf{x}) + I_2(\mathbf{x}) + 2\sqrt{I_1(\mathbf{x})I_2(\mathbf{x})}\, |\gamma(\mathbf{x}_1, \mathbf{x}_2, \tau)| \cos[\phi(\mathbf{x}_1, \mathbf{x}_2, \tau)], \qquad (2.1.12)$$

where the contributions from pinholes P_1 and P_2 to the space point Q are given by

$$I_j(\mathbf{x}) = |a_j(\nu_0)|^2 \Gamma(\mathbf{x}_j, \mathbf{x}_j, 0), \quad j = 1, 2, \qquad (2.1.13)$$

the degree of coherence is defined as

$$\gamma(\mathbf{x}_1, \mathbf{x}_2, \tau) \equiv \gamma_{12}(\tau) = \frac{\Gamma_{12}(\tau)}{\sqrt{\Gamma_{11}(0)\Gamma_{22}(0)}}, \qquad (2.1.14)$$

and $\phi(\mathbf{x}_1, \mathbf{x}_2, \tau)$ is its phase. We note that the mean intensities $\Gamma_{jj}(0) = \int_0^\infty G_{jj}(\nu)\, d\nu$, $j = 1, 2$, can be expressed in terms of the spectral densities $G_{jj}(\nu)$. The degree of coherence fulfills the inequality

$$0 \leq |\gamma_{12}(\tau)| \leq 1, \qquad (2.1.15)$$

in analogy to (2.1.10). The optical beams are mutually coherent in real space if $|\gamma_{12}(\tau)| = 1$, they are mutually incoherent if $|\gamma_{12}(\tau)| = 0$, and they are partially coherent if $0 < |\gamma_{12}(\tau)| < 1$. We refer the reader to further discussion of complementarity of interference in real space–time and that in spectral region to papers by Friberg and Wolf (1995) and by Agarwal (1995) (and references therein). This phenomenon can be related to nonlocal properties of quantum theory [Peřina, Hradil and Jurčo (1994)].

The visibility of fringes in real space is then expressed, similarly as in (2.1.9) for spectral region, in the form

$$
\begin{aligned}
C(\mathbf{x}, \tau) &= \frac{I_{\max}(\mathbf{x}) - I_{\min}(\mathbf{x})}{I_{\max}(\mathbf{x}) + I_{\min}(\mathbf{x})} \\
&= \frac{2}{\sqrt{\frac{I_1(\mathbf{x})}{I_2(\mathbf{x})}} + \sqrt{\frac{I_2(\mathbf{x})}{I_1(\mathbf{x})}}}\, |\gamma_{12}(\tau)|,
\end{aligned}
\qquad (2.1.16)
$$

where $I_{\max}(\mathbf{x})$ and $I_{\min}(\mathbf{x})$ are the maximum and minimum values of intensity (envelops of white and dark fringes), respectively. Complete coherence is defined in terms of the maximum visibility of interference pattern. From the point of view of classical waves the interference fringes occur provided that the interference terms are not zero, i. e., if there is a fixed difference of phases between beams. Also a simple quantum interpretation of the interference terms can be provided. Assuming for simplicity that the beams are fully coherent and defining the probabilities $p_j(\mathbf{x}) = I_j(\mathbf{x})/[I_1(\mathbf{x}) + I_2(\mathbf{x})]$, $j = 1, 2$, we see from (2.1.16) that

$$[p_1(\mathbf{x}) - p_2(\mathbf{x})]^2 = 1 - [C(\mathbf{x}, \tau)]^2, \quad p_1(\mathbf{x})p_2(\mathbf{x}) = \frac{[C(\mathbf{x}, \tau)]^2}{4}. \tag{2.1.17}$$

Hence, if it is certain that a photon belongs to the first or the second beam ($p_1(\mathbf{x}) = 1$ or $p_2(\mathbf{x}) = 1$), the visibility of interference fringes $C(\mathbf{x}, \tau) = 0$ and the interference pattern is not observable. However, if it is impossible to distinguish to which beam the photon belongs, then $p_1(\mathbf{x}) = p_2(\mathbf{x}) = 1/2$ and $C(\mathbf{x}, \tau) = 1$, i. e., the interference fringes having the maximum visibility are observed. This is related to the impossibility to distinguish the paths of photons as a consequence of the Heisenberg uncertainty principle $\Delta t \Delta E \geq h$, i. e., $\tau \Delta \nu \geq 1$, in contradiction with the condition $\tau \leq \tau_c$ for observing interference fringes, which means that only photons inside the time interval τ along the beam can interfere which cannot be distinguished. It has been shown by Mandel (1991) that the modulus of degree of coherence just equals the degree of indistinguishability of beams. Thus the modulus of the degree of coherence and the function of mutual coherence can directly be measured in such an interference experiment. The phase of the degree of coherence and that of the function of mutual coherence can in principle be determined from positional measurements.

If the spectral properties of optical beams are conserved during propagation, the beams are called to be cross-spectrally pure, which means that the temporal coherence, specified by the degree of coherence $\gamma(\mathbf{x}, \mathbf{x}, \tau)$, and spatial coherence, specified by the degree of coherence $\gamma(\mathbf{x}_1, \mathbf{x}_2, 0)$, are independent, i. e., $\gamma(\mathbf{x}_1, \mathbf{x}_2, \tau) = \gamma(\mathbf{x}_1, \mathbf{x}_1, \tau)\gamma(\mathbf{x}_1, \mathbf{x}_2, 0)$. This can be seen dividing (2.1.14) by (2.1.7), which gives

$$\frac{\gamma_{12}(\tau)}{g_{12}(\nu)} = \frac{\Gamma_{12}(\tau)}{G_{12}(\nu)}\sqrt{g_{11}(\nu)g_{22}(\nu)},$$

where $g_{jj}(\nu), j = 1, 2$, are normalized spectral functions $G_{jj}(\nu)/\Gamma_{jj}(0)$. Expressing this equation in the form

$$\gamma_{12}(\tau)\frac{G_{12}(\nu)}{\Gamma_{12}(\tau)} = g_{12}(\nu)\sqrt{g_{11}(\nu)g_{22}(\nu)}$$

and performing the Fourier transformation, we arrive at

$$\gamma_{12}(\tau) = \int_0^\infty \sqrt{g_{11}(\nu)g_{22}(\nu)}\, g_{12}(\nu)\exp(-i2\pi\nu\tau)d\nu.$$

Assuming the spectrum conservation during propagation, $g_{11}(\nu) = g_{22}(\nu) = g(\nu)$, and $g_{12}(\nu) = \gamma_{12}(\tau_0)\exp(i2\pi\nu\tau_0)$, $\gamma_{12}(\tau_0)$ being a degree of spatial coherence and τ_0 a fixed time delay, we finally arrive at the cross-spectral purity condition $\gamma_{12}(\tau) = \gamma_{12}(\tau_0)\gamma_{11}(\tau - \tau_0)$ separating spatial and temporal coherence.

Note that any partially coherent beam can be considered a mixture of coherent beams and incoherent beams, since we can write that

$$I(\mathbf{x}) = |\gamma(\mathbf{x}_1, \mathbf{x}_2, \tau)| \left\{ I_1(\mathbf{x}) + I_2(\mathbf{x}) + 2\sqrt{I_1(\mathbf{x})I_2(\mathbf{x})} \, \cos[\phi(\mathbf{x}_1, \mathbf{x}_2, \tau)] \right\}$$
$$+ [1 - |\gamma(\mathbf{x}_1, \mathbf{x}_2, \tau)|][I_1(\mathbf{x}) + I_2(\mathbf{x})], \tag{2.1.18}$$

where the first term represents the interference law for two completely coherent beams and the second term sums the partial intensities of incoherent beams.

For a more general treatment involving also nonstationary fields, e. g., light pulses, we can use the second-order correlation function

$$\Gamma(x_1, x_2) = \langle V^*(x_1)V(x_2) \rangle, \tag{2.1.19}$$

where the brackets mean the ensemble average over all field realizations, $V(x)$ represents a scalar component of the optical field and $x \equiv (\mathbf{x}, t)$ denotes a space–time point.

The maximum visibility of the interference pattern for the whole field requires now the equality

$$|\Gamma(x_1, x_2)|^2 = \Gamma(x_1, x_1)\Gamma(x_2, x_2). \tag{2.1.20}$$

Writing this factorization in the form

$$\Gamma(x_1, x_2) = A(x_1)B(x_2) = \Gamma^*(x_2, x_1) = A^*(x_2)B^*(x_1), \tag{2.1.21}$$

we see that a real constant $K = A(x_1)/B^*(x_1) = A^*(x_2)/B(x_2)$ exists. Therefore $A(x) = K B^*(x)$ and we can introduce the field $V(x) = \sqrt{K} \, B(x)$, which means that the necessary and sufficient condition for the second-order coherence of the field is the factorization of the correlation function,

$$\Gamma(x_1, x_2) = V^*(x_1)V(x_2). \tag{2.1.22}$$

Assuming optical fields from natural sources, we may, more generally, write the correlation function in terms of its eigenvalues λ_j and eigenfunctions $\psi_j(x)$ as follows

$$\Gamma(x_1, x_2) = \sum_j \lambda_j \psi_j^*(x_1)\psi_j(x_2), \tag{2.1.23}$$

where the eigenfunctions and eigenvalues fulfill the Fredholm integral equation

$$\lambda_j \psi_j(x) = \int \Gamma^*(x, x')\psi_j(x') \, d^4x'. \tag{2.1.24}$$

This interpretation can be generalized if we use higher-order correlations, adopting the set of higher-order correlation functions

$$\Gamma^{(m,n)}(x_1, ..., x_{m+n}) = \langle V^*(x_1) \ldots V^*(x_m)V(x_{m+1}) \ldots V(x_{m+n}) \rangle, \tag{2.1.25}$$

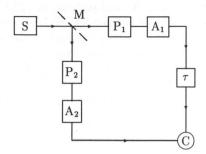

Figure 2.2: Scheme for measurements of Hanbury Brown–Twiss correlations between intensity fluctuations; S is a natural source, M is a splitter, P_1 and P_2 are photomultipliers, A_1 and A_2 are amplifiers, C is an electric correlator or a coincidence device, τ is a time delay element.

involving n complex field amplitudes V and m complex conjugate amplitudes V^*. For $m = n$ such correlation functions can be measured by a set of n photodetectors [e. g., Peřina (1985)] and, in particular, by using two photodetectors the Hanbury Brown–Twiss correlations can be measured (figure 2.2),

$$
\begin{aligned}
\Gamma^{(2,2)}(\mathbf{x}_1, \mathbf{x}_2, \mathbf{x}_2, \mathbf{x}_1, t, t+\tau, t+\tau, t) &= \langle I_1(t) I_2(t+\tau)\rangle \\
&= \langle I_1\rangle\langle I_2\rangle + \langle \Delta I_1(t)\Delta I_2(t+\tau)\rangle \\
&= \langle I_1\rangle\langle I_2\rangle[1 + |\xi_{12}(\tau)|^2], \quad (2.1.26)
\end{aligned}
$$

where the normalized fourth-order correlations of fluctuations are expressed as

$$
|\xi_{12}(\tau)|^2 = \frac{\langle \Delta I_1(t)\Delta I_2(t+\tau)\rangle}{\langle I_1\rangle\langle I_2\rangle}, \quad (2.1.27)
$$

provided that light beams are stationary, $\tau = t_2 - t_1$, and $\langle I_j\rangle = \Gamma_{jj}(0)$, $j = 1, 2$. For chaotic light of natural sources $|\xi_{12}(\tau)| = |\gamma_{12}(\tau)|$.

Now we can assume two light beams, with the wave vectors $\mathbf{k}_1$ and $\mathbf{k}_2$ and circular frequencies ω_1 and ω_2, respectively, having mutually random phases so that no observable interference effect of the second order can be obtained. However, using the above intensity correlations, we can obtain the fourth-order interference and beats as follows

$$
\begin{aligned}
\Gamma^{(2,2)}(x_1, x_2, x_2, x_1) &= \langle I(x_1)I(x_2)\rangle = \langle [I_1(x_1) + I_2(x_2)]^2\rangle + 2\langle I_1(x_1)\rangle\langle I_2(x_1)\rangle \\
&\times \cos[(\omega_1 - \omega_2)(t_1 - t_2) + (\mathbf{k}_1 - \mathbf{k}_2)\cdot(\mathbf{x}_2 - \mathbf{x}_1)], \quad (2.1.28)
\end{aligned}
$$

because some random phases can be mutually cancelled in the fourth-order moments creating the interference fringes (in space) and beats (in time) in (2.1.28). We have further assumed that $|t_1 - t_2| \ll 1/\Delta\nu$ and $|\mathbf{x}_1 - \mathbf{x}_2| \ll 1/\Delta k$, Δk being the spread of wave vectors. Hence, although the independent phase-averaged coherent fields cannot produce second-order interference, they can produce the fourth-order interference, thus enabling us, for instance, to observe interferences of optical fields from independent sources. Thus the phase plays a different role in these higher-order phenomena than in the second-order ones.

Finally we mention some results demonstrating interesting phase effects in optical fibres in the spatio–temporal and spectral domains [Hlubina (1993, 1995, 1997) and references therein].

2.2   Diffraction of light

Another nice demonstration of the meaning of the classical phase is the effect of diffraction of light, which may be considered a special case of scattering of light. The diffraction can be understood as a deviation of direction of propagation of light caused by its interaction with obstacles, such as apertures in screens, edges, and various object boundaries. When the resulting macroscopic field is treated as arising by superposition of many elementary waves from apertures with defined phase relations, this phenomenon can be reduced to the phenomenon of interference. Various propagation-invariant beams can be considered to which the superposition principle and interference phenomenon can be applied again. We will restrict our considerations to scalar theory of diffraction, which will be sufficient for demonstration of effects of classical phase.

Considering the light propagation in free space (figure 2.3) for simplicity, we could start from the wave equation for the complex amplitude of a wave field $V(\mathbf{x}, t) = U(\mathbf{x})\exp(-i2\pi\nu t)$,

$$\Delta V(\mathbf{x}, t) - \frac{1}{c^2}\frac{\partial^2 V(\mathbf{x}, t)}{\partial t^2} = 0, \qquad (2.2.1)$$

provided the field is fully coherent, so that the complex amplitude is sufficient to be examined. Here Δ means the Laplace operator with respect to the spatial coordinates. For the monochromatic field the problem is thus reduced to the boundary problem for the Helmholtz equation for the space amplitude $U(\mathbf{x})$,

$$\Delta U(\mathbf{x}) + k^2 U(\mathbf{x}) = 0, \qquad (2.2.2)$$

where $k = 2\pi\nu/c$ is the wave number. This is the usual way of treating diffraction problems in standard books and monographs [see, e. g., Nieto-Vesperinas (1991)]. In this section we will provide more general solution based on propagation of partial coherence [Born and Wolf (1959), Beran and Parrent (1964), Peřina (1972), Mandel and Wolf (1995)]. In this case we have two wave equations for the function of partial coherence as follows

$$\Delta_j \Gamma_{12}(\tau) - \frac{1}{c^2}\frac{\partial^2 \Gamma_{12}(\tau)}{\partial\tau^2} = 0, \quad j = 1, 2, \qquad (2.2.3)$$

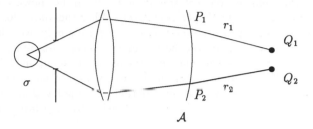

Figure 2.3: Propagation law for partial coherence; σ is a primary or secondary polychromatic extended source, light propagates from a wavefront $\mathcal{A}$ with typical points P_1 and P_2 to points Q_1 and Q_2 outside the wavefront; r_1, r_2 are the corresponding distances.

where the Laplace operator acts with respect to the first or the second set of spatial coordinates. Applying the Wiener–Khinchin theorem (2.1.11), we arrive at two Helmholtz equations for the spectral correlation function

$$\Delta_j G_{12}(\nu) + k^2 G_{12}(\nu) = 0, \quad j = 1, 2. \tag{2.2.4}$$

The diffraction of light can now be solved as a boundary problem of these partial differential equations involving partial coherence.

The standard technique of solving this problem is to use Green's formula. Assuming a closed boundary surface S of a volume V free of sources, the solution for the coherent field of the Helmholtz equation (2.2.2) inside V can be expressed in the Helmholtz–Kirchhoff integral form

$$U(Q) = \frac{1}{4\pi} \int_S \left[\frac{\partial U(P)}{\partial \mathbf{n}} \mathcal{G}(Q, P) - U(P) \frac{\partial \mathcal{G}(Q, P)}{\partial \mathbf{n}} \right] dP, \tag{2.2.5}$$

where $\mathcal{G}(Q, P)$ is the Green function fulfilling the following equation

$$(\Delta + k^2) \mathcal{G}(Q, P) = -\delta(Q - P), \tag{2.2.6}$$

where δ is the Dirac delta function, with the boundary condition

$$\mathcal{G}(Q, P')|_{P' \equiv P} = 0, \quad P \in S; \tag{2.2.7}$$

here Q is a point inside the volume V, P is a typical point on the surface S, the expression $\partial/\partial \mathbf{n}$ represents the derivative along the normal $\mathbf{n}$ at the point P and the Green function $\mathcal{G}(Q, P)$ in vacuum is represented by the outgoing spherical wave $\exp(ik|Q - P|)/4\pi|Q - P|$.

Usually it is assumed that Sommerfeld's radiation condition

$$\lim_{r \to \infty} r \left(\frac{\partial U}{\partial r} - ikU \right) = 0 \tag{2.2.8}$$

is fulfilled by the complex coherent amplitude U. It can be shown that the solution (2.2.5) is unique [e. g., Nieto-Vesperinas (1991)]. When the diffraction by aperture $\mathcal{A}$ is considered, we can express the solution by integrating only over the plane of this aperture, because the integral over the boundary surface with dimensions increasing to infinity tends to zero as a consequence of the radiation condition (2.2.8). This is the so-called Kirchhoff approximation or the approximation of physical optics. Finally we can consider that the dominant contribution to the point Q arises only from the aperture $\mathcal{A}$, which leads to the following Kirchhoff diffraction formula,

$$U(Q) = \frac{1}{4\pi} \int_{\mathcal{A}} \left[\frac{\partial U_i}{\partial \mathbf{n}} \mathcal{G} - U_i \frac{\partial \mathcal{G}}{\partial \mathbf{n}} \right] dP, \qquad (2.2.9)$$

where U_i is the complex amplitude of the incident coherent beam. Although this approximate solution can be criticized from a rigorous point of view, for instance, this solution cannot be analytic everywhere, it is a good solution for physical applications (for further discussion of this point we refer the reader to the book by Nieto-Vesperinas (1991)). It was shown by Kottler (1965) that if U satisfies the homogeneous Helmholtz equation (2.2.2) in V except the surface constituting the screen, then Kirchhoff's diffraction formula (2.2.9) represents the rigorous solution of the problem of diffraction by the screen. Also other solutions of this inconsistency can be suggested (see the above book by Nieto-Vesperinas).

In fact the problem in (2.2.9) is overspecified and it is sufficient to give only the values of U and the values of derivative $\partial U / \partial \mathbf{n}$ are not needed. Thus choosing the Green function at the aperture to be zero and considering partially coherent fields in general, we obtain the propagation law for the spectral correlation function

$$G(Q_1, Q_2, \nu) = \int_{\mathcal{A}} \int_{\mathcal{A}} \frac{\partial \mathcal{G}_1(Q_1, P_1, \nu)}{\partial \mathbf{n}_1} \frac{\partial \mathcal{G}_2(Q_2, P_2, \nu)}{\partial \mathbf{n}_2} G(P_1, P_2, \nu) \, dP_1 \, dP_2, \qquad (2.2.10)$$

where $\partial / \partial \mathbf{n}_j$ denotes the derivative with respect to the normal to the surface $\mathcal{A}$ at the point P_j, $j = 1, 2$. From $G^*(Q_1, Q_2, \nu) = G(Q_2, Q_1, \nu)$ it follows that $\mathcal{G}_1 = \mathcal{G}_2^*$, and defining the diffraction function $K(Q, P, \nu) \equiv \partial \mathcal{G}(Q, P, \nu) / \partial \mathbf{n}$ (the complex amplitude at Q caused by a unit point source of frequency ν situated at P), we arrive at the general propagation law of partial coherence

$$G(Q_1, Q_2, \nu) = \int_{\mathcal{A}} \int_{\mathcal{A}} K^*(Q_1, P_1, \nu) K(Q_2, P_2, \nu) G(P_1, P_2, \nu) \, dP_1 \, dP_2. \qquad (2.2.11)$$

The power spectrum $G(Q, \nu) \equiv G(Q, Q, \nu)$ is then obtained as a particular case; it is important to mention that this quantity is not determined by its values over the surface $\mathcal{A}$, but those of the spectral correlation function $G(P_1, P_2, \nu)$ have to be available over $\mathcal{A}$.

In free space the Green functions can be chosen in the form

$$\mathcal{G}_j = \frac{\exp(ikr_j)}{4\pi r_j} - \frac{\exp(ikr_j'')}{4\pi r_j''}, \quad j = 1, 2, \qquad (2.2.12)$$

where we have used the well-known method of images, extensively used in electro-statics, so that

$$
\begin{aligned}
r_j &= \sqrt{(x_j - x_j')^2 + (y_j - y_j')^2 + (z_j - z_j')^2}, \\
r_j'' &= \sqrt{(x_j - x_j')^2 + (y_j - y_j')^2 + (z_j + z_j')^2};
\end{aligned}
\tag{2.2.13}
$$

here (x_j, y_j, z_j) are the coordinates of the point Q_j and (x_j', y_j', z_j') are the coordinates of the point P_j on the surface $\mathcal{A}$. Performing the derivative of the Green functions with respect to the normal, we obtain

$$
\frac{\partial \mathcal{G}_j}{\partial n_j} = 2 \left(ik - \frac{1}{r_j} \right) \frac{\exp(ikr_j)}{4\pi r_j} \cos(\theta_j),
\tag{2.2.14}
$$

because

$$
\frac{\partial r_j}{\partial n_j} = -\frac{\partial r_j''}{\partial n_j} = -\left. \frac{\partial r_j}{\partial z_j'} \right|_A = \left. \frac{\partial r_j''}{\partial z_j'} \right|_A = \left. \frac{z_j}{r_j} \right|_A = \cos(\theta_j),
\tag{2.2.15}
$$

where θ_j is the inclination angle of light at the point P_j. Substituting (2.2.14) into the propagation law (2.2.10) for the spectral correlation function, we arrive at

$$
\begin{aligned}
G(Q_1, Q_2, \nu) = {} & \frac{1}{4\pi^2} \int_A \int_A (1 + ikr_1)(1 - ikr_2) \frac{\exp[-ik(r_1 - r_2)]}{r_1^2 r_2^2} \\
& \times \cos(\theta_1) \cos(\theta_2) G(P_1, P_2, \nu)\, dP_1\, dP_2.
\end{aligned}
\tag{2.2.16}
$$

Using the Wiener–Khinchin theorem (2.1.11), we can derive the function of mutual coherence arising as a result of general propagation

$$
\Gamma(Q_1, Q_2, \tau) = \frac{1}{4\pi^2} \int_A \int_A \frac{\cos(\theta_1)\cos(\theta_2)}{r_1^2 r_2^2} \mathcal{D}\Gamma \left(P_1, P_2, \tau + \frac{r_1 - r_2}{c} \right) dP_1\, dP_2,
\tag{2.2.17}
$$

where the operator $\mathcal{D}$ is expressed in the form

$$
\mathcal{D} = 1 - \frac{r_1 - r_2}{c} \frac{\partial}{\partial \tau} - \frac{r_1 r_2}{c^2} \frac{\partial^2}{\partial \tau^2}.
\tag{2.2.18}
$$

Equation (2.2.17) represents the general propagation law for mutual coherence of an optical field produced by a plane polychromatic partially coherent primary or secondary source.

It usually holds for the wave number that

$$
k = \frac{2\pi}{\lambda} \gg \frac{1}{r},
\tag{2.2.19}
$$

which provides the following simplified propagation law corresponding to the use of the Huygens–Fresnel principle,

$$
\Gamma(Q_1, Q_2, \tau) = \frac{1}{4\pi^2} \int_A \int_A \frac{\Gamma \left(P_1, P_2, \tau + \frac{r_1 - r_2}{c} \right)}{r_1 r_2} \bar{\Lambda}_1^* \bar{\Lambda}_2\, dP_1\, dP_2,
\tag{2.2.20}
$$

where $\Lambda_j = ik\cos(\theta_j)$ are the inclination factors describing the change of direction of secondarily emitted light on $\mathcal{A}$ and $\bar{\Lambda}_j$ are the values of the inclination factors at the mean frequency, respecting smoothed dependence of the inclination factors on frequency. Of course, under the assumption (2.2.19) we also can have the exact solution as follows

$$
\begin{aligned}
\Gamma(Q_1, Q_2, \tau) = & -\frac{1}{4\pi^2 c^2} \int_{\mathcal{A}} \int_{\mathcal{A}} \frac{\cos(\theta_1)\cos(\theta_2)}{r_1 r_2} \\
& \times \frac{\partial^2}{\partial \tau^2}\Gamma\left(P_1, P_2, \tau + \frac{r_1 - r_2}{c}\right) dP_1 \, dP_2.
\end{aligned} \tag{2.2.21}
$$

The mean intensity at the point Q is then expressed, on introducing the degree of coherence γ, as

$$
I(Q) = \int_{\mathcal{A}} \int_{\mathcal{A}} \frac{\sqrt{I(P_1)I(P_2)}}{4\pi^2 r_1 r_2} \gamma\left(P_1, P_2, \frac{r_1 - r_2}{c}\right) \bar{\Lambda}_1^* \bar{\Lambda}_2 \, dP_1 \, dP_2. \tag{2.2.22}
$$

We can point out that only measurable quantities occur in these propagation laws, compared to directly non-measurable vectors of the Maxwell theory or coherent complex amplitudes. Applying the propagation law (2.2.21) to two-slit screen of the Young experiment, we obtain the interference law (2.1.12) for two partially coherent beams.

These propagation laws have to be used in general cases including laser sources, where the correlation state on the source area is to be prescribed. On the other hand, we can consider propagation of light from the area of a primary natural source exhibiting spatial incoherence in this area, expressed by the conditions

$$
\begin{aligned}
G(P_1, P_2, \nu) &= G(P_1, \nu)\delta(P_1 - P_2), \\
\Gamma(P_1, P_2, \tau) &= \Gamma(P_1, P_1, \tau)\delta(P_1 - P_2).
\end{aligned} \tag{2.2.23}
$$

Now the propagation law (2.2.20) has the form

$$
\begin{aligned}
\Gamma(Q_1, Q_2, \tau) = & \frac{1}{4\pi^2} \int_0^\infty \exp(-i2\pi\nu\tau) \\
& \times \int_\sigma G(P, \nu) \frac{\exp[-ik(r_1 - r_2)]}{r_1 r_2} \bar{\Lambda}_1^* \bar{\Lambda}_2 \, dP \, d\nu,
\end{aligned} \tag{2.2.24}
$$

where the source has been denoted by σ and the mean intensities for calculation of the degree of coherence are given by

$$
\Gamma(Q_j, Q_j, 0) \equiv I(Q_j) = \frac{1}{4\pi^2} \int_\sigma \frac{I(P)}{r_j^2} |\bar{\Lambda}_j|^2 \, dP. \tag{2.2.25}
$$

It is often the case that the overall sum of time delay and path difference with respect to the light velocity c is small compared with the coherence time τ_c, i. e.,

$$
\left| \tau + \frac{r_1 - r_2}{c} \right| \ll \frac{1}{\Delta\nu}, \tag{2.2.26}
$$

which means that the exponential functions in (2.2.24) can be considered at a mean frequency $\bar{\nu}$ only and we thus arrive at

$$
\begin{aligned}
\Gamma(Q_1, Q_2, \tau) &= \sqrt{I(Q_1)I(Q_2)}\,\gamma(Q_1, Q_2, \tau) \\
&= \left(\frac{\bar{k}}{2\pi}\right)^2 \exp(-i2\pi\bar{\nu}\tau) \int_\sigma I(P)\frac{\exp[-i\bar{k}(r_1 - r_2)]}{r_1 r_2}\,dP, \quad (2.2.27)
\end{aligned}
$$

where we consider only small transversal dimensions compared to longitudinal ones, so that $\cos(\theta_j) \approx 1$. Hence the degree of coherence

$$
\gamma(Q_1, Q_2, \tau) \equiv \gamma_{12}(\tau) = \gamma_{12}(0)\exp(-i2\pi\bar{\nu}\tau) \quad (2.2.28)
$$

exhibits full temporal coherence, whereas the spatial coherence is given by the degree of spatial coherence

$$
\gamma_{12}(0) \equiv \gamma_{12} = \frac{1}{\sqrt{I(Q_1)I(Q_2)}} \int_\sigma I(P)\frac{\exp[-i\bar{k}(r_1 - r_2)]}{r_1 r_2}\,dP, \quad (2.2.29)
$$

where

$$
I(Q_j) = \int_\sigma \frac{I(P)}{r_j^2}\,dP, \quad j = 1, 2. \quad (2.2.30)
$$

This is the van Cittert–Zernike theorem which enables us to calculate, under the above assumptions, the degree of coherence of an extended quasimonochromatic spatially incoherent (primary) source in terms of the intensity distribution over the source. It says that the degree of spatial coherence of light from this source is equal to the normalized complex amplitude at the point Q_2 obtained by the diffraction of a spherical wave centred at the point Q_1, at an aperture of the form σ with the real amplitude numerically equal to the source intensity. This is a generalization of the standard diffraction problem assuming coherent complex field amplitudes, respecting states of partial coherence.

　　As a result of the effect of optical phase matching we can observe that an incoherent source gives rise to partially coherent field, which means that multiple propagation and diffraction can improve the degree of coherence of light. This can be expressed in terms of the entropy of light beams. Using the eigenvalues of the correlation function (2.1.23), we can define the probability

$$
p_n = \frac{\lambda_n}{\sum_j \lambda_j} \quad (2.2.31)
$$

that a photon belongs to the nth coherent elementary beam. Then the entropy of a light beam is defined as

$$
S = -\sum_n p_n \ln p_n \quad (2.2.32)
$$

and it becomes $S = 0$ for coherent beams and $S = \ln N$ for N incoherent beams ($p_n = 1/N$ for $n = 1, ..., N$, $p_n = 0$ for $n = N+1, ...$). Thus the multiple propagation and diffraction cause an increase of entropy. However, this is not in contradiction with the second law of thermodynamics, because the system is open and the decrease of entropy is followed by the decrease of energy of the system, as a consequence of light scattering and other losses.

Further we can conclude from the above results that even if, in general, temporal coherence and spatial coherence cannot be separated and are related by the wave equation for the function of partial coherence, they can be separated under the condition

$$\frac{|r_1 - r_2|}{c} \ll \frac{1}{\Delta\nu}, \tag{2.2.33}$$

leading to the condition of cross-spectrally pure field

$$\gamma(Q_1, Q_2, \tau) = \gamma(\tau)\gamma(Q_1, Q_2, 0), \tag{2.2.34}$$

where

$$\gamma(\tau) = \frac{\Gamma(\tau)}{\Gamma(0)}, \quad \gamma(Q_1, Q_2, 0) = \frac{\Gamma(Q_1, Q_2, 0)}{\sqrt{\Gamma(Q_1, Q_1, 0)\Gamma(Q_2, Q_2, 0)}}, \tag{2.2.35}$$

and

$$\Gamma(\tau) = \int_0^\infty G(\nu)\exp(-i2\pi\nu)\,d\nu,$$

$$\Gamma(Q_1, Q_2, 0) = \frac{1}{4\pi^2}\int_\sigma \frac{\exp[-i\bar{k}(r_1 - r_2)]}{r_1 r_2}\,dP, \tag{2.2.36}$$

on assuming independence of the spectral density $G(P, \nu)$ on P. In this case temporal coherence is connected with the spectral properties of light beams and spatial coherence is connected with geometrical properties of the optical system.

If quasimonochromatic light from a laser source is considered instead of a primary natural source, we can introduce the mutual intensity $\Gamma(Q_1, Q_2) \equiv \Gamma(Q_1, Q_2, 0)$ and we obtain the propagation law

$$\Gamma(Q_1, Q_2) = \sqrt{I(Q_1)I(Q_2)}\,\gamma(Q_1, Q_2)$$

$$= \left(\frac{\bar{k}}{2\pi}\right)^2 \int_\sigma\int_\sigma \frac{\exp[-i\bar{k}(r_1 - r_2)]}{r_1 r_2}\Gamma(P_1, P_2)\,dP_1\,dP_2. \tag{2.2.37}$$

Making use of the coordinates of the points $Q(x_j, y_j)$, $j = 1, 2$, and $P(\xi, \eta)$, we obtain, in the Fraunhofer approximation of far field, the degree of spatial coherence in the form of the normalized Fourier transformation of the intensity distribution over the source

$$\gamma(Q_1, Q_2) = \exp(i\psi)\frac{\int_\sigma I(\xi, \eta)\exp[i\bar{k}(p\xi + q\eta)]d\xi d\eta}{\int_\sigma I(\xi, \eta)\,d\xi\,d\eta}, \tag{2.2.38}$$

where we have replaced $r_1 r_2$ by the squared distance r^2 between the plane of the source and the screen containing the points Q_1 and Q_2 and the normalized coordinates p, q and the phase factor ψ are defined as

$$p = \frac{x_1 - x_2}{r}, \quad q = \frac{y_1 - y_2}{r}, \quad \psi = -\frac{\bar{k}[(x_1^2 + y_1^2) - (x_2^2 + y_2^2)]}{2r}. \tag{2.2.39}$$

Applying this theorem to a central, quasimonochromatic, uniform, spatially incoherent, circular source of the radius ρ, we obtain

$$\gamma(Q_1, Q_2) = \frac{2J_1(v)}{v} \exp(i\psi), \tag{2.2.40}$$

where $v = \bar{k}\rho(p^2 + q^2)^{1/2}$ and $J_1(v)$ is the Bessel function. For the same source of the rectangular form we obtain similarly

$$\gamma(Q_1, Q_2) = \frac{\sin(\bar{k}pa)}{\bar{k}pa}\frac{\sin(\bar{k}qb)}{\bar{k}qb} \exp(i\psi). \tag{2.2.41}$$

Equation (2.2.41) can serve as a basis for definition of the coherence area A_c. Using the first zero of the function $\sin x$, we have $\bar{k}pa = \pi$, $\bar{k}qb = \pi$ and for the coherence area we obtain $A_c = (r\bar{\lambda})^2/S$ where $S = 4ab$ is the area of the rectangular source. Similarly we can define the coherence area for the circular source.

Thus we have demonstrated using both the phenomena of optical interference and diffraction that the mutual classical phase relations can result in modulations of intensities at the point of observation occurring as a consequence of constructive and destructive interaction of field amplitudes. These amplitudes fully develop provided that the field is fully coherent, whereas they partially develop for partially coherent fields. For these demonstrations the scalar formulation of the theory is sufficient. However, more general formulations based on the vector theory, involving effects of polarization of light, can be developed [Nieto-Vesperinas (1991), Mandel and Wolf (1995)].

An interesting problem deals with propagation-invariant (self-imaging) properties of optical beams [Durnin (1987), Turunen, Vasara and Friberg (1991), Bouchal (1993), Bouchal and Olivík (1995), Kowarz and Agarwal (1995), and references therein]. This can be formulated along the beam by the condition

$$G(x_1, y_1, z, x_2, y_2, z, \nu) = G(x_1, y_1, 0, x_2, y_2, 0, \nu), \tag{2.2.42}$$

which expresses the concept of nonmodified propagation of a beam including partial coherence. This condition requires that spectral correlation function and, consequently spectral intensity function, is the same in every plane perpendicular to the propagation direction (assumed to be along the positive-z axis). The simplest partially coherent propagation-invariant field can be expressed in the Bessel form

$$G(\mathbf{x}_1, \mathbf{x}_2, \nu) = \exp[i\beta(z_1 - z_2)]J_0\left(\alpha\sqrt{(x_1 - x_2)^2 + (y_1 - y_2)^2}\right), \tag{2.2.43}$$

where for numbers α and β it holds that $\alpha^2 + \beta^2 = k^2$. Then optical fields with a longitudinal periodicity may be considered as self-imaging. The self-imaging property with partially coherent light can be defined as follows

$$G(x_1, y_1, z + d, x_2, y_2, z + d, \nu) = G(x_1, y_1, z, x_2, y_2, z, \nu), \qquad (2.2.44)$$

which implies that the spectral correlation function is the same at any two planes separated by a distance d. Also more general definitions of self-imaging property can be given [Lohmann, Ojeda-Castaneda and Streibl (1983)]. This definition leads to conclusion that the self-imaging coherent fields are superpositions of propagation-invariant fields expressed in the form

$$U(\mathbf{x}, \nu) = \exp(i\beta z) \int_0^{2\pi} F(\theta) \exp[i\alpha(x \cos\theta + y \sin\theta)] \, d\theta, \qquad (2.2.45)$$

where the form-factor $F(\theta) = 1/2\pi$ for the fundamental Bessel beam

$$U(\mathbf{x}, \nu) = \exp(i\beta z) J_0[\alpha(x^2 + y^2)]. \qquad (2.2.46)$$

Also in these cases the universal superposition principle valid for linear fields provides phase relations leading to interesting effects observable in macroscopic electromagnetic fields.

More detail about the diffraction of light can be found, for instance, in books [Nieto-Vesperinas (1991), Solimeno, Crosignani and Di Porto (1986), Mandel and Wolf (1995)].

2.3   Holography

Nice application of influence of classical optical phase on optical imaging is optical holography discovered by Gabor in 1948 [Gabor (1948, 1949, 1951)], originally in the field of electron microscopy. The method represents a two-stage imaging involving a coherent reference beam superimposed, before detection is realized by means of a photographic plate, on the signal beam coming from the object. In this way an interference pattern is created on the photographic plate, which can serve as diffraction grating during the reconstruction of the original object and the conjugated one, using again a coherent reference wave. Thus with respect to the phase of the reference wave, the reconstructed object includes the phase information and it represents a spatial object in three dimensions. Making use of the holographic principle on lower intensity levels, combination of the object and the conjugated object was also suggested for generation of squeezed light [Liu and Chen (1995)], however this suggestion was criticized [Shapiro (1996)]. If the holographic principle is adopted mixing a signal wave with a coherent reference wave on a beamsplitter and if both the outcomes from the beamsplitter are detected by photodetectors and then their outcomes are subtracted,

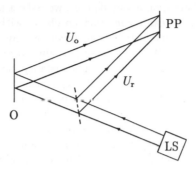

Figure 2.4: Holographic principle; LS is a laser source of high spatial coherence, O is an object to be imaged, U_o and U_r are object and reference complex field amplitudes, respectively, the interference pattern is registered by a photographic plate PP.

we obtain the basis for homodyne detection scheme (section 4.12), which is important for the detection of nonclassical (squeezed) light. The important assumption for holography is a large spatial coherence (coherence area) of the light source.

We can demonstrate the holographic principle (figure 2.4) assuming a scalar coherent wave U_o coming to a photographic plate PP from an object O and making the superposition of this signal beam with a complex scalar field U_r of a reference beam; in this way we obtain the interference pattern on the plate called the hologram as follows,

$$I_{\text{plate}} = |U_o + U_r|^2 = I_o + I_r + U_o U_r^* + U_o^* U_r$$
$$= I_o + I_r + 2|U_o||U_r| \cos(\phi_o - \phi_r), \qquad (2.3.1)$$

where the object and reference beam intensities are given as $I_o = |U_o|^2$ and $I_r = |U_r|^2$, respectively, and ϕ_o and ϕ_r are the corresponding phases of the object and reference waves.

This is the first step of holographic imaging. The second step of this imaging is the reconstruction of the object. It is performed by means of the reference wave. Illuminating the hologram by the reference wave, we obtain from (2.3.1)

$$U_r I_{\text{plate}} = U_r(I_o + I_r) + U_o |U_r|^2 + U_o^* U_r^2, \qquad (2.3.2)$$

which means that the object has been reconstructed. The first term in (2.3.2) represents unimportant background, the second term represents just the reconstructed object provided that the reference intensity $|U_r|^2$ is uniform, and the last term represents the complex-conjugated image, which is spatially separated from the direct image. We see from (2.3.1) and (2.3.2) that the object is registered and reconstructed including the phase information about it, i. e., we have a faithful three-dimensional copy. Thus in this way one can reconstruct an object with high degree of accuracy from a record of the intensity distribution, taken on any plane behind the object. Holography is therefore a method of imaging capable to reconstruct wavefronts.

There are various types of holograms. For instance, using a lens in between the object structure and the photographic plate registering the hologram under the assumption of larger distances to justify the Fraunhofer approximation, we can obtain the Fourier holograms using the same principle. The reconstructed wave is then represented by the inverse Fourier transformation realized again by a lens. Using the Fourier holography, we can construct the so-called Vander Lugt filters important for pattern recognition. It is sufficient for this to use an object structure $f(x,y)$ for preparation of the Fourier hologram, which is reconstructed using the Fourier transformation of another structure $h(x,y)$ performed by a lens. The reconstructed image is the convolution of both the structures, which may be used as a holographic spatial filter for pattern recognition. Very important class of holograms is formed by volume holograms enabling us to use white light for reconstruction instead of monochromatic light. Consider the simplest case that the object and reference waves are plane waves with wave vectors $\mathbf{k}_o$ and $\mathbf{k}_r$, respectively. Then the intensity registered in the hologram is

$$I_{\text{plate}} = I_o + I_r + 2\sqrt{I_o I_r}\,\cos[(\mathbf{k}_o - \mathbf{k}_r)\cdot \mathbf{x}]. \qquad (2.3.3)$$

In this way a sinusoidal spatial pattern is created having the period $\Lambda = 2\pi/|\mathbf{k}_o - \mathbf{k}_r|$. If the reference wave propagates along the z axis and the object wave makes an angle θ with the z axis, then $|\mathbf{k}_o - \mathbf{k}_r| = 2k\sin(\theta/2)$, $k = 2\pi/\lambda$, and the period is

$$\Lambda = \frac{\lambda}{2\sin(\frac{\theta}{2})}. \qquad (2.3.4)$$

In the emulsion the interference pattern forms a thick diffraction grating—a volume hologram. When illuminating this hologram by a reference wave, the parallel planes of the grating reflect the wave only when the Bragg condition $\sin\phi = \lambda/2\Lambda$ is satisfied, where ϕ is the angle between the planes of the grating and the incident reference wave. The object wave will be reconstructed only if the wavelength of the reconstruction source coincides with the wavelength of the recording source. If white light is used for the reconstruction, only the correct wavelength will be reflected and the reconstruction with it will be successful. Of course, the recording process must be done with monochromatic light. The possibility to use white light for reconstruction provides great advantage in many applications of holography.

The above considerations clearly show the role of the classical field phase (more accurately of difference of phases) in the holographic imaging, which leads to conservation of complete information about the object. Such a demonstration is not so simple if partially coherent light beams are involved [DeVelis and Reynolds (1967), Soroko (1971)]. Nevertheless the principal conclusions are the same, only more general interference law (2.1.12) must be adopted for the description of the interference

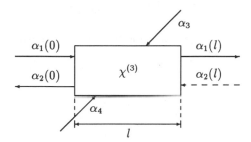

Figure 2.5: Four-wave mixing of counterpropagating waves; a nonlinear crystal characterized by the cubic susceptibility $\chi^{(3)}$ is used, α_3, α_4 are complex amplitudes of strong pumping beams, α_1, α_2 are complex amplitudes of signal beams, l is length of the crystal; the process of phase conjugation can be spontaneous or stimulated.

of the object and reference waves and the van Cittert–Zernike theorem (2.2.29) is appropriate for the description of diffraction effects in the Fourier holography.

Some disadvantage of the classical holographic imaging is its two-stage procedure. Using the nonlinear process of four-wave mixing, both the stages can be realized in real time (section 2.4). For this also various electronic devices combined with computers can be used.

The purpose of this section is to point out the meaning of classical phase for the holographic method. There are many books explaining the principles and applications of holography in great detail, e. g., [Born and Wolf (1959), DeVelis and Reynolds (1967), Soroko (1978), Saleh and Teich (1991)].

2.4   Phase conjugation

The above examples have demonstrated the role of phase in linear classical optics. In this section we will demonstrate the role of classical phase in the process of phase conjugation, which is a third-order nonlinear optical process. We assume nonlinear dielectric medium characterized by the third-order susceptibility $\chi^{(3)}$ and pumped by counterpropagating coherent classical waves of the complex amplitudes α_3 and α_4, on which a signal of the coherent complex amplitude α_1 is incident (figure 2.5). Further we consider the degenerate four-wave mixing process where frequencies of all beams are the same. Denoting the complex amplitude of the conjugate beam as α_2, the basic equations of motion of this process, taking into account the counterpropagation of the conjugate beam, are

$$\frac{d\alpha_1}{dz} \;=\; i\kappa\alpha_2^*,$$

$$\frac{d\alpha_2}{dz} = -i\kappa\alpha_1^*, \tag{2.4.1}$$

where the constant $\kappa = g^*\alpha_3\alpha_4$ and g is the coupling constant proportional to the susceptibility $\chi^{(3)}$.

From the equations of motion (2.4.1) the wave equation

$$\frac{d^2\alpha_1}{dz^2} = -|\kappa|^2\alpha_1 \tag{2.4.2}$$

follows, which can be solved in the form

$$\alpha_1(z) = C_1\exp(i|\kappa|z) + C_2\exp(-i|\kappa|z), \tag{2.4.3}$$

under the boundary conditions

$$\begin{aligned} C_1 + C_2 &= \alpha_1(0), \\ i|\kappa|[C_1\exp(i|\kappa|l) - C_2\exp(-i|\kappa|l)] &= i\kappa\alpha_2^*(l), \end{aligned} \tag{2.4.4}$$

where l is the length of the phase-conjugating medium. The solution is expressed as follows

$$\begin{aligned} \alpha_1(z) &= \frac{1}{\cos(|\kappa|l)}\left[\alpha_1(0)\cos(|\kappa|(l-z)) + i\frac{\kappa}{|\kappa|}\alpha_2^*(l)\sin(|\kappa|z)\right], \\ \alpha_2(z) &= \frac{1}{\cos(|\kappa|l)}\left[\alpha_2(l)\cos(|\kappa|z) + i\frac{\kappa}{|\kappa|}\alpha_1^*(0)\sin(|\kappa|(l-z))\right]. \end{aligned} \tag{2.4.5}$$

If $|\kappa|l = (2k+1)\pi/2$, k integer, then the process is in the regime of generator. In the spontaneous counterpropagation process it holds that $\alpha_2(l) = 0$ and we have the following solutions,

$$\begin{aligned} \alpha_1(l) &= \frac{\alpha_1(0)}{\cos(|\kappa|l)}, \\ \alpha_2(0) &= i\frac{\kappa}{|\kappa|}\alpha_1^*(0)\tan(|\kappa|l). \end{aligned} \tag{2.4.6}$$

From these relations of the incoming and outgoing fields, we can conclude that the incident signal beam is amplified for $|\kappa|l < \pi/2$, whereas for $|\kappa|l = \pi/2$ the regime of generator is obtained. From the second equation in (2.4.6) one can see that the signal α_2 is phase conjugated with respect to the incident signal α_1, additional phase shifts of the amount $\pi/2$ (the factor i) and from the coupling constant κ can arise, and the phase-conjugated wave can be amplified provided that $\pi/4 < |\kappa|l < \pi/2$. Such a phase-conjugating mirror does not obey the standard law of reflection, but the beam is reversed to the same direction.

This principle can serve for performing holographic imaging in real time avoiding two stages of the imaging (registration of the hologram and its reconstruction). If the signal beam α_1 is assumed to interfere with the pumping (reference) beam α_3 (α_4) to create a hologram, then the pumping (reference) beam α_4 (α_3) can serve to reconstruct the object in the form of phase-conjugated wave α_2, using transmitted light in the former case and reflected light in the latter case. This is the method of the so-called dynamical holography. Finally we can mention that this principle can be used for compensation of distorsions of wavefronts in adaptive optics. When a wave propagates through distorsive random medium, such as turbulent atmosphere, deformation of the wavefront arises. When this wave propagates back through the same medium before the regime of fluctuations is substantially changed, the distorsions are compensated and the optical beam is cleaned. For realistic description of the process finite apertures in the system are to be taken into account [Bouchal and Peřina (1992) and references therein].

The four-wave mixing is a very effective process for the generation of nonclassical (squeezed) light, which needs its quantum description (for a review, see [Peřina (1991), section 10.5]). To obtain such light, one has to mix the incident signal wave together with a phase-conjugated wave. The phase conjugated wave alone cannot exhibit nonclassical behaviour and it is typically noisier than the incident signal field, provided that the process is spontaneous, i. e., $\alpha_2(l) = 0$, which reflects, from the quantum point of view, that only vacuum fluctuations are present at this port [Gaeta and Boyd (1988)]. If nonclassical light is injected into this port, phase-conjugated nonclassical light can be obtained exhibiting squeezing of vacuum fluctuations and/or sub Poissonian photon statistics [Bajer and Peřina (1991), Lanzerotti, Gaeta and Boyd (1995)], which may be used for precise quantum-optical measurements [Lanzerotti and Gaeta (1995)]. The use of phase conjugation of weak continuous-wave optical signals for abberation corrections was demonstrated by Lanzerotti, Schirmer, Gaeta and Agarwal (1996).

Relations of the initial phases of optical beams interacting in a nonlinear medium are decisive for the direction of development of the interaction. For instance, they fully determine whether the second-harmonic or second-subharmonic radiation is generated in the process of degenerate optical parametric process, or whether such radiation is sub- or super-Poissonian, etc., [Peřina (1991), chapter 10].

2.5   Phase retrieval methods

The so-called phase problem arises in various branches of physics, let us name only quantum theory of decay [Khalfin (1960)], scattering theory [Goldberger, Lewis and Watson (1963), Nieto-Vesperinas (1991)], diffraction theory of image formation [Walther (1963), Peřina (1963), Nieto-Vesperinas and Hignette (1979), Nieto-Vesperinas (1991)], inverse source problems [Baltes (1978)], inverse scattering and diffraction tomography problems [Baltes (1980), Wolf (1996)], and coherence theory [Peřina (1985), Mandel and Wolf (1995)].

This problem can be formulated as follows: The modulus of an analytic function

$\gamma(\tau)$ is given and we desire to determine its phase. In the coherence theory this function can be the degree of temporal coherence and if the procedure is successful, we can determine the spectrum of radiation from the measurement of the visibility of interference fringes, making use of the inverse to Wiener–Khinchin theorem (2.1.11). Note that in this case the function $\gamma(\tau)$ is analytic in the lower half-plane τ. This is expressed by the following pair of direct and inverse Fourier transformations of the degree of temporal coherence $\gamma(\tau)$ and normalized spectral density $g(\nu)$,

$$
\begin{aligned}
\gamma(\tau) &= \int_0^\infty g(\nu) \exp(-i2\pi\nu\tau)\, d\nu, \\
g(\nu) &= \int_{-\infty}^\infty \gamma(\tau) \exp(i2\pi\nu\tau)\, d\tau, \quad \nu \geq 0,
\end{aligned}
\tag{2.5.1}
$$

and $g(\nu) = 0$ for $\nu < 0$.

The phase problem is trivial if it is known *a priori* that the spectrum is symmetrical and quasimonochromatic light is involved, i. e., $g(\nu_0 + \nu) = g(\nu_0 - \nu)$, and the spectral bandwidth $\Delta\nu$ is much less than the mean frequency ν_0. Performing a simple substitution in the first integral (2.5.1) and making use of this symmetry, we arrive at the simple phase $-2\pi\nu_0\tau$ and the degree of temporal coherence is fully determined from the visibility measurements,

$$
\gamma(\tau) = |\gamma(\tau)| \exp(-i2\pi\nu_0\tau).
\tag{2.5.2}
$$

If the spectrum of radiation is not symmetrical, analytic properties of $\gamma(\tau)$ must be used, because positional measurements of the interference fringes, giving the possibility to determine the phase of $\gamma(\tau)$, are difficult to perform in the visible region. Denoting the phase $\phi(\tau) = \arg[\gamma(\tau)]$, it is true that also the function $\ln[\gamma(\tau)] = \ln[|\gamma(\tau)|] + i\phi(\tau)$ is analytic in the lower half $\Pi^{(-)}$ of the complex τ-plane, but it will have logarithmic branch points at zeros of $\gamma(\tau)$. Hence if there are no zeros of $\gamma(\tau)$ in $\Pi^{(-)}$, we can apply the Hilbert transformation to obtain relations between the modulus of $\gamma(\tau)$ and its phase $\phi(\tau)$ [Toll (1956), Wolf (1962), Roman and Marathay (1963)]

$$
\begin{aligned}
\phi(\tau) &= \frac{1}{\pi}\text{V. p.} \int_{-\infty}^\infty \frac{\ln[|\gamma(\tau')|]}{\tau' - \tau}\, d\tau', \\
\ln[|\gamma(\tau)|] &= -\frac{1}{\pi}\text{V. p.} \int_{-\infty}^\infty \frac{\phi(\tau')}{\tau' - \tau}\, d\tau',
\end{aligned}
\tag{2.5.3}
$$

where V. p. denotes the Cauchy principal value of the integral.

If the function $\gamma(\tau)$ has zeros in $\Pi^{(-)}$, we can write for the analytic and regular function $\gamma(\tau)$ the Cauchy integral with the contour of integration composed of the real axis and semi-circle of infinite radius lying in $\Pi^{(-)}$, taking into account that the integral vanishes over the semi-circle as a consequence of the assumption that $|\gamma(\tau)|$ tends to zero at least as fast as $|\tau|^{-1}$ for $|\tau|$ going to infinity,

$$
\gamma(\tau) = -\frac{1}{2\pi i} \int_{-\infty}^\infty \frac{\gamma(\tau')}{\tau' - \tau}\, d\tau', \quad \text{Im}\,\tau < 0.
\tag{2.5.4}
$$

Letting $\text{Im}\,\tau \to 0$ in (2.5.4), using the symbolic identity

$$\lim_{\epsilon \to 0} \frac{1}{\tau \mp i\epsilon} = \mathcal{P}\frac{1}{\tau} \pm i\pi\delta(\tau), \quad \epsilon > 0, \tag{2.5.5}$$

where $\mathcal{P}$ refers to the Cauchy principal value and $\delta(\tau)$ is the Dirac delta function, and separating the real and the imaginary parts, we arrive at the dispersion relations, or the Hilbert transformations,

$$\text{Im}[\gamma(\tau)] = \frac{1}{\pi}\text{V. p.} \int_{-\infty}^{\infty} \frac{\text{Re}[\gamma(\tau')]}{\tau' - \tau}\,d\tau',$$

$$\text{Re}[\gamma(\tau)] = -\frac{1}{\pi}\text{V. p.} \int_{-\infty}^{\infty} \frac{\text{Im}[\gamma(\tau')]}{\tau' - \tau}\,d\tau'. \tag{2.5.6}$$

Although the problem of positions of zeros of $\gamma(\tau)$ is complicated and not fully solved, a superposition of a constant background can remove this problem [Burge, Fiddy, Greenaway and Ross (1974)]. Then the most general solution of the phase problem can be obtained using a method of solution of a singular integral equation [Peřina (1985)], which may be useful also in other branches of physics. We can assume the validity of the Lipschitz condition $|\gamma(\tau_1) - \gamma(\tau_2)| \leq \alpha|\tau_1 - \tau_2|^\beta$, where α and β are positive constants. The modulus and the phase of $\gamma(\tau)$ can be expressed as

$$|\gamma(\tau)| = \sqrt{(\text{Re}[\gamma(\tau)])^2 + (\text{Im}[\gamma(\tau)])^2},$$

$$\tan[\phi(\tau)] = \frac{\text{Im}[\gamma(\tau)]}{\text{Re}[\gamma(\tau)]}. \tag{2.5.7}$$

First assume that $\tan[\phi(\tau)] \neq 0$ and $\text{Re}[\gamma(\tau)] \neq 0$ for all τ. This is equivalent to the assumption that $\gamma(\tau)$ has no zeros in $\Pi^{(-)}$, since then $\text{Im}[\gamma(\tau)] \neq 0$ for all τ and the function $\gamma(\tau)$ indeed has no zeros for real τ. Applying a theorem given by Landau and Lifshitz (1964) (section 125), the function $\gamma(\tau)$ indeed has no zeros in $\Pi^{(-)}$. Denoting $\text{Im}[\gamma(\tau)] = f(\tau)$ and substituting from (2.5.7) into (2.5.6), we obtain the singular integral equation

$$\frac{f(\tau)}{\tan[\phi(\tau)]} = -\frac{1}{\pi}\text{V. p.} \int_{-\infty}^{\infty} \frac{f(\tau')}{\tau' - \tau}\,d\tau'. \tag{2.5.8}$$

To solve this equation we use the Sokhotski–Plemelj formulae. Let $\psi(\tau)$ be defined by the Cauchy integral

$$\psi(\tau) = \frac{1}{2\pi i} \int_{-\infty}^{\infty} \frac{f(\tau')}{\tau' - \tau}\,d\tau', \quad \text{Im}\,\tau \neq 0. \tag{2.5.9}$$

Using the identity (2.5.5), we arrive at the Sokhotski–Plemelj formulae

$$\psi^+(\tau) - \psi^-(\tau) = f(\tau),$$

$$\psi^+(\tau) + \psi^-(\tau) = \frac{1}{\pi i}\text{V. p.} \int_{-\infty}^{\infty} \frac{f(\tau')}{\tau' - \tau}\,d\tau', \tag{2.5.10}$$

where ψ^+ and ψ^- are boundary values of $\psi(\tau)$ on the real τ axis from the upper and lower half-planes, respectively.

Denoting $A(\tau) = i/\tan[\phi(\tau)]$, we obtain from (2.5.8)

$$A(\tau)f(\tau) - \frac{1}{\pi i}\text{V. p.}\int_{-\infty}^{\infty}\frac{f(\tau')}{\tau'-\tau}\,d\tau' = 0, \tag{2.5.11}$$

and using (2.5.10), we arrive at the equation

$$A(\tau)[\psi^+(\tau) - \psi^-(\tau)] - [\psi^+(\tau) + \psi^-(\tau)] = 0. \tag{2.5.12}$$

Since $\gamma(\tau)$ has no zeros in $\Pi^{(-)}$ (that is, $\ln[\gamma(\tau)]/2\pi i|_C = \phi(\tau)/2\pi|_C = 0$ for any closed contour C in $\Pi^{(-)}$), the following relation follows from (2.5.12),

$$\{\ln[\psi(\tau)]\}^+ - \{\ln[\psi(\tau)]\}^- = \ln\left[\frac{A(\tau)+1}{A(\tau)-1}\right] = -i2\phi(\tau), \tag{2.5.13}$$

which determines the unique function $\ln[\psi(\tau)]$, for which, according to (2.5.9),

$$\ln[\psi(\tau)] = \frac{1}{2\pi i}\int_{-\infty}^{\infty}\frac{-i2\phi(\tau')}{\tau'-\tau}\,d\tau' = -\frac{1}{\pi}\int_{-\infty}^{\infty}\frac{\phi(\tau')}{\tau'-\tau}\,d\tau', \quad \text{Im}\,\tau \neq 0. \tag{2.5.14}$$

This means that

$$\psi(\tau) = \exp\left[-\frac{1}{\pi}\int_{-\infty}^{\infty}\frac{\phi(\tau')}{\tau'-\tau}\,d\tau'\right], \quad \text{Im}\,\tau \neq 0. \tag{2.5.15}$$

Making use of the first equation of (2.5.10) once again, we have

$$
\begin{aligned}
f(\tau) &= \text{Im}[\gamma(\tau)] = \psi^+(\tau) - \psi^-(\tau) \\
&= i2\sin[\phi(\tau)]\exp\left[-\frac{1}{\pi}\text{V. p.}\int_{-\infty}^{\infty}\frac{\phi(\tau')}{\tau'-\tau}\,d\tau'\right].
\end{aligned}
\tag{2.5.16}
$$

Making use of the fact that the solution of a linear homogeneous equation is determined up to a multiplicative constant and thus omitting the constant 2π, we obtain from (2.5.7)

$$\text{Re}[\gamma(\tau)] = \cos[\phi(\tau)]\exp\left[-\frac{1}{\pi}\text{V.p.}\int_{-\infty}^{\infty}\frac{\phi(\tau')}{\tau'-\tau}\,d\tau'\right] \tag{2.5.17}$$

and again from (2.5.7)

$$|\gamma(\tau)| = \exp\left[-\frac{1}{\pi}\text{V. p.}\int_{-\infty}^{\infty}\frac{\phi(\tau')}{\tau'-\tau}\,d\tau'\right]. \tag{2.5.18}$$

From this relation the modulus of the degree of coherence can be calculated if its phase is known. Using the forward and backward Hilbert transformations for two functions G and H as follows

$$
\begin{aligned}
G(\tau) &= \frac{1}{\pi i}\text{V. p.}\int_{-\infty}^{\infty}\frac{H(\tau')}{\tau'-\tau}\,d\tau', \\
H(\tau) &= \frac{1}{\pi i}\text{V. p.}\int_{-\infty}^{\infty}\frac{G(\tau')}{\tau'-\tau}\,d\tau',
\end{aligned}
\tag{2.5.19}
$$

we obtain the final solution for the phase ϕ from (2.5.18),

$$i \ln[|\gamma(\tau)|] = \frac{1}{\pi i} \text{V. p.} \int_{-\infty}^{\infty} \frac{\phi(\tau')}{\tau' - \tau} d\tau', \qquad (2.5.20)$$

in the form

$$\phi(\tau) = \frac{1}{\pi} \text{V. p.} \int_{-\infty}^{\infty} \frac{\ln[|\gamma(\tau')|]}{\tau' - \tau} d\tau', \qquad (2.5.21)$$

which is the desired expression relating the phase $\phi(\tau)$ to $|\gamma(\tau)|$.

Consider now that the function $\gamma(\tau)$ has zero in $\Pi^{(-)}$ at the points τ_j, $j = 1, 2, \ldots$. In this case the most general form of analytic function can be written in the form [van Kampen (1953), Edwards and Parrent (1959)]

$$\gamma(\tau) = \gamma_0(\tau) \exp(-i2\pi\nu_0\tau) \prod_j \frac{\tau - \tau_j}{\tau - \tau_j^*}, \qquad (2.5.22)$$

where the number ν_0 represents the frequency of light and in the lower half of the τ plane it holds that $\text{Im}\,\tau_j < 0$; $\gamma_0(\tau)$ represents the part of the degree of coherence containing no zeros in $\Pi^{(-)}$. The positions of zeros of $\gamma(\tau)$ are then described by the Blaschke factors $(\tau - \tau_j)/(\tau - \tau_j^*)$ and it holds that $|\gamma(\tau)| = |\gamma_0(\tau)|$ on the real axis. Therefore the complete phase $\phi(\tau)$ of $\gamma(\tau)$ is obtained on the basis of (2.5.22) in the form [Toll (1956)]

$$\phi(\tau) = \frac{1}{\pi} \text{V. p.} \int_{-\infty}^{\infty} \frac{\ln[|\gamma(\tau')|]}{\tau' - \tau} d\tau' + \sum_j \arg\left(\frac{\tau - \tau_j}{\tau - \tau_j^*}\right) - 2\pi\nu_0\tau. \qquad (2.5.23)$$

The physical significance of the last term is that it represents a shift of the spectrum to ν_0 and therefore we can omit it. It can further be shown that the second term in (2.5.23), expressing the so-called Blaschke phase, is non-negative. Hence, the first term giving the part of the phase in terms of the Hilbert transformation of $\ln[|\gamma(\tau)|]$ may be called the minimum phase.

Thus for the unique determination of the phase $\phi(\tau)$ of the degree of coherence $\gamma(\tau)$ from its modulus (visibility of interference pattern) and hence for the unique determination of the spectrum of light from the visibility measurement, one must know the positions of zeros of $\gamma(\tau)$ in $\Pi^{(-)}$. The knowledge of the visibility $|\gamma(\tau)|$ is sufficient only to determine the minimum phase. Using the residue theorem, when calculating the spectrum $g(\nu)$ from (2.5.23) by the Wiener–Khinchin theorem, we arrive at [Peřina and Tillich (1966)]

$$g(\nu) = g_0(\nu) + 2\pi i \sum_j (\tau_j^* - \tau_j) \exp(i2\pi\nu\tau_j^*)$$

$$\times \prod_{k \neq j} \frac{\tau_j^* - \tau_k}{\tau_j^* - \tau_k^*} \int_0^{\nu} g_0(\mu) \exp(-i2\pi\mu\tau_j^*) \, d\mu, \qquad (2.5.24)$$

where the minimum spectrum $g_0(\nu)$ is the Fourier transform of $\gamma_0(\tau)$ (the exponential factor in (2.5.22), leading to the frequency shift, has been omitted).

It is possible to restrict location of zeros of $\gamma(\tau)$ by various physical conditions. For example, since the spectrum is real, i. e., $g^*(\nu) = g(\nu)$, the cross-symmetry condition $\gamma(\tau) = \gamma^*(-\tau)$ must hold, and consequently the zeros must be distributed symmetrically with respect to the imaginary τ axis, because with every factor $(\tau - \tau_j)/(\tau - \tau_j^*)$ the factor $(\tau + \tau_k^*)/(\tau + \tau_k)$ occurs (to every j a k exists such that $\tau_j = -\tau_k^*$) [Roman and Marathay (1963)]. The condition $g(\nu) \geq 0$ of the non-negative values of the spectrum leads to elimination of the zeros from the imaginary τ axis, because if $\tau = ia$ ($a < 0$ and real), then $\gamma(ia) = \int_0^\infty g(\nu) \exp(2\pi\nu a)\, d\nu > 0$, i. e., $\gamma(ia) \neq 0$. Little is known about other physical conditions restricting location of zeros of $\gamma(\tau)$ in $\Pi^{(-)}$. Some physical conditions can be used to eliminate zeros at all. For instance, it can be shown [Peřina (1985), p. 51] that the reconstruction of the phase under the additional condition of the minimum value of the first moment of the spectrum is unique, because the zeros are eliminated in this case. Also a stationary coherent field must be monochromatic and the Blaschke factors have to be absent in order to satisfy the wave equation, i. e., zeros cannot be present in the solution [Peřina (1985), p. 42]. It is possible to show [Kano and Wolf (1962)] that for blackbody radiation the function $\gamma(\tau)$ has really no zeros in $\Pi^{(-)}$ so that its phase as well as the spectrum can be reconstructed uniquely from the visibility of interference fringes in spite of the fact that the spectrum of blackbody radiation is not symmetrical. In general, the contribution of zeros to the phase $\phi(\tau)$ may be so large that the minimum phase cannot be considered as a good approximation to the actual phase.

The solution of the phase problem in bright-field electron microscopy [Misell (1973a, b), Misell, Burge and Greenaway (1974), Misell and Greenaway (1974a, b), Spence (1974), Ferwerda and Hoenders (1975a, b), Ferwerda (1978)] enables us to consider a constant bright background of a level $C > 0$, so that the analytic function $\gamma(\tau) + C \equiv \gamma'(\tau)$ has no zeros in $\Pi^{(-)}$ and the above method of the singular integral equations can be applied giving for the phase [Burge, Fiddy, Greenaway and Ross (1974)]

$$\tan[\phi(\tau)] = \frac{|\gamma'(\tau)| \sin[\alpha(\tau)]}{|\gamma'(\tau)| \cos[\alpha(\tau)] - C}, \qquad (2.5.25)$$

where $\alpha(\tau)$ is the phase of $\gamma'(\tau)$, which is equal to the minimum phase determined from the dispersion relations, since $\gamma'(\tau)$ has no zeros in $\Pi^{(-)}$. The most complex study of the phase problem from this point of view has been provided by Burge, Fiddy, Greenaway and Ross (1976), who also employed the dispersion relations with subtractions [Nussenzveig (1972)] and suggested some algorithms for determination of positions of zeros.

There are other possibilities how to recover the phase $\phi(\tau)$ from the modulus of the degree of coherence $|\gamma(\tau)|$. For instance, Gamo (1963) suggested the triple intensity correlation measurements for recovering the phase. Another method was proposed by Mehta (1965) on the basis of Cauchy–Riemann conditions

$$\frac{\partial |\gamma(\tau_r, \tau_i)|}{\partial \tau_r} = -|\gamma(\tau_r, \tau_i)| \frac{\partial \phi(\tau_r, \tau_i)}{\partial \tau_i},$$

$$\frac{\partial |\gamma(\tau_r, \tau_i)|}{\partial \tau_i} = |\gamma(\tau_r, \tau_i)| \frac{\partial \phi(\tau_r, \tau_i)}{\partial \tau_r}, \qquad (2.5.26)$$

which must hold in $\Pi^{(-)}$; here τ_r and τ_i are the real and imaginary parts of τ, i. e., $\tau = \tau_r + i\tau_i$. The phase may be obtained by integrating the second equation in (2.5.26). In order to determine the derivative $\partial \phi(\tau_r, \tau_i)/\partial \tau_r|_{\tau_i=0}$ along the real axis, we must also know values of $|\gamma(\tau)|$ in $\Pi^{(-)}$ for $\tau_i < 0$, but the physical significance of the degree of coherence $\gamma(\tau_r, \tau_i)$ in the complex τ plane is obvious from the Wiener–Khinchin theorem giving

$$\gamma(\tau_r, \tau_i) = \int_0^\infty g(\nu) \exp(2\pi\nu\tau_i) \exp(-i2\pi\nu\tau_r) \, d\nu, \qquad (2.5.27)$$

i. e., $\gamma(\tau_i, \tau_i)$ for $\tau_i < 0$ can be obtained from the spectrum $g(\nu)$ by using the exponential filter $\exp[-2\pi\nu(-\tau_i)]$. The use of Cauchy–Riemann conditions for the experimental reconstruction of the phase was demonstrated by Kohler and Mandel (1973) and Ablekov, Zubkov and Frolov (1976). The use of exponential filters has been discussed by Nakajima and Asakura (1982) on the basis of computer simulations. Also holographic approach to the phase problem can be developed [Kohler and Mandel (1970)].

Yet another method was suggested by Mehta (1968), which is based on using a reference beam with the known degree of coherence $\gamma_r(\tau)$, including its phase ϕ_r, and with the spectrum $g_r(\nu)$. If we superimpose the light from the reference source on light with an unknown spectrum $g(\nu)$, we obtain light with the resulting spectrum

$$\bar{g}(\nu) = g(\nu) + g_r(\nu), \qquad (2.5.28)$$

since both the sources are statistically independent. Thus we have

$$\bar{\gamma}(\tau) = \gamma(\tau) + \gamma_r(\tau) \qquad (2.5.29)$$

for the corresponding degree of coherence. If two separate experiments are realized with light of the unknown spectrum and that of the superimposed spectrum, we obtain $|\gamma(\tau)|$ and $|\bar{\gamma}(\tau)|$, respectively. Then taking the squared modulus of (2.5.29), we arrive at

$$|\bar{\gamma}(\tau)|^2 = |\gamma(\tau)|^2 + |\gamma_r(\tau)|^2 + 2|\gamma(\tau)||\gamma_r(\tau)| \cos[\phi(\tau) - \phi_r(\tau)], \qquad (2.5.30)$$

where all quantities are known except the phase $\phi(\tau)$ which can be determined.

Other approaches to classical phase problem can be found in [Peřina (1985), section 4.4], where also additional references are available. A brief interesting discussion of the phase problem from the point of view of inverse light diffraction can be found in [Nieto-Vesperinas (1991), section 9.10]. In this connection two cases are practically important—the function under discussion is periodic, like in crystalography, electron microscopy, neutron diffraction, etc., or it represents an object of finite extent and so it is a function of finite support. Some discussions of uniqueness of the solution can be found in chapters [Ferwerda (1978), Hoenders (1978), Ross, Fiddy and Nieto-Vesperinas (1980)]. Great progress was achieved in the solution of phase problem when various computational algorithms were developed [Fienup (1978, 1982, 1984), Bates and Fright (1983), Fienup and Wakerman (1986), Perez-Ilzarbe, Nieto-Vesperinas and Navarro (1990), Negrete-Regagnon (1996), Miura and Baba (1996), Gureyev and Nugent (1996)]. Futher detail can be found in works by Dainty (1984), Saxton (1978), and Bates and McDonnell (1986). Such algorithms can be effective in phase recovery in optical imaging, e. g., [Dong, Zhuang and Ersoy (1994), and references therein, Rathjen (1995)] and the use of Zernike polynomials can be helpful [Gureyev, Roberts and Nugent (1995a, b), Voitsekhovich (1995), Gureyev and Nugent (1996)]. Sometimes iterative computational methods can lead to spurious phases and therefore they must be applied with great care [Wedberg and Stamnes (1995)]. Determination of the amplitude and the phase of scattered fields by means of holographic principle was discussed by Wolf (1970) and a review of the phase retrieval in crystallography was provided by Millane (1990). Also the method of homodyne tomography (section 4.13) recently developed in quantum optics can help to reconstruct the classical optical phase [McAlister, Beck, Clarke, Mayer and Raymer (1995)] or the phase of the wave function in quantum mechanics [Orlowski and Paul (1995)] via the reconstruction of the Wigner quasidistribution

$$\Phi_{\mathrm{W}}(x,k) = \frac{1}{\pi} \int_{-\infty}^{\infty} \langle V^*(x+x')V(x-x') \rangle \exp(-i2kx')\,dx'. \tag{2.5.31}$$

Particular problems are related to determination of the amplitude and phase of ultrashort optical pulses [Nibbering, Franco, Prade, Grillon, Chambaret and Mysyrowicz (1996), Fittinghoff, Bowie, Sweetser, Jennings, Krumbügel, DeLong and Trebino (1996)]. Phase retrieval in nonlinear optical spectroscopy using the maximum-entropy method was demonstrated by Vartiainen, Peiponen, Kishida and Koda (1996). There are effective methods of obtaining super-resolution by data inversion, as reviewed by Bertero and De Mol (1996). Recently interesting review appeared about dispersion relations and phase retrieval in optical spectroscopy [Peiponen, Vartiainen and Asakura (1997)].

As we have seen in this chapter the classical phase can be determined from simple interference and diffraction measurements. The holographic principle is very fruitful in this case. Also the method of nonlinear phase conjugation based on the use of a third-order nonlinear medium provides useful phase information.

Chapter 3

Phase-space description of light field

The notion of operator-valued measure occurs as a positive operator-valued measure in quantum detection and estimation theory and quantum theory of measurement, which here are treated in chapters 4 and 5. But this notion can be used also in the treatment of some of the phase-space quasidistributions. To show this, we define a generalized Wigner function for two arbitrary operators, even though it is still an open problem, which pair of operators—besides the position and momentum ones—is appropriate here.

In quantum optics the orderings of field operators have important physical meaning and the operator generating functions should be appropriately ordered. In this connection the resolutions of the identity in terms of the coherent states have proved to be very useful and generalizations of the corresponding phase-space descriptions have included the Wigner function for two conjugate quadratures. Nonclassical features of light beams can be analyzed using the phase-space methods. These can be completed with the formalism of the multimode Gaussian state (i. e., the generalized superposition of coherent fields and quantum noise). Using the phase-space formalism, radiation interacting with matter can be described. Also losses can be taken into account by this description and decay of quantum macroscopic superposition states can be illustrated.

3.1 Wigner function for two arbitrary operators

Quantum optics is understood here as applied quantum mechanics, whose fundamental concepts are only introduced in a slightly modified way. Here we assume a finite number of modes J, which are equivalent to harmonic oscillators. The underlying Hilbert space $\mathcal{H}$ is a completion of the span of $\{|n_1, ..., n_J\rangle\}$, the basis formed by the Fock or number states. As usual in quantum theory, we consider operators on the space $\mathcal{H}$ and, particularly, we remember the trace-class operators [Davies (1976)],

because, of these operators, the Hermitian and positive ones are related to the state
of a quantum optical system. They are the so-called statistical operators. Single Hermitian operators can be measured, a random variable can be associated with each
and their distributions can be connected to a statistical operator $\hat{\rho}$.

3.1.1 Operator-valued measures

Considering an Hermitian operator $\hat{B}$, $\hat{B}^\dagger = \hat{B}$, we are interested in its eigenvectors
(eigenkets), i. e., vectors $|B\rangle$ such that

$$\hat{B}|B\rangle = B|B\rangle, \tag{3.1.1}$$

where B is the so-called eigenvalue. The set of eigenvalues is known as the spectrum
of the operator $\hat{B}$ and will be denoted as Spec $\hat{B}$. The random variable resulting from
the measurement is just a random eigenvalue. In some important cases of quantum
mechanical Hamiltonians [Dirac (1958)] the spectrum consists of point and continuous
parts

$$\text{Spec } \hat{B} = \{b_n\} \cup \{B\}, \tag{3.1.2}$$

where b_n are the elements of the point part and, as well as from now on, B means
only an element of the continuous part. Discrete and continuous eigenvectors are
orthogonal and can be normalized,

$$\begin{aligned}
\langle b_n|b_m\rangle &= \delta_{nm}, \\
\langle b_n|B\rangle &= 0, \\
\langle B'|B''\rangle &= \delta(B' - B''),
\end{aligned} \tag{3.1.3}$$

where $\delta(B' - B'')$ is the Dirac delta function. We assume that they form a complete
set so that arbitrary vectors $|\psi\rangle$ can be expressed in terms of them,

$$|\psi\rangle = \sum_n c_n|b_n\rangle + \int_{\{B\}} \psi(B)|B\rangle \, dB, \tag{3.1.4}$$

where

$$c_n = \langle b_n|\psi\rangle, \quad \psi(B) = \langle B|\psi\rangle. \tag{3.1.5}$$

The relation of completeness can be written as a resolution of the identity

$$\sum_n |b_n\rangle\langle b_n| + \int_{\{B\}} |B\rangle\langle B| \, dB = \hat{1}. \tag{3.1.6}$$

Applying the operator $\hat{B}$ from the left or the right to the relation (3.1.6), we obtain
an expression for the operator in terms of the eigenvectors

$$\hat{B} = \sum_n b_n|b_n\rangle\langle b_n| + \int_{\{B\}} B|B\rangle\langle B| \, dB. \tag{3.1.7}$$

Just as kets are represented by column vectors made of their components c_n, $\psi(B)$, where also a continuous "subscript" is understood, operators are represented by matrices. Thus the operator $\hat{X}$ corresponds to a matrix $\mathbf{X}$ with elements

$$
\begin{aligned}
X_{nm}^{\mathrm{dd}} &= \langle b_n | \hat{X} | b_m \rangle, \\
X_n^{\mathrm{dc}}(B) &= \langle b_n | \hat{X} | B \rangle, \\
X_m^{\mathrm{cd}}(D) &= \langle B | \hat{X} | b_m \rangle, \\
X^{\mathrm{cc}}(B, B') &= \langle B | \hat{X} | B' \rangle.
\end{aligned}
\tag{3.1.8}
$$

The generalized sum of the diagonal elements of the matrix $\mathbf{X}$ is called the trace of the operator $\hat{X}$,

$$
\mathrm{Tr}\,\hat{X} = \sum_n X_{nn}^{\mathrm{dd}} + \int_{\{B\}} X^{\mathrm{cc}}(B, B)\, dB.
\tag{3.1.9}
$$

Especially, for $\hat{X} = \hat{B}$ we obtain the generalized sum of the eigenvalues

$$
\mathrm{Tr}\,\hat{B} = \sum_n b_n + \int_{\{B\}} B\, dB.
\tag{3.1.10}
$$

We remember the cyclic property of the trace,

$$
\mathrm{Tr}(\hat{X}\hat{Y}) = \mathrm{Tr}(\hat{Y}\hat{X}),
\tag{3.1.11}
$$

both the sides of this equation being equal to

$$
\sum_n \sum_m \langle b_n | \hat{X} | b_m \rangle \langle b_m | \hat{Y} | b_n \rangle + \sum_n \int_{\{B\}} \langle b_n | \hat{X} | B \rangle \langle B | \hat{Y} | b_n \rangle\, dB
$$

$$
+ \int_{\{B\}} \sum_m \langle B | \hat{X} | b_m \rangle \langle b_m | \hat{Y} | B \rangle\, dB + \int_{\{B\}} \int_{\{B'\}} \langle B | \hat{X} | B' \rangle \langle B' | \hat{Y} | B \rangle\, dB\, dB'.
\tag{3.1.12}
$$

The importance of Hermitian operators in quantum detection and estimation theory and in quantum theory of measurement is demonstrated by the projection operators [Helstrom (1976), Bush, Lahti and Mittelstaedt (1991)]. These operators describe orthogonal projections of the Hilbert space $\mathcal{H}$ onto a linear subspace $\mathcal{H}' \subset \mathcal{H}$. Any vector $|\psi\rangle$ in $\mathcal{H}$ can be written uniquely as the sum of a vector $|\psi'\rangle$ in $\mathcal{H}'$ and a vector $|\psi''\rangle$ that is orthogonal to $|\psi'\rangle$ and to all the other vectors in $\mathcal{H}'$,

$$
|\psi\rangle = |\psi'\rangle + |\psi''\rangle, \quad \langle \psi''' | \psi'' \rangle = 0, \quad \text{all } |\psi'''\rangle \in \mathcal{H}'.
\tag{3.1.13}
$$

The operator $\hat{E}$ taking each ket $|\psi\rangle$ into its component $|\psi'\rangle$ in $\mathcal{H}'$,

$$
|\psi'\rangle = \hat{E} |\psi\rangle,
\tag{3.1.14}
$$

is a linear operator known as a projection operator. Projection operators satisfy the equation

$$
\hat{E}^2 = \hat{E}.
\tag{3.1.15}
$$

The vectors $|\psi''\rangle$ from (3.1.13) form the orthogonal complement $\mathcal{H}'^{\perp}$,

$$\mathcal{H}'^{\perp} = (\hat{1} - \hat{E})\mathcal{H} = \{(\hat{1} - \hat{E})|\psi\rangle; |\psi\rangle \in \mathcal{H}\}. \tag{3.1.16}$$

It is possible to build a projection operator by picking out any subset I of the real line $\mathbb{R}$ (of the spectrum Spec $\hat{B}$). The appropriate operator is

$$\hat{E}(I) = \sum_{\substack{n \\ b_n \in I}} |b_n\rangle\langle b_n| + \int_{I \cap \{B\}} |B\rangle\langle B|\, dB. \tag{3.1.17}$$

Particularly, by (3.1.6) $\hat{E}(\mathbb{R}) = \hat{1}$. For the disjoint sets I', I'', i. e., such that $I' \cap I'' = \emptyset$, the introduced projection operators are orthogonal in the sense that

$$\hat{E}(I')\hat{E}(I'') = \hat{0}. \tag{3.1.18}$$

Introducing the shorthand notation

$$\hat{E}(dB) = |B\rangle\langle B|\, dB \tag{3.1.19}$$

and assuming that the operator $\hat{B}$ has a continuous spectrum, we may use the integral notation for (3.1.17) and (3.1.7),

$$\hat{E}(I) = \int_I \hat{E}(dB) \tag{3.1.20}$$

and

$$\hat{B} = \int_{-\infty}^{\infty} B\hat{E}(dB). \tag{3.1.21}$$

In general, the relations (3.1.17) and (3.1.7) involve sums, but we use (3.1.20) and (3.1.21) as their equivalents, too. By means of the resolutions of the identity (3.1.6) we can define functions of operators. In the shorthand notation, the corresponding integral and/or ordinary sum is written as

$$f(\hat{B}) = \int_{-\infty}^{\infty} f(B)\hat{E}(dB). \tag{3.1.22}$$

Measurement of two commuting operators associates with these physical quantities a two-dimensional random vector. Its distribution can be connected to the statistical operator $\hat{\rho}$. We are interested in the simultaneous eigenvalue problems for two commuting operators $\hat{B}_1$, $\hat{B}_2$, i. e., such that $[\hat{B}_1, \hat{B}_2] = \hat{0}$, and parametrize the eigenvectors with two numbers,

$$\hat{B}_j|B_1, B_2\rangle = B_j|B_1, B_2\rangle, \quad j = 1, 2. \tag{3.1.23}$$

The random vector resulting from the measurement is composed just of random eigenvalues. Now the spectrum consists of vectors and, in principle, its decomposition into point, curvilinear, and continuous parts may be rather involved,

$$\text{Spec}(\hat{B}_1, \hat{B}_2) = \{(b_{1n_1}, b_{2n_2})\} \cup \{(b_{1n}(B), b_{2n}(B))\} \cup \{(B_1, B_2)\}, \tag{3.1.24}$$

where (b_{1n_1}, b_{2n_2}) are the elements of the point part, $(b_{1n}(B), b_{2n}(B))$, with B a natural parameter (length), are the elements of the curvilinear part, (B_1, B_2) is a generic element of the continuous part. In analogy to (3.1.3) we assume that the eigenvectors are normalized,

$$
\begin{aligned}
\langle b_{1n_1}, b_{2n_2} | b_{1m_1}, b_{2m_2} \rangle &= \delta_{n_1,m_1} \delta_{n_2,m_2}, \\
\langle b_{1n_1}, b_{2n_2} | b_{1n}(B), b_{2n}(B) \rangle &- 0, \\
\langle b_{1n_1}, b_{2n_2} | B_1, B_2 \rangle &= 0, \\
\langle b_{1n}(B'), b_{2n}(B') | b_{1m}(B''), b_{2m}(B'') \rangle &= \delta_{nm} \delta(B' - B''), \\
\langle b_{1n}(B), b_{2n}(B) | B_1, B_2 \rangle &= 0, \\
\langle B_1', B_2' | B_1'', B_2'' \rangle &= \delta(B_1' - B_1'') \delta(B_2' - B_2'').
\end{aligned} \tag{3.1.25}
$$

In this situation we assume that $I \subset \mathbb{R}^2$ and we define the projection operators

$$
\hat{E}(I) = \sum_{\substack{n_1, n_2 \\ (b_{1n_1}, b_{2n_2}) \in I}} |b_{1n_1}, b_{2n_2}\rangle\langle b_{1n_1}, b_{2n_2}|
$$

$$
+ \sum_n \int_{\{(b_{1n}(B), b_{2n}(B))\} \cap I} |b_{1n}(B), b_{2n}(B)\rangle\langle b_{1n}(B), b_{2n}(B)|\, dB
$$

$$
+ \iint_{I \cap \{(B_1, B_2)\}} |B_1, B_2\rangle\langle B_1, B_2|\, dB_1\, dB_2. \tag{3.1.26}
$$

Using the abstract integration, we write a resolution of the identity as

$$
\int_{-\infty}^{\infty} \int_{-\infty}^{\infty} \hat{E}(dB_1 dB_2) = \hat{1}. \tag{3.1.27}
$$

By means of (3.1.27) we can form the original operators

$$
\hat{B}_j = \int_{-\infty}^{\infty} \int_{-\infty}^{\infty} B_j \hat{E}(dB_1 dB_2), \quad j = 1, 2, \tag{3.1.28}
$$

along with all of the operators that are functions of $\hat{B}_1$, $\hat{B}_2$,

$$
f(\hat{B}_1, \hat{B}_2) = \int_{-\infty}^{\infty} \int_{-\infty}^{\infty} f(B_1, B_2) \hat{E}(dB_1 dB_2). \tag{3.1.29}
$$

We may remember the meaning of the shorthand notation when we rewrite also (3.1.26),

$$
\hat{E}(I) = \iint_I \hat{E}(dB_1 dB_2). \tag{3.1.30}
$$

Assuming a continuous spectrum, we have

$$
\hat{E}(dB_1 dB_2) = |B_1, B_2\rangle\langle B_1, B_2| dB_1 dB_2. \tag{3.1.31}
$$

Generalizations to an arbitrary number of parameters connected, of course, to a number of consistent operators are apparent. Dividing both sides of (3.1.19) and (3.1.31) by dB and $dB_1 dB_2$, respectively, we obtain the interpretation of the projection operators as derivatives of operator-valued measures,

$$|B\rangle\langle B| = \frac{\hat{E}(dB_{|B})}{dB_{|B}} \tag{3.1.32}$$

and

$$|B_1, B_2\rangle\langle B_1, B_2| = \frac{\hat{E}\left(dB_1 dB_{2|(B_1, B_2)}\right)}{dB_1 dB_{2|(B_1, B_2)}}. \tag{3.1.33}$$

Measurement of the operator $\hat{B}$ or of the operators $\hat{B}_j$, $j = 1, ..., J$, associates with this physical quantity (these physical quantities) a random variable B (J-dimensional random vector $(B_1, ..., B_J)$). Using the mathematical terminology, we recognize the distribution of B (the joint distribution of $B_1, ..., B_J$) as the probability of the quantity B (the vector $(B_1, ..., B_J)$) lying in I,

$$\text{Prob}[B \in I] = \text{Tr}\{\hat{\rho}\hat{E}(I)\}, \quad I \subset \mathbb{R}, \tag{3.1.34}$$

$$\text{Prob}[(B_1, ..., B_J) \in I] = \text{Tr}\{\hat{\rho}\hat{E}(I)\}, \quad I \subset \mathbb{R}^J. \tag{3.1.35}$$

Let us recall, restricting ourselves to the case of $\hat{B}$, that the following ways of describing the distributions are sufficient:
(i) the probabilities of eigenvalues in the point part of the spectrum

$$p_n = \langle b_n|\hat{\rho}|b_n\rangle \tag{3.1.36}$$

and
(ii) the probability density of eigenvalues in the continuous part of the spectrum

$$P(B) = \langle B|\hat{\rho}|B\rangle. \tag{3.1.37}$$

In the theory of quantum observation and measurement it is useful to consider classes of operators obeying certain requirements. These requirements or postulates enable one to obtain probability distributions of a random vector $(u_1, ..., u_J)$ describing an ideal observing instrument. The role of the operators $\hat{E}(I)$ is taken over by the operators $\hat{\Pi}(I)$ and the relations (3.1.34), (3.1.35) become

$$\text{Prob}[(u_1, ..., u_J) \in I] = \text{Tr}\{\hat{\rho}\hat{\Pi}(I)\}, \quad I \subset \mathbb{R}^J. \tag{3.1.38}$$

Because a probability must be a nonnegative real number, the operators $\hat{\Pi}(I)$ must be positive and Hermitian. As

$$\text{Prob}[(u_1, ..., u_J) \in \emptyset] = 0, \tag{3.1.39}$$

it is required that
$$\hat{\Pi}(\emptyset) = \hat{0}. \tag{3.1.40}$$

As
$$\text{Prob}[(u_1, ..., u_J) \in \mathbb{R}^J] = 1, \tag{3.1.41}$$

it is postulated that
$$\hat{\Pi}(\mathbb{R}^J) = \hat{1}. \tag{3.1.42}$$

Another property of ordinary probability motivates that the operator associated with the union of disjoint regions in $\mathbb{R}^J$ is determined by addition,

$$\hat{\Pi}\left(\overset{\infty}{\underset{k=1}{\cup}} I_k\right) = \sum_{k=1}^{\infty} \hat{\Pi}(I_k), \quad I_j \cap I_k = \emptyset, \quad \text{all } j \neq k. \tag{3.1.43}$$

In analogy to the case when a number is associated with the so-called measurable sets, the mapping of sets into operators is called a normalized positive operator-valued measure or a probability-operator measure in $\mathbb{R}^J$. A resolution of the identity (3.1.6) is a special type of probability-operator measure known as an orthogonal or projection-valued probability-operator measure.

For quasidistributions of two noncommuting operators, it may be useful to resign just the positivity of the operator-valued measure and, among others, to retain the postulate (3.1.42).

3.1.2 Generalized Wigner function as a measure

The phase-space formulation of classical mechanics was connected to the stochastic considerations in the statistical physics so closely that the phase-space formalism of quantum mechanics became a significant contribution to quantum statistical physics [Weyl (1927), Wigner (1932)]. Both classical and quantum formalisms provide joint distributions of canonically conjugate variables. Whereas in the classical case the simultaneous measurement with arbitrary accuracy of such variables is possible, in quantum theory, the simultaneous measurement with arbitrary accuracy of noncommuting and particularly canonically conjugate operators is excluded, in principle. In this section we would like not only remember the striking possibility of defining joint quasidistributions for canonically conjugate operators but also that of writing joint quantum quasidistributions for arbitrary two noncommuting operators. We have in mind the application to the photon-number operator and the momentum-like quadrature operator, which can frequently be used for the description of phase properties. The details will not be included here [Lukš and Peřinová (1997a)].

The prototypes of canonically conjugate operators are the position-coordinate and the linear momentum operators, whose commutator is known to be the Planck constant reduced by 2π and multiplied by i. In quantum theory, canonical transformations of classical mechanics are replaced by unitary and antiunitary transformations

[Wigner (1931), Messiah (1961, 1962)], which preserve commutators of evolving operators in the Heisenberg picture, and thus, the canonical conjugacy in the evolution. On the contrary, several canonical transformations cannot be realized as unitary or antiunitary transformations in quantum theory. This maybe an indicator of a pitfall in the treatment of some pairs of operators. Especially, the action–phase-angle pair has been connected to the same commutator as the position-coordinate–momentum pair, which is not true. A conventional elimination of the reduced Planck constant by either setting it a unity or preferring another physical quantity (for example, number to action) means that a complementarity is usual where a commutator is any c-number. In the case of two operators which exhibit finite point spectra it is advantageous to approach the canonical conjugacy problem in a rather different way [Pegg, Vaccaro and Barnett (1990)]. The property usually required of the quasidistribution is that it be even nonnegative. Nevertheless, the existing formalisms show that it cannot be achieved along with the accomplishment of another natural postulate. This postulate is that the marginal distributions for the individual operators be quantum mechanically intact.

To perform the outlined program, we may use the method of quantum characteristic function [Scully and Cohen (1987)]. We relate this concept to the operators $\hat{A}$ and $\hat{B}$ and denote

$$C^{AB}(\theta, \tau) = \langle \exp(i\theta\hat{A} + i\tau\hat{B}) \rangle = \mathrm{Tr}\{\hat{\rho}\exp(i\theta\hat{A} + i\tau\hat{B})\}. \qquad (3.1.44)$$

We take $I = I_A \times I_B$, where I_A, I_B are the closed intervals $[a_1, a_2]$, $[b_1, b_2]$, and we express the filter $\tilde{j}_{-I}(\theta, \tau)$,

$$\tilde{j}_{-I}(\theta, \tau) = \int\int_{-I} \exp(i\theta A + i\tau B)\, dA\, dB$$

$$= \frac{4}{\theta\tau} \exp\left[-i\theta\frac{(a_1 + a_2)}{2} - i\tau\frac{(b_1 + b_2)}{2}\right] \sin\left[\frac{(a_2 - a_1)}{2}\theta\right] \sin\left[\frac{(b_2 - b_1)}{2}\tau\right], \quad (3.1.45)$$

where $-I = \{(-a, -b); (a, b) \in I\}$. Using the properties of the Fourier transformation, we define the quasiprobability measure

$$\Pi^{AB}(I) = \frac{1}{4\pi^2} \int_{-\infty}^{\infty} \int_{-\infty}^{\infty} \mathrm{Re}\left[\tilde{j}_{-I}(\theta, \tau) C^{AB}(\theta, \tau)\right] d\theta\, d\tau. \qquad (3.1.46)$$

Let us note that a similar derivation leads to the quasiprobability generalized function. Introducing the operators

$$\hat{\Pi}^{AB}(I) = \frac{1}{4\pi^2} \int_{-\infty}^{\infty} \int_{-\infty}^{\infty} \mathrm{Re}\left[\tilde{j}_{-I}(\theta, \tau) \exp(i\theta\hat{A} + i\tau\hat{B})\right] d\theta\, d\tau, \qquad (3.1.47)$$

where Re means the Hermitian part, we write the quasiprobability measure as

$$\Pi^{AB}(I) = \mathrm{Tr}\{\hat{\rho}\hat{\Pi}^{AB}(I)\}. \qquad (3.1.48)$$

The properties of marginal distributions can be found. We derive that

$$\hat{\Pi}^{AB}(I_A \times \mathbb{R}) = \lim_{B \to \infty} \frac{1}{2\pi^2} \int_{-\infty}^{\infty} \int_{-\infty}^{\infty} \mathrm{Re}\left[\tilde{j}_{-I_A}(\theta) \frac{\sin(B\tau)}{\tau} \exp(i\theta\hat{A} + i\tau\hat{B})\right] d\theta \, d\tau$$

$$= \frac{1}{2\pi} \int_{-\infty}^{\infty} \mathrm{Re}\left[\tilde{j}_{-I_A}(\theta) \exp(i\theta\hat{A})\right] d\theta = \hat{E}^A(I_A) \tag{3.1.49}$$

and similarly

$$\hat{\Pi}^{AB}(\mathbb{R} \times I_B) = \hat{E}^B(I_B). \tag{3.1.50}$$

Here

$$\tilde{j}_{-I_A}(\theta) = \int_{-I_A} \exp(i\theta A) \, dA = \frac{2}{\theta} \exp\left[-i\theta\frac{(a_1 + a_2)}{2}\right] \sin\left[\frac{(a_2 - a_1)}{2}\theta\right],$$

$$\tilde{j}_{-I_B}(\tau) = \int_{-I_B} \exp(i\tau B) \, dB = \frac{2}{\tau} \exp\left[-i\tau\frac{(b_1 + b_2)}{2}\right] \sin\left[\frac{(b_2 - b_1)}{2}\tau\right] \tag{3.1.51}$$

and $\hat{E}^B(I_B) \equiv \hat{E}(I)$ in (3.1.17), and $\hat{E}^A(I_A)$ is the analogue. From any of the relations (3.1.49) and (3.1.50) it follows that (3.1.42) for $J - 2$ is fulfilled.

We use the operator-valued measure $\hat{\Pi}^{AB}$ to assign operators $\hat{M}$ to ordinary functions $f(A, B)$, e. g., the monomials $A^k B^l$,

$$\hat{M} = \int \int f(A, B) \hat{\Pi}^{AB}(dAdB). \tag{3.1.52}$$

A mapping theorem holds for expectations,

$$\mathrm{Tr}\{\hat{\rho}\hat{M}\} = \int \int f(A, B) \mathrm{Tr}\left\{\hat{\rho}\hat{\Pi}^{AB}(dAdB)\right\} = \int \int f(A, B) \Pi^{AB}(dAdB). \tag{3.1.53}$$

Using (3.1.49) in the integral form, we obtain for $f(A, B) = f_A(A)$ that

$$\hat{M} = \int f(A) \int_{B=-\infty}^{\infty} \hat{\Pi}^{AB}(dAdB)$$

$$= \int f(A) \hat{E}^A(dA) = f_A(\hat{A}). \tag{3.1.54}$$

Similarly by (3.1.50) for $f(A, B) = f_B(B)$,

$$\hat{M} = f_B(\hat{B}). \tag{3.1.55}$$

Assuming the absolute continuity of the quasiprobability measure $\Pi^{AB}(I)$ with respect to the Lebesgue measure, we introduce the Wigner function $W^{AB}(A, B)$ for the operators $\hat{A}$, $\hat{B}$ as the derivative

$$W^{AB}(A, B) = \frac{\Pi^{AB}\left(dAdB_{|(A,B)}\right)}{dAdB_{|(A,B)}}. \tag{3.1.56}$$

Under the assumption of weak absolute continuity of the operator-valued measure $\hat{\Pi}^{AB}(I)$ with respect to the Lebesgue measure, we introduce the Wigner operators

$$\hat{W}^{AB}(A, B) = \frac{\hat{\Pi}^{AB}\left(dA\,dB_{|(A,B)}\right)}{dA\,dB_{|(A,B)}}. \tag{3.1.57}$$

The relations (3.1.49) and (3.1.50) can be written in the forms

$$\int_{-\infty}^{\infty} \hat{W}^{AB}(A, B)\,dB = \frac{\hat{E}^A\left(dA_{|A}\right)}{dA_{|A}} = |A\rangle\langle A| \tag{3.1.58}$$

and

$$\int_{-\infty}^{\infty} \hat{W}^{AB}(A, B)\,dA = \frac{\hat{E}^B\left(dB_{|B}\right)}{dB_{|B}} = |B\rangle\langle B|, \tag{3.1.59}$$

respectively. The relation $\hat{\Pi}^{AB}(\mathbb{R}^2) = \hat{I}$ reads as

$$\int_{-\infty}^{\infty}\int_{-\infty}^{\infty} \hat{W}^{AB}(A, B)\,dA\,dB = \hat{I}. \tag{3.1.60}$$

The counterpart of the relation (3.1.48) reads

$$W^{AB}(A, B) = \text{Tr}\{\hat{\rho}\hat{W}^{AB}(A, B)\}. \tag{3.1.61}$$

The formula (3.1.52) and the mapping theorem (3.1.53) can be rewritten in the forms

$$\hat{M} = \int\int f(A, B)\hat{W}^{AB}(A, B)\,dA\,dB \tag{3.1.62}$$

and

$$\text{Tr}\{\hat{\rho}\hat{M}\} = \int\int f(A, B)W^{AB}(A, B)\,dA\,dB. \tag{3.1.63}$$

To obtain the usual Wigner function for one degree of freedom [Wigner (1932)], we must set $\hat{A} = \hat{q}$, $\hat{B} = \hat{p}$ (see (3.2.11) below). From the commutation relation $[\hat{q}, \hat{p}] = i\hbar\hat{I}$, a great number of further properties and simplifications follow. The usual Wigner function as applied in the framework of quantum optics can be found in section 3.3.

Let us address the question how to proceed when at least one of the physical quantities A and B cannot be associated with an Hermitian operator and its measurement must be treated according to the formulae (3.1.38)–(3.1.43). It can be seen immediately that the definition (3.1.44) of the quantum characteristic function for A and B is not at our disposal. Assuming that the operator $\hat{A}$ exists and resorting to a more general description of the physical quantity B in terms of the operators $\hat{\Pi}(I_B)$, $I_B \subset \mathbb{R}$, which resolve the identity (cf. (3.1.42)),

$$\hat{I} = \int_{-\infty}^{\infty} \hat{\Pi}(dB), \tag{3.1.64}$$

we can define a new quantum characteristic function

$$C_{\text{MEM}}^{AB}(\theta, \tau) = \text{Tr}\left\{\hat{\rho}\exp\left(i\frac{\theta}{2}\hat{A}\right)\widehat{\exp}(i\tau B)\exp\left(i\frac{\theta}{2}\hat{A}\right)\right\}, \qquad (3.1.65)$$

where

$$\widehat{\exp}(i\tau B) = \int_{-\infty}^{\infty}\exp(i\tau B)\hat{\Pi}(dB). \qquad (3.1.66)$$

Here the subscript MEM stands for the matrix elements method and the matrix elements are understood in the basis, where the operator $\hat{A}$ is diagonal. As above we define the quasiprobability measure

$$\Pi_{\text{MEM}}^{AB}(I) = \text{Tr}\left\{\hat{\rho}\hat{\Pi}_{\text{MEM}}^{AB}(I)\right\}, \qquad (3.1.67)$$

where the operators

$$\hat{\Pi}_{\text{MEM}}^{AB}(I)$$

$$= \frac{1}{4\pi^2}\int_{-\infty}^{\infty}\int_{-\infty}^{\infty}\text{Re}\left[\hat{j}_{-I}(\theta, \tau)\exp\left(i\frac{\theta}{2}\hat{A}\right)\widehat{\exp}(i\tau B)\exp\left(i\frac{\theta}{2}\hat{A}\right)\right]d\theta\,d\tau. \qquad (3.1.68)$$

We also use the operator-valued measure $\hat{\Pi}_{\text{MEM}}^{AB}$ to assign operators $\hat{M}_{\text{MEM}}$ to ordinary functions $f(A, B)$,

$$\hat{M}_{\text{MEM}} = \int\int f(A, B)\hat{\Pi}_{\text{MEM}}^{AB}(dAdB). \qquad (3.1.69)$$

Especially, in the case of the products $A^k B^l$,

$$\left[\hat{A}^k\widehat{B^l}\right]_{\text{MEM}}^{AB} = \frac{1}{2^k}\sum_{j=0}^{k}\binom{k}{j}\hat{A}^j\widehat{B^l}\hat{A}^{k-j}, \qquad (3.1.70)$$

where

$$\widehat{B^l} = \int_{-\infty}^{\infty}B^l\hat{\Pi}(dB). \qquad (3.1.71)$$

To include more quantum optical concepts, we set $\hat{A} = \hat{n} = \hat{a}^\dagger\hat{a}$, where $\hat{a}$ is the photon annihilation operator and $\hat{n}$ is the photon-number operator (see section 3.2). It can easily be seen that the method of characteristic function yields the quasidistribution for the number and the quantity B in the form

$$\overline{\Phi}_{\text{MEM}}^{nB}(n, B) = \text{Tr}\left\{\hat{\rho}\hat{\Phi}_{\text{MEM}}^{nB}(n, B)\right\}, \qquad (3.1.72)$$

where

$$\hat{\Phi}_{\text{MEM}}^{nB}(n, B) = \sum_{k=-n}^{n}\left\langle n-k\left|\frac{\hat{\Pi}(dB_{|B})}{dB_{|B}}\right|n+k\right\rangle|n-k\rangle\langle n+k|, \qquad (3.1.73)$$

with $|n \pm k\rangle$ the eigenkets of the operator $\hat{n}$, and $n = 0, \pm\frac{1}{2}, \pm1, \pm\frac{3}{2}, \ldots$. The appropriate quasiprobability measure is found in the form

$$\Pi_{\text{MEM}}^{nB}(I) = \sum_{n=-\infty}^{\infty}{}' \int_{I_B(n)} \overline{\Phi}_{\text{MEM}}^{nB}(n, B) \, dB, \tag{3.1.74}$$

where $I_B(n) = \{B : (n, B) \in I\}$ and $\sum'$ means that the summation step is $\frac{1}{2}$. Our instance of the quantity B is the quantum phase [Lukš and Peřinová (1997a)] (see the relation (4.6.97) in section 4.6). Let us remark that there is no problem, in principle, when the operator $\hat{B}$ exists. In this case one may criticize only our introducing different quasiprobability measures according to (3.1.46) and (3.1.67) simultaneously. Not without a connection with the quantum phase problem, we set $\hat{B} = \hat{P} = -i(\hat{a} - \hat{a}^\dagger)$, $\hat{P}$ is the momentum-like quadrature operator (cf. (3.5.30)). Then we see a pragmatic distinction between the formulae (3.1.46) and (3.1.67), because when determined in a Glauber coherent state, the formula (3.1.67) results in a simple closed formula, whereas the original probability measure $\hat{\Pi}^{nP}(I)$ ((3.1.46)) cannot be obtained. Nevertheless, there exists a sequence of probability measures $\hat{\Pi}_r^{nP}(I)$, $r = 2, 4, 6, \ldots$, which in a sense approximates the desired measure [Peřinová, Lukš and Křepelka (1997b)]. The case $\hat{B} = \hat{a}$ has been related to the study of quantum phase by Dupertuis (1988), who has applied, however, completely different technique of ordering.

3.2   Basic properties of coherent states

The fundamental states enabling us to formulate a description of any quantum system closely to a classical description are coherent states (more extended review of the technique of coherent states can be found in [Peřina (1985), Peřina, Hradil and Jurčo (1994), Walls and Milburn (1994), Vogel and Welsch (1994), Mandel and Wolf (1995)]. In order to obtain such a quantum-classical correspondence a phase space is used. For the optical field in coherent states the normal quantum correlation functions are factorized as follows,

$$\begin{aligned}
\Gamma_{\mathcal{N}}^{(m,n)}(x_1, \ldots, x_{m+n}) &= \text{Tr}\{\hat{\rho}\hat{A}^{(-)}(x_1)\ldots\hat{A}^{(-)}(x_m)\hat{A}^{(+)}(x_{m+1})\ldots\hat{A}^{(+)}(x_{m+n})\} \\
&= V^*(x_1)\ldots V^*(x_m)V(x_{m+1})\ldots V(x_{m+n}), \tag{3.2.1}
\end{aligned}$$

in relation to the same factorization of classical correlation function (2.1.22), expressing that the classical completely coherent field is noiseless (deterministic); here $\hat{\rho}$ is the statistical operator of the optical field, and $\hat{A}^{(+)}(x)$ and $\hat{A}^{(-)}(x)$ are the annihilation and creation operators of a photon at the space–time point $x \equiv (\mathbf{x}, t)$, respectively. This means that the coherent states fulfill the condition of complete coherence of an optical field and they may therefore contain only fluctuations of the physical vacuum. Furthermore the coherent states form an overcomplete set of states and consequently they can be used as a basis for decompositions of vectors and operators.

For simplicity we consider a single-mode field, because the generalization of the corresponding notions to multimode fields is straightforward. There are three ways of the definition of the coherent state $|\alpha\rangle$, which are equivalent for the coherent states of the harmonic oscillator. The first way adopts the displacement operator

$$\hat{D}(\alpha) = \exp(\alpha\hat{a}^\dagger - \alpha^*\hat{a}) \qquad (3.2.2)$$

to define the coherent state

$$|\alpha\rangle = \hat{D}(\alpha)|0\rangle, \qquad (3.2.3)$$

where α is a complex field amplitude, $\hat{a}$ and $\hat{a}^\dagger$ are the photon annihilation and creation operators, respectively, and $|0\rangle$ is the vacuum state. Using the Baker–Hausdorff identity (see, e. g., Louisell (1964))

$$\exp(\hat{A} + \hat{B}) = \exp(\hat{A})\exp(\hat{B})\exp\left(-\frac{1}{2}[\hat{A}, \hat{B}]\right), \qquad (3.2.4)$$

which is valid for any operators $\hat{A}$ and $\hat{B}$ whose commutator $[\hat{A}, \hat{B}] = \hat{A}\hat{B} - \hat{B}\hat{A}$ is a c-number, i. e., $[[\hat{A}, \hat{B}], \hat{A}] = [[\hat{A}, \hat{B}], \hat{B}] = \hat{0}$, we obtain from (3.2.2) and (3.2.3)

$$\begin{aligned} |\alpha\rangle &= \exp(-\frac{1}{2}|\alpha|^2)\exp(\alpha\hat{a}^\dagger)\exp(-\alpha^*\hat{a})|0\rangle \\ &= \exp(-\frac{1}{2}|\alpha|^2)\sum_{n=0}^{\infty}\frac{\alpha^n}{\sqrt{n!}}|n\rangle, \end{aligned} \qquad (3.2.5)$$

where we have expanded both $\exp(\alpha\hat{a}^\dagger)$ and $\exp(-\alpha^*\hat{a})$ in the Taylor series, used the vacuum stability condition $\hat{a}|0\rangle = 0$ and the definition of the Fock (number) state

$$|n\rangle = \frac{\hat{a}^{\dagger n}}{\sqrt{n!}}|0\rangle. \qquad (3.2.6)$$

Taking into account that the annihilation and creation operators act on the Fock state as follows

$$\begin{aligned} \hat{a}|n\rangle &= \sqrt{n}|n-1\rangle, \\ \hat{a}^\dagger|n\rangle &= \sqrt{n+1}|n+1\rangle, \end{aligned} \qquad (3.2.7)$$

we arrive at the second definition of the coherent state as an eigenstate of the annihilation operator,

$$\begin{aligned} \hat{a}|\alpha\rangle &= \alpha|\alpha\rangle, \\ \langle\alpha|\hat{a}^\dagger &= \langle\alpha|\alpha^*, \end{aligned} \qquad (3.2.8)$$

where $\langle\alpha|$ is the Hermitian conjugate state to the state $|\alpha\rangle$. Since the annihilation operator $\hat{a}$ is not Hermitian, the number α is complex. The equations (3.2.8) show that the coherent state $|\alpha\rangle$ must contain an uncertain and indefinite number of photons since the action of the annihilation operator does not change this state except its norm. Whereas the Fock state $|n\rangle$ has a definite number of photons and completely uncertain phase, the coherent state $|\alpha\rangle$, $\alpha \neq 0$, is characterized by the complex number with a definite phase arg α, however the number of photons is uncertain.

In a straightforward way we see from (3.2.5) that the coherent state is normalized, since $\langle\alpha|\alpha\rangle = 1$ taking into account orthonormality of the Fock states, $\langle n|m\rangle = \delta_{nm}$, and also that the probability of n photons being in the coherent state $|\alpha\rangle$ is given by the Poisson probability distribution

$$|\langle n|\alpha\rangle|^2 = \frac{|\alpha|^{2n}}{n!}\exp(-|\alpha|^2), \tag{3.2.9}$$

which means that the photon-number distribution of the coherent state $|\alpha\rangle$ is given by the Poisson distribution with the mean number of photons $\langle\hat{n}\rangle = |\alpha|^2$, where the photon-number operator $\hat{n} = \hat{a}^\dagger\hat{a}$. In the same way we can verify that the coherent states are not orthogonal, since

$$|\langle\alpha|\beta\rangle|^2 = \exp(-|\alpha - \beta|^2). \tag{3.2.10}$$

The coherent states can be considered as approximately orthonormal provided that their distance in the complex plane is large, i. e., if $|\alpha - \beta| \gg 1$.

The third definition of the coherent states demonstrates that these states are minimum-uncertainty wave packets. Defining the operators of the generalized coordinate $\hat{q}$ and momentum $\hat{p}$ in the standard way

$$\hat{q} = \sqrt{\frac{\hbar}{2\omega}}(\hat{a} + \hat{a}^\dagger),$$

$$\hat{p} = -i\sqrt{\frac{\hbar\omega}{2}}(\hat{a} - \hat{a}^\dagger), \tag{3.2.11}$$

fulfilling the commutation rule

$$[\hat{q}, \hat{p}] = i\hbar\hat{1}, \tag{3.2.12}$$

we obtain the following equality for the variances of $\hat{q}$ and $\hat{p}$ in the coherent state $|\alpha\rangle$

$$\langle(\Delta\hat{q})^2\rangle\langle(\Delta\hat{p})^2\rangle = \frac{\hbar^2}{4}, \tag{3.2.13}$$

since for the variances in the coherent state we have

$$\begin{aligned}\langle(\Delta\hat{q})^2\rangle &= \langle\hat{q}^2\rangle - \langle\hat{q}\rangle^2 = \frac{\hbar}{2\omega},\\ \langle(\Delta\hat{p})^2\rangle &= \langle\hat{p}^2\rangle - \langle\hat{p}\rangle^2 = \frac{\hbar\omega}{2}.\end{aligned} \tag{3.2.14}$$

In general, any state has to fulfill the Heisenberg inequality,

$$\langle(\Delta\hat{q})^2\rangle\langle(\Delta\hat{p})^2\rangle \geq \frac{\hbar^2}{4}, \tag{3.2.15}$$

and therefore the coherent state $|\alpha\rangle$ possesses minimum-uncertainty fluctuations, which are related to the vacuum fluctuations. This can be seen as follows. For the fluctuating part of the energy of this mode we obtain

$$\frac{1}{2}[\langle(\Delta\hat{p})^2\rangle + \omega^2\langle(\Delta\hat{q})^2\rangle] = \frac{1}{2}[\langle\hat{p}^2\rangle + \omega^2\langle\hat{q}^2\rangle] - \frac{1}{2}[\langle\hat{p}\rangle^2 + \omega^2\langle\hat{q}\rangle^2] = \frac{\hbar\omega}{2}. \tag{3.2.16}$$

This expression is just the difference of the mean energy and the coherent energy and therefore it gives the incoherent energy equal to the zero-point fluctuations of the value $\hbar\omega/2$.

Taking into account the resolution of the identity in terms of the Fock states

$$\sum_{n=0}^{\infty} |n\rangle\langle n| = \hat{1} \tag{3.2.17}$$

due to the completeness of the Fock states, we can derive the following resolution of the identity in terms of coherent states by substituting from (3.2.5)

$$\frac{1}{\pi}\int|\alpha\rangle\langle\alpha|\,d^2\alpha = \hat{1}, \tag{3.2.18}$$

where the integration is taken over the whole complex-α plane. This condition holds due to the overcompleteness of the coherent states. It can be used to decompose any vector $|\psi\rangle$ in terms of the coherent states,

$$|\psi\rangle = \frac{1}{\pi}\int|\alpha\rangle\langle\alpha|\psi\rangle\,d^2\alpha. \tag{3.2.19}$$

Similarly any operator $\hat{M}$ can be decomposed in the form

$$\hat{M} = \int\int|\alpha\rangle\langle\alpha|\hat{M}|\beta\rangle\langle\beta|\,d^2\alpha\,d^2\beta. \tag{3.2.20}$$

There are some interesting properties of the displacement operator which is unitary,

$$\hat{D}^{\dagger}(\alpha) = \hat{D}^{-1}(\alpha) = \hat{D}(-\alpha). \tag{3.2.21}$$

Using the identities

$$[\hat{a}, \hat{M}] = \frac{\partial\hat{M}}{\partial\hat{a}^{\dagger}}, \quad [\hat{M}, \hat{a}^{\dagger}] = \frac{\partial\hat{M}}{\partial\hat{a}} \tag{3.2.22}$$

valid for any operator $\hat{M}$, which can be decomposed in the Taylor series, we can demonstrate the displacement property of the operator $\hat{D}$,

$$\hat{D}^{-1}(\alpha)\hat{M}(\hat{a}^{\dagger}, \hat{a})\hat{D}(\alpha) = \hat{M}(\hat{a}^{\dagger} + \alpha^*\hat{1}, \hat{a} + \alpha\hat{1}). \tag{3.2.23}$$

Further, using the Baker–Hausdorff identity, we can prove a group-theoretical property

$$\hat{D}(\alpha)\hat{D}(\beta) = \hat{D}(\alpha + \beta)\exp\left[\frac{1}{2}(\alpha\beta^* - \alpha^*\beta)\right].$$ (3.2.24)

The displacement operators form a complete set of operators, because they are eigenoperators $\hat{U}$ of a superoperator, the eigenvalue problem being $[\hat{a}, \hat{U}] = \lambda\hat{U}$, $\lambda = \alpha$, $\hat{U} = \hat{D}(\alpha)$. They generalize the Fourier exponential function in the complex form $\exp(\alpha\beta^* - \alpha^*\beta)$ and they are orthogonal, since it holds that

$$\text{Tr}\{\hat{D}^\dagger(\alpha)\hat{D}(\beta)\} = \pi\delta(\alpha - \beta),$$ (3.2.25)

where we have introduced the Dirac δ function in the complex form

$$\delta(\alpha) = \frac{1}{\pi^2}\int \exp(\alpha\beta^* - \alpha^*\beta)\,d^2\beta.$$ (3.2.26)

If we assume to have an operator $\hat{M}^{(N)}(\hat{a}^\dagger, \hat{a})$ in the normal form (all creation operators stand to the left of all annihilation operators), we obtain its expectation value in the coherent state

$$\langle\alpha|\hat{M}^{(N)}(\hat{a}^\dagger, \hat{a})|\alpha\rangle = M^{(N)}(\alpha^*, \alpha).$$ (3.2.27)

This is a very useful rule enabling us to replace the quantum expectation value by a classical expression in the complex α plane, replacing the annihilation and creation operators by the complex numbers, $\hat{a} \to \alpha$, $\hat{a}^\dagger \to \alpha^*$. As a special case, when detecting the single-mode electromagnetic field by photodetection, we obtain the measured correlation functions in the monomial form and it holds that

$$\langle\alpha|\hat{a}^{\dagger k}\hat{a}^l|\alpha\rangle = \alpha^{*k}\alpha^l.$$ (3.2.28)

It is often helpful to use the following rules when calculating with the coherent states [Lukš and Peřinová (1987)],

$$\hat{a}^\dagger|\alpha\rangle\langle\alpha| = |\alpha\rangle\langle\alpha|\left(\alpha^* + \frac{\overleftarrow{\partial}}{\partial\alpha}\right),$$

$$|\alpha\rangle\langle\alpha|\hat{a} = \left(\alpha + \frac{\partial}{\partial\alpha^*}\right)|\alpha\rangle\langle\alpha|,$$ (3.2.29)

or more generally for an operator function $\hat{M}$

$$\hat{M}(\hat{a}^\dagger, \hat{a})|\alpha\rangle\langle\alpha| = |\alpha\rangle\langle\alpha|M\left(\alpha^* + \frac{\overleftarrow{\partial}}{\partial\alpha}, \alpha\right),$$

$$|\alpha\rangle\langle\alpha|\hat{M}(\hat{a}^\dagger, \hat{a}) = M\left(\alpha^*, \alpha + \frac{\partial}{\partial\alpha^*}\right)|\alpha\rangle\langle\alpha|.$$ (3.2.30)

Taking the coherent-state average of (3.2.30), we arrive at the c-number expressions

$$\langle\alpha|\hat{M}(\hat{a}^\dagger,\hat{a})|\alpha\rangle = 1M\left(\alpha^* + \overleftarrow{\frac{\partial}{\partial\alpha}},\alpha\right) = M\left(\alpha^*,\alpha + \frac{\partial}{\partial\alpha^*}\right)1. \qquad (3.2.31)$$

Sometimes even and odd coherent states are useful (section 3.10) [Malkin and Man'ko (1970)] defined as

$$\begin{aligned}|\alpha_+\rangle &= \frac{1}{2}\exp\left(\frac{1}{2}|\alpha|^2\right)\cosh^{-1/2}(|\alpha|^2)(|\alpha\rangle + |-\alpha\rangle) \\ &= \cosh^{-1/2}(|\alpha|^2)\sum_{n=0}^{\infty}\frac{\alpha^{2n}}{\sqrt{(2n)!}}|2n\rangle\end{aligned} \qquad (3.2.32)$$

and

$$\begin{aligned}|\alpha_-\rangle &= \frac{1}{2}\exp\left(\frac{1}{2}|\alpha|^2\right)\sinh^{-1/2}(|\alpha|^2)(|\alpha\rangle - |-\alpha\rangle) \\ &= \sinh^{-1/2}(|\alpha|^2)\sum_{n=0}^{\infty}\frac{\alpha^{2n+1}}{\sqrt{(2n+1)!}}|2n+1\rangle,\end{aligned} \qquad (3.2.33)$$

in correspondence to the two-photon displacement operators

$$\frac{1}{2}[\hat{D}(\alpha) + \hat{D}(-\alpha)] = \cosh(\alpha\hat{a}^\dagger - \alpha^*\hat{a}) \qquad (3.2.34)$$

and

$$\frac{1}{2}[\hat{D}(\alpha) - \hat{D}(-\alpha)] = \sinh(\alpha\hat{a}^\dagger - \alpha^*\hat{a}), \qquad (3.2.35)$$

respectively. These are eigenstates of $\hat{a}^2$,

$$\hat{a}^2|\alpha_\pm\rangle = \alpha^2|\alpha_\pm\rangle. \qquad (3.2.36)$$

3.3 Quantum characteristic functions and quasidistributions

In quantum optics we deal more with mixed states than with pure states and therefore we have to use the statistical operator to describe general optical fields and systems. For this we need to develop the statistical operator $\hat{\rho}$ in terms of the coherent states. In general, we can apply the decomposition (3.2.20) to the statistical operator to obtain

$$\hat{\rho} = \frac{1}{\pi^2}\int\int|\alpha\rangle\langle\alpha|\hat{\rho}|\beta\rangle\langle\beta|\,d^2\alpha\,d^2\beta, \qquad (3.3.1)$$

which is an expansion of the statistical operator in terms of the coherent-state matrix elements $\langle\alpha|\hat{\rho}|\beta\rangle$ and the dyadic products $|\alpha\rangle\langle\beta|$. Adopting the nondiagonal decomposition of the statistical operator in terms of the Fock states

$$\hat{\rho} = \sum_{n=0}^{\infty} \sum_{m=0}^{\infty} \rho_{nm} |n\rangle\langle m|, \tag{3.3.2}$$

where the matrix elements are defined as $\rho_{nm} = \langle n|\hat{\rho}|m\rangle$, we arrive at the following relation of the coherent-state and Fock-state matrix elements of the statistical operator

$$\langle\alpha|\hat{\rho}|\beta\rangle = \sum_{n=0}^{\infty} \sum_{m=0}^{\infty} \rho_{nm} \frac{\alpha^{*n}\beta^m}{\sqrt{n!m!}} \exp\left(-\frac{1}{2}|\alpha|^2 - \frac{1}{2}|\beta|^2\right). \tag{3.3.3}$$

Since $\text{Tr}\{\hat{\rho}^2\} = \sum_{n,m} |\rho_{nm}|^2 \leq 1$, as a consequence of the fact that $|\rho_{nm}|^2 \leq \rho_{nn}\rho_{mm}$, i. e., $\text{Tr}\{\hat{\rho}^2\} \leq (\text{Tr } \hat{\rho})^2 = 1$, the series in (3.3.3) converges for all finite values of $|\alpha|, |\beta|$ and so $\langle\alpha|\hat{\rho}|\beta\rangle$ is an entire function in α and β, i. e., it is a well-behaved function. However, exploiting the resolution of the identity given in (3.2.18), we can expand the statistical operator in the diagonal form called the Glauber–Sudarshan representation [Glauber (1963a, b), Sudarshan (1963)]

$$\hat{\rho} = \int \Phi_{\mathcal{N}}(\alpha)|\alpha\rangle\langle\alpha|\, d^2\alpha, \tag{3.3.4}$$

which represents a mixture of projection operators $|\alpha\rangle\langle\alpha|$ onto the coherent states $((|\alpha\rangle\langle\alpha|)^2 = |\alpha\rangle\langle\alpha|)$, where $\Phi_{\mathcal{N}}(\alpha)$ is a weighting function. From the normalization condition $\text{Tr }\hat{\rho} = 1$, this function is also normalized,

$$\int \Phi_{\mathcal{N}}(\alpha)\, d^2\alpha = 1, \tag{3.3.5}$$

and it follows from the Hermiticity $\hat{\rho}^\dagger = \hat{\rho}$ that it is a real function of the complex variable, $[\Phi_{\mathcal{N}}(\alpha)]^* = \Phi_{\mathcal{N}}(\alpha)$.

   The diagonal form (3.3.4) of the statistical operator is very suitable for calculations of quantum expectation values of normally ordered operators, e. g., as given in (3.2.28) in the case of photodetection,

$$\begin{aligned}
\text{Tr}\{\hat{\rho}\hat{a}^{\dagger k}\hat{a}^l\} &= \text{Tr}\left\{\int \Phi_{\mathcal{N}}(\alpha)|\alpha\rangle\langle\alpha|\, d^2\alpha\, \hat{a}^{\dagger k}\hat{a}^l\right\} \\
&= \int \Phi_{\mathcal{N}}(\alpha)\langle\alpha|\hat{a}^{\dagger k}\hat{a}^l|\alpha\rangle\, d^2\alpha = \int \Phi_{\mathcal{N}}(\alpha)\alpha^{*k}\alpha^l\, d^2\alpha \\
&= \langle\alpha^{*k}\alpha^l\rangle_{\mathcal{N}}, \tag{3.3.6}
\end{aligned}$$

using the eigenvalue properties (3.2.8) of the coherent states. The suffix $\mathcal{N}$ at Φ expresses the fact that this function is related to averaging of the normally ordered operators. Consequently, the quantum expectation value of a normally ordered operator can be expressed as a "classical" expectation value in a generalized phase space with the quasiprobability density, in brief, quasidistribution $\Phi_{\mathcal{N}}(\alpha)$, if the integration is carried out over the whole complex-α plane. Thus the Glauber–Sudarshan representation of the statistical operator provides a basis for a formal equivalence between quantum and classical descriptions of optical beams. For the field in the coherent state $|\beta\rangle$ we have

$$\Phi_{\mathcal{N}}(\alpha) = \delta(\alpha - \beta) \tag{3.3.7}$$

and the statistical operator $\hat{\rho}$ is just the projection operator $|\beta\rangle\langle\beta|$ onto the coherent state. Although the weighting function $\Phi_{\mathcal{N}}(\alpha)$ has some properties of a probability density function (e. g., it is a real-valued function fulfilling the normalization), it cannot be generally interpreted as a real probability density function, because it can take on negative values and has singularities stronger than the Dirac generalized function for all pure states except the coherent ones as a reflection of quantum properties of described systems. As a consequence of the fact that $\operatorname{Re}\alpha$ and $\operatorname{Im}\alpha$ are eigenvalues of the noncommuting Hermitian operators $(\hat{a} + \hat{a}^\dagger)/2$ and $(\hat{a} - \hat{a}^\dagger)/2i$ in the coherent state and therefore they cannot be measured simultaneously with arbitrary accuracy, the quasidistribution $\Phi_{\mathcal{N}}(\alpha)$ cannot be measured directly. However, as discussed in section 4.12, this function is accessible indirectly by homodyne tomography measurements. It represents one of the set of quasidistributions.

Using the diagonal representation (3.3.4), we can obtain relations between the Fock-state matrix elements or coherent-state matrix elements of the statistical operator and the quasidistribution $\Phi_{\mathcal{N}}(\alpha)$ as follows

$$
\begin{aligned}
\langle n|\hat{\rho}|m\rangle &= \rho_{nm} = \int \Phi_{\mathcal{N}}(\alpha)\langle n|\alpha\rangle\langle\alpha|m\rangle \, d^2\alpha \\
&= \int \Phi_{\mathcal{N}}(\alpha)\frac{\alpha^n \alpha^{*m}}{\sqrt{n!m!}} \exp(-|\alpha|^2) \, d^2\alpha.
\end{aligned} \tag{3.3.8}
$$

Taking the diagonal coherent-state matrix element of the statistical operator, we can define another quasidistribution denoted as

$$\Phi_{\mathcal{A}}(\alpha) = \frac{1}{\pi}\langle\alpha|\hat{\rho}|\alpha\rangle = \frac{1}{\pi}\int \Phi_{\mathcal{N}}(\beta)\exp(-|\alpha - \beta|^2) \, d^2\beta. \tag{3.3.9}$$

Thus this quasidistribution is obtained as the convolution of the quasidistribution $\Phi_{\mathcal{N}}(\alpha)$ related to normally ordered operators and the exponential function $\exp(-|\beta|^2) = |\langle 0|\beta\rangle|^2$ according to the commutation rule $\hat{a}\hat{a}^\dagger = \hat{a}^\dagger\hat{a} + \hat{1}$ and $\langle|\alpha|^2\rangle_{\mathcal{A}} = \langle|\alpha|^2\rangle_{\mathcal{N}} + 1$. The quasidistribution function $\Phi_{\mathcal{A}}(\alpha)$, called also the Husimi quasidistribution, is really related to the antinormal ordering of field operators (all annihilation operators are to the left of all creation operators),

$$
\begin{aligned}
\operatorname{Tr}\{\hat{\rho}\hat{a}^l\hat{a}^{\dagger k}\} &= \operatorname{Tr}\{\hat{a}^{\dagger k}\hat{\rho}\hat{a}^l\} = \operatorname{Tr}\left\{\frac{1}{\pi}\int |\alpha\rangle\langle\alpha| \, d^2\alpha \, \hat{a}^{\dagger k}\hat{\rho}\hat{a}^l\right\} \\
&= \frac{1}{\pi}\int \alpha^{*k}\alpha^l\langle\alpha|\hat{\rho}|\alpha\rangle \, d^2\alpha,
\end{aligned} \tag{3.3.10}
$$

where we have substituted the identity operator in terms of coherent states and carried out a cyclic permutation of the operators, which does not change the trace. Putting $k = l = 0$ in (3.3.10), we see again that the quasidistribution $\Phi_{\mathcal{A}}(\alpha)$ is normalized,

$$\frac{1}{\pi} \int \Phi_{\mathcal{A}}(\alpha) \, d^2\alpha = 1. \tag{3.3.11}$$

Moreover, the quasidistribution $\Phi_{\mathcal{A}}(\alpha)$ related to the antinormal ordering of field operators is always nonnegative, similarly as classical distributions. This can be seen as follows, if $|\psi_n\rangle$ is an orthonormal basis in which the statistical operator is diagonal with eigenvalues λ_n,

$$\hat{\rho} = \sum_n \lambda_n |\psi_n\rangle\langle\psi_n|, \tag{3.3.12}$$

then

$$\Phi_{\mathcal{A}}(\alpha) = \frac{1}{\pi} \sum_n \lambda_n |\langle\alpha|\psi_n\rangle|^2 \geq 0, \tag{3.3.13}$$

and also

$$\Phi_{\mathcal{A}}(\alpha) \leq \frac{1}{\pi} \sum_n |\langle\alpha|\psi_n\rangle|^2 = \frac{1}{\pi}, \tag{3.3.14}$$

because $0 \leq \lambda_n \leq 1$ and the coherent state $|\alpha\rangle$ is normalized ($|\alpha\rangle = \sum_n |\psi_n\rangle\langle\psi_n|\alpha\rangle$, and consequently $\langle\alpha|\alpha\rangle = \sum_n |\langle\alpha|\psi_n\rangle|^2 = 1$). We can also express this quasidistribution in terms of the Fock-state matrix elements of the statistical operator as follows

$$\Phi_{\mathcal{A}}(\alpha) = \frac{1}{\pi} \sum_{n,m} \rho_{nm} \frac{\alpha^{*n}\alpha^m}{\sqrt{n!m!}} \exp(-|\alpha|^2). \tag{3.3.15}$$

It is obvious from (3.3.9) that the vacuum fluctuations smooth behaviour when one goes from $\Phi_{\mathcal{N}}(\alpha)$ to $\Phi_{\mathcal{A}}(\alpha)$.

In analogy to a classical characteristic function we can define its quantum analogue

$$C(\beta) = \mathrm{Tr}\{\hat{\rho}\hat{D}(\beta)\} = \mathrm{Tr}\{\hat{\rho}\exp(\beta\hat{a}^\dagger - \beta^*\hat{a})\}, \tag{3.3.16}$$

where β is a parameter of the quantum characteristic function, and making use of the Baker–Hausdorff identity, we can obtain

$$C(\beta) = \exp\left(-\frac{1}{2}|\beta|^2\right) C_{\mathcal{N}}(\beta) = \exp\left(\frac{1}{2}|\beta|^2\right) C_{\mathcal{A}}(\beta), \tag{3.3.17}$$

where the normal characteristic function is given by

$$C_{\mathcal{N}}(\beta) = \mathrm{Tr}\{\hat{\rho}\exp(\beta\hat{a}^\dagger)\exp(-\beta^*\hat{a})\} = \int \Phi_{\mathcal{N}}(\alpha)\exp(\alpha^*\beta - \alpha\beta^*) \, d^2\alpha \tag{3.3.18}$$

and the antinormal characteristic function is given by

$$C_{\mathcal{A}}(\beta) = \mathrm{Tr}\{\hat{\rho}\exp(-\beta^*\hat{a})\exp(\beta\hat{a}^\dagger)\} = \int \Phi_{\mathcal{A}}(\alpha)\exp(\alpha^*\beta - \alpha\beta^*) \, d^2\alpha. \tag{3.3.19}$$

Employing the complex form (3.2.26) of the Dirac function, we can use the inverse Fourier transformation to obtain the quasidistributions in terms of the characteristic functions

$$\Phi_{\mathcal{N}}(\alpha) = \frac{1}{\pi^2} \int C_{\mathcal{N}}(\beta) \exp(\alpha\beta^* - \alpha^*\beta) \, d^2\beta \qquad (3.3.20)$$

and

$$\Phi_{\mathcal{A}}(\alpha) = \frac{1}{\pi^2} \int C_{\mathcal{A}}(\beta) \exp(\alpha\beta^* - \alpha^*\beta) \, d^2\beta. \qquad (3.3.21)$$

The Wigner quasidistribution (function) $\Phi_S(\alpha)$ is determined in terms of the characteristic function (3.3.16) by

$$\Phi_S(\alpha) = \frac{1}{\pi^2} \int C(\beta) \exp(\alpha\beta^* - \alpha^*\beta) \, d^2\beta. \qquad (3.3.22)$$

The moments of the Wigner function can be calculated as follows

$$\int \alpha^{*k}\alpha^l \Phi_S(\alpha) \, d^2\alpha = \frac{\partial^k}{\partial\beta^k} \frac{\partial^l}{\partial(-\beta^*)^l} C(\beta) \bigg|_{\beta=\beta^*=0} = \langle \alpha^{*k}\alpha^l \rangle_S, \qquad (3.3.23)$$

where the suffix at the last bracket in (3.3.23) denotes the Weyl symmetrical ordering of the annihilation and creation operators, for instance, $\langle \alpha^{*2}\alpha^2 \rangle_S$ relates to

$$\{\hat{a}^{\dagger 2}\hat{a}^2\}_S = \frac{1}{3!} \left[\hat{a}^{\dagger 2}\hat{a}^2 + \hat{a}^\dagger\hat{a}\hat{a}^\dagger\hat{a} + \hat{a}^\dagger\hat{a}^2\hat{a}^\dagger + \hat{a}\hat{a}^{\dagger 2}\hat{a} + \hat{a}\hat{a}^\dagger\hat{a}\hat{a}^\dagger + \hat{a}^2\hat{a}^{\dagger 2} \right]. \qquad (3.3.24)$$

The quantum characteristic function $C(\beta)$ is expressed in terms of the Wigner function as

$$C(\beta) = \int \Phi_S(\alpha) \exp(\alpha^*\beta - \alpha\beta^*) \, d^2\alpha. \qquad (3.3.25)$$

It is evident from (3.3.23), putting $k = l = 0$, that also the Wigner function is normalized,

$$\int \Phi_S(\alpha) \, d^2\alpha = C(0) = 1. \qquad (3.3.26)$$

Even if the Wigner function is regular and bounded, $|\Phi_S(\alpha)| \leq 2/\pi$, it can take on negative values as a consequence of quantum behaviour of physical systems. Of course, the normally and antinormally ordered moments $\langle \alpha^{*k}\alpha^l \rangle_{\mathcal{N}}$ and $\langle \alpha^{*k}\alpha^l \rangle_{\mathcal{A}}$ can be derived from the normal and antinormal characteristic functions $C_{\mathcal{N}}(\beta)$ and $C_{\mathcal{A}}(\beta)$ with the help of derivatives in the same way as in (3.3.23).

Taking into account the relation (3.3.17) between characteristic functions, we can derive, using the Fourier transformation, the following relations between the introduced quasidistributions,

$$\Phi_{\mathcal{A}}(\alpha) = \frac{2}{\pi} \int \exp(-2|\alpha - \beta|^2)\Phi_S(\beta) \, d^2\beta \qquad (3.3.27)$$

and

$$\Phi_S(\alpha) = \frac{2}{\pi} \int \exp(-2|\alpha - \beta|^2)\Phi_{\mathcal{N}}(\beta) \, d^2\beta, \qquad (3.3.28)$$

in addition to (3.3.9). In general, the quasidistributions $\Phi_A(\alpha)$ and $\Phi_S(\alpha)$ on the left-hand side represent averages of the quasidistributions $\Phi_S(\beta)$ and $\Phi_N(\beta)$ on the right-hand side with exponential weighting factors corresponding to vacuum fluctuations. This averaging process going successively from Φ_N through Φ_S to Φ_A tends to smooth out any unruly behaviour of Φ_N and to transform it into a smooth function Φ_A, which is regular and nonnegative. Whereas the quasidistributions Φ_A and Φ_S exist for all quantum states, the cases in which quasidistribution Φ_N cannot be defined in the ordinary sense are precisely those in which the integral equations (3.3.9) and (3.3.28) cannot be solved.

Adopting the normal characteristic function (3.3.18), we can easily specify the type of interaction of radiation field with matter leading to the field in the coherent state. Considering a multimode field interacting with matter, we have for the Hamiltonian

$$\hat{H} = \sum_{j,k} f_{jk}(t)\hat{a}_j^\dagger \hat{a}_k + \sum_k [g_k(t)\hat{a}_k^\dagger + g_k^*(t)\hat{a}_k] + h(t), \qquad (3.3.29)$$

where $f_{jk}(t) = f_{kj}^*(t)$, $h^*(t) = h(t)$, and $g_k(t)$ are arbitrary deterministic functions of time. The diagonal elements in the first term of this Hamiltonian describe free oscillations, the off-diagonal elements represent the exchange of energy between modes, the second term in (3.3.29) is the interaction term between classical deterministic currents and the radiation modes, and the last term in (3.3.29) is a classical energy background. When writing the Heisenberg equations of motion for annihilation operators $\hat{a}_j(t)$ using (3.3.29), we obtain a set of equations for the annihilation operators, which does not contain any creation operators. Solving this system in principle and substituting this solution into the normal characteristic function (3.3.18) for $\hat{a}_j(t)$ and $\hat{a}_j^\dagger(t)$, this characteristic function will also be normally ordered in the initial operators $\hat{a}_k(0)$ and $\hat{a}_k^\dagger(0)$ and therefore the initial field statistics will be conserved. This means that also the initial coherent state remains coherent for all times (only the complex amplitude $\alpha_j(t)$ develops) and the vacuum state develops into a coherent state. Consequently the optical field in the coherent state is generated by a system of nonrandom prescribed classical currents.

The quasidistributions $\Phi_N(\alpha)$ and $\Phi_A(\alpha)$ can be obtained directly from the statistical operator $\hat{\rho}$. We have

$$\Phi_A(\alpha) = \frac{1}{\pi}\langle\alpha|\hat{\rho}^{(N)}|\alpha\rangle = \frac{1}{\pi}\rho^{(N)}(\alpha^*, \alpha). \qquad (3.3.30)$$

Writing the statistical operator in the antinormal form

$$\hat{\rho}^{(A)} = \sum_{n,m} G_{nm}\hat{a}^n \hat{a}^{\dagger m}, \qquad (3.3.31)$$

with some decomposition coefficients G_{nm}, substituting the identity operator in terms of the coherent states and comparing the result with the Glauber–Sudarshan representation of the statistical operator, we arrive at

$$\hat{\rho}^{(A)} = \sum_{n,m} G_{nm} \hat{a}^n \frac{1}{\pi} \int |\alpha\rangle\langle\alpha| \, d^2\alpha \, \hat{a}^{\dagger m}$$

$$= \frac{1}{\pi} \int \sum_{n,m} G_{nm} \alpha^n \alpha^{*m} |\alpha\rangle\langle\alpha| \, d^2\alpha = \int \Phi_{\mathcal{N}}(\alpha)|\alpha\rangle\langle\alpha| \, d^2\alpha, \qquad (3.3.32)$$

from which it follows that

$$\Phi_{\mathcal{N}}(\alpha) = \frac{1}{\pi}\rho^{(A)}(\alpha^*, \alpha). \qquad (3.3.33)$$

Thus we conclude that the quasidistribution $\Phi_{\mathcal{N}}(\alpha)$ ($\Phi_{\mathcal{A}}(\alpha)$) can be obtained from the antinormal (normal) form of the statistical operator (making use of the commutation rules for the field operators) by means of the substitutions $\hat{a} \to \alpha$, $\hat{a}^{\dagger} \to \alpha^*$; vice versa substituting $\alpha^{*m}\alpha^n \to \hat{a}^n\hat{a}^{\dagger m}$ ($\alpha^{*m}\alpha^n \to \hat{a}^{\dagger m}\hat{a}^n$) in $\Phi_{\mathcal{N}}(\alpha)$ ($\Phi_{\mathcal{A}}(\alpha)$), we obtain the antinormal (normal) form of the statistical operator.

As an example we can assume the optical field being the superposition of the coherent state $|\beta\rangle$ with noise specified by the mean number of photons $\langle\hat{n}\rangle$, which is described by the statistical operator

$$\hat{\rho} = \frac{\langle\hat{n}\rangle^{\hat{b}^{\dagger}\hat{b}}}{(1 + \langle\hat{n}\rangle)^{\hat{b}^{\dagger}\hat{b}+\hat{1}}}, \qquad (3.3.34)$$

where $\hat{b} = \hat{a} - \beta\hat{1}$ and $[\hat{b}, \hat{b}^{\dagger}] = [\hat{a}, \hat{a}^{\dagger}] = \hat{1}$. After simple algebraic manipulations [Peřina (1985)] and using the above rules (3.3.33) and (3.3.30), we arrive at the quasidistributions describing the single-mode superposition of the coherent and chaotic fields

$$\Phi_{\mathcal{N}}(\alpha) = \frac{1}{\pi\langle\hat{n}\rangle} \exp\left(-\frac{|\alpha - \beta|^2}{\langle\hat{n}\rangle}\right) \qquad (3.3.35)$$

and

$$\Phi_{\mathcal{A}}(\alpha) = \frac{1}{\pi(\langle\hat{n}\rangle + 1)} \exp\left(-\frac{|\alpha - \beta|^2}{\langle\hat{n}\rangle + 1}\right). \qquad (3.3.36)$$

If the chaotic field is in thermal equilibrium, then

$$\langle\hat{n}\rangle = \frac{1}{e^{\Theta} - 1}, \qquad (3.3.37)$$

where $\Theta = \hbar\omega/KT$, K being the Boltzmann constant, and we obtain three equivalent forms of the statistical operator for this field,

$$\begin{aligned}
\hat{\rho} &= (1 - e^{-\Theta})\exp[-\Theta(\hat{a}^{\dagger} - \beta^*\hat{1})(\hat{a} - \beta\hat{1})], \\
\hat{\rho}^{(\mathcal{N})} &= (1 - e^{-\Theta})\{\exp[-(1 - e^{-\Theta})(\hat{a}^{\dagger} - \beta^*\hat{1})(\hat{a} - \beta\hat{1})]\}_{\mathcal{N}}, \\
\hat{\rho}^{(\mathcal{A})} &= (e^{\Theta} - 1)\{\exp[-(e^{\Theta} - 1)(\hat{a}^{\dagger} - \beta^*\hat{1})(\hat{a} - \beta\hat{1})]\}_{\mathcal{A}},
\end{aligned} \qquad (3.3.38)$$

where the subscripts $\mathcal{N}$ and $\mathcal{A}$ mean operations of the normal and antinormal orderings, respectively.

If the field α is a result of the superposition of independent fields α_j, $j = 1, 2, ...$, specified by partial quasidistributions $\Phi_{\mathcal{N}}^{(j)}(\alpha_j)$, then the resulting field is described by the convolution law

$$\Phi_{\mathcal{N}}(\alpha) = \int ... \int \delta\Big(\alpha - \sum_j \alpha_j\Big) \prod_k \Phi_{\mathcal{N}}^{(k)}(\alpha_k) \, d^2\alpha_k, \qquad (3.3.39)$$

in close analogy to the classical convolution law for the probability distributions of statistically independent quantities and in agreement with the superposition principle.

3.4   Operator orderings

In the previous treatment we have concentrated our attention to particular orderings of field operators, i. e., to the normal ordering related to photodetection process detecting the electromagnetic field by means of the detection of photons, to the antinormal ordering related to the detection of field by emitting a stimulated photon to the field [Mandel (1966), Peřina (1985), section 16.4, Mandel and Wolf (1995), section 12.9.1] and to the Weyl (symmetrical) ordering related, for instance, to the detection of field by means of a scattering process, in which one photon is absorbed and another one emitted [Crosignani, Di Porto and Bertolotti (1975)]. With respect to general phase-space descriptions [Klauder and Sudarshan (1968), Haken (1970), Louisell (1973), Loudon (1973), Schubert and Wilhelmi (1986), Peřina (1985, 1991), Meystre and Sargent III (1991), Gardiner (1991), Vogel and Welsch (1994), Walls and Milburn (1994), Peřina, Hradil and Jurčo (1994), Mandel and Wolf (1995), Wünsche (1996b)] of free optical fields and optical fields in interaction with matter, in particular in nonlinear media, and their detection, especially homodyne detection, a generalization of the operator-ordering technique is appropriate, enabling us, for instance, to interpolate between the normal and antinormal orderings. Such a theory was basically developed by Cahill and Glauber (1969a, b) and Agarwal and Wolf (1970a, b, c) (for a review, see [Peřina (1985), chapter 16], which also includes a multimode formulation). This theory is not only related to modern quantum optical measurements, but it also makes clear how singularities, unruly behaviour, and negative values appear in quasidistributions.

The completeness property of the displacement operator $\hat{D}(\alpha)$ enables us to decompose any bounded operator $\hat{A}$, for which the trace $\mathrm{Tr}\{\hat{A}^\dagger \hat{A}\}$ is finite, i. e., the Hilbert–Schmidt operator, in terms of the displacement operator $\hat{D}(\alpha)$ in the form

$$\hat{A} = \frac{1}{\pi} \int \tilde{a}(\beta) \hat{D}^\dagger(\beta) \, d^2\beta, \qquad (3.4.1)$$

where the coefficient of decomposition is

$$\tilde{a}(\beta) = \mathrm{Tr}\{\hat{D}(\beta)\hat{A}\}. \qquad (3.4.2)$$

Having another operator $\hat{B}$, we obtain similarly

$$\text{Tr}\{\hat{A}\hat{B}\} = \frac{1}{\pi} \int \tilde{a}(-\beta)\tilde{b}(\beta)\, d^2\beta. \tag{3.4.3}$$

If we choose $\hat{B} = \hat{A}^\dagger$, we obtain for the norm of the operator

$$||\hat{A}||^2 = \text{Tr}\{\hat{A}^\dagger\hat{A}\} = \frac{1}{\pi} \int |\tilde{a}(\beta)|^2\, d^2\beta. \tag{3.4.4}$$

Thus, if $\tilde{a}(\beta)$ is a square integrable function, then the Hilbert–Schmidt norm $||\hat{A}|| = \sqrt{\text{Tr}\{\hat{A}^\dagger\hat{A}\}}$ is finite and vice versa. In particular, for the statistical operator $\hat{\rho} = \hat{\rho}^\dagger = \hat{A}$ we have $\text{Tr}\{\hat{\rho}^2\} = \int |C(\beta)|^2\, d^2\beta/\pi \leq 1$ and consequently $0 \leq |C(\beta)| \leq 1$. If the statistical operator $\hat{\rho}$ is used in (3.4.1) instead of $\hat{A}$, we have the representation of the statistical operator in terms of the displacement operator $\hat{D}(\beta)$. Then the decomposition coefficient $\tilde{a}(\beta)$ in (3.4.2) is equal to the characteristic function $C(\beta)$.

Considering a filter function $\Omega(\beta^*, \beta)$, we can introduce the Ω-ordered displacement operator as follows

$$\hat{D}(\beta, \Omega) = \Omega(\beta^*, \beta)\hat{D}(\beta) \equiv \{\hat{D}(\beta)\}_\Omega. \tag{3.4.5}$$

Further we can introduce a quantization of the displaced Dirac δ function performing the Fourier transformation of (3.4.5)

$$\begin{aligned}
\hat{\Lambda}(\alpha, \Omega) &= \frac{1}{\pi} \int \hat{D}(\beta, \Omega) \exp(\alpha\beta^* - \alpha^*\beta)\, d^2\beta \\
&= \frac{1}{\pi} \int \Omega(\beta^*, \beta)\hat{D}(\beta) \exp(\alpha\beta^* - \alpha^*\beta)\, d^2\beta \\
&= \frac{1}{\pi} \int \Omega(\beta^*, \beta) \exp[\beta(\hat{a}^\dagger - \alpha^*\hat{1}) - \beta^*(\hat{a} - \alpha\hat{1})]\, d^2\beta \\
&\equiv \pi\{\delta(\hat{a} - \alpha\hat{1})\}_\Omega, \tag{3.4.6}
\end{aligned}$$

which means that the operator $\pi^{-1}\hat{\Lambda}(\alpha, \Omega)$ represents the Ω-ordered form of the displaced Dirac function $\delta(\hat{a} - \alpha\hat{1})$.

The filter function $\Omega(\alpha, \beta)$ may be an entire analytic function, $\Omega(0, 0) = 1$, if it possesses no zeros, it represents a minimum filter, which can be symmetrical, $\Omega(-\alpha, -\beta) = \Omega(\alpha, \beta)$, etc. Using the Weierstrass theorem on entire functions, the filter function $\Omega(\alpha, \beta)$ has the exponential form $\exp[\omega(\alpha, \beta)]$, where $\omega(\alpha, \beta)$ is an entire analytic function with $\omega(0, 0) = 0$. The most fruitful choice is $\Omega(\beta^*, \beta) = \exp(s|\beta|^2/2)$, because it includes the normal ordering ($s = 1$), the symmetrical ordering ($s = 0$), and the antinormal ordering ($s = -1$), as can be seen from the following forms of the Ω-ordered displacement operator,

$$\begin{aligned}
\hat{D}(\beta, \Omega) &\equiv \hat{D}(\beta, s) = \exp\left(\frac{s}{2}|\beta|^2\right) \exp(\beta\hat{a}^\dagger - \beta^*\hat{a}) \\
&= \exp\left(\frac{s-1}{2}|\beta|^2\right) \exp(\beta\hat{a}^\dagger) \exp(-\beta^*\hat{a}) \\
&= \exp\left(\frac{s+1}{2}|\beta|^2\right) \exp(-\beta^*\hat{a}) \exp(\beta\hat{a}^\dagger). \tag{3.4.7}
\end{aligned}$$

Here we have used the Baker–Hausdorff identity.

If we define the $\tilde{\Omega}$ ordering as the Ω ordering with $\Omega(\beta^*, \beta) \to \Omega^{-1}(\beta^*, \beta)$, we see that $\hat{D}^\dagger(\beta, \Omega) = \hat{D}(-\beta, \Omega^*) = \hat{D}^{-1}(\beta, \Omega^{*-1}) = \hat{D}^{-1}(\beta, \tilde{\Omega}^*)$ and $\hat{\Delta}^\dagger(\alpha, \Omega) = \hat{\Delta}(\alpha, \Omega^*)$. In particular, for the s-ordering, we have $\hat{D}^\dagger(\beta, s) = \hat{D}(-\beta, s^*) = \hat{D}^{-1}(\beta, -s^*)$ and $\hat{\Delta}^\dagger(\alpha, s) = \hat{\Delta}(\alpha, s^*)$.

The system of Ω-ordered displacement operators and that of $\tilde{\Omega}$-ordered displacement operators are biorthogonal,

$$\text{Tr}\{\hat{D}(\beta, \Omega)\hat{D}(\gamma, \tilde{\Omega})\} = \text{Tr}\{\hat{D}(\beta, s)\hat{D}(\gamma, -s)\} = \text{Tr}\{\hat{D}(\beta)\hat{D}(\gamma)\} = \pi\delta(\beta + \gamma). \quad (3.4.8)$$

Then the operator $\hat{A}$ can be decomposed in the more general form

$$\hat{A} = \frac{1}{\pi} \int \tilde{a}(\beta, \Omega)\hat{D}^{-1}(\beta, \Omega)\, d^2\beta, \quad (3.4.9)$$

where the coefficient of decomposition is determined by

$$\tilde{a}(\beta, \Omega) = \text{Tr}\{\hat{D}(\beta, \Omega)\hat{A}\}. \quad (3.4.10)$$

Applying (3.4.9) and (3.4.10) to the statistical operator $\hat{\rho}$ with $\hat{D}(\beta, \Omega) \equiv \hat{D}(\beta, s)$, we regain the characteristic functions $C_\mathcal{N}(\beta) = \tilde{a}(\beta, 1)$ and $C_\mathcal{A}(\beta) = \tilde{a}(\beta, -1)$. For the symmetrical ordering $s = 0$ and we have equations (3.4.1) and (3.4.2).

From (3.4.6) and (3.4.8) it follows that

$$\text{Tr}\{\hat{\Delta}(\alpha, \Omega)\hat{\Delta}(\beta, \tilde{\Omega})\} = \pi\delta(\alpha - \beta), \quad (3.4.11)$$

which means that also the operators $\hat{\Delta}(\alpha, \Omega)$ and $\hat{\Delta}(\alpha, \tilde{\Omega})$ form biorthogonal systems. There are other interesting properties of this operator following from its definition, for example,

$$\int \hat{\Delta}(\alpha, \Omega)\, d^2\alpha = \pi\hat{D}(0, \Omega) = \pi \quad (3.4.12)$$

and

$$\text{Tr}\{\hat{\Delta}(\beta, \Omega)\} = 1. \quad (3.4.13)$$

Further, using the definition (3.4.6) in which the normally ordered form of $\hat{D}(\beta, s)$ from (3.4.7) is substituted and performing the integration, the following expression for the operator $\hat{\Delta}(\alpha, s)$ can be obtained on applying the relation between the photon-number characteristic function and its normally ordered form (see (3.4.54)) [Cahill and Glauber (1969a, b)]

$$\hat{\Delta}(\alpha, s) = \frac{2}{1 - s}\hat{D}(\alpha)\left(\frac{s + 1}{s - 1}\right)^{\hat{a}^\dagger \hat{a}}\hat{D}^{-1}(\alpha). \quad (3.4.14)$$

As a special case for $s = -1$ we have the important result that

$$\hat{\Delta}(\alpha, -1) = |\alpha\rangle\langle\alpha|, \quad (3.4.15)$$

i. e., this is the projection operator onto the coherent state $|\alpha\rangle$.

The operators $\hat{\Delta}(\alpha, 0)$ form a complete set of operators, because they are eigenoperators of a superoperator, the eigenvalue problem being $(\hat{a}\hat{V} + \hat{V}\hat{a})/2 = \lambda\hat{V}$, $\lambda = \alpha$, $\hat{V} = \hat{\Delta}(\alpha, 0)$. At least for $\Omega(\beta^*, \beta) = 1$ the completeness property makes it possible to express any bounded operator $\hat{A}$, for which the trace $\text{Tr}\{\hat{A}^\dagger\hat{A}\}$ is finite, in the form

$$\hat{A} - \frac{1}{\pi} \int a(\alpha, \tilde{\Omega})\hat{\Delta}(\alpha, \Omega)\, d^2\alpha, \tag{3.4.16}$$

where the decomposition coefficient is given as

$$a(\alpha, \tilde{\Omega}) = \text{Tr}\{\hat{\Delta}(\alpha, \tilde{\Omega})\hat{A}\}, \tag{3.4.17}$$

which provides, with respect to $\hat{\Delta}(\alpha, \Omega) = \pi\{\delta(\hat{a} - \alpha\hat{1})\}_\Omega$, the Ω-ordered form of the operator $\hat{A}$. For the product of two operators $\hat{A}$ and $\hat{B}$ we obtain

$$\text{Tr}\{\hat{A}\hat{B}\} = \frac{1}{\pi} \int a(\alpha, \Omega)b(\alpha, \tilde{\Omega})\, d^2\alpha. \tag{3.4.18}$$

If the operator $\hat{B}$ has the meaning of the statistical operator $\hat{\rho}$, we can decompose it as

$$\hat{\rho} = \frac{1}{\pi} \int \Phi(\alpha, \tilde{\Omega})\hat{\Delta}(\alpha, \Omega)\, d^2\alpha, \tag{3.4.19}$$

where the quasidistribution multiplied by π is

$$\Phi(\alpha, \tilde{\Omega}) = \text{Tr}\{\hat{\rho}\hat{\Delta}(\alpha, \tilde{\Omega})\} = \pi\langle\{\delta(\hat{a} - \alpha\hat{1})\}_{\tilde{\Omega}}\rangle. \tag{3.4.20}$$

The expectation value of the operator $\hat{A}$ is equal to

$$\text{Tr}\{\hat{\rho}\hat{A}\} = \frac{1}{\pi} \int \Phi(\alpha, \tilde{\Omega})a(\alpha, \Omega)\, d^2\alpha. \tag{3.4.21}$$

The corresponding equations for the s-ordering are

$$\hat{\rho} = \frac{1}{\pi} \int \Phi(\alpha, -s)\hat{\Delta}(\alpha, s)\, d^2\alpha, \tag{3.4.22}$$

where

$$\Phi(\alpha, -s) = \text{Tr}\{\hat{\rho}\hat{\Delta}(\alpha, -s)\}, \tag{3.4.23}$$

and the expectation value is obtained as

$$\text{Tr}\{\hat{\rho}\hat{A}\} = \frac{1}{\pi} \int \Phi(\alpha, -s)a(\alpha, s)\, d^2\alpha. \tag{3.4.24}$$

Hence, the expectation value of the operator $\hat{A}$ can be calculated as the "classical" expectation value in the complex plane of the classical function $a(\alpha, \Omega)$ corresponding to the operator $\hat{A}$ via the $\tilde{\Omega}$ ordering with the quasiprobability $\Phi(\alpha, \tilde{\Omega})/\pi$ corresponding to the statistical operator $\hat{\rho}$ via the Ω ordering, as follows from (3.4.16) and (3.4.19). This agrees with the particular rules discussed above for obtaining the Glauber–Sudarshan and Husimi quasidistributions from the statistical operator. Putting $s = -1$ in (3.4.22) and taking into account (3.4.15), we obtain the Glauber–Sudarshan representation of the statistical operator, because $\Phi_{\mathcal{N}}(\alpha) = \Phi(\alpha, 1)/\pi$.

Putting $\hat{A} \equiv \hat{1}$, we see that $a(\alpha, \Omega) = 1$ and we also have the normalization condition for quasidistribution

$$\text{Tr}\,\hat{\rho} = 1 = \frac{1}{\pi} \int \Phi(\alpha, \tilde{\Omega})\, d^2\alpha; \tag{3.4.25}$$

the Hermiticity of the statistical operator, $\hat{\rho} = \hat{\rho}^\dagger$, implies that $\Phi^*(\alpha, \Omega) = \Phi(\alpha, \Omega^*)$ or $\Phi^*(\alpha, s) = \Phi(\alpha, s^*)$.

Writing (3.4.20) in the form

$$\Phi(\alpha, \tilde{\Omega}) = \frac{1}{\pi} \int \text{Tr}\{\hat{\rho}\hat{D}(\beta, \tilde{\Omega})\} \exp(\alpha\beta^* - \alpha^*\beta)\, d^2\beta, \tag{3.4.26}$$

we see that $\text{Tr}\{\hat{\rho}\hat{D}(\beta, \tilde{\Omega})\}$ is the characteristic function $C(\beta, \tilde{\Omega})$ of the quasidistribution $\Phi(\alpha, \tilde{\Omega})/\pi$ and so we have the following pair of the Fourier transformations in the complex form

$$\begin{aligned}
\frac{1}{\pi}\Phi(\alpha, \tilde{\Omega}) &= \frac{1}{\pi^2} \int C(\beta, \tilde{\Omega}) \exp(\alpha\beta^* - \alpha^*\beta)\, d^2\beta, \\
C(\beta, \tilde{\Omega}) &= \frac{1}{\pi} \int \Phi(\alpha, \tilde{\Omega}) \exp(-\alpha\beta^* + \alpha^*\beta)\, d^2\alpha.
\end{aligned} \tag{3.4.27}$$

The characteristic function is bounded as follows

$$|C(\beta, \tilde{\Omega})| = |\tilde{\Omega}|\text{Tr}\{\hat{\rho}\hat{D}(\beta)\}| \leq |\tilde{\Omega}|. \tag{3.4.28}$$

The Ω-ordered moments calculated with the help of the quasidistribution $\Phi(\alpha, \Omega)/\pi$ are obtained in the following way

$$\begin{aligned}
\frac{\partial^{k+l}}{\partial\beta^k\partial(-\beta^*)^l}C(\beta, \Omega)\bigg|_{\beta=\beta^*=0} &= \text{Tr}\{\hat{\rho}\{\hat{a}^{\dagger k}\hat{a}^l\}_\Omega\} \\
&= \frac{1}{\pi} \int \Phi(\alpha, \Omega)\alpha^{*k}\alpha^l\, d^2\alpha = \langle \alpha^{*k}\alpha^l \rangle_\Omega. \tag{3.4.29}
\end{aligned}$$

Using the inverse Fourier transformation to (3.4.6), we obtain the Ω-ordered product

$$\{\hat{a}^{\dagger k}\hat{a}^l\}_\Omega = \frac{\partial^{k+l}}{\partial\beta^k\partial(-\beta^*)^l}\hat{D}(\beta, \Omega)\bigg|_{\beta=\beta^*=0} = \frac{1}{\pi} \int \alpha^{*k}\alpha^l\hat{\Delta}(\alpha, \Omega)\, d^2\alpha. \tag{3.4.30}$$

Considering $\Omega(\beta^*, \beta) = \exp(s|\beta|^2/2)$, we obtain the corresponding expressions for the normal, antinormal, and symmetrical orderings for $s = 1, -1$, and 0, respectively, from equations (3.4.27), (3.4.29), and (3.4.30), introducing $\Phi(\alpha, 1)/\pi \equiv \Phi_{\mathcal{N}}(\alpha)$,

$\Phi(\alpha, -1)/\pi \equiv \Phi_A(\alpha)$, $\Phi(\alpha, 0)/\pi \equiv \Phi_S(\alpha)$, more generally $\Phi(\alpha, s)/\pi \equiv \Phi_s(\alpha)$, $C(\beta, 1)$ $\equiv C_N(\beta)$, $C(\beta, -1) \equiv C_A(\beta)$, and $C(\beta, 0) \equiv C(\beta)$.

It can be shown [Cahill and Glauber (1969a, b), Peřina (1985), p. 206] in general that the broadest class of operators possess normally ordered expansions, whereas a relatively smaller class of operators possess antinormally ordered expansions. This means that the class of optical fields each of which possesses the Glauber–Sudarshan representation of the statistical operator with the weighting function $\Phi_N(\alpha) = \pi^{-1}$ $\rho^{(A)}(\alpha^*, \alpha)$ as an ordinary function is substantially smaller than the class of all optical fields with the quasidistribution function $\Phi_A(\alpha) = \pi^{-1}\rho^{(N)}(\alpha^*, \alpha)$, which belongs to the class of ordinary functions. More precisely, the Hilbert–Schmidt operators can be expanded in convergent s-ordered power series for $\mathrm{Re}\, s > 1/2$, i. e., when the ordering is closer to normal than to symmetrical. A characteristic feature of (3.4.22) is that both s and $-s$ appear in it. Hence, because the projection operator (3.4.15) is regular ($s = 1$), the quasidistribution $\Phi_N(\alpha)$ may have singularities, whereas regular and nonnegative quasidistributions $\Phi_A(\alpha)$ require that the decomposition operator $\hat{\Delta}(\alpha, 1)$ be singular ($s = -1$). Thus extreme smoothness of one quantity leads to singular behaviour of the other. Only for symmetrical ordering ($s = 0$) are both the quantities $\Phi_S(\alpha) = \pi^{-1}\Phi(\alpha, 0)$ and $\hat{\Delta}(\alpha, 0)$ regular (the operator $\hat{\Delta}(\alpha, 0)$ is not Hilbert–Schmidt, $\mathrm{Tr}\{[\hat{\Delta}(\alpha, 0)]^2\} = +\infty$, it is bounded and finite in the sense that the maximum value over β of the expectation value $\langle\beta|\hat{\Delta}(\alpha, 0)|\beta\rangle$ is finite for β finite).

A measure of the degree of nonclassical behaviour can be defined for pure and mixed states [Lee (1991), Lütkenhaus and Barnett (1995)]. By Hadamard's theorem [Saks and Zykmund (1971)] the critical value of s dividing well-behaved from not so well-behaved quasiprobability distributions is -1 for all pure states except those with Gaussian quasidistributions $\Phi_s(\alpha)$ [Lütkenhaus and Barnett (1995)].

It is also possible to establish connecting relations for different orderings. Considering the relation (3.4.6) for two orderings Ω_1 and Ω_2, expressing the displacement operator $\hat{D}(\beta)$ from the first equation by means of the inverse Fourier transformation and substituting it into the second equation, we arrive at the connecting relation

$$\hat{\Delta}(\alpha, \Omega_2) = \int K_{21}(\alpha - \beta)\hat{\Delta}(\beta, \Omega_1)\, d^2\beta, \tag{3.4.31}$$

where

$$K_{21}(\beta) = \frac{1}{\pi^2}\int \tilde{\Omega}_1(\gamma^*, \gamma)\Omega_2(\gamma^*, \gamma)\exp(\beta\gamma^* - \beta^*\gamma)\, d^2\gamma. \tag{3.4.32}$$

Multiplying (3.4.31) by the statistical operator $\hat{\rho}$ and taking the trace, we obtain the relation between quasidistributions

$$\Phi(\alpha, \Omega_2) = \int K_{21}(\alpha - \beta)\Phi(\beta, \Omega_1)\, d^2\beta, \tag{3.4.33}$$

with the corresponding relation between the characteristic functions

$$C(\beta, \Omega_2) = \tilde{\Omega}_1(\beta^*, \beta)\Omega_2(\beta^*, \beta)C(\beta, \Omega_1). \tag{3.4.34}$$

In the particular case of the s-ordering we obtain the relation between the characteristic functions

$$C(\beta, s_2) = \exp\left(\frac{1}{2}(s_2 - s_1)|\beta|^2\right) C(\beta, s_1), \tag{3.4.35}$$

and performing the Fourier transformation, we arrive at the relation of quasidistributions as follows

$$\Phi(\alpha, s_2) = \frac{2}{\pi(s_1 - s_2)} \int \Phi(\beta, s_1) \exp\left(-\frac{2|\alpha - \beta|^2}{s_1 - s_2}\right) d^2\beta, \quad \mathrm{Re}(s_1) > \mathrm{Re}(s_2). \tag{3.4.36}$$

Equations (3.3.9), (3.3.27), and (3.3.28) are obtained as special cases for $s_1 = 1, s_2 = -1$, $s_1 = 0, s_2 = -1$, and $s_1 = 1, s_2 = 0$, respectively. We see that the Gaussian convolution (3.4.36) tends to smooth out any unruly behaviour of the quasidistribution $\Phi(\beta, s_1)/\pi$ as a result of the vacuum fluctuations. For example, for the coherent state $|\gamma\rangle$ and $s_1 = 1$ we have $\pi^{-1}\Phi(\beta, 1) \equiv \Phi_N(\beta) = \delta(\beta - \gamma)$ and the s-ordered quasidistribution $\Phi_s(\alpha) = 2/[\pi(1 - s)]\exp[-2|\alpha - \gamma|^2/(1 - s)]$, which is a regular nonnegative function for $s \neq 1$; for $s = -1$ we have the Husimi quasidistribution $\pi^{-1}\Phi(\alpha, -1) \equiv \Phi_A(\alpha) = \pi^{-1}\exp(-|\alpha - \gamma|^2)$ and for $s = 0$ we obtain the Wigner quasidistribution $\pi^{-1}\Phi(\alpha, 0) \equiv \Phi_S(\alpha) = 2\pi^{-1}\exp(-2|\alpha - \gamma|^2)$.

The convolution integral equation (3.4.36) can in principle be solved in the space of generalized functions [Gelfand and Shilov (1964)]. Formally the solution has the operator form

$$\Phi(\alpha, s_1) = \exp\left(-\frac{s_1 - s_2}{2}\frac{\partial^2}{\partial\alpha\partial\alpha^*}\right) \Phi(\alpha, s_2). \tag{3.4.37}$$

The corresponding relation between moments can be written as

$$\begin{aligned}
\langle \hat{a}^{\dagger k}\hat{a}^l\rangle_{s_2} &= k!\left(\frac{s_1 - s_2}{2}\right)^k \left\langle \hat{a}^{l-k}L_k^{l-k}\left(\frac{2\hat{a}^\dagger\hat{a}}{s_2 - s_1}\right)\right\rangle_{s_1}, \quad l \geq k, \\
&= l!\left(\frac{s_1 - s_2}{2}\right)^l \left\langle \hat{a}^{\dagger k-l}L_l^{k-l}\left(\frac{2\hat{a}^\dagger\hat{a}}{s_2 - s_1}\right)\right\rangle_{s_1}, \quad l \leq k,
\end{aligned} \tag{3.4.38}$$

where $L_n^\gamma(x)$ is the Laguerre polynomial [Vilenkin (1968)]

$$L_n^\gamma(x) = \Gamma(n + \gamma + 1)\sum_{j=1}^{n}\frac{(-x)^j}{j!(n - j)!\Gamma(j + \gamma + 1)}, \tag{3.4.39}$$

where $\Gamma(x)$ is the gamma function.

In particular cases we obtain for $s_1 = -1$ and $s_2 = 1$ the normal moments in terms of the antinormal ones,

$$\langle \hat{a}^{\dagger k}\hat{a}^l\rangle_N = k!(-1)^k\langle \hat{a}^{l-k}L_k^{l-k}(\hat{a}^\dagger\hat{a})\rangle_A, \quad l \geq k, \tag{3.4.40}$$

for $s_1 = 1$ and $s_2 = -1$ we express the antinormal moments in terms of the normal ones,

$$\langle \hat{a}^{\dagger k} \hat{a}^l \rangle_{\mathcal{A}} = k! \langle \hat{a}^{l-k} L_k^{l-k}(-\hat{a}^\dagger \hat{a}) \rangle_{\mathcal{N}}, \quad l \geq k, \tag{3.4.41}$$

and when $s_1 = 1$ and $s_2 = 0$, we express the symmetrically ordered moments in terms of the normally ordered ones,

$$\langle a^{\dagger k} a^l \rangle_{\mathcal{S}} = k! \left(\frac{1}{2}\right)^k \langle \hat{a}^{l-k} L_k^{l-k}(-2\hat{a}^\dagger \hat{a}) \rangle_{\mathcal{N}}, \quad l \geq k. \tag{3.4.42}$$

The standard multimode description is straightforward, therefore we will pay some attention to a multimode formulation of the operator ordering based on the following relation of the photon-number characteristic function and its s-ordered forms [Peřina (1985)]

$$\langle \exp(ix\hat{a}^\dagger \hat{a}) \rangle = \frac{2}{1 + e^{ix} + s(1 - e^{ix})} \left\langle \exp\left[\frac{2(1 - e^{ix})\hat{a}^\dagger \hat{a}}{1 + e^{ix} + s(1 - e^{ix})}\right] \right\rangle_s, \tag{3.4.43}$$

where ix denotes a parameter of the characteristic function. Defining the M-mode operator of the number of photons as $\hat{n} = \sum_{\lambda=1}^M \hat{n}_\lambda$, where λ is the mode index and $\hat{n}_\lambda$ are the single-mode number operators, we arrive at the multimode characteristic function

$$\begin{aligned}
\langle \exp(ix\hat{n}) \rangle &= \left\langle \prod_\lambda^M \exp(ix\hat{a}_\lambda^\dagger \hat{a}_\lambda) \right\rangle \\
&= \left[1 - \frac{1-s}{2}(1 - e^{ix})\right]^{-M} \left\langle \exp\left[-\frac{\hat{n}(1 - e^{ix})}{1 - \frac{1-s}{2}(1 - e^{ix})}\right] \right\rangle_s.
\end{aligned} \tag{3.4.44}$$

We note that this result is valid inside any volume of the electromagnetic field, dimensions of which are much larger than the wavelength.

Making the substitution

$$\frac{2(e^{ix} - 1)}{1 + e^{ix} + s(1 - e^{ix})} = iy \tag{3.4.45}$$

in (3.4.44), we obtain the following expression for the s-ordered characteristic function

$$\langle \exp(iy\hat{n}) \rangle_s \equiv \langle \exp(iyW) \rangle_s = \left(1 - \frac{1-s}{2}iy\right)^{-M} \left\langle \left(\frac{1 + \frac{1+s}{2}iy}{1 - \frac{1-s}{2}iy}\right)^{\hat{n}} \right\rangle, \tag{3.4.46}$$

where $W = \sum_\lambda |\alpha_\lambda|^2$ is the integrated intensity. Considering this characteristic function for two orderings $s = s_1$ and $s = s_2$ and applying the substitution (3.4.45) once again, we arrive at the relation of the s_1- and s_2-ordered multidimensional characteristic functions

$$\langle \exp(iy\hat{n}) \rangle_{s_2} = \left(1 + \frac{s_2 - s_1}{2}iy\right) \left\langle \exp\left(\frac{iy\hat{n}}{1 + \frac{s_2 - s_1}{2}iy}\right) \right\rangle_{s_1}. \tag{3.4.47}$$

Performing the Fourier transformation, we obtain the following relation between the integrated intensity distributions related to the s_1- and s_2-orderings

$$
\begin{aligned}
P(W, s_2) &= \frac{2}{s_1 - s_2} \int_0^\infty \left(\frac{W}{W'}\right)^{(M-1)/2} \exp\left[-\frac{2(W + W')}{s_1 - s_2}\right] \\
&\times I_{M-1}\left(4\frac{(WW')^{1/2}}{s_1 - s_2}\right) P(W', s_1)\, dW', \quad \mathrm{Re}(s_1) > \mathrm{Re}(s_2), \quad (3.4.48)
\end{aligned}
$$

where $I_M(x)$ is the modified Bessel function.

Taking the Fourier transformation of (3.4.46), we obtain the integrated intensity distribution related to the s-ordering in terms of the photon-number distribution

$$
\begin{aligned}
P(W, s) &= \left(\frac{2W}{1 - s}\right)^M \frac{\exp\left(-\frac{2W}{1-s}\right)}{W} \\
&\times \sum_{n=0}^\infty \frac{n!p(n)}{\Gamma(n + M)} \left(\frac{s + 1}{s - 1}\right)^n L_n^{M-1}\left(\frac{4W}{1 - s^2}\right), \quad (3.4.49)
\end{aligned}
$$

which can be inverted, using the orthogonality condition for the Laguerre polynomials [Vilenkin (1968)], in the form of the generalized photodetection equation [Peřina (1985)]

$$
\begin{aligned}
p(n) &= \left(\frac{2}{1 + s}\right)^M \left(\frac{s - 1}{s + 1}\right)^n \\
&\times \int_0^\infty P(W, s) L_n^{M-1}\left(\frac{4W}{1 - s^2}\right) \exp\left(-\frac{2W}{1 + s}\right) dW. \quad (3.4.50)
\end{aligned}
$$

For the normal ordering $s \to 1$ and using the asymptotic expression for the Laguerre polynomial, we arrive at the standard photodetection equation for the photon-number distribution [Mandel (1963), section 3.7]

$$
p(n) = \int_0^\infty P_\mathcal{N}(W) \frac{W^n}{n!} \exp(-W)\, dW, \quad (3.4.51)
$$

where $P_\mathcal{N}(W) = P(W, 1)$. The useful relation between the s_1- and s_2-ordered multimode moments is

$$
\langle \hat{n}^k \rangle_{s_2} = k! \left(\frac{s_1 - s_2}{2}\right)^k \left\langle L_k^{M-1}\left(\frac{2\hat{n}}{s_2 - s_1}\right) \right\rangle_{s_1}, \quad (3.4.52)
$$

because the generating function of the Laguerre polynomials enters the right-hand side of (3.4.47). The photon-number moments are expressed in terms of the s-ordered moments in the form

$$
\langle \hat{n}^k \rangle = \sum_{j=0}^k \left(\frac{1 - s}{2}\right)^j \sum_{r=0}^j \frac{(-1)^r j! r^k}{r!(j - r)!} \left\langle L_j^{M-1}\left(\frac{2\hat{n}}{1 - s}\right) \right\rangle_s. \quad (3.4.53)
$$

Many other relations can be obtained in particular cases, for instance, for $s = 1, 0, -1$ we have from (3.4.46)

$$\langle \exp(iy\hat{n}) \rangle_{\mathcal{N}} = \langle (1 + iy)^{\hat{n}} \rangle, \tag{3.4.54}$$

$$\langle \exp(iy\hat{n}) \rangle_{\mathcal{S}} = \left(1 - \frac{iy}{2}\right)^{-M} \left\langle \left(\frac{1 + \frac{iy}{2}}{1 - \frac{iy}{2}}\right)^{\hat{n}} \right\rangle, \tag{3.4.55}$$

$$\langle \exp(iy\hat{n}) \rangle_{\mathcal{A}} = \langle (1 - iy)^{-\hat{n}-M\hat{1}} \rangle, \tag{3.4.56}$$

respectively. From (3.4.44) we obtain the inverse relations for these cases

$$\begin{aligned} \langle \exp(ix\hat{n}) \rangle &= \langle \exp[\hat{n}(e^{ix} - 1)] \rangle_{\mathcal{N}} \\ &= \left(\frac{2}{1 + e^{ix}}\right)^{M} \left\langle \exp\left[\frac{2\hat{n}(e^{ix} - 1)}{1 + e^{ix}}\right] \right\rangle_{\mathcal{S}} \\ &= \langle \exp[-ixM\hat{1} + \hat{n}(1 - e^{-ix})] \rangle_{\mathcal{A}}. \end{aligned} \tag{3.4.57}$$

The first relation in (3.4.57) and (3.4.54) are characteristic of the photodetection (section 3.7).

From (3.4.47) we can obtain the mutual relations between the normal and anti normal multimode characteristic functions

$$\langle \exp(iy\hat{n}) \rangle_{\mathcal{A}} = (1 - iy)^{-M} \left\langle \exp\left(\frac{iy\hat{n}}{1 - iy}\right) \right\rangle_{\mathcal{N}} \tag{3.4.58}$$

putting $s_1 = 1$, $s_2 = -1$ and

$$\langle \exp(iy\hat{n}) \rangle_{\mathcal{N}} = (1 + iy)^{-M} \left\langle \exp\left(\frac{iy\hat{n}}{1 + iy}\right) \right\rangle_{\mathcal{A}} \tag{3.4.59}$$

putting $s_1 = -1$, $s_2 = 1$.

The antinormal moments will contain the contribution of vacuum fluctuations through the number of degrees of freedom M,

$$\langle \hat{n}^k \rangle_{\mathcal{A}} = \langle [\hat{n} + M\hat{1}][\hat{n} + (M + 1)\hat{1}] \ldots [\hat{n} + (M + k - 1)\hat{1}] \rangle. \tag{3.4.60}$$

From (3.4.49) putting $s = -1$, we obtain useful relations

$$P_{\mathcal{A}}(W) = \exp(-W) \sum_{n=0}^{\infty} \frac{W^{n+M-1}}{\Gamma(n + M)} p(n), \tag{3.4.61}$$

so that

$$p(n) = \frac{d^{n+M-1}}{dW^{n+M-1}} [P_{\mathcal{A}}(W) \exp W] \bigg|_{W=0}. \tag{3.4.62}$$

In the classical limit, when the average photon number per mode becomes large, the distinction between different orderings of the field operators vanishes as a consequence of the correspondence principle. Thus all quasidistributions $P(W, s)$ for various s and photon-number distribution $p(n)$ identify and so do their moments. This can be seen, for instance, from (3.4.51), because for a strong field the Poisson function $W^n \exp(-W)/n!$ under the integral sign gives its main contribution to the integral in the neighbourhood of the point $W = n$, where it has its maximum. However this function tends to the function $\delta(W - n)$ for large W. Thus $p(n) \approx P_N(n)$ for n large ($n \approx W$), $\langle \hat{n}^k \rangle \approx \langle \hat{n}^k \rangle_N$, etc.

3.5   Squeezed states

The concept of squeezed states has been introduced in connection with a reduction of quantum noise in measurements of an observable, at the cost of information about the conjugate observable. More generally, it is possible to consider uncertainty relations for two noncommuting observables, which need not be canonically conjugate [Messiah (1961)]. In spite of the prominent role of the uncertainty relations, their mathematical proof rests merely upon the Schwartz inequality, which can be applied as follows. Let us consider a concrete physical system and two observables $\hat{A}$, $\hat{B}$ therein. On the assumption that the system is in the pure state $|\psi\rangle$, $\langle \psi | \psi \rangle = 1$, the observables $\hat{A}$, $\hat{B}$ have the expectations

$$\langle \hat{A} \rangle = \langle \psi | \hat{A} | \psi \rangle, \quad \langle \hat{B} \rangle = \langle \psi | \hat{B} | \psi \rangle, \tag{3.5.1}$$

respectively. A physical interpretation of the following mathematical derivation is based on the mean squares of the operators

$$\Delta \hat{A} = \hat{A} - \langle \hat{A} \rangle \hat{1}, \quad \Delta \hat{B} = \hat{B} - \langle \hat{B} \rangle \hat{1}. \tag{3.5.2}$$

Introducing the vectors

$$|\varphi\rangle = \Delta \hat{A} |\psi\rangle, \quad |\chi\rangle = \Delta \hat{B} |\psi\rangle \tag{3.5.3}$$

and considering the Schwartz inequality in the form

$$|\langle \varphi | \chi \rangle|^2 \leq \langle \varphi | \varphi \rangle \langle \chi | \chi \rangle, \tag{3.5.4}$$

we obtain the inequality

$$|\langle \Delta \hat{A} \Delta \hat{B} \rangle|^2 \leq \langle (\Delta \hat{A})^2 \rangle \langle (\Delta \hat{B})^2 \rangle. \tag{3.5.5}$$

On the right-hand side there is a product of variances. To interpret the left-hand side more easily, we resolve the constituent product

$$\begin{aligned} \Delta \hat{A} \Delta \hat{B} &= \frac{1}{2} \left(\{ \Delta \hat{A}, \Delta \hat{B} \} + [\Delta \hat{A}, \Delta \hat{B}] \right) \\ &= \frac{1}{2} \left(\{ \Delta \hat{A}, \Delta \hat{B} \} + i \hat{C} \right), \end{aligned} \tag{3.5.6}$$

where
$$\hat{C} = -i[\Delta\hat{A}, \Delta\hat{B}] = -i[\hat{A}, \hat{B}]. \tag{3.5.7}$$

Then the relation (3.5.5) becomes

$$\langle(\Delta\hat{A})^2\rangle\langle(\Delta\hat{B})^2\rangle \geq \text{cov}^2(\hat{A}, \hat{B}) + \frac{1}{4}\langle\hat{C}\rangle^2, \tag{3.5.8}$$

where the covariance of the observables $\hat{A}$, B [Lévy-Leblond (1986)]

$$\text{cov}(\hat{A}, \hat{B}) = \frac{1}{2}\langle\{\Delta\hat{A}, \Delta\hat{B}\}\rangle. \tag{3.5.9}$$

Regardless of the above modifications, the equality in the relation (3.5.8) is attained on the assumption that either

$$\Delta\hat{B}|\psi\rangle = 0, \tag{3.5.10}$$

or a suitable complex number κ can be found that

$$(\Delta\hat{A} + \kappa\Delta\hat{B})|\psi\rangle = 0. \tag{3.5.11}$$

We rewrite this relation in the form

$$(\hat{A} + \kappa\hat{B})|\psi\rangle - \lambda|\psi\rangle, \tag{3.5.12}$$

where

$$\lambda = \langle\hat{A}\rangle + \kappa\langle\hat{B}\rangle. \tag{3.5.13}$$

If the equality sign in the relation (3.5.8) applies, then

$$\kappa = -\frac{\langle\Delta\hat{B}\Delta\hat{A}\rangle}{\langle(\Delta\hat{B})^2\rangle} = \frac{-\text{cov}(\hat{A}, \hat{B}) + \frac{i}{2}\langle\hat{C}\rangle}{\langle(\Delta\hat{B})^2\rangle}. \tag{3.5.14}$$

For the states $|\psi\rangle$ fulfilling

$$\langle\hat{C}\rangle = 0, \tag{3.5.15}$$

the relation (3.5.8) and formula (3.5.14) simplify.

The relation (3.5.8) simplifies to the form

$$\text{cov}^2(\hat{A}, \hat{B}) \leq \langle(\Delta\hat{A})^2\rangle\langle(\Delta\hat{B})^2\rangle, \tag{3.5.16}$$

which means only that the variances of the operators and the covariance between them obey the same relations as the quantities in the classical theory of statistics. This proves that these characteristics have been consistently defined. Equation (3.5.14) simplifies to the form

$$\kappa = \gamma, \quad \gamma = -\frac{\text{cov}(\hat{A}, \hat{B})}{\langle(\Delta\hat{B})^2\rangle} = -\text{sgn}[\text{cov}(\hat{A}, \hat{B})]\frac{\Delta A}{\Delta B}, \tag{3.5.17}$$

where

$$\Delta A = \sqrt{\langle (\Delta \hat{A})^2 \rangle}, \quad \Delta B = \sqrt{\langle (\Delta \hat{B})^2 \rangle}. \tag{3.5.18}$$

From the properties of inequalities it follows that the relation (3.5.16) holds also regardless of the condition (3.5.15). The equality sign in the relation (3.5.16) takes place when the correlation between the operators of observables $\hat{A}$, $\hat{B}$ is as strong as possible. Equation (3.5.12) takes on the form

$$(\hat{A} + \gamma \hat{B})|\psi\rangle = \lambda|\psi\rangle. \tag{3.5.19}$$

So far we assumed the property (3.5.15), which unsatisfactorily emphasizes the quantum nature of the relation for the product of uncertainties. Now we restrict ourselves to the states $|\psi\rangle$ fulfilling the condition

$$\text{cov}(\hat{A}, \hat{B}) = 0. \tag{3.5.20}$$

Then the relation (3.5.8) reads

$$\langle (\Delta \hat{A})^2 \rangle \langle (\Delta \hat{B})^2 \rangle \geq \frac{1}{4} \langle \hat{C} \rangle^2, \tag{3.5.21}$$

or equivalently

$$\Delta A \Delta B \geq \frac{1}{2} |\langle \hat{C} \rangle|, \tag{3.5.22}$$

which we call the Heisenberg uncertainty relation. But the pioneers of the quantum theory restricted themselves first to the case $\hat{C} = \hbar \hat{1}$. The relation (3.5.14) becomes, under the condition (3.5.20),

$$\kappa = i\gamma, \quad \gamma = \frac{\langle \hat{C} \rangle}{2 \langle (\Delta \hat{B})^2 \rangle} = \text{sgn}\langle \hat{C} \rangle \frac{\Delta A}{\Delta B}. \tag{3.5.23}$$

The uncertainty relations (3.5.21) and (3.5.22) hold generally regardless of the condition (3.5.20). Equation (3.5.12) takes on the form

$$(\hat{A} + i\gamma \hat{B})|\psi\rangle = \lambda|\psi\rangle. \tag{3.5.24}$$

With respect to the equality attained in the relation (3.5.21) we call the eigenstates the minimum-uncertainty states [Messiah (1961)]. When the uncertainty product $\Delta A \Delta B = \frac{1}{2} |\langle \hat{C} \rangle|$ is state dependent, some authors speak of the $\hat{A}$–$\hat{B}$ intelligent states in place of the $\hat{A}$–$\hat{B}$ minimum-uncertainty ones [Aragone, Chalbaud and Salamó (1976)].

Following Wódkiewicz and Eberly (1985), we shall say that the variances of the operators $\hat{A}$ and $\hat{B}$ are squeezed if

$$\langle (\Delta \hat{A})^2 \rangle < \frac{1}{2} |\langle \hat{C} \rangle| \tag{3.5.25}$$

or

$$\langle (\Delta \hat{B})^2 \rangle < \frac{1}{2} |\langle \hat{C} \rangle|. \qquad (3.5.26)$$

To measure the degree of squeezing with respect to the pair of operators $\hat{A}$ and $\hat{B}$, Bužek, Wilson-Gordon, Knight and Lai (1992) have introduced the parameters

$$S_A = \frac{\langle (\Delta \hat{A})^2 \rangle - \frac{1}{2}|\langle \hat{C} \rangle|}{\frac{1}{2}|\langle \hat{C} \rangle|}, \qquad (3.5.27)$$

$$S_B = \frac{\langle (\Delta \hat{B})^2 \rangle - \frac{1}{2}|\langle \hat{C} \rangle|}{\frac{1}{2}|\langle \hat{C} \rangle|}. \qquad (3.5.28)$$

The squeezing condition now takes on the simple form

$$S_A < 0 \quad \text{or} \quad S_B < 0, \qquad (3.5.29)$$

while $S_B > 0$ or $S_A > 0$, respectively. Borrowing the terminology introduced by Glauber and Lewenstein (1991), we call the variable $\hat{A}$ subfluctuant (superfluctuant) when $S_A < 0$ $(S_A > 0)$.

Mathematically well-defined physical quantum states enable us to study a reduction of the quantum noise in the measurement. In the following we recall the basic types of squeezed states. For a single-mode electromagnetic field of frequency ω described by the annihilation (creation) operator $\hat{a}$ $(\hat{a}^\dagger)$, we consider slowly varying operators $\hat{a} \rightarrow \exp(i\omega t)\hat{a}$, $\hat{a}^\dagger \rightarrow \exp(-i\omega t)\hat{a}^\dagger$ and we define in the interaction picture the position-like and momentum-like quadrature operators

$$\hat{Q} = \hat{a} + \hat{a}^\dagger, \quad \hat{P} = -i(\hat{a} - \hat{a}^\dagger), \quad [\hat{Q}, \hat{P}] = 2i\hat{1}, \qquad (3.5.30)$$

which are proportional to the generalized position operator and the generalized momentum operator, respectively. They obey the appropriate uncertainty relation

$$\langle (\Delta \hat{Q})^2 \rangle \langle (\Delta \hat{P})^2 \rangle \geq 1, \qquad (3.5.31)$$

where

$$\begin{aligned}
\langle (\Delta \hat{Q})^2 \rangle &= 1 + 2[\langle \Delta \hat{a}^\dagger \Delta \hat{a} \rangle + \text{Re}\langle (\Delta \hat{a})^2 \rangle], \\
\langle (\Delta \hat{P})^2 \rangle &= 1 + 2[\langle \Delta \hat{a}^\dagger \Delta \hat{a} \rangle - \text{Re}\langle (\Delta \hat{a})^2 \rangle].
\end{aligned} \qquad (3.5.32)$$

For $\kappa = i$ the solution of the appropriate problem (3.5.11) are the coherent states of the harmonic oscillator (section 3.2). If the variance of one of the quadrature operators $\hat{Q}$ and $\hat{P}$ is smaller than the corresponding value for the coherent state,

$$\min(\langle (\Delta \hat{Q})^2 \rangle, \langle (\Delta \hat{P})^2 \rangle) < 1, \qquad (3.5.33)$$

then the field is in a squeezed state according to the standard definition [Walls (1983)]. In this case we speak of the squeezed state in a strict sense. The notion of principal squeezing will be introduced in section 3.10.

As an example of the squeezed states the two-photon coherent states $|\beta\rangle_g$ can serve [Stoler (1970, 1971a, b, 1975), Yuen (1976)]. They are defined as eigenstates of the boson operator

$$\hat{b} = \mu\hat{a} + \nu\hat{a}^\dagger \quad (\hat{a} = \mu^*\hat{b} - \nu\hat{b}^\dagger), \tag{3.5.34}$$

in terms of the photon annihilation and creation operators $\hat{a}$, $\hat{a}^\dagger$. As

$$[\hat{b}, \hat{b}^\dagger] = \hat{1}, \tag{3.5.35}$$

it holds that

$$|\mu|^2 - |\nu|^2 = 1. \tag{3.5.36}$$

Hence we have the eigenvalue properties of the two-photon coherent states

$$\hat{b}|\beta\rangle_g = \beta|\beta\rangle_g, \quad _g\langle\beta|\hat{b}^\dagger = {}_g\langle\beta|\beta^*. \tag{3.5.37}$$

Consequently a number of properties of the coherent states, such as their overcompleteness and diagonal representation of the statistical operator, can be transferred directly to the two-photon coherent states. Further it holds that

$$\begin{aligned}
_g\langle\beta|\hat{a}^\dagger\hat{a}|\beta\rangle_g &= |\mu^*\beta - \nu\beta^*|^2 + |\nu|^2, \\
_g\langle\beta|(\Delta\hat{Q})^2|\beta\rangle_g &= |\mu - \nu|^2, \\
_g\langle\beta|(\Delta\hat{P})^2|\beta\rangle_g &= |\mu + \nu|^2,
\end{aligned} \tag{3.5.38}$$

which shows the squeezing properties of two-photon coherent states. The two-photon coherent states involve the vacuum fluctuations, since $_g\langle 0|\hat{a}^\dagger\hat{a}|0\rangle_g = |\nu|^2$, and their quantum noise is reflected by the expression

$$_g\langle\beta|\Delta\hat{a}^\dagger\Delta\hat{a}|\beta\rangle_g = |\nu|^2. \tag{3.5.39}$$

The coherent state is obtained for $\mu = 1$, $\nu = 0$. Alternatively the two-photon coherent states can be defined with the help of squeeze and displacement operators [Yuen (1976)]

$$|\beta\rangle_g \equiv |\beta, \mu, \nu\rangle = \hat{S}(\mu, \nu)\hat{D}(\beta)|0\rangle, \tag{3.5.40}$$

or equivalently [Caves (1981)]

$$|\alpha, \zeta\rangle = \hat{D}(\alpha)\hat{S}(\zeta)|0\rangle, \tag{3.5.41}$$

where the squeeze operator is

$$\hat{S}(\zeta) = \exp\left[\frac{1}{2}(\zeta^*\hat{a}^2 - \zeta\hat{a}^{\dagger 2})\right]; \tag{3.5.42}$$

here $\zeta = r \exp(i\theta)$ is a complex squeeze parameter, $\beta = \mu\alpha + \nu\alpha^*$, $\mu = \cosh r$, $\nu = \sinh r \exp(i\theta)$. The difference between both the definitions consists in that in the former case the vacuum fluctuations in the coherent state are squeezed, whereas in the latter case the vacuum is squeezed and then the state is displaced. In this case we have

$$\langle \alpha, \zeta | \hat{a}^\dagger \hat{a} | \alpha, \zeta \rangle = |\alpha|^2 + \sinh^2 r,$$
$$\langle \alpha, \zeta | (\Delta \hat{Q})^2 | \alpha, \zeta \rangle = \exp(-2r),$$
$$\langle \alpha, \zeta | (\Delta \hat{P})^2 | \alpha, \zeta \rangle = \exp(2r), \qquad (3.5.43)$$

provided that $\theta = 0$. The second term in the first equation expresses the effect of vacuum fluctuations. The second equation shows squeezing of vacuum fluctuations in the Q quadrature. Fluctuations in the P quadrature are increased to have the minimum-uncertainty state.

The photon-number distribution of the two-photon coherent state is given by

$$p(n) = |\langle n|\beta\rangle_g|^2,$$
$$\langle n|\beta\rangle_g = (n!\mu)^{-1/2} \left(\frac{\nu}{2\mu}\right)^{n/2} \exp\left(-\frac{1}{2}|\beta|^2 + \frac{\nu^*}{2\mu}\beta^2\right) H_n\left(\frac{\beta}{\sqrt{2\mu\nu}}\right), \quad (3.5.44)$$

where $H_n(x)$ is the Hermite polynomial.

The two-photon coherent states are generated by the two-photon interaction described by the Hamiltonian

$$\hat{H} = \hbar\omega + g\hat{a}^2 + g^*\hat{a}^{\dagger 2} + f\hat{a} + f^*\hat{a}^\dagger, \qquad (3.5.45)$$

where f, g are arbitrary deterministic functions. More efficiently they are generated in the optical parametric processes with classical pumping [Peřina (1991), chapter 10]. In this case the squeeze parameter r is equal to gt, where g is the coupling constant.

Also other types of squeezing of vacuum fluctuations can be defined. Hong and Mandel (1985a, b) defined the $2N$th-order squeezing of vacuum fluctuations by the condition

$$\langle (\Delta \hat{Q})^{2N} \rangle < (2N-1)!!. \qquad (3.5.46)$$

Another form of higher-order squeezing can be defined in terms of real and imaginary parts of higher powers or of products of the field amplitudes [Hillery (1987a, b)]. For instance, we can write

$$\hat{a}^2 = \hat{Y}_1 + i\hat{Y}_2, \qquad (3.5.47)$$

where

$$\hat{Y}_1 = \frac{1}{2}(\hat{a}^2 + \hat{a}^{\dagger 2}), \quad \hat{Y}_2 = \frac{1}{2i}(\hat{a}^2 - \hat{a}^{\dagger 2}). \qquad (3.5.48)$$

These components obey the following commutation rules

$$[\hat{Y}_1, \hat{Y}_2] = i(2\hat{n} + \hat{1}), \qquad (3.5.49)$$

where $\hat{n}$ is the number operator and for their variances we have

$$\langle(\Delta\hat{Y}_{1,2})^2\rangle = \frac{1}{2}[1 - 2\langle\hat{a}\hat{a}^\dagger\rangle + \langle\Delta\hat{a}^2\Delta\hat{a}^{\dagger 2}\rangle \pm \text{Re}\langle(\Delta\hat{a}^2)^2\rangle]; \qquad (3.5.50)$$

they satisfy the uncertainty relation

$$\langle(\Delta\hat{Y}_1)^2\rangle\langle(\Delta\hat{Y}_2)^2\rangle \geq \left\langle \hat{n} + \frac{1}{2}\hat{1} \right\rangle^2. \qquad (3.5.51)$$

We speak of the amplitude-squared squeezing when it holds that

$$\langle(\Delta\hat{Y}_1)^2\rangle \text{ or } \langle(\Delta\hat{Y}_2)^2\rangle < \left\langle \hat{n} + \frac{1}{2}\hat{1} \right\rangle. \qquad (3.5.52)$$

The uncertainty product (3.5.51) is minimized by the generalized coherent states $|\psi\rangle$ as the solutions of the eigenvalue problem (3.5.24), for which the relation $\langle(\Delta\hat{Y}_1)^2\rangle = \langle(\Delta\hat{Y}_2)^2\rangle = \langle\hat{n} + \hat{1}/2\rangle$ is valid. Such a type of squeezing is appropriate for the description of quantum effects in the second-harmonic generation. It can be shown that the fundamental (sub-frequency) mode exhibits this type of squeezing, while the second-harmonic mode is squeezed in the standard way and both types of squeezing are related. This concept is more general, because the right-hand side of (3.5.49) is dependent on the state of the field. Also sum- and difference-amplitude squeezing can be introduced [Hillery (1989)] suitable for the description of nondegenerate optical parametric processes and Raman scattering [Hillery, Yu and Bergou (1992), Chizhov, Haus and Yeong (1995), Kumar and Gupta (1997)].

From a number of reviews devoted to squeezed states we mention reviews by Loudon and Knight (1987), Teich and Saleh (1989), Zhang, Feng and Gilmore (1990), Zaheer and Zubairy (1991), Klyshko (1996), and the corresponding chapters or sections in books by Walls and Milburn (1994), Peřina, Hradil and Jurčo (1994), Vogel and Welsch (1994), and Mandel and Wolf (1995).

3.6   Atomic coherent states

Radcliffe (1971) and Arecchi, Courtens, Gilmore and Thomas (1972) have shown that a system of N two-level atoms can be described in terms of the so-called atomic coherent states, which have many properties in common with coherent states of the harmonic oscillator. They are eigenstates of an angular-momentum operator. They correspond to a set of classical dipoles, additionally involving physical vacuum fluctuations, and they are generated by a classical radiation field, much as the coherent states are generated by classical currents.

We introduce the angular-momentum operators $\hat{J}_1$, $\hat{J}_2$, $\hat{J}_3$, which obey the commutation rules

$$\left[\hat{J}_1, \hat{J}_2\right] = i\hat{J}_3 \text{ (and cyclic permutations)},$$

$$\left[\hat{J}_3, \hat{J}_\pm\right] = \pm\hat{J}_\pm, \quad \hat{J}_\pm = \hat{J}_1 \pm i\hat{J}_2,$$
$$\left[\hat{J}_+, \hat{J}_-\right] = 2\hat{J}_3. \tag{3.6.1}$$

Now we can introduce the so-called Dicke states $|l, m\rangle$,

$$\hat{J}^2|l, m\rangle = l(l+1)|l, m\rangle,$$
$$\hat{J}_3|l, m\rangle = m|l, m\rangle, \tag{3.6.2}$$

where $\hat{J}^2 = \hat{J}_1^2 + \hat{J}_2^2 + \hat{J}_3^2$. From the theory of angular momentum it is known that the number l can be integer or half-odd, and the number m, for a given l, varies from $-l$ to l. The maximum value of l is $N/2$, that is

$$|m| \le l \le \frac{N}{2}. \tag{3.6.3}$$

The quantum number l is called the cooperation number and the ground state is defined as $\hat{J}_-|l, -l\rangle = 0$. Similarly as in (3.2.7)

$$\hat{J}_+|l, m\rangle = \sqrt{(l-m)(l+m+1)}|l, m+1\rangle,$$
$$\hat{J}_-|l, m\rangle = \sqrt{(l+m)(l-m+1)}|l, m-1\rangle. \tag{3.6.4}$$

Hence, $\hat{J}_\pm$ are raising and lowering operators, respectively.

The corresponding Hamiltonian of the radiation and the atom system in the interaction is given by

$$\hat{H} = \hbar\omega\hat{a}^\dagger\hat{a} + \hbar\omega_a\hat{J}_3 + \hbar G(\hat{a}\hat{J}_+ + \hat{a}^\dagger\hat{J}_-), \tag{3.6.5}$$

where ω_a is the atom transition frequency and G is the coupling constant. At resonance, $\omega = \omega_a$, and writing $g = G/\omega$, we have, when omitting ω,

$$\hat{H} = \hbar\hat{a}^\dagger\hat{a} + \hbar\hat{J}_3 + \hbar g(\hat{a}\hat{J}_+ + \hat{a}^\dagger\hat{J}_-). \tag{3.6.6}$$

Following Jordan (1935) and Schwinger (1965), we can introduce the boson annihilation operators

$$\hat{a}_1 = \sum_{l=0}^{\infty}{}' \sum_{m=-l}^{l} \sqrt{(l+m+1)}|l, m\rangle\langle l+\tfrac{1}{2}, m+\tfrac{1}{2}|,$$
$$\hat{a}_2 = \sum_{l=0}^{\infty}{}' \sum_{m=-l}^{l} \sqrt{(l-m+1)}|l, m\rangle\langle l+\tfrac{1}{2}, m-\tfrac{1}{2}|, \tag{3.6.7}$$

so that $[\hat{a}_j, \hat{a}_k^\dagger] = \delta_{jk}\hat{1}$, $j, k = 1, 2$, and

$$\hat{J}_+ = \hat{a}_1^\dagger\hat{a}_2, \quad \hat{J}_- = \hat{a}_2^\dagger\hat{a}_1, \quad \hat{J}_3 = \frac{1}{2}(\hat{a}_1^\dagger\hat{a}_1 - \hat{a}_2^\dagger\hat{a}_2). \tag{3.6.8}$$

These operators satisfy the commutation rules (3.6.1) and it holds that

$$\hat{J}^2 = \hat{L}(\hat{L} + \hat{1}), \quad \hat{L} = \frac{1}{2}(\hat{a}_1^\dagger \hat{a}_1 + \hat{a}_2^\dagger \hat{a}_2). \tag{3.6.9}$$

Using this we can write the Hamiltonian in the form of the Hamiltonian for optical parametric processes [Walls and Barakat (1970), Nussenzveig (1973)]

$$\hat{H} = \hbar\omega\hat{a}^\dagger\hat{a} + \hbar\omega_1\hat{a}_1^\dagger\hat{a}_1 + \hbar\omega_2\hat{a}_2^\dagger\hat{a}_2 + \hbar g(\hat{a}\hat{a}_1\hat{a}_2^\dagger + \hat{a}^\dagger\hat{a}_1^\dagger\hat{a}_2), \tag{3.6.10}$$

provided that $\omega = 1$, $\omega_1 = -1/2$, and $\omega_2 = 1/2$.

The atomic coherent states or the Bloch states can be defined from the ground state $|l, -l\rangle$, in analogy to the displacement operator definition of the coherent state $|\alpha\rangle$, as

$$\begin{aligned} |\theta, \phi\rangle &= \exp(\xi\hat{J}_+ - \xi^*\hat{J}_-)|l, -l\rangle \\ &= \cos^{2l}\left(\frac{\theta}{2}\right) \exp\left[e^{-i\phi}\tan\left(\frac{\theta}{2}\right)\hat{J}_+\right]|l, -l\rangle, \end{aligned} \tag{3.6.11}$$

where $\xi = \theta\exp(-i\phi)/2$ [θ and ϕ being polar and azimuthal angles, respectively, in the angular-momentum space (the Bloch sphere)]. In terms of the Dicke states we obtain

$$\begin{aligned} |\theta, \phi\rangle &= (1 + |\tau|^2)^{-l} \exp(\tau\hat{J}_+)|l, -l\rangle \\ &= (1 + |\tau|^2)^{-l} \sum_{m=-l}^{l} \frac{(2l)!}{(l+m)!(l-m)!} \tau^{l+m}|l, m\rangle, \end{aligned} \tag{3.6.12}$$

where $\tau = \tan(\theta/2)\exp(-i\phi)$. This expression is analogous to the decomposition (3.2.5) valid for coherent states of the harmonic oscillator.

The atomic coherent states have the following eigenvalue property

$$(\mathbf{n} \cdot \hat{\mathbf{J}})|\theta, \phi\rangle = -l|\theta, \phi\rangle, \tag{3.6.13}$$

$\mathbf{n}$ being the unit vector in the direction (θ, ϕ). They also represent the intelligent states, because

$$\langle(\Delta\hat{J}_1)^2\rangle\langle(\Delta\hat{J}_2)^2\rangle = \frac{1}{4}\langle\hat{J}_3\rangle^2. \tag{3.6.14}$$

The atomic coherent states furnish the following resolution of the identity (see subsection 4.8.2)

$$\frac{2l+1}{4} \int |\theta, \phi\rangle\langle\theta, \phi| \sin\theta \, d\theta \, d\phi = \sum_{m=-l}^{l} |l, m\rangle\langle l, m| = \hat{1}_{2l}^{SU(2)}, \tag{3.6.15}$$

which can serve for representing vectors and operators in terms of the atomic coherent states. On introducing new numbers n_1, n_2 such that

$$l = \frac{1}{2}(n_1 + n_2), \quad m = \frac{1}{2}(n_1 - n_2), \tag{3.6.16}$$

the number states $|n_1, n_2\rangle$ will correspond to the Dicke states $|l, m\rangle$ and it holds that $\hat{a}_j^\dagger \hat{a}_j |n_1, n_2\rangle = n_j |n_1, n_2\rangle$, $j = 1, 2$.

We may conclude that under the influence of a classical deterministic current, an electromagnetic field initially in a coherent state, or in its ground state, will evolve into a coherent state, and under the influence of a classical deterministic electromagnetic field, an atomic system initially in an atomic coherent state, or in its ground state, will evolve into an atomic coherent state. The coherent state corresponds to a deterministic classical field and the atomic coherent state corresponds to a set of classical dipoles, involving additionally fluctuations of the physical vacuum. It is obvious that the Dicke states correspond to the Fock states and the atomic coherent states (the Bloch states) correspond to the coherent states (the Glauber states), and the following correspondence is appropriate $\hat{a} \leftrightarrow \hat{J}_-$, $\hat{a}^\dagger \leftrightarrow \hat{J}_+$.

Interesting applications can be obtained with the binomial states [Stoler, Saleh and Teich (1985)] and squeezed and displaced number states [Kim, de Oliveira and Knight (1989a), de Oliveira, Kim, Knight and Bužek (1990), Král (1990a, b)]. More about the generalized coherent states can be found in the literature [Malkin and Man'ko (1979), Perelomov (1986), Klauder and Skagerstam (1985), Bužek and Knight (1995)].

A logical completion of the investigations in the atomic coherent states is a phase-space method. This comprises quasidistributions on the appropriate phase space, for example, the Stratonovich–Wigner function [Stratonovich (1956), Agarwal (1981)]. It is worth noting that in the branch of spin systems continuous quasidistributions have been introduced. They are based on the Stratonovitch–Weyl correspondence [Várilly and Gracia-Bondía (1989)] and a positive distribution function of rotated spin variables [Dodonov and Man'ko (1997)]. The dynamics of the radiation and the atom system in the interaction can be described using the joint Wigner function for atom–field interactions [Czirják and Benedict (1996)].

3.7 Photocount statistics

In this section we review the basic notions of photocount statistics of radiation field. For more detail we refer the reader to recent monographs [Peřina (1991), Walls and Milburn (1994), Vogel and Welsch (1994), Mandel and Wolf (1995)]. The photodetection equation relates the photocount distribution $p(n, T, t)$ of the detection of n counts in the time interval $(t, t + T)$ with the probability distribution $P_\mathcal{N}(W)$ of the integrated intensity W,

$$p(n, T, t) = \int_0^\infty \frac{(\eta W)^n}{n!} \exp(-\eta W) P_\mathcal{N}(W) \, dW, \qquad (3.7.1)$$

$$P_\mathcal{N}(W) = \int \ldots \int \Phi_\mathcal{N}(\{\alpha_\lambda\}, T, t) \delta\left(\sum_\lambda |\alpha_\lambda|^2 - W\right) \prod_\lambda d^2\alpha_\lambda,$$

$$W = \int_t^{t+T} I(t')\, dt',$$ (3.7.2)

where η is the photodetection efficiency and $I(t)$ is the intensity of light incident on a photocathode. This distribution is related to quantum characteristic functions (3.4.54) and (3.4.57) and the photon-number distribution (3.4.51) is obtained for $\eta = 1$. The photocount distribution $p(n, T, t)$ and the photon-number distribution $p(m) = \langle m|\hat\rho|m\rangle$ are related by the Bernoulli transformation

$$p(n, T, t) = \sum_{m=n}^{\infty} \frac{m!}{n!(m-n)!}\eta^n(1-\eta)^{m-n}p(m).$$ (3.7.3)

In particular, we obtain for the mean number of counts $\langle n\rangle = \eta\langle W\rangle_{\mathcal{N}}$ and for the variance

$$\langle(\Delta n)^2\rangle = \langle n\rangle + \eta^2\langle(\Delta W)^2\rangle_{\mathcal{N}}.$$ (3.7.4)

This formula has an important physical interpretation. It shows that the variance in the number of ejected photoelectrons is the sum of the variance $\langle n\rangle$ in the number of classical particles obeying the Poisson distribution and the variance $\eta^2\langle(\Delta W)^2\rangle_{\mathcal{N}}$ of the generalized wave field. This relation represents the wave–particle duality. Introducing the so-called Fano factor $F_n = \langle(\Delta n)^2\rangle/\langle n\rangle$, the fluctuation formula (3.7.4) can be written in terms of the photocount number n and the photon number m in the form of the quantum Burgess variance theorem [Peřina, Saleh and Teich (1983)]

$$F_n - 1 = \eta(F_m - 1).$$ (3.7.5)

It can be seen from the variance theorem (3.7.4) that the generalized wave fluctuations are calculated with the use of the Glauber–Sudarshan quasiprobability, which can be singular and negative, so that it is also possible that $\langle(\Delta W)^2\rangle_{\mathcal{N}} < 0$ and in this case the sub-Poisson photon statistics are detected with $\langle(\Delta n)^2\rangle < \langle n\rangle$, i. e., for the Fano factor we have $F_n < 1$; the field has no classical analogue. If $F_n = 1$, the field is coherent and, if $F_n > 1$, it is super-Poissonian. For chaotic light of natural sources, $F_m = 1 + \langle\hat m\rangle$, whereas for the Fock state $F_m = 0$. The photocount distribution can be obtained from the normal generating function through derivatives

$$p(n, T, t) = \frac{1}{n!}\frac{d^n}{d(is)^n}\langle\exp(is\eta W)\rangle_{\mathcal{N}}\bigg|_{is=-1}.$$ (3.7.6)

Also factorial moments can be used to specify the photocount statistics,

$$\left\langle\frac{n!}{(n-k)!}\right\rangle = \eta^k\langle W^k\rangle_{\mathcal{N}} = \frac{d^k}{d(is)^k}\langle\exp(is\eta W)\rangle_{\mathcal{N}}\bigg|_{is=0},$$ (3.7.7)

equally as cumulants, which are defined as

$$\kappa_j^{(W)} = \frac{d^j}{d(is)^j}\ln\langle\exp(is\eta W)\rangle_{\mathcal{N}}\bigg|_{is=0}, \quad j = 1, 2, \ldots;$$ (3.7.8)

it holds that $\kappa_0^{(W)} = 0$, $\kappa_1^{(W)} = \eta \langle W \rangle_{\mathcal{N}}$, $\kappa_2^{(W)} = \eta^2 \langle (\Delta W)^2 \rangle_{\mathcal{N}}$, etc., and we have that $\kappa_j^{(W)} = 0$, $j > 1$, for a coherent field and $\kappa_j^{(W)} = (j-1)! \eta^j \langle W \rangle_{\mathcal{N}}^j$ for chaotic light of natural sources.

The bunching, unbunching, and antibunching properties of photons in light beams are described by the fourth-order correlation functions, which can be normalized as follows,

$$\gamma_{\mathcal{N}}^{(2)}(\tau) = \frac{\langle I(t) I(t+\tau) \rangle_{\mathcal{N}}}{\langle I(t) \rangle_{\mathcal{N}}^2} = \frac{\Gamma_{\mathcal{N}}^{(2,2)}(t, t+\tau, t+\tau, t)}{\left[\Gamma_{\mathcal{N}}^{(1,1)}(t, t) \right]^2}, \tag{3.7.9}$$

where τ is a time delay. From the Schwartz inequality it follows that for classical fields $\gamma_{\mathcal{N}}^{(2)}(\tau) \leq \gamma_{\mathcal{N}}^{(2)}(0)$, which expresses the bunching property of photons of classical beams. Putting $\tau = 0$, we have $\gamma_{\mathcal{N}}^{(2)}(0) = 1$ for coherent beams, $\gamma_{\mathcal{N}}^{(2)}(0) \geq 1$ for bunched beams, and $\gamma_{\mathcal{N}}^{(2)}(0) \leq 1$ for antibunched beams (the classical inequality $\langle I^2 \rangle_{\mathcal{N}} \geq \langle I \rangle_{\mathcal{N}}^2$ is violated in this case). As far as we studied the integrated intensity, we were close to the detection process and despite this or therefore the results we have obtained are relevant for usual quantum optical processes. Since the definition based on the correlation functions (3.7.9) is connected immediately with the intensity $I(t)$, the relation to the usual nonlinear processes is not so clear [Miranowicz, Bajer, Ekert and Leoński (1997)].

Whereas two-level systems are sufficient for the detection of phase-independent moments of radiation fields, Dalton and Knight (1990) have demonstrated that three-level photon counting systems can be made sensitive to the phase-dependent properties of radiation fields.

3.8   Generalized superposition of coherent fields and quantum noise

In this section we provide a brief review of quantum generalization of the superposition of signal and noise, which is able to describe simply nonclassical behaviour of optical beams. We give only a single-mode formulation, a general multimode formulation can be found in [Peřina, Hradil and Jurčo (1994), section 4.8]; in greater detail in [Peřina (1991)]. This formulation is equivalent to the two-photon coherent state technique.

In solving problems of interaction of radiation with matter, very often nonclassical states of radiation arise, which cannot be simply described by the Glauber–Sudarshan quasidistribution related to the normal ordering of field operators, because this quasidistribution exists only as a generalized function. Moreover, sometimes we also include optical beams which are initially in nonclassical states. On the other hand, the quasidistribution related to the antinormal ordering always exists as a nonnegative and regular function, therefore we adopt the antinormal ordering in the following. The antinormal characteristic function for the generalized superposition of a coherent field and quantum noise can be written in the Gaussian form averaged over the initial statistical behaviour of the field

$$
\begin{aligned}
C_A(\beta, t) &= \operatorname{Tr}\left\{\hat{\rho}\exp[-\beta^*\hat{a}(t)]\exp[\beta\hat{a}^\dagger(t)]\right\} \\
&= \left\langle \exp\left\{-B_A(t)|\beta|^2 + \frac{1}{2}[C^*(t)\beta^2 + \text{c. c.}] + [\beta\xi^*(t) - \text{c. c.}]\right\}\right\rangle, \quad (3.8.1)
\end{aligned}
$$

where the angular brackets mean the average over the initial complex amplitude $\xi(0)$ and the Heisenberg picture has been adopted. The quantum noise coefficients are expressed as

$$
B_A(t) = \langle\Delta\hat{a}(t)\Delta\hat{a}^\dagger(t)\rangle = \langle\Delta\hat{a}^\dagger(t)\Delta\hat{a}(t)\rangle - 1, \quad C(t) = \langle(\Delta\hat{a}(t))^2\rangle. \quad (3.8.2)
$$

The corresponding quasidistribution is

$$
\begin{aligned}
\Phi_A(\alpha, t) &= \left\langle \frac{1}{\sqrt{K(t)}}\exp\left\{-\frac{B_A(t)}{K(t)}|\alpha - \xi(t)|^2\right.\right. \\
&\quad\left.\left. + \left[\frac{C^*(t)}{2K(t)}(\alpha - \xi(t))^2 + \text{c. c.}\right]\right\}\right\rangle, \quad (3.8.3)
\end{aligned}
$$

where $K(t) = [B_A(t)]^2 - |C(t)|^2$ is the determinant of the quantum noise matrix. The corresponding normal generating function is given by

$$
\begin{aligned}
\langle\exp(-\lambda W(t))\rangle_N &= \left\langle \frac{1}{\sqrt{1 + \lambda[E(t) - 1]}}\frac{1}{\sqrt{1 + \lambda[F(t) - 1]}}\right. \\
&\quad\left. \times \exp\left\{-\frac{\lambda A_1(t)}{1 + \lambda[E(t) - 1]} - \frac{\lambda A_2(t)}{1 + \lambda[F(t) - 1]}\right\}\right\rangle, \quad (3.8.4)
\end{aligned}
$$

where the parameter $\lambda = -is$ and

$$
E(t) = B_A(t) - |C(t)|, \quad F(t) = B_A(t) + |C(t)|,
$$

$$
A_{1,2}(t) = \frac{1}{2}\left\{|\xi(t)|^2 \mp \frac{1}{2|C(t)|}\left[[\xi(t)]^2 C^*(t) + \text{c. c.}\right]\right\}. \quad (3.8.5)
$$

The expression in the angular brackets in (3.8.4) has the form of the product of generating functions for the superposition of signals $A_{1,2}(t) \geq 0$ and the principal normally ordered quantum-noise components $E(t) - 1$, $F(t) - 1$ (it holds that $F(t) - 1 \geq 0$, however, $E(t) - 1$ can also be negative). The deviation of the expression within the angular brackets from the Poisson statistics with the generating function $\exp\{-\lambda[A_1(t) + A_2(t)]\} = \exp[-\lambda|\xi(t)|^2]$ generally reflects the change of the statistics caused by the interaction, whereas the average over the initial complex amplitude, represented by the angular brackets, produces additional change of statistics respecting initial states. The quantum features are reflected by negative values of the principal normally ordered quantum-noise component $E(t) - 1 = B_A(t) - |C(t)| - 1$.

The corresponding factors in the angular brackets in (3.8.4) are also generating functions for the Laguerre polynomials $L_n^{-1/2}(x)$ and consequently we obtain for the photon-number distribution $(\eta = 1)$

$$
p(n, t) = \left\langle \frac{1}{\sqrt{E(t)F(t)}} \left[1 - \frac{1}{F(t)}\right]^n \exp\left[-\frac{A_1(t)}{E(t)} - \frac{A_2(t)}{F(t)}\right] \sum_{k=0}^{n} \left[\frac{1 - 1/E(t)}{1 - 1/F(t)}\right]^k \right.
$$
$$
\left. \times L_k^{-1/2}\left(-\frac{A_1(t)}{E(t)[E(t) - 1]}\right) L_{n-k}^{-1/2}\left(-\frac{A_2(t)}{F(t)[F(t) - 1]}\right) \right\rangle \quad (3.8.6)
$$

and for its factorial moments

$$
\langle [W(t)]^k \rangle_N = k! \left\langle [F(t) - 1]^k \sum_{l=0}^{k} \left[\frac{E(t) - 1}{F(t) - 1}\right]^l \right.
$$
$$
\left. \times L_l^{-1/2}\left(-\frac{A_1(t)}{E(t) - 1}\right) L_{k-l}^{-1/2}\left(-\frac{A_2(t)}{F(t) - 1}\right) \right\rangle. \quad (3.8.7)
$$

The mean photon number equals

$$
\langle \hat{n}(t) \rangle = \langle W(t) \rangle_N = \langle A_1(t) \rangle + \langle A_2(t) \rangle + \frac{1}{2}[E(t) + F(t) \quad 2] \quad (3.8.8)
$$

and the integrated-intensity variance is

$$
\langle [\Delta W(t)]^2 \rangle_N = 2\langle A_1(t) \rangle [E(t) - 1] + 2\langle A_2(t) \rangle [F(t) - 1]
$$
$$
+ \frac{1}{2}\{[E(t) - 1]^2 + [F(t) - 1]^2\}. \quad (3.8.9)
$$

The last terms in (3.8.8) and (3.8.9) represent quantum noise contributions related to physical vacuum fluctuations. They are not zero even if the input light intensity is zero $(A_1(t) = A_2(t) = 0)$. For nonclassical fields $E(t) - 1 < 0$, the first interference term in (3.8.9) between the signal $A_1(t)$ and the normally ordered quantum noise $E(t) - 1$ may be dominant at least for some time intervals and the fields exhibit the sub-Poissonian photon statistics $\langle [\Delta\hat{n}(t)]^2 \rangle < \langle \hat{n}(t) \rangle$ $(\langle [\Delta W(t)]^2 \rangle_N < 0)$.

As a simple demonstration of physical meaning of this description we can use the second-subharmonic generation with strong and coherent classical pumping, which is determined by the Hamiltonian

$$
\hat{H} = \hbar\omega\left(\hat{a}^\dagger\hat{a} + \frac{1}{2}\hat{1}\right) - \frac{1}{2}[\hbar g\hat{a}^2 \exp(i2\omega t - i\psi) + \text{H. c.}], \quad (3.8.10)
$$

where g is the coupling constant including the real pump amplitude and 2ω is a frequency of a pump beam of a phase ψ. The solution of the corresponding Heisenberg equation

$$
\frac{d\hat{a}}{dt} = -i\omega\hat{a} + ig^*\hat{a}^\dagger \exp(-i2\omega t + i\psi) \quad (3.8.11)
$$

together with its Hermitian conjugate is

$$\hat{a}(t) = u(t)\hat{a}(0) + v(t)\hat{a}^\dagger(0), \tag{3.8.12}$$

where

$$
\begin{aligned}
u(t) &= \exp(-i\omega t)\cosh(|g|t), \\
v(t) &= i\exp(-i\omega t + i\psi)\frac{g}{|g|}\sinh(|g|t),
\end{aligned}
\tag{3.8.13}
$$

so that $|u(t)|^2 - |v(t)|^2 = 1$. Assuming now g real (phase matching) and the initial coherent state $|\xi\rangle$, we observe that the time-dependent complex amplitude $\xi(t) = u(t)\xi + v(t)\xi^*$, the determinant $K(t) = \cosh^2(gt)$, and we obtain that

$$
\begin{aligned}
A_{1,2}(t) &= \frac{1}{2}|\xi|^2\exp(\mp 2gt)[1 \mp \sin(2\varphi - \psi)] \geq 0, \\
E(t) - 1 &= \frac{1}{2}[\exp(-2gt) - 1] \leq 0, \\
F(t) - 1 &= \frac{1}{2}[\exp(2gt) - 1] \geq 0, \\
B_A(t) &= |u(t)|^2 = \cosh^2(gt), \\
C(t) &= u(t)v(t) = \frac{i}{2}\sinh(2gt)\exp(i\psi),
\end{aligned}
\tag{3.8.14}
$$

where φ is the phase of the initial complex amplitude ξ. We see the correspondence with the two-photon coherent states, if the squeeze parameter $r = gt$. The squeezed vacuum state ($\xi = 0$) is super-chaotic, because $\langle[\Delta\hat{n}(t)]^2\rangle = 2\langle\hat{n}(t)\rangle[1 + \langle\hat{n}(t)\rangle]$, $\langle\hat{n}(t)\rangle = \sinh^2(gt)$. The normal generating function (3.8.4) is a product of two generating functions describing the superposition of the signal $A_1(t)$ with the normally ordered noise $E(t) - 1$ and that of the signal $A_2(t)$ and the normally ordered noise $F(t) - 1$. Thus the resulting photon-number distribution is a discrete convolution of photon-number quasidistributions, one of which can take on negative values as a consequence of $E(t) - 1 < 0$ and therefore the resulting distribution oscillates [Peřina and Bajer (1990), Peřina, Hradil and Jurčo (1994), section 5.4]. This is again a fully quantum phenomenon, which can also be interpreted in terms of the interference in phase space [Schleich and Wheeler (1987a, b), Vogel and Schleich (1992)]. Dissipations generally degrade all these nonclassical effects.

3.9   Irreversible quantum dynamics

The status of irreversible evolution in physics has been doubted since the beginning of the statistical physics. In quantum optics the irreversibility of the master equation is an approximation to the exact reversible evolution of the system and its environment.

We consider N radiation modes described by the photon annihilation and creation operators $\hat{a}_j$ and $\hat{a}_j^\dagger$, $j = 1, 2, ..., N$, respectively, having frequencies ω_j, which are in interaction with large reservoirs described by the boson annihilation (creation) operators $\hat{b}_l^{(j)}$ ($\hat{b}_l^{(j)\dagger}$) having frequencies $\psi_l^{(j)}$. We assume broad dense spectra of reservoir frequencies. We will review only basic results suitable for further considerations. Greater detail about the Heisenberg–Langevin and Schrödinger approaches to quantum statistics of systems can be found in a number of monographs and monographic chapters [Lax (1968), Haken (1970), Scully and Whitney (1972), Louisell (1973), Agarwal (1973), Haake (1973), van Kampen (1981), Risken (1984), Meystre and Sargent III (1991), Gardiner (1991), Peřina (1991), Carmichael (1993), Walls and Milburn (1994), Vogel and Welsch (1994)], and in special issue of Quantum and Semiclassical Optics devoted to stochastic quantum optics [Carmichael (1996)].

The Hamiltonian of this system can be written in the form

$$\hat{H} = \hat{H}_{0,\text{rad}} + \hat{H}_{\text{res}} + \hat{H}_{\text{int}}, \tag{3.9.1}$$

where

$$\hat{H}_{0,\text{rad}} = \sum_{j=1}^{N} \hbar\omega_j \left(\hat{a}_j^\dagger \hat{a}_j + \frac{1}{2}\hat{1} \right) \tag{3.9.2}$$

is the free radiation Hamiltonian,

$$\hat{H}_{\text{res}} = \sum_{j,l} \hbar\psi_l^{(j)} \left(\hat{b}_l^{(j)\dagger} \hat{b}_l^{(j)} + \frac{1}{2}\hat{1} \right) \tag{3.9.3}$$

is the free reservoir Hamiltonian, and according to Peřina and Peřinová (1976)

$$\hat{H}_{\text{int}}(t) = \sum_{j,l} \hbar \left\{ \kappa_l^{(j)} \hat{b}_l^{(j)}(t,0) \left[\mu_j^* \hat{a}_j^\dagger(t) - \nu_j^* \hat{a}_j(t) \exp(2i\omega_j t) \right] + \text{H. c.} \right\} \tag{3.9.4}$$

is the interaction Hamiltonian between the radiation modes and the large reservoir system $\{\hat{b}_l^{(j)}(t,0)\}$; here $\kappa_l^{(j)}\mu_j^*$, $\kappa_l^{(j)}\nu_j^*$ are coupling constants between the radiation modes and the reservoir. We assume that $|\mu_j|^2 - |\nu_j|^2 = 1$, or that

$$\mu_j = \cosh(|\zeta_j|),$$

$$\nu_j = -\frac{\zeta_j}{|\zeta_j|} \sinh(|\zeta_j|), \quad j = 1, ..., N, \tag{3.9.5}$$

with an appropriate ζ_j. The second parameter of $\hat{b}_l^{(j)}$ expresses the fact that this reservoir operator is standard. We index the reservoir operator $\hat{b}_l^{(j)}(t,0)$ with positive and negative subscripts so that

$$\psi_l^{(j)} + \psi_{-l}^{(j)} = 2\omega_j \tag{3.9.6}$$

and we assume that

$$\kappa_{-l}^{(j)*} = \kappa_l^{(j)}. \tag{3.9.7}$$

Using (3.9.7) (still not (3.9.6)), we rewrite (3.9.4) in the form

$$\hat{H}_{\text{int}} = \sum_{j,l} \hbar \left[\kappa_l^{(j)} \hat{b}_l^{(j)}(t, \zeta(t)) \hat{a}_j^\dagger(t) + \text{H. c.} \right], \tag{3.9.8}$$

where for the squeezed reservoir operators it holds that

$$\hat{b}_l^{(j)}(t, \zeta(t)) = \mu_j^* \hat{b}_l^{(j)}(t,0) - \nu_j \hat{b}_{-l}^{(j)\dagger}(t,0) \exp(-2i\omega_j t), \tag{3.9.9}$$

with

$$\zeta_j(t) = \zeta_j \exp(-2i\omega_j t). \tag{3.9.10}$$

Ideally squeezed operators are characterized by the equality $|\nu_j| = 1$ and then $|\zeta_j| \to \infty$. Introducing an alternative parameter

$$\eta_j = \frac{\zeta_j}{|\zeta_j|} \tanh(|\zeta_j|) = -\frac{\nu_j}{\mu_j^*}, \tag{3.9.11}$$

we see that also $|\eta_j| = 1$ for ideally squeezed operators. In the following we provide a brief review of both the approach to this problem based on the Heisenberg–Langevin equations and that based on the master and generalized Fokker–Planck equations. The Heisenberg equations for radiation and reservoir operators are obtained in the form

$$\frac{d}{dt}\hat{a}_j(t) = -i\omega_j \hat{a}_j(t) - i \sum_l \kappa_l^{(j)} \hat{b}_l^{(j)}(t, \zeta_j(t)),$$

$$\frac{d}{dt}\hat{b}_l^{(j)}(t, \zeta_j(t)) = -i\psi_l^{(j)} \hat{b}_l^{(j)}(t, \zeta_j(t)) - i\kappa_l^{(j)*} \hat{a}_j(t). \tag{3.9.12}$$

If the second equation in (3.9.12) is solved under the assumption of quasimonochromatic radiation, we can eliminate the evolved reservoir operators in the first equation in the Wigner–Weisskopf approximation arriving at the Heisenberg–Langevin equations for damped radiation modes as follows

$$\frac{d}{dt}\hat{a}_j(t) = -\left(i\omega_j' + \frac{\gamma_j}{2}\right) \hat{a}_j(t) + \hat{L}_j(t, \zeta_j(t)), \tag{3.9.13}$$

where $\omega_j' = \omega_j + \Delta\omega_j$ are shifted radiation frequencies, with the frequency shifts $\Delta\omega_j$ being given by

$$\Delta\omega_j = \text{V. p.} \int_{-\infty}^{\infty} \frac{|\kappa^{(j)}(\psi^{(j)})|^2 \rho_j(\psi^{(j)})}{\omega_j - \psi^{(j)}} d\psi^{(j)} \tag{3.9.14}$$

and the damping constants are defined as

$$\gamma_j = 2\pi |\kappa^{(j)}(\omega_j)|^2 \rho_j(\omega_j), \tag{3.9.15}$$

with $\rho_j(\psi^{(j)})$ being density functions of the frequencies of the reservoir oscillators, after the sum over l has been replaced by the integration over reservoir frequencies

$\psi^{(j)}$ respecting high reservoir oscillator density assumed. Here the squeezed Langevin force operators have been obtained in terms of the initial squeezed reservoir operators in the form

$$\hat{L}_j(t, \zeta_j(t)) = -i \sum_l \kappa_l^{(j)} \hat{b}_l^{(j)}(0, \zeta_j(0)) \exp\left(-i\psi_l^{(j)}t\right).$$

(3.9.16)

If we additionally assume that the reservoir is a Gaussian system, then

$$\hat{\rho}_{\text{th}} = \prod_{j,l} \frac{\langle \hat{n}_{\text{R}l}^{(j)} \rangle^{\hat{b}_l^{(j)\dagger}\hat{b}_l^{(j)}}}{\left(1 + \langle \hat{n}_{\text{R}l}^{(j)} \rangle\right)^{\hat{b}_l^{(j)}\hat{b}_l^{(j)\dagger}}},$$

(3.9.17)

where $\langle \hat{n}_{\text{R}l}^{(j)} \rangle = \langle \hat{b}_l^{(j)\dagger}(0,0)\hat{b}_l^{(j)}(0,0) \rangle_{\text{R}}$ represents the mean photon number in the reservoir oscillators in the reservoir mode l coupled to the radiation mode j and $\hat{b}_l^{(j)} = \hat{b}_l^{(j)}(0,0)$. The squeezed Langevin force operators (3.9.16) have the following properties

$$\langle \hat{L}_j(t, \zeta_j(t)) \rangle_{\text{R}} = \langle \hat{L}_j^\dagger(t, \zeta_j(t)) \rangle_{\text{R}} = 0,$$
$$\langle \hat{L}_j^\dagger(t, \zeta_j(t)) \hat{L}_k(t', \zeta_k(t')) \rangle_{\text{R}} = \gamma_j B_{\text{R}\mathcal{N}}^{(j)} \delta_{jk} \delta(t - t'),$$
$$\langle \hat{L}_j(t, \zeta_j(t)) \hat{L}_k^\dagger(t', \zeta_k(t')) \rangle_{\text{R}} = \gamma_j B_{\text{R}\mathcal{A}}^{(j)} \delta_{jk} \delta(t - t'),$$
$$\langle \hat{L}_j(t, \zeta_j(t)) \hat{L}_k(t', \zeta_k(t')) \rangle_{\text{R}} = -\gamma_j C_{\text{R}}^{(j)} \exp(-2i\omega_j t) \delta_{jk} \delta(t - t'),$$
$$\langle \hat{L}_j^\dagger(t, \zeta_j(t)) \hat{L}_k^\dagger(t', \zeta_k(t')) \rangle_{\text{R}} = -\gamma_j C_{\text{R}}^{(j)*} \exp(2i\omega_j t) \delta_{jk} \delta(t - t'),$$

(3.9.18)

where

$$B_{\text{R}\mathcal{N}}^{(j)} = |\mu_j|^2 \langle n_{\text{R}}^{(j)} \rangle + |\nu_j|^2 (\langle n_{\text{R}}^{(j)} \rangle + 1),$$
$$B_{\text{R}\mathcal{A}}^{(j)} = |\mu_j|^2 (\langle n_{\text{R}}^{(j)} \rangle + 1) + |\nu_j|^2 \langle n_{\text{R}}^{(j)} \rangle,$$
$$C_{\text{R}}^{(j)} = -\mu_j^* \nu_j (2\langle n_{\text{R}}^{(j)} \rangle + 1).$$

(3.9.19)

Here $\langle \ldots \rangle_{\text{R}}$ means the average over the reservoir with respect to the statistical operator (3.9.17) and we assume broad-band reservoir spectra, so that $\langle n_{\text{R}}^{(j)} \rangle \equiv \langle \hat{n}_{\text{R}l}^{(j)} \rangle$ independently of l. At a temperature T it holds that $\langle n_{\text{R}}^{(j)} \rangle = [\exp(\hbar\psi^{(j)}/KT) - 1]^{-1}$, K being the Boltzmann constant.

The definition (3.9.16) indicates that the properties (3.9.18) can be derived from those of the squeezed reservoir operators,

$$\langle \hat{b}_l^{(j)\dagger}(0, \zeta_j) \hat{b}_m^{(k)}(0, \zeta_k) \rangle = B_{\text{R}\mathcal{N}}^{(j)} \delta_{jk} \delta_{lm},$$
$$\langle \hat{b}_l^{(j)}(0, \zeta_j) \hat{b}_m^{(k)\dagger}(0, \zeta_k) \rangle = B_{\text{R}\mathcal{A}}^{(j)} \delta_{jk} \delta_{lm},$$
$$\langle \hat{b}_l^{(j)}(0, \zeta_j) \hat{b}_m^{(k)}(0, \zeta_k) \rangle = C_{\text{R}}^{(j)} \delta_{jk} \delta_{l,-m}.$$

(3.9.20)

With this in mind we are changing the picture. In the new picture the reservoir operators are not squeezed and the pertinent quantum statistical property is attributed to a squeezed reservoir state,

$$\hat{\rho}_{\mathrm{th}}(\{\zeta_j\}) = \exp\left[\sum_{j,\pm l}\left(\zeta_j\hat{b}_l^{(j)\dagger}\hat{b}_{-l}^{(j)\dagger} - \zeta_j^*\hat{b}_l^{(j)}\hat{b}_{-l}^{(j)}\right)\right]$$

$$\times \hat{\rho}_{\mathrm{th}}\exp\left[\sum_{j,\pm l}\left(\zeta_j^*\hat{b}_l^{(j)}\hat{b}_{-l}^{(j)} - \zeta_j\hat{b}_l^{(j)\dagger}\hat{b}_{-l}^{(j)\dagger}\right)\right]. \tag{3.9.21}$$

The shorthand $\sum_{j,\pm l} = \frac{1}{2}\sum_{j,l}$ is used. The statistical operator of the squeezed reservoir state (3.9.21) is a generalization of the zero-temperature case treated by Gilson, Barnett and Stenholm (1987), where the replacements $l \to r + l$, $-l \to r - l$ enable one to index the reservoir modes with nonnegative integers. The picture adopted leads to simplification in the observables, because the ζ argument can be left out in $\hat{b}_l^{(j)}(t, \zeta_j(t))$ and $\hat{L}_j(t, \zeta_j(t))$. Of course, the ζ dependence is respected in (3.9.21).

The solution of equations (3.9.13) has the form

$$\hat{a}_j(t) = u_j(t)\hat{a}_j(0) + \sum_l w_{jl}(t)\hat{b}_l^{(j)}(0), \tag{3.9.22}$$

where the time-dependent functions $u_j(t)$ and $w_{jl}(t)$ must fulfill the following identities as a consequence of the boson commutation rules $[\hat{a}_j(t), \hat{a}_j^\dagger(t)] = \hat{1}$,

$$|u_j(t)|^2 + \sum_l |w_{jl}(t)|^2 = 1. \tag{3.9.23}$$

Here

$$u_j(t) = \exp\left(-i\omega_j' t - \frac{\gamma_j}{2}t\right),$$

$$w_{jl}(t) = \kappa_l^{(j)}\exp\left(-i\psi_l^{(j)}t\right)\frac{1 - \exp\left[i\left(\psi_l^{(j)} - \omega_j'\right) - \frac{\gamma_j}{2}t\right]}{\psi_l^{(j)} - \omega_j' + i\frac{\gamma_j}{2}}. \tag{3.9.24}$$

From the commutation rules we obtain

$$\sum_l |w_{jl}(t)|^2 = 1 - \exp(-\gamma_j t). \tag{3.9.25}$$

For the squeezed reservoir state we have from (3.9.22)

$$\langle \hat{n}_j(t)\rangle_{\mathrm{R}} = \langle \hat{a}_j^\dagger(t)\hat{a}_j(t)\rangle_{\mathrm{R}} = \exp(-\gamma_j t)\langle \hat{a}_j^\dagger(0)\hat{a}_j(0)\rangle_{\mathrm{R}} + B_{\mathrm{RN}}^{(j)}(t)\hat{1}, \tag{3.9.26}$$

$$\langle \hat{a}_j^2(t)\rangle_{\mathrm{R}} = \exp(-\gamma_j t)\langle \hat{a}_j^2(0)\rangle_{\mathrm{R}} + C_{\mathrm{R}}^{(j)}(t)\exp(-2i\omega_j t)\hat{1}, \tag{3.9.27}$$

where $\langle...\rangle_R$ means the average over the reservoir with respect to the statistical operator (3.9.21) and

$$
\begin{aligned}
B^{(j)}_{R\mathcal{N}}(t) &= B^{(j)}_{R\mathcal{N}}[1 - \exp(-\gamma_j t)], \\
C^{(j)}_R(t) &= C^{(j)}_R[1 - \exp(-\gamma_j t)].
\end{aligned}
\tag{3.9.28}
$$

The first terms represent the time evolution of the initial radiation and the second terms are contributions of the reservoir. These operators obey the following equations of motion

$$
\frac{d}{dt}\langle \hat{n}_j(t)\rangle_R = -\gamma_j \langle \hat{n}_j(t)\rangle_R + \gamma_j B^{(j)}_{R\mathcal{N}}(t)\hat{1},
\tag{3.9.29}
$$

$$
\frac{d}{dt}\langle \hat{a}_j^2(t)\rangle_R = -(2i\omega_j + \gamma_j)\langle \hat{a}_j^2(t)\rangle_R + \gamma_j C^{(j)}_R(t) \exp(-2i\omega_j t)\hat{1}.
\tag{3.9.30}
$$

Making use of the equations of motion (3.9.13), we can arrive at the important identities

$$
\langle \hat{L}_j^\dagger(t)\hat{a}_j(t) + \text{H. c.}\rangle_R = \gamma_j B^{(j)}_{R\mathcal{N}}(t)\hat{1},
\tag{3.9.31}
$$

$$
\langle \hat{L}_j(t)\hat{a}_j(t) + \hat{a}_j(t)\hat{L}_j(t)\rangle_R = \gamma_j C^{(j)}_R(t) \exp(-2i\omega_j t)\hat{1}.
\tag{3.9.32}
$$

The properties of the Langevin force operator $\hat{L}_j$ ensure the validity of the boson commutation rules for all times and make the quantum description fully consistent. If the Langevin forces are neglected, the solution $\hat{a}_j(t) = \hat{a}_j(0)\exp(-i\omega_j' t - \gamma_j \frac{t}{2})$ violates the commutation rules, since $[\hat{a}_j(t), \hat{a}_j^\dagger(t)] = \exp(-\gamma_j t)\hat{1} \to \hat{0}$ for $t \to \infty$, which may be approximately correct for $\gamma_j t \ll 1$ only.

The time development of the statistical properties of radiation can be described in the framework of the Heisenberg–Langevin approach assuming the time-independent statistical operator $\hat{\rho}_e$,

$$
\hat{\rho}_e = \hat{\rho} \otimes \hat{\rho}_{th}(\{\zeta_j\}),
\tag{3.9.33}
$$

where the statistical operator of the state of radiation

$$
\hat{\rho} = \int ... \int \Phi_{\mathcal{N}}(\{\xi_j\}, 0)|\{\xi_j\}\rangle\langle\{\xi_j\}| \prod_{j=1}^N d^2\xi_j,
\tag{3.9.34}
$$

with a Glauber–Sudarshan quasidistribution $\Phi_{\mathcal{N}}(\{\xi_j\}, 0)$. To this aim we use the normal quantum characteristic function

$$
\begin{aligned}
C_{\mathcal{N}}(\{\beta_j\}, t) &= \text{Tr}\left\{ \hat{\rho}(0) \prod_j \exp\left[\beta_j \hat{a}_j^\dagger(t)\right] \exp\left[-\beta_j^* \hat{a}_j(t)\right] \right\} \\
&= \left\langle \prod_j \exp\left\{ - B^{(j)}_{R\mathcal{N}}(t)|\beta_j|^2 + \left[\frac{1}{2}C^{(j)*}_R(t)\right.\right.\right. \\
&\quad \left.\left.\left. \times \exp(2i\omega_j t)\beta_j^2 + \text{c. c.}\right] + \xi_j^*(t)\beta_j - \xi_j(t)\beta_j^* \right\}\right\rangle,
\end{aligned}
\tag{3.9.35}
$$

where β_j are complex parameters of the characteristic function and we have used (3.9.17), (3.9.21), and (3.9.22); here the time-developed complex amplitude is $\xi_j(t) = \xi_j \exp(-i\omega'_j t - \gamma_j \frac{t}{2})$ and the brackets $\langle ... \rangle$ mean the average over the initial radiation complex amplitudes $\{\xi_j\}$ described by the quasidistribution $\Phi_{\mathcal{N}}(\{\xi_j\}, 0)$. For the time-developed Glauber–Sudarshan quasidistribution we obtain [Peřinová (1981)]

$$
\begin{aligned}
\Phi_{\mathcal{N}}(\{\alpha_j\}, t) &= \frac{1}{\pi^{2N}} \int \cdots \int C_{\mathcal{N}}(\{\beta_j\}, t) \prod_j \exp(\alpha_j \beta_j^* - \alpha_j^* \beta_j) d^2 \beta_j \\
&= \Bigg\langle \prod_j \frac{1}{\pi \sqrt{K_j(t)}} \exp\Bigg\{ -\frac{1}{K_j(t)} B_{\mathrm{R}\mathcal{N}}^{(j)}(t) |\alpha_j - \xi_j(t)|^2 \\
&\quad + \frac{1}{2K_j(t)} \Big[C_{\mathrm{R}}^{(j)}(t) \exp(-2i\omega_j t)(\alpha_j^* - \xi_j^*(t))^2 + \text{c. c.} \Big] \Bigg\} \Bigg\rangle, \quad (3.9.36)
\end{aligned}
$$

where

$$
K_j(t) = [B_{\mathrm{R}\mathcal{N}}^{(j)}(t)]^2 - |C_{\mathrm{R}}^{(j)}(t)|^2, \quad j = 1, \ldots, N. \tag{3.9.37}
$$

If the radiation is initially in the coherent state described by $\hat{\rho} = |\{\xi_j\}\rangle\langle\{\xi_j\}|$, then the average in (3.9.36) is performed with the use of the Dirac delta functions, i. e., the angular brackets can be omitted; the peak of the Glauber–Sudarshan quasidistribution moves along the spiral $\xi_j(t) = \xi_j \exp(-i\omega'_j t - \gamma_j \frac{t}{2})$ with the increasing quantum noise characteristics $B_{\mathrm{R}\mathcal{N}}^{(j)}(t)$ and $C_{\mathrm{R}}^{(j)}(t)$. The latter is responsible for the possible non-classical properties. The photocount distribution $p(n, t)$ and its factorial moments $\langle [W(t)]^k \rangle_{\mathcal{N}}$ can be expressed with the help of the generalized superposition of the signal $\{\xi_j(t)\}$ and squeezed noise characterized by $B_{\mathrm{R}\mathcal{N}}^{(j)}(t)$ and $C_{\mathrm{R}}^{(j)}(t) \exp(-2i\omega_j t)$ using the Laguerre polynomials [Peřina, Peřinová and Horák (1973)].

The Heisenberg picture is not so explicitly statistical. This drawback is not possessed by the alternative treatment based on the Schrödinger approach. In this picture the equation of motion for the "original" statistical operator reads

$$
i\hbar \frac{\partial}{\partial t} \hat{\rho}_{\mathrm{e,orig}} = [\hat{H}_{\mathrm{orig}}, \hat{\rho}_{\mathrm{e,orig}}], \tag{3.9.38}
$$

where

$$
\hat{H}_{\mathrm{orig}} = \hat{H}_{0,\mathrm{rad}} + \hat{H}_{\mathrm{res}} + \hat{H}_{\mathrm{int,orig}}, \tag{3.9.39}
$$

with the Hamiltonians $\hat{H}_{0,\mathrm{rad}}$ and $\hat{H}_{\mathrm{res}}$ given in (3.9.2) and (3.9.3), respectively, and

$$
\hat{H}_{\mathrm{int,orig}} = \sum_{j,l} \hbar \left(\kappa_l^{(j)} \hat{b}_l^{(j)} \hat{a}_j^\dagger + \text{H. c.} \right). \tag{3.9.40}
$$

Adopting the partial interaction picture with respect to the reservoir, i. e., defining

$$
\hat{\rho}_{\mathrm{e}}(t) = \exp\left(\frac{i}{\hbar} \hat{H}_{\mathrm{res}} t \right) \hat{\rho}_{\mathrm{e,orig}}(t) \exp\left(-\frac{i}{\hbar} \hat{H}_{\mathrm{res}} t \right), \tag{3.9.41}
$$

where

$$\hat{H}_{\text{int}}(t) = \exp\left(\frac{i}{\hbar}\hat{H}_{\text{res}}t\right)\hat{H}_{\text{int,orig}}\exp\left(-\frac{i}{\hbar}\hat{H}_{\text{res}}t\right),\qquad(3.9.42)$$

we obtain the equation of motion

$$i\hbar\frac{\partial}{\partial t}\hat{\rho}_{\text{e}}(t) = \left[\hat{H}_{\text{rad}} + \hat{H}_{\text{int}}(t), \hat{\rho}_{\text{e}}(t)\right]\qquad(3.9.43)$$

and the initial condition

$$\hat{\rho}_{\text{e}}(t)|_{t=0} = \hat{\rho}_{\text{e}}.\qquad(3.9.44)$$

If we perform the second iteration, we obtain

$$\begin{aligned}
\hat{\rho}_{\text{e}}(t+\Delta t) &= \hat{\rho}_{\text{e}}(t) - \frac{i}{\hbar}\int_t^{t+\Delta t}[\hat{H}(t'), \hat{\rho}_{\text{e}}(t')]\,dt' \\
&\quad - \frac{1}{\hbar^2}\int_t^{t+\Delta t}\int_t^{t'}[\hat{H}(t'), [\hat{H}(t''), \hat{\rho}_{\text{e}}(t'')]]\,dt''\,dt',\qquad(3.9.45)
\end{aligned}$$

where

$$\hat{H}(t) = \hat{H}_{\text{rad}} + \hat{H}_{\text{int}}(t).\qquad(3.9.46)$$

Using the definition of the time derivative and introducing the reduced statistical operator $\hat{\rho}(t) = \text{Tr}_{\text{R}}\{\hat{\rho}_{\text{e}}(t)\}$, we derive the formula

$$\frac{\partial}{\partial t}\hat{\rho}(t) = \lim_{\Delta t \to 0}\frac{1}{\Delta t}\text{Tr}_{\text{R}}\{\hat{\rho}_{\text{e}}(t+\Delta t) - \hat{\rho}_{\text{e}}(t)\}.\qquad(3.9.47)$$

In this kind of the interaction picture, the interaction Hamiltonian can be written as

$$\hat{H}_{\text{int}}(t) = \sum_{j=1}^N \hbar\left[\hat{G}^{(j)}(t)\hat{a}_j^\dagger + \text{H. c.}\right],\qquad(3.9.48)$$

where

$$\hat{G}^{(j)}(t) = \sum_l \kappa_l^{(j)}\hat{b}_l^{(j)}\exp(-i\psi_l^{(j)}t).\qquad(3.9.49)$$

Apart from the factor $-i$, the noise operators $\hat{G}^{(j)}(t)$ are the unsqueezed Langevin force operators.

We decompose the trace functional over the reservoir as

$$\text{Tr}_{\text{R}} = \text{Tr}_{>t}\text{Tr}_{<t},\qquad(3.9.50)$$

where $\text{Tr}_{<t}$ ($\text{Tr}_{>t}$) means the trace over the history (the prospect) of the quantum noise, and observe that (cf. (3.9.18))

$$\begin{aligned}
\langle\hat{G}^{(j)}(t')\rangle_{>t} &= \langle\hat{G}^{(j)\dagger}(t')\rangle_{>t} = 0, \\
\langle\hat{G}^{(j)\dagger}(t')\hat{G}^{(k)}(t'')\rangle_{>t} &= \gamma_j B_{\text{RN}}^{(j)}\delta_{jk}\delta(t'-t''), \\
\langle\hat{G}^{(j)}(t')\hat{G}^{(k)\dagger}(t'')\rangle_{>t} &= \gamma_j B_{\text{RA}}^{(j)}\delta_{jk}\delta(t'-t''), \\
\langle\hat{G}^{(j)}(t')\hat{G}^{(k)}(t'')\rangle_{>t} &= \gamma_j C_{\text{R}}^{(j)}\exp(-2i\omega_j t')\delta_{jk}\delta(t'-t''), \\
\langle\hat{G}^{(j)\dagger}(t')\hat{G}^{(k)}(t'')\rangle_{>t} &= \gamma_j C_{\text{R}}^{(j)*}\exp(2i\omega_j t')\delta_{jk}\delta(t'-t'').\qquad(3.9.51)
\end{aligned}$$

According to the rule

$$\lim_{\Delta t \to +0} \frac{1}{\Delta t} \text{Tr}_{\text{R}} \{\hat{\rho}_e(t + \Delta t) - \hat{\rho}_e(t)\}$$

$$= \lim_{\Delta t \to +0} \frac{1}{\Delta t} \text{Tr}_{>t} \{\hat{\rho}_{e+}^{\text{appr}}(t + \Delta t) - \hat{\rho}_{e+}(t)\}, \qquad (3.9.52)$$

where

$$\hat{\rho}_{e+}(t) = \text{Tr}_{<t} \{\hat{\rho}_e(t)\}, \qquad (3.9.53)$$

$$\hat{\rho}_{e+}^{\text{appr}}(t + \Delta t) = \hat{\rho}_{e+}(t) - \frac{i}{\hbar} \int_t^{t+\Delta t} [\hat{H}(t'), \hat{\rho}_{e+}(t)] \, dt'$$

$$- \frac{1}{\hbar^2} \int_t^{t+\Delta t} \int_t^{t'} [\hat{H}(t'), [\hat{H}(t''), \hat{\rho}_{e+}(t)]] \, dt'' \, dt', \qquad (3.9.54)$$

and using the statistical independence of radiation and the prospect of the reservoir

$$\hat{\rho}_{e+}(t) = \hat{\rho}(t) \otimes \hat{\rho}_{\text{R}+}(t), \qquad (3.9.55)$$

where

$$\hat{\rho}_{\text{R}+}(t) = \text{Tr}_{<t} \{\hat{\rho}_{\text{R}}(0)\}, \qquad (3.9.56)$$

we arrive at the master equation

$$\frac{\partial}{\partial t} \hat{\rho}(t) = \sum_{j=1}^{N} \left\{ - i\omega_j [\hat{a}_j^\dagger \hat{a}_j \hat{\rho}(t) - \hat{\rho}(t)\hat{a}_j^\dagger \hat{a}_j] \right.$$

$$+ \frac{\gamma_j}{2} C_{\text{R}}^{(j)*} \exp(2i\omega_j t)[2\hat{a}_j \hat{\rho}(t)\hat{a}_j - \hat{a}_j^2 \hat{\rho}(t) - \hat{\rho}(t)\hat{a}_j^2]$$

$$+ \frac{\gamma_j}{2} B_{\text{RA}}^{(j)} [2\hat{a}_j \hat{\rho}(t)\hat{a}_j^\dagger - \hat{a}_j^\dagger \hat{a}_j \hat{\rho}(t) - \hat{\rho}(t)\hat{a}_j^\dagger \hat{a}_j]$$

$$+ \frac{\gamma_j}{2} B_{\text{R}N}^{(j)} [2\hat{a}_j^\dagger \hat{\rho}(t)\hat{a}_j - \hat{a}_j \hat{a}_j^\dagger \hat{\rho}(t) - \hat{\rho}(t)\hat{a}_j \hat{a}_j^\dagger]$$

$$\left. + \frac{\gamma_j}{2} C_{\text{R}}^{(j)} \exp(-2i\omega_j t)[2\hat{a}_j^\dagger \hat{\rho}(t)\hat{a}_j^\dagger - \hat{a}_j^{\dagger 2} \hat{\rho}(t) - \hat{\rho}(t)\hat{a}_j^{\dagger 2}] \right\}. \qquad (3.9.57)$$

The foregoing derivation is valid on the assumption that $t \ll \frac{1}{\gamma_j}$ and t is much larger than a typical correlation time of the reservoir. Particularly, for a single mode of radiation ($N = 1$), $C_{\text{R}}^{(1)} = 0$, $B_{\text{RA}}^{(1)} = 1$, $B_{\text{R}N}^{(1)} = 0$, the matrix elements of the statistical operator in a number-state basis have been studied [Arnoldus (1994)].

Let us consider squeezed number states [Kim, de Oliveira and Knight (1989a, b), Král (1990a, b)]

$$|\zeta, n\rangle_j = \exp \left[\frac{1}{2}(\zeta \hat{a}_j^{\dagger 2} - \zeta^* \hat{a}_j^2) \right] |n\rangle_j. \qquad (3.9.58)$$

The diagonal elements of the statistical operator $\hat{\rho}$ are the probabilities

$$P_{\{n_j\}}(\{-\zeta_j(t)\}, t) = \prod_{j=1}^{N} p_{n_j}(-\zeta_j(t), t). \qquad (3.9.59)$$

Here the factors

$$p_{n_j}(-\zeta_j(t),t) = {}_j\langle -\zeta_j(t), n_j|\hat{\rho}(t)| - \zeta_j(t), n_j\rangle_j \qquad (3.9.60)$$

obey the differential equation

$$
\begin{aligned}
\frac{d}{dt}p_{n_j}(-\zeta_j(t),t) &= -\gamma_j(\langle n_R^{(j)}\rangle + 1)[n_j p_{n_j}(-\zeta_j(t),t) - (n_j+1) \\
&\quad \times p_{n_j+1}(-\zeta_j(t),t)] - \gamma_j\langle n_R^{(j)}\rangle[(n_j+1)p_{n_j}(-\zeta_j(t),t) \\
&\quad - n_j p_{n_j-1}(-\zeta_j(t),t)], \quad j=1,...,N.
\end{aligned} \qquad (3.9.61)
$$

Since in the steady state $\frac{d}{dt}p_{n_j}(-\zeta_j(t),t) = 0$, we can easily obtain these factors in the form of the Bose–Einstein distribution

$$\lim_{t\to\infty} p_{n_j}(-\zeta_j(t),t) = \frac{\langle n_R^{(j)}\rangle^{n_j}}{(\langle n_R^{(j)}\rangle + 1)^{n_j+1}}. \qquad (3.9.62)$$

When also detailed balance between absorption and emission processes is assumed and thermal equilibrium considered, then

$$\langle n_R^{(j)}\rangle = \left[1 - \exp\left(-\frac{\hbar\omega_j}{KT_j}\right)\right]^{-1} \exp\left(-\frac{\hbar\omega_j}{KT_j}\right). \qquad (3.9.63)$$

Using the coherent state technique, we can adopt the s-ordered quasidistribution $\Phi_s(\{\alpha_j\},t) = \frac{1}{\pi^N}\Psi(\{\alpha_j\},t,s)$ to obtain the generalized Fokker–Planck equation, applying the corresponding operator ordering in the master equation (3.9.57),

$$
\begin{aligned}
\frac{\partial}{\partial t}\Phi_s(\{\alpha_j\},t) &= \sum_{j=1}^{N}\left\{\left[\left(\frac{\gamma_j}{2} + i\omega_j\right)\frac{\partial}{\partial\alpha_j}(\alpha_j\Phi_s(\{\alpha_j\},t))\right.\right. \\
&\quad \left.- \gamma_j C_R^{(j)}\exp(-2i\omega_j t)\frac{\partial^2}{\partial\alpha_j^2}\Phi_s(\{\alpha_j\},t) + \text{c. c.}\right] \\
&\quad \left.+ \gamma_j B_{Rs}^{(j)}\frac{\partial^2}{\partial\alpha_j\partial\alpha_j^*}\Phi_s(\{\alpha_j\},t)\right\}.
\end{aligned} \qquad (3.9.64)
$$

Here

$$B_{Rs}^{(j)} = B_{R\mathcal{S}}^{(j)} - \frac{s}{2}, \qquad (3.9.65)$$

with

$$B_{R\mathcal{S}}^{(j)} = B_{R\mathcal{N}}^{(j)} + \frac{1}{2} = B_{R\mathcal{A}}^{(j)} - \frac{1}{2}. \qquad (3.9.66)$$

The drift terms involving the first-order derivatives are responsible for a motion of the peak of the quasidistribution, whereas the diffusion term including the second-order derivative causes a broadening of the quasidistribution and an increase of the uncertainty. The rotating terms involving $\frac{\partial^2}{\partial\alpha_j^2}$ and their complex conjugates account for nonclassical properties. The Fourier transformation provides the equation of motion for the characteristic function $C_s(\{\beta_j\}, t)$,

$$
\frac{\partial}{\partial t} C_s(\{\beta_j\}, t) = \sum_{j=1}^{N} \left\{ \left[-\left(\frac{\gamma_j}{2} + i\omega_j \right) \beta_j^* \frac{\partial}{\partial \beta_j^*} - \gamma_j C_R^{(j)} \exp(-2i\omega_j t) \beta_j^{*2} \right. \right.
$$
$$
\left. \left. + \text{c. c.} \right] C_s(\{\beta_j\}, t) - \gamma_j B_{Rs}^{(j)} |\beta_j|^2 C_s(\{\beta_j\}, t) \right\}. \tag{3.9.67}
$$

In the interaction picture with respect to the radiation modes, the renormalizations $\omega_j \to 0$ simplify the description. In this usual picture the quantum noise is strictly stationary also in the case of the squeezed reservoir, because the reservoir oscillators are squeezed symmetrically about the zero frequency. On the contrary, the squeezing about the frequencies $\omega_j \gg 0$ leads to nonstationary quantum noise. As the loss of stationarity is due to a factor $\exp(-i\omega_j t)$, it is worth noting that this factor does not change the statistical properties of the unsqueezed (standard) noise. The effect of a reservoir on a signal beam can be simulated by an infinite array of beamsplitters and has been analyzed in the Heisenberg picture [Imoto, Jeffers and Loudon (1992), Jeffers, Imoto and Loudon (1993)] and in the Schrödinger picture [Kim and Imoto (1995)].

One can consider more general case of the interaction of multimode radiation with a nonlinear medium composed of two-level atoms, taking into account multiphoton absorption and emission events. The free Hamiltonian of the electron system and the interaction Hamiltonian can be written in the form [McNeil and Walls (1974)]

$$
\hat{H}_{\text{atom}} = \sum_{\lambda=1}^{2} E_\lambda \sum_l \hat{c}_{\lambda l}^\dagger \hat{c}_{\lambda l}, \tag{3.9.68}
$$

$$
\hat{H}_{\text{int}} = \sum_l \hbar \mu^{(m,n-m)}(\mathbf{x}_l) \hat{c}_{2l}^\dagger \hat{c}_{1l} \hat{O}^{(m,n-m)} + \text{H. c.}, \tag{3.9.69}
$$

where

$$
\hat{O}^{(m,n-m)} \equiv \hat{O} = \prod_{j=1}^{m} \hat{a}_j^\dagger \prod_{k=m+1}^{n} \hat{a}_k. \tag{3.9.70}
$$

Here E_λ are energy levels, $\hat{c}_{\lambda l}$ and $\hat{c}_{\lambda l}^\dagger$ are the annihilation and creation operators of the λth level of the lth atom, respectively, $\hat{a}_j$ and $\hat{a}_j^\dagger$ are the annihilation and creation operators of a photon in the radiation mode j, respectively, and $\mu^{(m,n-m)}(\mathbf{x}_l)$ is a coupling coefficient proportional to the n-photon transition matrix element of the atom situated at a point $\mathbf{x}_l$. This interaction Hamiltonian describes m emission and $n - m$ absorption events during one atom transition. We only consider resonant transitions occurring under the condition

$$
\sum_{j=1}^{m} \omega_j - \sum_{k=m+1}^{n} \omega_k = \Omega, \tag{3.9.71}
$$

where the atomic transition frequency Ω,

$$\Omega = \frac{1}{\hbar}(E_2 - E_1). \tag{3.9.72}$$

If we are interested in the properties of radiation rather than in atomic behaviour, as is usual in quantum optics, we can eliminate the atomic variables similarly as the reservoir variables and we can obtain Heisenberg–Langevin equations, master equation or generalized Fokker–Planck equation. If only virtual electronic transitions are taken into account and real transitions are neglected, an effective Hamiltonian can be derived [Peřina Jr. (1993) and references therein] and then the Heisenberg–Langevin and generalized Fokker–Planck equations directly follow. The derivation of the master equation proceeds in the Markov approximation under the condition of thermal equilibrium, when $\langle \hat{c}^\dagger_{\lambda l} \hat{c}_{\lambda l} \rangle = N_\lambda$, independent of l, and $\langle \hat{c}^\dagger_{1l} \hat{c}_{2l} \rangle = 0$. The master equation for the reduced statistical operator can be obtained in the form

$$\frac{\partial}{\partial t}\hat{\rho}(t) = K\left\{ N_1\left([\hat{O}\hat{\rho}(t),\hat{O}^\dagger] + [\hat{O},\hat{\rho}(t)\hat{O}^\dagger]\right) \right.$$
$$\left. - N_2\left([\hat{O},\hat{O}^\dagger\hat{\rho}(t)] + [\hat{\rho}(t)\hat{O},\hat{O}^\dagger]\right)\right\}, \tag{3.9.73}$$

where K is a constant related to $\mu^{(m,n-m)}$. We can also consider more general interaction of the radiation modes with dissipative system symbolically described by the expression $\hbar\left(\hat{G}^\dagger(t)\hat{O} + \text{H. c.}\right)$, $\hat{G}(t)$ being a random noise operator with the properties

$$\langle \hat{G}(t')\rangle_{>t} = \langle \hat{G}^\dagger(t')\rangle_{>t} = 0,$$
$$\langle \hat{G}^\dagger(t')\hat{G}(t'')\rangle_{>t} = 2D_{G^\dagger G}\delta(t'-t''),$$
$$\langle \hat{G}(t')\hat{G}^\dagger(t'')\rangle_{>t} = 2D_{GG^\dagger}\delta(t'-t''),$$
$$\langle \hat{G}(t')\hat{G}(t'')\rangle_{>t} = 2D_{GG}(t')\delta(t'-t''),$$
$$\langle \hat{G}^\dagger(t')\hat{G}^\dagger(t'')\rangle_{>t} = 2D_{G^\dagger G^\dagger}(t')\delta(t'-t''). \tag{3.9.74}$$

The function $D_{GG}(t)$ and its conjugate have a special form,

$$D_{GG}(t) = D_{FF}\exp(-2i\Omega t),$$
$$D_{G^\dagger G^\dagger}(t) = D_{F^\dagger F^\dagger}\exp(2i\Omega t). \tag{3.9.75}$$

The quantities $D_{G^\dagger G}$, $D_{GG^\dagger}$, D_{FF}, $D_{F^\dagger F^\dagger}$ are parameters of the quantum noise with the properties

$$D_{GG^\dagger} > 0, \quad D_{G^\dagger G} \geq 0, \quad D_{GG^\dagger} > D_{G^\dagger G}, \tag{3.9.76}$$

$$D_{GG^\dagger}D_{G^\dagger G} - |D_{FF}|^2 \geq 0, \quad D_{F^\dagger F^\dagger} = D_{FF}^*. \tag{3.9.77}$$

Using the technique exposed in the paragraph focusing on the rule (3.9.52), we express the limit values

$$\lim_{\Delta t \to +0} \frac{1}{\Delta t}\left\langle \int_t^{t+\Delta t}\int_t^{t'}\hat{G}^\dagger(t')\hat{G}(t'')\, dt''\, dt'\right\rangle_{>t} = D_{G^\dagger G},$$

$$\lim_{\Delta t \to +0} \frac{1}{\Delta t} \left\langle \int_t^{t+\Delta t} \int_t^{t'} \hat{G}(t')\hat{G}(t'') \, dt'' \, dt' \right\rangle_{>t} = D_{GG}(t), \qquad (3.9.78)$$

and arrive at the more general master equation comprising rotating terms

$$\begin{aligned}
\frac{\partial}{\partial t}\hat{\rho}(t) &= -\frac{i}{\hbar}[\hat{H}_{\text{rad}}, \hat{\rho}(t)] + D_{G^\dagger G^\dagger}(t)\left[2\hat{O}\hat{\rho}(t)\hat{O} - \hat{O}^2\hat{\rho}(t) - \hat{\rho}(t)\hat{O}^2\right] \\
&\quad + D_{G^\dagger G}\left[2\hat{O}^\dagger\hat{\rho}(t)\hat{O} - \hat{O}\hat{O}^\dagger\hat{\rho}(t) - \hat{\rho}(t)\hat{O}\hat{O}^\dagger\right] \\
&\quad + D_{GG^\dagger}\left[2\hat{O}\hat{\rho}(t)\hat{O}^\dagger - \hat{O}^\dagger\hat{O}\hat{\rho}(t) - \hat{\rho}(t)\hat{O}^\dagger\hat{O}\right] \\
&\quad + D_{GG}(t)\left[2\hat{O}^\dagger\hat{\rho}(t)\hat{O}^\dagger - \hat{O}^{\dagger 2}\hat{\rho}(t) - \hat{\rho}(t)\hat{O}^{\dagger 2}\right], \qquad (3.9.79)
\end{aligned}$$

where

$$\hat{H}_{\text{rad}} = \hat{H}_{0,\text{rad}} + \hat{H}_{\text{int,rad}}, \qquad (3.9.80)$$

with $\hat{H}_{0,\text{rad}}$ according to (3.9.2), $N = n$, and $\hat{H}_{\text{int,rad}}$ is the usual effective interaction Hamiltonian. For the case of one-photon absorption in a single mode, we have $\hat{O} = \hat{a}$ ($m = 0$, $n = 1$) and the rate coefficients

$$D_{G^\dagger G} = \begin{cases} K\langle \hat{c}_{2l}^\dagger \hat{c}_{2l} \rangle_{\text{atom}}, \\ \frac{\gamma}{2}\langle \hat{b}_l^\dagger \hat{b}_l \rangle_{\text{R}}, \end{cases} \qquad (3.9.81)$$

$$D_{GG^\dagger} = \begin{cases} K\langle \hat{c}_{1l}^\dagger \hat{c}_{1l} \rangle_{\text{atom}} = K\langle \hat{c}_{2l} \hat{c}_{2l}^\dagger \rangle_{\text{atom}}, \\ \frac{\gamma}{2}\langle \hat{b}_l \hat{b}_l^\dagger \rangle_{\text{R}}, \end{cases} \qquad (3.9.82)$$

$$D_{FF} = \begin{cases} K\langle \hat{c}_{1l}^\dagger \hat{c}_{2l} \rangle_{\text{atom}}, \\ \frac{\gamma}{2}\langle \hat{b}_l^2 \rangle_{\text{R}}. \end{cases} \qquad (3.9.83)$$

The second lines in the braces of (3.9.81)–(3.9.83) have been added to exhibit the analogy with a thermal reservoir. In the case of one-photon absorption and for

$$\hat{H}_{\text{rad}} = \hat{H}_{0,\text{rad}} = \hbar\omega\left(\hat{a}^\dagger\hat{a} + \frac{1}{2}\hat{1}\right), \qquad (3.9.84)$$

the Fokker–Planck equation appropriate to (3.9.79) is of the form [Peřinová, Lukš and Szlachetka (1989)]

$$\begin{aligned}
\frac{\partial}{\partial t}\Phi_s(\alpha, t) &= \left[\frac{\partial}{\partial\alpha}\left(\Gamma\alpha\Phi_s(\alpha, t)\right) - D_{\alpha\alpha(s)}(t)\frac{\partial^2}{\partial\alpha^2}\Phi_s(\alpha, t) + \text{c. c.}\right] \\
&\quad + 2D_{\alpha\alpha^*(s)}\frac{\partial^2}{\partial\alpha\partial\alpha^*}\Phi_s(\alpha, t), \qquad (3.9.85)
\end{aligned}$$

where

$$\begin{aligned}
\Gamma &= i\omega + D_{GG^\dagger} - D_{G^\dagger G}, \\
D_{\alpha\alpha^*(s)} &= -\frac{s}{2}(D_{GG^\dagger} - D_{G^\dagger G}) + \frac{1}{2}(D_{GG^\dagger} + D_{G^\dagger G}), \\
D_{\alpha\alpha(s)}(t) &= D_{GG}(t), \\
D_{\alpha^*\alpha^*(s)}(t) &= D_{\alpha\alpha(s)}^*(t).
\end{aligned} \qquad (3.9.86)$$

An exact irreversibility enters the physical descriptions only via von Neumann projection postulate. Of course, an opinion prevails that it is no fundamental physical truth that the irreversibility is present in the Nature. von Neumann pointed out the two fundamentally different dynamics for describing the change of a quantum state in quantum theory: The unitary continuous evolution of a closed system and the instantaneous, but unpredictable projection of its state during a measurement causing a reduction of information on the history.

The theory of continuous measurement based on the concept of the open system represents a way how to render the irreversible measurement dynamics as acceptable as the master equation models. The essentials of the ideal measurement will be illustrated by the measurement of the photon number in one of two modes. The Hilbert space $\mathcal{H}$ of a two-mode optical system is a completion of the span of the number states $|n_1, n_2\rangle = |n_1\rangle_1 \otimes |n_2\rangle_2$, where $|n_j\rangle_j$ are elements of the single-mode Hilbert space $\mathcal{H}_j, j = 1, 2$. Thus, $\mathcal{H} = \mathcal{H}_1 \otimes \mathcal{H}_2$. In terms of these states any statistical operator $\hat{\rho}$ can be expanded in a fourfold sum with c-number coefficients. Assuming a measurement of the operator $\hat{n}_2$, we consider a series with q-number coefficients

$$\hat{\rho} = \sum_{m_2=0}^{\infty} \sum_{n_2=0}^{\infty} {}_2\langle n_2|\hat{\rho}|m_2\rangle_2 \otimes |n_2\rangle_{22}\langle m_2|, \qquad (3.9.87)$$

where

$$_2\langle n_2|\hat{\rho}|m_2\rangle_2 = \sum_{m_1=0}^{\infty} \sum_{n_1=0}^{\infty} \langle n_1, n_2|\hat{\rho}|m_1, m_2\rangle|n_1\rangle_{11}\langle m_1|. \qquad (3.9.88)$$

The notation used on the left-hand side of (3.9.88) can be associated with the single-mode scalar product such that

$$_2\langle n_2|m_1, m_2\rangle = \delta_{n_2, m_2}|m_1\rangle. \qquad (3.9.89)$$

This scalar product yields elements of the first-mode Hilbert space. According to quantum theory, the consequence of the measurement of the operator $\hat{n}_2$ when the result of measurement is not taken into account is the reduction of the state described by $\hat{\rho}$ to the state described by the statistical operator

$$\hat{\rho}' = \sum_{M=0}^{\infty} {}_2\langle M|\hat{\rho}|M\rangle_2 \otimes |M\rangle_{22}\langle M| \qquad (3.9.90)$$

$$= \sum_{M=0}^{\infty} \hat{\rho}_{|M}^{(1)} \otimes p(M)|M\rangle_{22}\langle M|, \qquad (3.9.91)$$

where the operator

$$\hat{\rho}_{|M}^{(1)} = \frac{1}{p(M)} {}_2\langle M|\hat{\rho}|M\rangle_2 \qquad (3.9.92)$$

and the real number

$$p(M) = \text{Tr}_1 \{ {}_2\langle M|\hat{\rho}(t)|M\rangle_2 \} = {}_2\langle M|\text{Tr}_1\{\hat{\rho}\}|M\rangle_2. \qquad (3.9.93)$$

When the result M of the measurement is known, the operator (3.9.92) describes the reduced state conditional on the result M. The relation (3.9.93) determines the probability of this result. In a more usual notation, the projection operator $\hat{E}$ is used with the property

$$\hat{E}(M)|\psi\rangle = {}_2\langle M|\psi\rangle \otimes |M\rangle_2. \tag{3.9.94}$$

This enables one to simplify the relation (3.9.90) as

$$\hat{\rho}' = \sum_{M=0}^{\infty} \hat{E}(M)\hat{\rho}\hat{E}(M) \tag{3.9.95}$$

and suggests to work with the conditional operator

$$\hat{\rho}_{|M} = \frac{1}{p(M)}\hat{E}(M)\hat{\rho}\hat{E}(M). \tag{3.9.96}$$

Also the distribution (3.9.93) can be expressed as

$$p(M) = \mathrm{Tr}\left\{\hat{\rho}\hat{E}(M)\right\}. \tag{3.9.97}$$

It is attractive to have a model of nonideal photon-number measurement. A corresponding framework is provided by the theory of continuous photon-number measurement originating from the Srinivas–Davies model of destructive continuous measurement [Srinivas and Davies (1981, 1982), Ueda (1989, 1990), Ueda, Imoto and Ogawa (1990a, b), Imoto, Ueda and Ogawa (1990), Ueda and Kitagawa (1992)].

Recently, models of nondemolition continuous photon-number measurement have been presented [Ueda, Imoto, Nagaoka and Ogawa (1992), Brune, Haroche and Raimond (1992)]. We assume that the measurement is performed with a detector, which—although a macroscopic system—can be described quantum mechanically. Of course, this description need not be a simple task in all cases. We suppose that— if outlined—the description coincides with the formulae (3.9.68), (3.9.69), (3.9.70), (3.9.73) except that we admit the degeneracy of a multiphoton process.

To keep touch with the theory of continuous measurement, we put $N_2 = 0$, $KN_1 = R$ in (3.9.73). Just as in the case of the ideal measurement, where the second mode is measured, we consider $\hat{O} = \hat{a}_2$ ($\hat{O} = \hat{a}_2^{\dagger}\hat{a}_2$) in the case of destructive (nondemolition) photon-number measurement. The nonreferring measurement [Ueda (1990)], i. e., the process when the results are not being taken into account, is described by the master equation in the Lindblad form [Lindblad (1976)]

$$\frac{\partial}{\partial T}\hat{\rho}(T) = R\left[\hat{O}\hat{\rho}(T)\hat{O}^{\dagger} - \frac{1}{2}\hat{O}^{\dagger}\hat{O}\hat{\rho}(T) - \frac{1}{2}\hat{\rho}(T)\hat{O}^{\dagger}\hat{O}\right], \tag{3.9.98}$$

with an initial condition

$$\hat{\rho}(T)|_{T=0} = \hat{\rho}. \tag{3.9.99}$$

Contrary to the ideal measurement taken for instantaneous, T is chosen to denote the detection time here.

In the case of destructive measurement the probabilities

$$p_{n_2}^{(2\text{dem})}(T) = {}_2\langle n_2|\text{Tr}_1\{\hat{\rho}_{\text{dem}}(T)\}|n_2\rangle_2 \tag{3.9.100}$$

obey the ordinary differential equations

$$\frac{d}{dT}p_{n_2}^{(2\text{dem})}(T) = -R\left[n_2 p_{n_2}^{(2\text{dem})}(T) - (n_2+1)p_{n_2+1}^{(2\text{dem})}(T)\right], \tag{3.9.101}$$

under the initial conditions

$$p_{n_2}^{(2\text{dem})}(T)\Big|_{T=0} = p_{n_2}^{(2)}(0). \tag{3.9.102}$$

Although the relationship to the reservoir or atomic system has not been clarified, these equations can be interpreted as rate equations for the Markov process with independent decrements $\underline{n}_{2\text{dem}}(T)$,

$$\underline{n}_{2\text{dem}}(T) = \underline{n}_2(0) - \sum_{j=1}^{\infty}\theta\left(T - \underline{\tau}_j^{(\text{dem})}\right), \tag{3.9.103}$$

where θ is the Heaviside unit-step function, $\underline{n}_{2(\text{dem})}(0)$ is the initial value of the process, which is a random variable with the distribution $p_{n_2}^{(2)}(0)$, and $\underline{\tau}_j^{(\text{dem})}$ are the instants of the events $\underline{\tau}_1^{(\text{dem})} < \underline{\tau}_2^{(\text{dem})} < \underline{\tau}_3^{(\text{dem})} < ...$, which are stochastic variables. Typically, the Markovian process (3.9.103) is characterized by the transition probabilities ($\Delta T > 0$)

$$\text{Prob}\left(\underline{n}_{2\text{dem}}(T + \Delta T) = n_2|\underline{n}_{2\text{dem}}(T) = n_2\right) = 1 - Rn_2\Delta T + o(\Delta T), \tag{3.9.104}$$

$$\text{Prob}\left(\underline{n}_{2\text{dem}}(T + \Delta T) = n_2 - 1|\underline{n}_{2\text{dem}}(T) = n_2\right) = Rn_2\Delta T + o(\Delta T), \tag{3.9.105}$$

$$\text{Prob}\left(\underline{n}_{2\text{dem}}(T + \Delta T) \neq n_2, n_2 - 1|\underline{n}_{2\text{dem}}(T) = n_2\right) = o(\Delta T), \tag{3.9.106}$$

where $o(\Delta T)$ has the property that $\frac{o(\Delta T)}{\Delta T} \to 0$ for $\Delta T \to 0$.

The operation of the detector can be described by the related process $\underline{M}_{\text{dem}}(T)$,

$$\underline{M}_{\text{dem}}(T) = \sum_{j=1}^{\infty}\theta\left(T - \underline{\tau}_j^{(\text{dem})}\right). \tag{3.9.107}$$

Although this process is understood at another level than the process (3.9.103) under study, the following conservation law is worth noting

$$\underline{n}_{2\text{dem}}(T) + \underline{M}_{\text{dem}}(T) = \underline{n}_2(0). \tag{3.9.108}$$

In the case of nondemolition measurement the probabilities

$$p_{n_2}^{(2\text{nondem})}(T) = {}_2\langle n_2|\text{Tr}_1\{\hat{\rho}_{\text{nondem}}(T)\}|n_2\rangle_2 \tag{3.9.109}$$

do not evolve,

$$p_{n_2}^{(2\text{nondem})}(T) = p_{n_2}^{(2)}(0),$$ (3.9.110)

and the study of the process

$$\underline{n}_{2\text{nondem}}(T) = \underline{n}_2(0)$$ (3.9.111)

is of no use. It is necessary to introduce the detection process $\underline{M}_{\text{nondem}}(T)$,

$$\underline{M}_{\text{nondem}}(T) = \sum_{j=1}^{\infty} \theta(T - \underline{\tau}_j^{(\text{nondem})}),$$ (3.9.112)

where $\underline{\tau}_j^{(\text{nondem})}$ are analogues of $\underline{\tau}_j^{(\text{dem})}$.

Formally, we may treat the detector as another degree of freedom with a Hilbert space $\mathcal{H}_{\text{det}}$ enlarging the product Hilbert space $\mathcal{H}$, $\mathcal{H} = \mathcal{H}_1 \otimes \mathcal{H}_2 \to \mathcal{H}_{\text{m}} = \mathcal{H} \otimes \mathcal{H}_{\text{det}}$, where the subscript m stands for "modified". Macroscopically, the detector does not admit the superposition principle (the Schrödinger-cat paradox excluded) and we work only with the mixtures of the number states described by $|M\rangle_{\text{det}}\,_{\text{det}}\langle M|$. Nevertheless, it is useful to define the shift operators

$$\begin{aligned}
\widehat{\exp}(i\varphi_{\text{det}}) &= \hat{1} \otimes \sum_{M=0}^{\infty} |M\rangle_{\text{det}}\,_{\text{det}}\langle M+1|, \\
\widehat{\exp}(-i\varphi_{\text{det}}) &= [\widehat{\exp}(i\varphi_{\text{det}})]^{\dagger},
\end{aligned}$$ (3.9.113)

but we rely only on its pairwise use as in the characteristic relation

$$\widehat{\exp}(-i\varphi_{\text{det}})|M\rangle_{\text{det}}\,_{\text{det}}\langle M|\widehat{\exp}(i\varphi_{\text{det}}) = |M+1\rangle_{\text{det}}\,_{\text{det}}\langle M+1|.$$ (3.9.114)

Let us mention the analogue of the formula (3.9.89),

$$_{\text{det}}\langle M|m_1, m_2, M'\rangle = \delta_{MM'}|m_1, m_2\rangle.$$ (3.9.115)

In this treatment we pass from $\hat{\rho}(T)$ to the statistical operator $\hat{\rho}_{\text{m}}(T)$.

Introducing the master equation

$$\frac{\partial}{\partial T}\hat{\rho}_{\text{m}}(T) = R\left[\hat{O}_{\text{m}}\hat{\rho}_{\text{m}}(T)\hat{O}_{\text{m}}^{\dagger} - \frac{1}{2}\hat{O}_{\text{m}}^{\dagger}\hat{O}_{\text{m}}\hat{\rho}_{\text{m}}(T) - \frac{1}{2}\hat{\rho}_{\text{m}}(T)\hat{O}_{\text{m}}^{\dagger}\hat{O}_{\text{m}}\right],$$ (3.9.116)

where

$$\hat{O}_{\text{m}} = \hat{O}\widehat{\exp}(-i\varphi_{\text{det}}),$$ (3.9.117)

and the initial condition

$$\hat{\rho}_{\text{m}}(T)|_{T=0} = \hat{\rho} \otimes |0\rangle_{\text{det}}\,_{\text{det}}\langle 0|,$$ (3.9.118)

we derive the rate equations for the two-component stochastic process $\{\underline{n_2}(T), \underline{M}(T)\}$,

$$
\begin{aligned}
\frac{d}{dT} p_{n_2, M}(T) = & -R\Big[\langle n_2|\hat{O}^\dagger \hat{O}|n_2\rangle p_{n_2, M}(T) \\
& - \big|\langle n_2|\hat{O}|n_2+r\rangle\big|^2 p_{n_2+r, M-1}(T)\Big],
\end{aligned}
\tag{3.9.119}
$$

where

$$
r = \begin{cases} 1 & \text{for } \hat{O} = \hat{a}_2, \\ 0 & \text{for } \hat{O} = \hat{a}_2^\dagger \hat{a}_2. \end{cases}
\tag{3.9.120}
$$

The considered operators $\hat{O}$ have the property

$$
\big|\langle n_2|\hat{O}|n_2+r\rangle\big|^2 = \langle n_2+r|\hat{O}^\dagger \hat{O}|n_2+r\rangle.
\tag{3.9.121}
$$

The perturbative solution to the master equation (3.9.116) with the initial condition (3.9.118) is of the form

$$
\hat{\rho}_{\mathrm{m}}(T) = \sum_{M=0}^{\infty} \hat{\hat{u}}_2(M, T)\hat{\rho} \otimes |M\rangle_{\mathrm{det}}\,_{\mathrm{det}}\langle M|.
\tag{3.9.122}
$$

Here the superoperator $\hat{\hat{u}}_2(M, T)$ is defined by the property

$$
\begin{aligned}
\hat{\hat{u}}_2(M, T)\hat{\rho} = & \int_0^T \int_0^{\tau_M} \cdots \int_0^{\tau_2} \hat{\hat{u}}_2(M, \tau_1, \ldots, \tau_{M-1}, \tau_M, T) \\
& \times \hat{\rho}\, d\tau_1 \ldots d\tau_{M-1}\, d\tau_M,
\end{aligned}
\tag{3.9.123}
$$

where

$$
\hat{\hat{u}}_2(M, \tau_1, \ldots, \tau_M, T)\hat{\rho} = \hat{u}_2(M, \tau_1, \ldots, \tau_M, T)\hat{\rho}\hat{u}_2^\dagger(M, \tau_1, \ldots, \tau_M, T),
\tag{3.9.124}
$$

with

$$
\begin{aligned}
\hat{u}_2(M, \tau_1, \ldots, \tau_M, T) = & R^{\frac{M}{2}} \exp\Big[-\frac{R}{2}(T - \tau_M)\hat{O}^\dagger \hat{O}\Big] \hat{O} \exp\Big[-\frac{R}{2}(\tau_M - \tau_{M-1})\hat{O}^\dagger \hat{O}\Big] \\
& \times \ldots \hat{O} \exp\Big(-\frac{R}{2}\tau_1 \hat{O}^\dagger \hat{O}\Big).
\end{aligned}
\tag{3.9.125}
$$

Substituting

$$
\hat{\rho}(T) = \mathrm{Tr}_{\mathrm{det}}\{\hat{\rho}_{\mathrm{m}}(T)\}
\tag{3.9.126}
$$

into the left-hand side of equation (3.9.98), using the achieved property (3.9.116) and the cyclic property of $\mathrm{Tr}_{\mathrm{det}}$, we derive that (3.9.126) is the solution of (3.9.98). Substituting (3.9.122) into (3.9.126), we arrive at either a coarser expansion

$$
\hat{\rho}(T) = \sum_{M=0}^{\infty} p(M, T)\hat{\rho}_{|M}(T)
\tag{3.9.127}
$$

or a finer one

$$\hat{\rho}(T) = \sum_{M=0}^{\infty} \int_0^T \int_0^{\tau_M} \cdots \int_0^{\tau_2} p(M, \tau_1, \ldots, \tau_M, T)$$
$$\times \hat{\rho}_{|M,\tau_1,\ldots,\tau_M}(T)\, d\tau_1 \, \ldots \, d\tau_{M-1}\, d\tau_M, \qquad (3.9.128)$$

where $p(M, T)$ is the probability of M counts being registered in the interval $[0, T)$,

$$p(M, T) = \mathrm{Tr}\left\{\hat{\hat{u}}_2(M, T)\hat{\rho}\right\}, \qquad (3.9.129)$$

and $p(M, \tau_1, \ldots, \tau_M, T)$ is the probability density of M counts being registered at the times $\tau_1, \ldots, \tau_M$, in the interval $[0, T)$,

$$p(M, \tau_1, \ldots, \tau_M, T) = \mathrm{Tr}\left\{\hat{\hat{u}}_2(M, \tau_1, \ldots, \tau_M, T)\hat{\rho}\right\}. \qquad (3.9.130)$$

Here $\hat{\rho}_{|M}(T)$ describes the resulting two-mode state under the coarser condition,

$$\hat{\rho}_{|M}(T) = \frac{1}{p(M, T)}\hat{\hat{u}}_2(M, T)\hat{\rho}, \qquad (3.9.131)$$

and $\hat{\rho}_{|M,\tau_1,\ldots,\tau_M}(T)$ describes the resulting state under the finer condition,

$$\hat{\rho}_{|M,\tau_1,\ldots,\tau_M}(T) = \frac{1}{p(M, \tau_1, \ldots, \tau_M, T)}\hat{\hat{u}}_2(M, \tau_1, \ldots, \tau_M, T)\hat{\rho}. \qquad (3.9.132)$$

It is important that in the two considered cases of measurement, $\hat{O} = \hat{a}_2$, $\hat{O} = \hat{a}_2^\dagger \hat{a}_2$, the refinement does not affect the resulting state, i. e.,

$$\hat{\rho}_{|M,\tau_1,\ldots,\tau_M}(T) = \hat{\rho}_{|M}(T). \qquad (3.9.133)$$

Taking into account the statistical operator (3.9.122), we obtain coupled master equations for unnormalized statistical operators $\hat{\hat{u}}_2(M, T)\hat{\rho}$,

$$\frac{\partial}{\partial T}\left[\hat{\hat{u}}_2(M, T)\hat{\rho}\right] = -R\left\{\frac{1}{2}\hat{O}^\dagger\hat{O}[\hat{\hat{u}}_2(M, T)\hat{\rho}] + \frac{1}{2}[\hat{\hat{u}}_2(M, T)\hat{\rho}]\hat{O}^\dagger\hat{O}\right.$$
$$\left. -\hat{O}[\hat{\hat{u}}_2(M-1, T)\hat{\rho}]\hat{O}^\dagger\right\}. \qquad (3.9.134)$$

Tracing over the two modes under study in equation (3.9.134), using (3.9.129) and the fact that

$$\mathrm{Tr}\left\{\hat{O}^\dagger\hat{O}\hat{\hat{u}}_2(M, T)\hat{\rho}\right\} = \mathrm{Tr}\left\{\hat{\rho}_{|M}(T)\hat{O}^\dagger\hat{O}\right\}p_M^{(\mathrm{det})}(T), \qquad (3.9.135)$$

we arrive at equations resembling (3.9.119)

$$\frac{\partial}{\partial T}p_M^{(\mathrm{det})}(T) = -R\left[\mathrm{Tr}\left\{\hat{\rho}_{|M}(T)\hat{O}^\dagger\hat{O}\right\}p_M^{(\mathrm{det})}(T)\right.$$
$$\left. -\mathrm{Tr}\left\{\hat{\rho}_{|M-1}(T)\hat{O}^\dagger\hat{O}\right\}p_{M-1}^{(\mathrm{det})}(T)\right]. \qquad (3.9.136)$$

On a closer inspection of (3.9.119), we see that equations (3.9.119) can be obtained from equations (3.9.136) putting

$$\hat{\rho}(T)|_{T=0} = \hat{\rho}_{|0}(0) = \hat{\rho}^{(1)} \otimes |n_2(0)\rangle_2 \, {}_2\langle n_2(0)|, \qquad (3.9.137)$$

where $\hat{\rho}^{(1)}$ describes any single-mode state. In this particular case, the solution of the master equation (3.9.98) has a finite expansion

$$\hat{\rho}(T) = \sum_{M=0}^{n_2(0)} p(M,T)\hat{\rho}_{|M}(T), \qquad (3.9.138)$$

where $p(M,T)$ is a binomial distribution and

$$\hat{\rho}_{|M}(T) = \hat{\rho}^{(1)} \otimes |n_2(0) - rM\rangle_2 \, {}_2\langle n_2(0) - rM|. \qquad (3.9.139)$$

In other words, in the course of the referring measurement, the second mode is in the pure number state all times. Recalling that in this case the initial condition may read

$$p_{n_2,M}(T)|_{T=0} = \delta_{n_2,n_2(0)}\delta_{M,0}, \qquad (3.9.140)$$

we may verify that

$$p_{n_2,M}(T) = p_M^{(\text{det})}(T)\delta_{n_2,n_2(0)-rM} \qquad (3.9.141)$$

are solutions of equations (3.9.119). As here (3.9.139) represents quantum component of a stochastic process whose detection part is the process $\underline{M}(T)$, we easily understand the idea of the quantum stochastic process treated in [Davies (1976)]. The property of this kind of processes that the initial pure state remains pure all the times has attracted the attention and leads to the Monte Carlo wave-function approach [Dalibard, Castin and Mølmer (1992), Carmichael (1993)].

The continuous photon-number measurement is a quantum Markov process. Although the statement that this process does not obey the projection postulate says a great deal, microscopic models of the continuous photon-number measurement have revealed that the projection postulate is being applied in this process continuously, i. e., in models almost continuously, repeatedly after short interaction times of a light mode with additional degrees of freedom [Imoto, Ueda and Ogawa (1990)]. This is connected with the idea of a quantum measurement [von Neumann (1932)] that the measured system should be coupled to another one on which the projection is perfomed. Although the latter is called detector the relationship with the detection process is clear only in the microscopic models. The projections interfere with the measured state slightly for the most of time, because one of the projections affects the diagonal elements only from time to time. The projections suppress small off-diagonal matrix elements each time due to a short interaction period. The measured photon-number operator of the light mode appropriate to the projection postulate happens to be replaced by the measured energy operators of additional degrees of freedom in the microscopic models of the continuous photon-number measurement. In order that this situation is illustrated, it is possible to take resonant harmonic oscillators for the additional degrees of freedom and to observe that after the interaction the state of one of them has changed only so that the probability of finding one photon in this degree of freedom is infinitesimal of the order of the interaction time, but the probability of finding more photons is negligible and, of course, the probability of no photon count is almost unity. On the whole, the no-count events constitute continua, which are separated by discrete one-count events. The measuring apparatus was early explained microscopically [Imoto, Ueda and Ogawa (1990)] or made concrete as an atomic quantum device [Ueda and Kitagawa (1992)]. From the point of view of the continuous application of the projection postulate, the microscopic explanation of the photodetector uses simply the ground and the excited state of the additional degree of freedom and the "pure" projections on the no- and one-count events. The change caused by the projection on the ground state is slight, but the projection on the excited state means a collapse (a quantum jump). This does not contradict the above characteristics of the interference between the mixed projections and the measured state, because when the distinction between no- and one-count events is not made, the jumps are averaged out due to the projection on the excited state. The quantum effects on a cavity mode of the electromagnetic field caused by measuring one of its quadrature components have been analyzed [Wiseman and Milburn (1993)]. An intracavity quantum-nondemolition coupling to another mode, simple homodyne detection, and balanced homodyne detection have been considered.

Let us remark that the almost continuous application of the projection postulate leads to continuous measurement in the above mentioned limit and changed conditions for the application of the projection postulate may be obtained.

3.10   Generation and decay of superposition states

In the remote past the fathers of new quantum theory (namely, W. Heisenberg) answered the question why the position–momentum uncertainty of the centroid of a macroscopic body is not observable. Since the eighties specialists have answered a slightly more abstract question why superpositions of classically distinguishable states are not observable in the everyday world of classical physics. Put in a different way, the question reads why the so-called quantum coherence is unobserved. In short, we ask why the world we see is essentially classical.

At the beginning of quantum theory part of physicists perceived the necessity of a new mathematical formalism, which was worked out by the mathematician J. von Neumann [von Neumann (1932)] for quantum mechanics. In that book the relevance of the projection postulate (the collapse of a state vector) for quantum measurements is assumed. Consequences of this postulate for measurements of correlated quantum systems were discussed by Einstein, Podolsky and Rosen (1935). In von Neumann's book also the so-called von Neumann measurement has been introduced as the development of a quantum correlation between a quantum system and a measuring instrument off which one reads out the outcome as the projection postulate describes. Schrödinger (1935) formulated a thought experiment which ridiculously resembled the von Neumann measurement. The quantum system is an atom of radioactive substance which can be in two states, |atom not decayed⟩ and |atom decayed⟩, while the measuring instrument is a cat, which can be in two states, |cat alive⟩ and |cat dead⟩. After the entanglement (the quantum correlation) of the two systems develops, the observer reads out the state of the atom according to the rule: If the cat is dead, then the atom has decayed.

Quite paradoxically, a slight modification of this "procedure of measurement" results in a superposition of states |cat alive⟩ and |cat dead⟩. The paradox has often been dismissed as having no observable consequences, although the state of being alive is clearly distinguishable from the state of being dead. Similarly, in quantum optics when a standing wave is assumed, the coherent states of the same amplitude, but shifted in phase by π, have the same spatial distribution of the optical intensity, but distinct (namely, opposite) electric fields. For a suitable wavelength we have a mesoscopic scale. The superpositions have been realized experimentally in the context of cavity quantum electrodynamics by S. Haroche and collaborators [Davidovich, Brune, Raimond and Haroche (1996)] in Paris.

Quantum coherence is an extraordinarily fragile property. Destroying the integrity of superpositions (the decay of coherence) can be described also as randomly disturbing the phases of the components in the superposition.

The concept of a superposition state has become of a great importance in quantum optics and it belongs to the fundamentals of quantum theory. Now we attempt the general definition useful for the study of statistics of these states. In the study of the superposition states in a system of light modes we suppose the presence of another component R of a composite of physical systems. For simplicity, we consider an L-level system R. Similarly as in [Helstrom (1976)] we describe the composite system by a statistical operator $\hat{\rho}_{S+R}$. With respect to the basis of L orthogonal states from the space $\mathcal{H}_R$, the matrix elements $\hat{\rho}_{jk}$ can be defined, which are operators on $\mathcal{H}_S$. These operators have the property

$$\hat{\rho}_{jk}^{\dagger} = \hat{\rho}_{kj} \tag{3.10.1}$$

and particularly $\hat{\rho}_{jj}$ are statistical operators in the system S up to a normalization constant

$$\hat{\rho}_{jj} = p_j \hat{\rho}_{|j}, \tag{3.10.2}$$

where $\hat{\rho}_{|j}$ is a statistical operator and p_j is a probability.

Nonreferring measurement [Ueda 1990] of the operator $\hat{X}$ of the system R,

$$\hat{X} = \sum_{j=1}^{L} j|j\rangle\langle j|, \tag{3.10.3}$$

yields a mixed state of the system S,

$$\hat{\rho}' = \sum_{j=1}^{L} \hat{\rho}_{jj}, \tag{3.10.4}$$

$$\hat{\rho}' = \sum_{j=1}^{L} p_j \hat{\rho}_{|j}. \tag{3.10.5}$$

Let us assume for a moment that

$$\text{Tr}_S\{\hat{\rho}_{jk}\} = p_j \delta_{jk}. \tag{3.10.6}$$

Then it is natural to define a superposition state of the system S as follows

$$\hat{\rho} = \sum_{j=1}^{L} \sum_{k=1}^{L} \hat{\rho}_{jk}. \tag{3.10.7}$$

It is obvious that this operator has the correct trace and it is positive as follows from the equation

$$\hat{\rho} = L \, \text{Tr}_R \left\{ \hat{\rho}_{S+R} \frac{1}{L} \hat{1}_{\text{para}} \right\}, \tag{3.10.8}$$

where a projection operator

$$\frac{1}{L} \hat{1}_{\text{para}} = \hat{1}_S \otimes |0_{\text{para}}\rangle_{RR}\langle 0_{\text{para}}|, \tag{3.10.9}$$

with

$$|0_{\text{para}}\rangle_R = \frac{1}{\sqrt{L}} \sum_{j=1}^{L} |j\rangle_R, \tag{3.10.10}$$

and Tr_R means the trace over the L-level system. In general, the superposition state can be introduced without the assumption (3.10.6) as a paradoxical mixture

$$\hat{\rho} = \sum_{j=1}^{L} \sum_{k=1}^{L} Z_j Z_k^* \hat{\rho}_{jk}, \tag{3.10.11}$$

where Z_j are any complex numbers obeying the normalization condition

$$\sum_{j=1}^{L} \sum_{k=1}^{L} Z_j Z_k^* \text{Tr}_S\{\hat{\rho}_{jk}\} = 1. \tag{3.10.12}$$

Contrary to (3.10.6) let us assume that $\text{Tr}_S\{\hat{\rho}_{jk}\} \neq 0$ for all j, k. This situation may occur in the detailed study of the statistics as we will demonstrate below in the coherent state. In this case we introduce paradoxical probabilities

$$p_{jk} = \text{Tr}_S\{\hat{\rho}_{jk}\} \tag{3.10.13}$$

and for $p_{jk} \neq 0$ operators

$$\hat{\rho}_{|jk} = \frac{\hat{\rho}_{jk}}{p_{jk}}. \tag{3.10.14}$$

Now we explain the usual and paradoxical analyses of variance of frequently occurring operators. The state $\hat{\rho}_{S+R}$ is pure, if there exists a ket $|\psi_{S+R}\rangle \in \mathcal{H}_{S+R}$ such that

$$\hat{\rho}_{S+R} = |\psi_{S+R}\rangle\langle\psi_{S+R}|. \tag{3.10.15}$$

With respect to the basis consisting of L orthogonal states from the space $\mathcal{H}_R$ we have the "coefficients" $|\psi_j\rangle$, which themselves are vectors from $\mathcal{H}_S$. It holds that the matrix elements

$$\hat{\rho}_{jk} = |\psi_j\rangle\langle\psi_k|. \tag{3.10.16}$$

The property (3.10.6) then implies the orthogonality property. It cannot be supposed that the vectors $|\psi_j\rangle$ will be orthogonal in any case because it would mean very strong quantum correlation (a complete entanglement). The vectors $|\psi_j\rangle$ are distinct, when the entanglement occurs, whereas in the case of the independence of the physical systems the vector $|\psi_j\rangle$ is independent of j. The quantum superposition obeys the relation

$$\hat{\rho} = |\psi\rangle\langle\psi|, \tag{3.10.17}$$

with

$$|\psi\rangle = \sum_{j=1}^{L} \mathcal{Z}_j |\psi_j\rangle. \tag{3.10.18}$$

Returning to the usual mixture, we rewrite the formula (3.10.5) in the form

$$\hat{\rho}' = E_x\left(\hat{\rho}_{|x}\right), \tag{3.10.19}$$

where E_x means averaging over the random variable x taking the values of j, $j = 1, ..., L$, with the probabilities p_j. The operator E_x stands for the angular brackets $\langle...\rangle$, which we use everywhere else. For simplicity we assume that the system S consists of a single mode. From (3.10.19) we obtain the expression for the normally ordered moments (cf. section 3.3) in the form

$$\text{Tr}\{\hat{\rho}'\hat{a}^{\dagger k}\hat{a}^l\} = E_x\left(\text{Tr}_S\{\hat{\rho}_{|x}\hat{a}^{\dagger k}\hat{a}^l\}\right) \tag{3.10.20}$$

or in the abbreviated notation

$$E(\hat{a}^{\dagger k}\hat{a}^l) = E_x E_S(\hat{a}^{\dagger k}\hat{a}^l|x). \tag{3.10.21}$$

Assuming an appropriate definition of symbols $\Delta_x\hat{a}^{\dagger}$, $\Delta_x\hat{a}$,

$$\begin{aligned}
\Delta_x\hat{a}^{\dagger} &= \hat{a}^{\dagger} - E_S(\hat{a}^{\dagger}|x), \\
\Delta_x\hat{a} &= \hat{a} - E_S(\hat{a}|x),
\end{aligned} \tag{3.10.22}$$

we can derive the following decomposition of the central second-order moments

$$E(\Delta\hat{a}^\dagger\Delta\hat{a}) = E_x E_S(\Delta_x\hat{a}^\dagger\Delta_x\hat{a}|x) + E_x(E_S(\hat{a}^\dagger|x)E_S(\hat{a}|x)) - E(\hat{a}^\dagger)E(\hat{a}), \quad (3.10.23)$$

$$E((\Delta\hat{a})^2) = E_x E_S((\Delta_x\hat{a})^2|x) + E_x\left([E_S(\hat{a}|x)]^2\right) - [E(\hat{a})]^2. \quad (3.10.24)$$

In analogy we modify the formulae (3.10.21), (3.10.23), (3.10.24) to comprise the notions (3.10.13) and (3.10.14),

$$\begin{aligned}
E(\Delta\hat{a}^\dagger\Delta\hat{a}) &= \sum_{j=1}^{L}\sum_{k=1}^{L} p_{jk}\mathrm{Tr}_S\{\hat{\rho}_{|jk}\Delta_{jk}\hat{a}^\dagger\Delta_{jk}\hat{a}\} \\
&\quad + \sum_{j=1}^{L}\sum_{k=1}^{L} p_{jk}\mathrm{Tr}_S\{\hat{\rho}_{|jk}\hat{a}^\dagger\}\mathrm{Tr}_S\{\hat{\rho}_{|jk}\hat{a}\} - \mathrm{Tr}\{\hat{\rho}\hat{a}^\dagger\}\mathrm{Tr}\{\hat{\rho}\hat{a}\}, \quad (3.10.25)
\end{aligned}$$

$$\begin{aligned}
E((\Delta\hat{a})^2) &= \sum_{j=1}^{L}\sum_{k=1}^{L} p_{jk}\mathrm{Tr}_S\{\hat{\rho}_{|jk}(\Delta_{jk}\hat{a})^2\} \\
&\quad + \sum_{j=1}^{L}\sum_{k=1}^{L} p_{jk}\left[\mathrm{Tr}_S\{\hat{\rho}_{|jk}\hat{a}\}\right]^2 - [\mathrm{Tr}\{\hat{\rho}\hat{a}\}]^2, \quad (3.10.26)
\end{aligned}$$

where

$$\begin{aligned}
\Delta_{jk}\hat{a}^\dagger &= \hat{a}^\dagger - \mathrm{Tr}_S\{\hat{\rho}_{|jk}\hat{a}^\dagger\}, \\
\Delta_{jk}\hat{a} &= \hat{a} - \mathrm{Tr}_S\{\hat{\rho}_{|jk}\hat{a}\}. \quad (3.10.27)
\end{aligned}$$

It holds that

$$E(\Delta\hat{a}^\dagger\Delta\hat{a}) = E(\hat{a}^\dagger\hat{a}) - E(\hat{a}^\dagger)E(\hat{a}), \quad (3.10.28)$$

whereas

$$E_x E_S(\Delta_x\hat{a}^\dagger\Delta_x\hat{a}|x) = E_x E_S(\hat{a}^\dagger\hat{a}|x) - E_x(E_S(\hat{a}^\dagger|x)E_S(\hat{a}|x)). \quad (3.10.29)$$

This has been obtained by averaging a relation similar to (3.10.28),

$$E_S(\Delta\hat{a}^\dagger\Delta\hat{a}|x) = E_S(\hat{a}^\dagger\hat{a}|x) - E_S(\hat{a}^\dagger|x)E_S(\hat{a}|x). \quad (3.10.30)$$

We shall proceed with important classes of pure superposition states [Bužek and Knight (1995)].

(i) We assume that $L = 2$ and into the formula (3.10.18) we substitute

$$Z_j = e^{i\varphi_j}\sqrt{p_j}, \quad |\psi_j\rangle = |n_j\rangle, \quad j = 1, 2, \quad (3.10.31)$$

where $|n_j\rangle$ are number states and $n_1 < n_2$. The states in (3.10.31) are orthogonal. The appropriate superposition states

$$|\psi\rangle = \sum_{j=1}^{2} e^{i\varphi_j}\sqrt{p_j}|n_j\rangle \quad (3.10.32)$$

are called k-photon superposition states, $k = n_2 - n_1$. The Fano factor reads

$$d = \frac{\langle (\Delta \hat{n})^2 \rangle}{\langle \hat{n} \rangle} = \frac{p_1 p_2 (n_1 - n_2)^2}{p_1 n_1 + p_2 n_2}. \tag{3.10.33}$$

Supposing $k = n_2 - n_1$ fixed, we see easily that for $p_1 k < 1$, the superposition state is sub-Poissonian for any n_1 and n_2, which occurs for one-photon state under non-restrictive condition $p_1 < 1$. For $p_1 k > 1$ the state can exhibit the super-Poissonian property. The phase-space interference approach has been generalized considering the Bohr–Sommerfeld band of the state $|\psi\rangle = \frac{1}{\sqrt{2}}(|n_1\rangle + |n_2\rangle)$) [Gagen (1995)]. Any superposition of the first $(N+1)$ number states can be constructed in a single-mode cavity as shown by Vogel, Akulin and Schleich (1993).

(ii) We assume that $L = N$ and substitute

$$Z_j = |Z_j| e^{i\varphi_j}, \quad |\psi_j\rangle = |\alpha_j\rangle, \quad j = 1, ..., N, \tag{3.10.34}$$

where α_j are the complex amplitudes of the coherent states $|\alpha_j\rangle$, into (3.10.18). If these states enter with the same weight, only the phases φ_j determine the superposition and

$$|\psi\rangle = Z \sum_{j=1}^{N} e^{i\varphi_j} |\alpha_j\rangle, \tag{3.10.35}$$

where $Z = |Z_j|$, independent of j. The quantum interference among the coherent states is constructive or destructive according to the concrete phase factors [Schleich and Wheeler (1987a, b, c), Schleich (1988), Schleich, Walther and Wheeler (1988), Bužek, Vidiella-Barranco and Knight (1992), Bužek, Knight and Vidiella-Barranco (1992)]. The normalization constant Z is expressed as

$$Z = \left(N + \sum_{\substack{j,k=1 \\ j \neq k}}^{N} W_{jk} \right)^{-\frac{1}{2}}, \tag{3.10.36}$$

where

$$W_{jk} = \exp \left[i \left(\varphi_j - \varphi_k \right) + \frac{1}{2} \left(\alpha_j \alpha_k^* - \alpha_j^* \alpha_k \right) \right] \exp \left(-\frac{1}{2} |\alpha_k - \alpha_j|^2 \right). \tag{3.10.37}$$

The statistical operator $\hat{\rho}$ corresponding to the pure state (3.10.35) is using (3.10.14)

$$\hat{\rho} = \sum_{j=1}^{N} p_j |\alpha_j\rangle\langle\alpha_j| + \sum_{\substack{j,k=1 \\ j \neq k}}^{N} p_{jk} \frac{|\alpha_j\rangle\langle\alpha_k|}{\langle \alpha_k | \alpha_j \rangle}, \tag{3.10.38}$$

where by (3.10.13)

$$p_j = Z^2, \quad p_{jk} = Z^2 W_{jk}. \tag{3.10.39}$$

Adopting the s-ordering (section 3.4), we express the characteristic function (section 3.3) of the physical state (3.10.38) as

$$C_s(\beta) = \sum_{j=1}^{N} p_j C_s(\beta; \alpha_j) + \sum_{\substack{j,k=1 \\ j \neq k}}^{N} p_{jk} C_s(\beta; \alpha_j, \alpha_k), \qquad (3.10.40)$$

where $C_s(\beta; \alpha_j)$ is the s-characteristic function for the coherent state $|\alpha_j\rangle$,

$$C_s(\beta; \alpha_j) = \exp\left[-\frac{(1-s)}{2}|\beta|^2 + \beta \alpha_j^* - \beta^* \alpha_j\right], \qquad (3.10.41)$$

and $C_s(\beta; \alpha_j, \alpha_k)$ is the expression for the interference between the terms $e^{i\varphi_j}|\alpha_j\rangle$, $e^{i\varphi_k}|\alpha_k\rangle$,

$$C_s(\beta; \alpha_j, \alpha_k) = \exp\left[-\frac{(1-s)}{2}|\beta|^2 + \beta \alpha_k^* - \beta^* \alpha_j\right]. \qquad (3.10.42)$$

The corresponding quasidistribution $\Phi_s(\alpha)$ can be written in the form

$$\Phi_s(\alpha) = \sum_{j=1}^{N} p_j \Phi_s(\alpha; \alpha_j) + \sum_{\substack{j,k=1 \\ j \neq k}}^{N} p_{jk} \Phi_s(\alpha; \alpha_j, \alpha_k), \qquad (3.10.43)$$

where $\Phi_s(\alpha; \alpha_j)$ is the quasidistribution corresponding to the coherent state $|\alpha_j\rangle$

$$\Phi_s(\alpha; \alpha_j) = \frac{2}{\pi(1-s)} \exp\left(-\frac{2|\alpha - \alpha_j|^2}{1-s}\right) \qquad (3.10.44)$$

and $\Phi_s(\alpha; \alpha_j, \alpha_k)$ describes the interference between the terms with the subscripts j, k in (3.10.35)

$$\Phi_s(\alpha; \alpha_j, \alpha_k) = \frac{2}{\pi(1-s)} \exp\left[-\frac{2}{1-s}(\alpha - \alpha_j)(\alpha^* - \alpha_k^*)\right]. \qquad (3.10.45)$$

We recall that $s = 0$ means the Wigner function, $s = -1$ the Husimi (Glauber Q (Φ_A)) function. The notation like (3.10.41) and (3.10.44) is extended to concepts (3.10.42) and (3.10.45) here, which are not a characteristic function and not a quasidistribution, respectively. The polar decomposition of the right-hand side of (3.10.45) can be accomplished, yielding the formula

$$\Phi_s(\alpha; \alpha_j, \alpha_k) = AB, \qquad (3.10.46)$$

where

$$A = \frac{2}{\pi(1-s)} \exp\left[\frac{1}{2(1-s)}|\alpha_k - \alpha_j|^2\right] \exp\left[-\frac{2}{1-s}\left|\alpha - \frac{1}{2}(\alpha_k + \alpha_j)\right|^2\right], \qquad (3.10.47)$$

$$B = \exp\left\{\frac{1}{1-s}\left\{\left[\alpha - \frac{1}{2}(\alpha_k + \alpha_j)\right](\alpha_k^* - \alpha_j^*) - \text{c. c.}\right\}\right\}. \qquad (3.10.48)$$

The limiting behaviour $s \nearrow 1$ is interesting, especially, because the corresponding quasidistributions for coherent states $|\alpha_j\rangle$ become the Dirac delta functions in this limit. We observe that one of the α-dependent exponentials in (3.10.47) corresponds to the coherent state $\left|\frac{1}{2}(\alpha_k + \alpha_j)\right\rangle$ and we see that it tends to the Dirac delta function as $s \nearrow 1$. Nevertheless, the α-dependent exponential (3.10.48) corresponds to a rotating wave, wavelength of which shortens more rapidly than does the diameter of the Gaussian surface, which results in ever more pronounced oscillations in the product. The amplitude factor in (3.10.47), which is important for large separation $|\alpha_k - \alpha_j|$, also contributes to the intensity of the oscillations in the Glauber P (Φ_N)-function limit. The "wave vector" of the exponential appropriate to the oscillations can be represented by the complex number $\frac{2i}{1-s}(\alpha_j - \alpha_k)$. In a graphical representation, the oscillations resemble a snapshot of wave propagating perpendicular to the line connecting α_j, α_k.

The assumption of the superposition of two coherent states ($N = 2$) is important, because it is evident that such superpositions are easy to detect. We shall deal with distributions of chosen rotated quadratures, because the interference term was recognized first in this distribution, not in the quasidistributions. Nevertheless, to this aim the Wigner function ($s = 0$) is appropriate, because from it the desired quasidistribution can be computed by integration. We choose the quadratures rotated by the angle

$$\tau \equiv \overline{\tau} = \arg[i(\alpha_1 - \alpha_2)], \qquad (3.10.49)$$

namely $Q(\tau)$, $P(\tau)$, defined as follows

$$\begin{aligned}
Q(\tau) &= \alpha e^{-i\tau} + \alpha^* e^{i\tau}, \\
P(\tau) &= -i\left(\alpha e^{-i\tau} - \alpha^* e^{i\tau}\right),
\end{aligned} \qquad (3.10.50)$$

which are the expectation values of the rotated quadrature operators $\hat{Q}(\tau)$, $\hat{P}(\tau)$ in the coherent state $|\alpha\rangle$,

$$\begin{aligned}
\hat{Q}(\tau) &= \hat{a} e^{-i\tau} + \hat{a}^\dagger e^{i\tau}, \\
\hat{P}(\tau) &= -i\left(\hat{a} e^{-i\tau} - \hat{a}^\dagger e^{i\tau}\right),
\end{aligned} \qquad (3.10.51)$$

respectively. From the geometric picture of a transformation, we expect the results corresponding to the case $Q(0)$, $P(0)$, when initially $i(\alpha_1 - \alpha_2) > 0$. The distributions of quadratures are defined as

$$\Phi_S^Q(Q, \tau) = \frac{1}{2} \int_{-\infty}^{\infty} \Phi_S\left(e^{i\tau}\left(\frac{Q}{2} + iy\right)\right) dy, \qquad (3.10.52)$$

$$\Phi_S^P(P, \tau) = \frac{1}{2} \int_{-\infty}^{\infty} \Phi_S\left(e^{i\tau}\left(x + i\frac{P}{2}\right)\right) dx. \qquad (3.10.53)$$

Substituting from (3.10.43) with $N = 2$ into (3.10.53), we obtain

$$\Phi_S^P(P,\tau) = \sum_{j=1}^{2} p_j \Phi_S^P\left(P,\tau;\overline{P}_j(\tau)\right) + (p_{12} + p_{21})\Phi_S^P\left(P,\tau;\overline{P}_{12}(\tau)\right), \qquad (3.10.54)$$

where the Gaussian distribution

$$\Phi_S^P(P,\tau;\overline{P}) = \frac{1}{\sqrt{2\pi}}\exp\left[-\frac{1}{2}(P-\overline{P})^2\right] \qquad (3.10.55)$$

and

$$\begin{aligned}
\overline{P}_j(\tau) &= -i\left(\alpha_j e^{-i\tau} - \alpha_j^* e^{i\tau}\right), \quad j=1,2, \\
\overline{P}_{12}(\tau) &= \frac{1}{2}\left[\overline{P}_1(\tau) + \overline{P}_2(\tau)\right].
\end{aligned} \qquad (3.10.56)$$

Although derived from the formula (3.10.53) and from the Wigner function, the quadrature distribution can be obtained directly from the wave function

$$\Phi_S^P(P,\tau) = |\psi^P(P,\tau)|^2, \qquad (3.10.57)$$

where

$$\psi^P(P,\tau) = Z\sum_{j=1}^{2} e^{i\varphi_j}\psi_j^P(P,\tau), \qquad (3.10.58)$$

with

$$\psi_j^P(P,\tau) = \frac{1}{\sqrt[4]{2\pi}}\exp\left\{-\frac{1}{4}\left[P-\overline{P}_j(\tau)\right]^2 - \frac{i}{2}P\overline{Q}_j(\tau) + \frac{i}{4}\overline{P}_j(\tau)\overline{Q}_j(\tau)\right\}. \qquad (3.10.59)$$

Let us recall the definition of the coherent state and the wave function for the vacuum state ((3.10.59) for $\overline{P}_j(\tau) = \overline{Q}_j(\tau) = 0$), which can serve for the derivation of the wave function (3.10.59)

$$\langle P,\tau|\alpha_j\rangle = \langle P,\tau|\hat{D}(\alpha_j)|0\rangle$$

$$= \exp\left(\frac{i}{4}\overline{P}_j(\tau)\overline{Q}_j(\tau)\right)\left\langle P,\tau\left|\hat{D}\left(\frac{1}{2}e^{i\tau}\overline{Q}_j(\tau)\right)\hat{D}\left(\frac{i}{2}e^{i\tau}\overline{P}_j(\tau)\right)\right|0\right\rangle. \qquad (3.10.60)$$

Here the Baker–Hausdorff identity (3.2.4) has been applied. On substituting into (3.10.57), we take into account that $\overline{Q}_1(\tau) = \overline{Q}_2(\tau)$ for τ defined by the relation (3.10.49) and we rederive the formula (3.10.54) with its three Gaussian components.

Substituting from the expression for the Wigner function (3.10.43) with $N = 2$ into (3.10.52), we obtain

$$\begin{aligned}
\Phi_S^Q(Q,\tau) &= \sum_{j=1}^{2} p_j \Phi_S^Q\left(Q,\tau;\overline{Q}_j(\tau)\right) \\
&\quad + p_{12}\Phi_S^Q\left(Q,\tau;\overline{Q}_{12}^c(\tau)\right) + p_{21}\Phi_S^Q\left(Q,\tau;\overline{Q}_{21}^c(\tau)\right), \qquad (3.10.61)
\end{aligned}$$

where the Gaussian distribution

$$\Phi_S^Q(Q,\tau;\overline{Q}) = \frac{1}{\sqrt{2\pi}} \exp\left[-\frac{1}{2}(Q - \overline{Q})^2\right] \qquad (3.10.62)$$

and

$$
\begin{aligned}
\overline{Q}_j(\tau) &= \alpha_j e^{-i\tau} + \alpha_j^* e^{i\tau}, \quad j = 1, 2, \\
\overline{Q}_{12}^c(\tau) &= \alpha_1 e^{-i\tau} + \alpha_2^* e^{i\tau}, \\
\overline{Q}_{21}^c(\tau) &= \left[\overline{Q}_{12}^c(\tau)\right]^*.
\end{aligned}
\qquad (3.10.63)
$$

Let us present the wave function

$$
\begin{aligned}
\psi^Q(Q,\tau) &= \frac{1}{2\sqrt{\pi}} \int_{-\infty}^{\infty} \exp\left(\frac{i}{2}QP\right)\psi^P(P,\tau)dP \\
&= Z\sum_{j=1}^{2} e^{i\varphi_j}\psi_j^Q(Q,\tau),
\end{aligned}
\qquad (3.10.64)
$$

where

$$\psi_j^Q(Q,\tau) = \frac{1}{\sqrt[4]{2\pi}} \exp\left\{-\frac{1}{4}\left[Q - \overline{Q}_j(\tau)\right]^2 + \frac{i}{2}Q\overline{P}_j(\tau) - \frac{i}{4}\overline{P}_j(\tau)\overline{Q}_j(\tau)\right\}, (3.10.65)$$

$$- \frac{1}{\sqrt[4]{2\pi}} \exp\left\{-\frac{1}{4}\left(Q - 2\alpha_j e^{-i\tau}\right)^2 + \frac{1}{2}\alpha_j^2 e^{-i2\tau} - \frac{1}{2}|\alpha_j|^2\right\}. \qquad (3.10.66)$$

Let us remember that the chosen phase shift τ leads to $\overline{Q}_1(\tau) = \overline{Q}_2(\tau)$ and $\overline{P}_1(\tau) < \overline{P}_2(\tau)$. Hence, the wave functions differ from each other. Whereas the momentum-like quadrature has the wave function, which oscillates, with a unique "wave vector" $(-\overline{Q}_1(\tau)) = (-\overline{Q}_2(\tau))$, the position-like quadrature has the wave function, which oscillates with two different wave vectors $\overline{P}_j(\tau)$, $j = 1, 2$. From this, in the momentum-like case the interference does not occur and in the position-like case the interference does occur. Regarding the wave functions as the pair of Fourier transformations, we can say that the interference in the position-like quadrature occurs because the wave function of P quadrature is a superposition of two representations with two different localizations. The interference in the momentum-like quadrature wave function does not occur, because the wave function of position-like quadrature is not a superposition of two representations with two different localizations. The interference manifests itself by the oscillations of the position-like quadrature distribution, whereas two peaks in the momentum-like quadrature distribution, which reveal the localization, are accompanied by the third hump, which is the so-called interference term.

With regard to the study of quadrature squeezing, we consider the mean number of thermal photons

$$B_N = \langle\Delta\hat{a}^\dagger\Delta\hat{a}\rangle = |\alpha_1 - \alpha_2|^2(p_1 p_2 - p_{12}p_{21}) \qquad (3.10.67)$$

and the moment

$$C = \langle (\Delta \hat{a})^2 \rangle = (\alpha_1^2 + \alpha_2^2)(p_1 + p_{12})(p_2 + p_{21}) - 2\alpha_1 \alpha_2 (p_1 + p_{21})(p_2 + p_{12}). \quad (3.10.68)$$

Respecting the property $p_1 = p_2$ of the quantities introduced in (3.10.39), we can write

$$C = (\alpha_1 - \alpha_2)^2 |p_1 + p_{12}|^2. \quad (3.10.69)$$

The standard quadrature variances can be obtained from the formulae

$$\langle (\Delta \hat{Q}(0))^2 \rangle = 1 + 2(B_N + \text{Re } C), \quad (3.10.70)$$

$$\langle (\Delta \hat{P}(0))^2 \rangle = 1 + 2(B_N - \text{Re } C). \quad (3.10.71)$$

The extreme values of rotated quadrature variances, which can be called principal quadrature variances, are attained for

$$\tau \equiv \frac{1}{2}\arg(-C) \pmod{\pi} \quad (3.10.72)$$

or for $\tau \equiv \arg[i(\alpha_1 - \alpha_2)] \pmod{\pi}$. In other words, the condition (3.10.49) yields one of the two offered possibilities for the extreme variances of $\hat{Q}(\tau)$ and $\hat{P}(\tau)$. The condition (3.10.72) comprises the minimum for the position-like quadrature variance and the maximum for the momentum-like quadrature variance, which may be abbreviated as [Lukš, Peřinová and Hradil (1988), Lukš, Peřinová and Peřina (1988)]

$$\langle (\Delta \hat{Q}^{(p)})^2 \rangle = 1 + 2(B_N - |C|), \quad (3.10.73)$$

$$\langle (\Delta \hat{P}^{(p)})^2 \rangle = 1 + 2(B_N + |C|). \quad (3.10.74)$$

Substituting from (3.10.67) and (3.10.69), we get

$$\langle (\Delta \hat{Q}^{(p)})^2 \rangle = 1 - 2|\alpha_1 - \alpha_2|^2 (p_1 p_{21} + 2p_{12}p_{21} + p_2 p_{12}) = \langle (\Delta \hat{Q}(\bar{\tau}))^2 \rangle, \quad (3.10.75)$$

$$\langle (\Delta \hat{P}^{(p)})^2 \rangle = 1 + 2|\alpha_1 - \alpha_2|^2 (p_1 p_{21} + 2p_1 p_2 + p_2 p_{12}) = \langle (\Delta \hat{P}(\bar{\tau}))^2 \rangle. \quad (3.10.76)$$

The comparison of various properties exhibited by the superposition states in the case $N = 2$ is possible after another simplification of the pair superposition states. Here we assume that $\alpha_1 > 0$ and $\alpha_2 = -\alpha_1$. As a consequence $\bar{\tau} \equiv \frac{\pi}{2} \pmod{\pi}$ and the minor principal quadrature variance is attained for the quadrature $\hat{P}(0)$, while the extreme variances are attained for $\hat{P}(0), \hat{Q}(0)$. From the continuum of the paired coherent states, the even coherent states (3.2.32) are distinguished for $\varphi_1 = \varphi_2$ and the odd coherent states (3.2.33) are obtained for $\varphi_2 \equiv \varphi_1 + \pi \pmod{2\pi}$ and the Yurke–Stoler superposition states occur for $\varphi_2 \equiv \varphi_1 \pm \frac{\pi}{2} \pmod{2\pi}$. Here

$$\begin{aligned} W_{12} &= \exp[i(\varphi_1 - \varphi_2)] \exp\left(-2\alpha_1^2\right), \\ W_{21} &= W_{12}^*, \\ Z &= \left[2 + 2\exp\left(-2\alpha_1^2\right)\cos(\varphi_1 - \varphi_2)\right]^{-\frac{1}{2}}. \end{aligned} \quad (3.10.77)$$

Particularly for the Yurke–Stoler superposition states

$$p_1 = p_2 = \frac{1}{2}, \quad p_{12} = \frac{i}{2}\exp(-2\alpha_1^2), \quad p_{21} = p_{12}^*, \qquad (3.10.78)$$

whereas for even and odd coherent states

$$p_1 = p_2 = Z_{\pm}^2, \quad p_{12} = \pm Z_{\pm}^2 \exp(-2\alpha_1^2), \qquad (3.10.79)$$

where

$$Z_{\pm}^2 = \frac{1}{2}\left[1 \pm \exp(-2\alpha_1^2)\right]^{-1}. \qquad (3.10.80)$$

The study of quadrature squeezing is exemplified by the following consequences of the formulae (3.10.75) and (3.10.76)
(i) in the Yurke–Stoler superposition states

$$\langle(\Delta\hat{Q}^{(\mathrm{p})})^2\rangle = 1 - 4\alpha_1^2\exp(-4\alpha_1^2), \qquad (3.10.81)$$

(ii) in the even and odd coherent states

$$\langle(\Delta\hat{Q}^{(\mathrm{p})})^2\rangle = 1 \mp 4\alpha_1^2\frac{\exp(-2\alpha_1^2)}{1 \pm \exp(-2\alpha_1^2)}. \qquad (3.10.82)$$

From (3.10.82) it follows that the odd coherent states cannot exhibit squeezing, whereas the even coherent states are squeezed for any complex field amplitude α_1. From the very simple expression (3.10.81) it follows that the Yurke–Stoler superposition states are always squeezed. Considering the even coherent state an elementary prototype of the squeezed states, Domokos and Janszky (1994) have studied the appropriate rotated quadrature distributions and "classically produced superpositions" of these nonclassical superposition states on the SU(2) and SU(1,1) beamsplitters.

Turning to sub-Poissonian behaviour, we note, that we do not use any special assumption to represent the most general superposition ($L = 2$) as in the previous paragraph, but only to arrive at a simplification. Provided that $\mathrm{Re}[(\alpha_1+\alpha_2)(\alpha_1^*-\alpha_2^*)] = 0$, we may arrive at examples of sub-Poissonian superposition states [Schaufler, Freyberger and Schleich (1994)]. On the other hand, when $\arg\alpha_1 = \arg\alpha_2$, we may obtain examples of phase squeezed (optimized) superposition states, see below [Schaufler, Freyberger and Schleich (1994)] in section 4.11.

Similarly as with the number states we have seen superpositions of orthogonal states, we may consider superpositions of odd and even coherent states. Particularly, the ordinary coherent state is a superposition of this kind. The Wigner function for the photon number and the continuous phase describes the interference term arising from two coherent states (see section 4.11) in an unusual way and the coherent state arising from the even and odd coherent states is described with the half-odd numbers corresponding to a new interference term.

The superpositions of states from discrete or "continuous" bases need not be the superposition states. This needs no explanation in the number states, which are not named the Schrödinger cats even for $N = 2$. Nevertheless, also some discrete coherent-state superpositions are rather kitten states which approximate suitable quantum states with arbitrary precision [Janszky, Domokos and Adam (1992), Janszky, Domokos, Szabo and Adam (1995), Domokos, Janszky and Adam (1994), Szabo, Adam, Janszky and Domokos (1996)]. The wrapped Gaussian superpositions of coherent states lying on a circle have been considered by Adam, Janszky and Vinogradov (1991). The set of states has been specified which can be expanded into the superposition of coherent states lying on a circle in phase space [Domokos, Adam and Janszky (1994)]. Bužek, Keitel and Knight (1995b) have concluded that by performing an operational phase-space measurement over a nonclassical pure continuous superposition of coherent states, one can always distinguish between this nonclassical state and the corresponding mixture state.

Yurke and Stoler (1986) suggested the use of a nonlinear Kerr-like medium with low dissipation for a transformation of an initial coherent state into a quantum superposition of macroscopically distinguishable states. Yurke and Stoler modelled a $(2l - 1)$th-order Kerr-like medium as an anharmonic oscillator described by the Hamiltonian

$$\hat{H} = \hat{H}_0 + \hat{H}_I, \tag{3.10.83}$$

where

$$
\begin{aligned}
\hat{H}_0 &= \hbar\omega\left(\hat{a}^\dagger\hat{a} + \frac{1}{2}\hat{1}\right), \\
\hat{H}_I &= \hbar\kappa(\hat{a}^\dagger\hat{a})^l,
\end{aligned}
\tag{3.10.84}
$$

with $\hbar\omega$ the energy-level spacing for the harmonic part of the Hamiltonian, l being an integer and κ being proportional to the nonlinear susceptibility of the medium.

The Schrödinger equation for the $(2l - 1)$th-order nonlinear oscillator reads

$$i\hbar\frac{\partial}{\partial t}|\psi(t)\rangle = \hat{H}_I|\psi(t)\rangle. \tag{3.10.85}$$

Describing the evolution of the system in the interaction picture, we obtain for any initial state $|\psi(0)\rangle$,

$$|\psi(t)\rangle = \hat{U}(t)|\psi(0)\rangle, \tag{3.10.86}$$

where

$$\hat{U}(t) = \exp(-i\kappa t\hat{n}^l). \tag{3.10.87}$$

It can be proved that the evolution has the period $\frac{2\pi}{\kappa}$ and that for t commensurable with this period, $t = \frac{L}{N}\frac{2\pi}{\kappa}$, L and $\overline{N}$ being mutually prime integers, the operator

$$\hat{U}\left(t = \frac{L}{N}\frac{2\pi}{\kappa}\right) = \sum_{n=0}^{\infty} u_n\left(t = \frac{L}{N}\frac{2\pi}{\kappa}\right)|n\rangle\langle n|, \tag{3.10.88}$$

where the diagonal matrix elements

$$u_n\left(t = \frac{1}{\overline{N}}\frac{2\pi}{\kappa}\right) = \langle n|\hat{U}\left(t = \frac{1}{\overline{N}}\frac{2\pi}{\kappa}\right)|n\rangle \tag{3.10.89}$$

form an $\overline{N}$-periodic sequence [Peřinová and Křepelka (1993)]. From this it follows that

$$\hat{U}^{l} = \sum_{k=0}^{\overline{N}-1} c_k^* \hat{U}_{\text{har}}^{h}, \tag{3.10.90}$$

where

$$\hat{U}_{\text{har}} = \exp\left(-i\frac{2\pi}{\overline{N}}\hat{n}\right),$$

$$c_k^* = \frac{1}{\overline{N}}\sum_{k'=0}^{\overline{N}-1} u_{k'}\left(t = \frac{1}{\overline{N}}\frac{2\pi}{\kappa}\right)\exp\left(i\frac{2\pi kk'}{\overline{N}}\right). \tag{3.10.91}$$

As a consequence of the decomposition (3.10.90), we obtain that (3.10.86) is a superposition of phase-shifted (rotated) initial states. Hence, the anharmonic oscillator arrives at these states in the process of amplitude dispersion.

Many authors recently analyzed the influence of reservoirs on quantum-mechanical superposition states of light. In particular, the influence of damping at zero temperature on quantum coherence was analyzed by Walls and Milburn (1985), Milburn and Holmes (1986), Vourdas and Wiener (1987), Kennedy and Drummond (1988), Milburn and Walls (1988), Vourdas and Bishop (1989), Agarwal and Adam (1989), Phoenix (1990), Vourdas (1992), Braunstein (1992a), Albrecht (1992), and Bužek, Vidiella-Barranco and Knight (1992). These authors showed that the off-diagonal terms in the field statistical operator expressed in the coherent-state basis are weighted with a time-dependent factor proportional to the distance $|\alpha|^2$ between the component states. The decay rate of the quantum coherence is proportional to $\gamma|\alpha|^2$, where γ is the damping constant of the cavity. Rapid decay of quantum coherence results in rapid fading of oscillations in the photon-number distribution.

To illustrate the decay of superposition states into the quiet reservoir, we assume the field mode initially prepared in a superposition of coherent states. The appropriate master equation is equivalent to that for an attenuator. This analogy emerges from the solution of the master equation in the normal ordering form (section 3.4). The description in terms of the normal characteristic function is even simpler

$$C_{\mathcal{N}}(\beta, t) = C_{\mathcal{N}}(e^{-\gamma t}\beta, 0), \tag{3.10.92}$$

where $C_{\mathcal{N}}(\beta, 0) \equiv C_{\mathcal{N}}(\beta)$ is given in (3.10.40). It is obvious that the evolved interference terms enter the expansion

$$C_{\mathcal{N}}(\beta, t) = \sum_{j=1}^{N} p_j(t) C_{\mathcal{N}}(\beta; \alpha_j(t)) + \sum_{\substack{j,k=1 \\ j\neq k}}^{N} p_{jk}(t) C_{\mathcal{N}}(\beta; \alpha_j(t), \alpha_k(t)), \tag{3.10.93}$$

where

$$p_j(t) = p_j, \quad p_{jk}(t) = p_{jk}, \quad j, k = 1, ..., N, \tag{3.10.94}$$

$$\alpha_j(t) = e^{-\frac{\gamma}{2}t}\alpha_j, \quad j = 1, ..., N. \tag{3.10.95}$$

So we have found the time-dependent expression for the statistical operator for the superposition of N coherent states [Walls and Milburn (1985), Phoenix (1990)]

$$\hat{\rho}(t) = \sum_{j=1}^{N} p_j(0)|\alpha_j(t)\rangle\langle\alpha_j(t)| + \sum_{\substack{j,k=1 \\ j \neq k}}^{N} \left[p_{jk}(0)\frac{|\alpha_j(t)\rangle\langle\alpha_k(t)|}{\langle\alpha_j(t)|\alpha_k(t)\rangle} + \text{H. c.} \right]. \tag{3.10.96}$$

To have the basic situation, we restrict ourselves to $N = 2$ and we observe that the weight of the evolved interference term is expressed by the ratio

$$S_{12} = \frac{\langle\alpha_2|\alpha_1\rangle}{\langle\alpha_2(t)|\alpha_1(t)\rangle} \tag{3.10.97}$$

or by its real part. In fact, S_{12} measures the decay of the interference term and also the decay (decoherence) of the superposition state. Because the decay of the superposition state is associated with the origin of statistical mixtures in most of the literature, we use the zero temperature reservoir to dispel the idea that the decay is linked solely to the loss of the orthogonality of the diagonal components.

Using the linearized form for the von Neumann entropy,

$$\text{Tr}\{\hat{\rho}(t)[\hat{1} - \hat{\rho}(t)]\} = 2\left\{ p_1(0)p_2(0)[1 - |\langle\alpha_2(t)|\alpha_1(t)\rangle|^2] \right.$$
$$\left. -p_{12}(0)p_{21}(0)\left[\frac{1}{|\langle\alpha_2(t)|\alpha_1(t)\rangle|^2} - 1 \right] \right\}, \tag{3.10.98}$$

we may observe the possibility of a mixture originating, but we also remember that in the long-time limit the mixture decays and a pure state (the vacuum) arises.

Taking into account the higher-order quadrature squeezing [Hong and Mandel (1985a, b)] (section 3.5), we may observe that the thermal behaviour does not preclude the origin of nonclassical behaviour. Even if a field mode does not exhibit fourth-order squeezing initially, it can be fourth-order squeezed sometimes during the decay into the thermal reservoir [Kim, Bužek and Kim (1994)].

Correlated (phase-sensitive) multimode reservoirs were studied extensively [Caves (1981, 1982), Milburn (1984), Caves and Schumaker (1985), Dupertuis and Stenholm (1987), Dupertuis, Barnett and Stenholm (1987a, b), Ekert and Knight (1990), Bužek, Knight and Kudryavtsev (1991)]. These are also called "rigged" reservoirs and are described in terms of broad-band squeezed light. The correlated reservoirs are characterized by the mean photon number $B_{d\mathcal{N}}$ of a quantum oscillator of the reservoir at a frequency and by the correlation C_d between oscillators, whose frequencies are symmetrically offset from some career frequency. An ideally squeezed reservoir [Gardiner

(1986)] is characterized by the equality $|C_d|^2 = B_{dN}B_{dA}$, whereas for a nonideally correlated reservoir we have $|C_d|^2 < B_{dN}B_{dA}$. For an uncorrelated reservoir we have $C_d = 0$. Each oscillator separately is in a thermal state with a number of thermal photons equal to $\langle \hat{n}_j(0) \rangle = B_{dN}$ provided that the expectation values of the reservoir oscillators are zero. As a measure of the quantum coherence the height of the central peak of the function

$$2\mathrm{Re}\,[p_{12}(t)\Phi_S(\alpha,t;\alpha_1,\alpha_2)]|_{\alpha=\frac{1}{2}(\alpha_1+\alpha_2)} \tag{3.10.99}$$

can be used. The decay rate of the quantum coherence is measured by the first derivative of the function (3.10.99) with respect to the time considered at $t = 0$. This leads to the quantity (for $\alpha_1 - \alpha_2 > 0$)

$$\gamma_{\mathrm{sq}} = \gamma \left[B_{dN} + \frac{1}{2}(\alpha_1 - \alpha_2)^2 \left(B_{dN} - C_d + \frac{1}{2} \right) \right]. \tag{3.10.100}$$

In the case of the zero temperature reservoir,

$$\gamma_{\mathrm{vac}} = \frac{\gamma}{4}(\alpha_1 - \alpha_2)^2. \tag{3.10.101}$$

In the more general case of the thermal reservoir

$$\gamma_{\mathrm{th}} = \gamma \left[B_{dN} + \frac{1}{2}(\alpha_1 - \alpha_2)^2 B_{dS} \right], \tag{3.10.102}$$

the quantum coherence is more sensitive to the influence of this reservoir and the decay rate of the quantum coherence increases for $(\alpha_1 - \alpha_2)^2$ fixed with the mean number of the reservoir quanta B_{dN}. If on the contrary B_{dN} is fixed and the correlation C_d increases, γ_{sq} falls to its minimum value

$$\gamma \left[B_{dN} + \frac{1}{2}(\alpha_1 - \alpha_2)^2 \left(B_{dS} - \sqrt{B_{dN}B_{dA}} \right) \right]$$

$$\underset{B_{dN} \to \infty}{\simeq} \gamma \left[B_{dN} - \frac{1}{16} \frac{(\alpha_1 - \alpha_2)^2}{B_{dS}} \right], \tag{3.10.103}$$

in other words the decay rate depends on $(\alpha_1 - \alpha_2)^2$ weakly for large B_{dN}. For details see [Bužek and Knight (1995) and references therein]. It has been predicted that transient macroscopic quantum superposition states will arise in degenerate parametric oscillator using a squeezed reservoir [Munro and Reid (1995)].

It is possible to consider multimode superposition of coherent states. Let us assume that the system S comprises J modes. As a generalization of the formula (3.10.35) we present

$$|\psi\rangle = Z \sum_{j=1}^{N} e^{i\varphi_j} |\{\alpha_{jl}\}\rangle, \tag{3.10.104}$$

where $\{\alpha_{jl}\} = (\alpha_{j1}, \ldots, \alpha_{jJ})$. The normalization constant Z is expressed again by the formula (3.10.36), where however

$$W_{jk} = \exp[i(\varphi_j - \varphi_k)] \prod_{l=1}^{J} W_{jkl}, \tag{3.10.105}$$

with

$$W_{jkl} = \exp\left[\frac{1}{2}(\alpha_{jl}\alpha_{kl}^* - \alpha_{jl}^*\alpha_{kl})\right] \exp\left(-\frac{1}{2}|\alpha_{kl} - \alpha_{jl}|^2\right). \tag{3.10.106}$$

The modification of the the the relation (3.10.38) reads as

$$\hat{\rho} = \sum_{j=1}^{N} p_j |\{\alpha_{jl}\}\rangle\langle\{\alpha_{jl}\}| + \sum_{\substack{j,k=1 \\ j \neq k}}^{N} p_{jk} \frac{|\{\alpha_{jl}\}\rangle\langle\{\alpha_{kl}\}|}{\langle\{\alpha_{kl}\}|\{\alpha_{jl}\}\rangle}, \tag{3.10.107}$$

whereas the relation (3.10.39) remains unchanged.

Multimode superposition states constructed from the multimode coherent states have been studied from the viewpoint of nonclassical properties [Chai (1992), Ansari and Man'ko (1994), Dodonov, Man'ko and Nikov (1995)]. Gerry and Grobe (1995a) have studied superpositions of the pair-coherent states (see subsection 4.11.3). Such superposition states can be generated from pair-coherent states by a method based on atomic interference [Zheng and Guo (1997a)]. The possibility of generation of four-mode superpositions of coherent states via a fully quantized nondegenerate four-wave mixing process has been investigated by Mogilevtsev and Kilin (1996). A simple scheme for realizing single- and two-mode Raman coupled models has been proposed in the context of trapped ion [Gerry (1997)]. In the one-mode case this scheme generates superpositions of coherent states. In the two-mode case, an entanglement of coherent states between the modes is obtained. A summary of work on the preparation of a quantum superposition of classically distinguishable states can be found in special issue of Journal of Modern Optics [Schleich and Raymer (1997)].

Chapter 4

Phase in quantum optics

The action–angle variables were important in classical mechanics and in the "old" quantum theory using the Bohr–Sommerfeld quantization condition. The quantization of phase-space trajectories was successful in the case of harmonic oscillator and that of plane rotator.

The superposition principle lies at the heart of present quantum theory [Schleich, Pernigo and Fam Le Kien (1991)] where the role of the phase factors is as crucial as in classical optics. There is great work concerning geometric phase or frequent ramifications of this concept (some of them being important in classical optics).

After the fathers of quantum theory realized the necessity of a reinterpretation of concepts of classical mechanics, it emerged that the action–angle variables could not be interpreted along the lines set. Soon thereafter in quantum field theory P. A. M. Dirac spoke about the phase and warned that the phase of a field in a given spatial point would not be able to be determined, but that a phase of a "modal component" (as we say today) would be possible. In the sixties this idea was revised and various proofs of non-existence or impossibility of a phase operator of an optical mode had been provided. Since nobody ever liked difficulties, people have used the quantum phase operator despite these proofs.

Of course, this led to various solutions of the quantum phase problem using the method of enlargement of the original Hilbert space of the optical mode, or—later— using the method of restriction of the original Hilbert space.

The well-known proof of the non-existence of the phase operator seems to be too strong from this perspective—there is no operator which would be, in the original sense, canonically conjugate to the number operator. But it is interesting to see how the appropriate canonical conjugation looks out and try also an appropriate non-existence proof.

The statistical treatment of angle or phase can be based on any solution of the quantum phase problem. One can substitute here "the classical phase" for "a quantum phase" and choose either a "right" phase distribution which is supported by an interval or a "convenient" phase distribution which is supported by a unit circle.

There are many measures of phase variability.

With the "convenient" phase distribution the idea of two optical phase operators (the cosine and sine operators) is connected, which unfortunately are not compatible in the basic Susskind–Glogower formalism. Orderings of exponential phase operators lead, due to this intrinsically quantum property, to "phase" distributions inside the unit circle. Only the antinormal ordering yields the expected "convenient" optical phase distribution.

The solution of the quantum phase problem using the method of enlargement of the original Hilbert space of the optical mode pays not only in the definition of the optical phase distribution, but also in definitions of various optical phase representations. In the same vein we derive a "new" Wigner function for a "positive" number and phase from a "new" Wigner function for signed number and phase. The solution of the quantum phase problem using the method of restriction of the original Hilbert space is appealing, even though one must use a limiting procedure, to arrive at a definition of the optical phase distribution. Also here, one can try definitions of various optical phase representations and derivation of a "new" Wigner function for a (positive) number and phase. These must comprise limiting procedure, but they are equivalent to the definitions related to the solution of the quantum phase problem using the method of enlargement and are more acceptable physically, because "one need not wait for a plane rotator becoming part of the model description", etc. The phase sum and phase difference as well as the $SU(2)$ and $SU(1,1)$ group formalisms have peculiar properties. All this must be taken into account when the contributions to the special states of the optical oscillator are evaluated.

We must not forget the quantum phase from quasidistributions of the complex amplitude or of two canonically conjugate quadratures (phase-space quasidistributions). We do not encounter here a solution of the quantum phase problem in the sense of definition of well-behaved optical phase operators. The formalisms are still based on the photon annihilation and creation operators of an optical mode. In spite of this observation the quantum phase from quasidistributions has made some "less well-behaved" optical phase operators very popular. The phase properties of real states of a single-mode optical field have been determined in this formalism. Nevertheless, the canonical phase properties of real states of optical field have been studied not only in the case of single mode, but also in that of two modes.

The eight-port scheme for simultaneous detection of canonically conjugate quadratures has contributed to understanding the quantum phase. The sampling of joint distribution of conjugate quadratures has grown to various detection techniques for the statistical operator of an optical mode. The latter problem differs from the former: Repeated measurements of the operators "tracking" the canonically conjugate quadratures yield data for a reconstruction of the statistical operator, but the reconstruction of the statistical operator does not require any "simultaneous measurements" in general.

4.1   Phase and the "old" quantum theory

Planck's and Einstein's investigations ingeneously using the hypothesis of the discrete energy levels (the so-called quantum hypothesis) found their counterpart in Bohr's model of atom, whose Kepler orbits were quantized according to the Bohr–Sommerfeld quantization condition still popular with the mathematicians of present [Landau and Lifshitz (1977), p. 170, Tabor (1989), p. 233]. As this old quantum theory has rested on the phase space concept of classical mechanics [Saletan and Cromer (1971)], we will touch on the parts thereof which, in our opinion, relate to the quantum phase problem.

Whereas the Lagrangian formalism is not mentioned here, we do remember that in the Hamilton theory we approach mechanics in the following way. A mechanical system of one degree of freedom is completely specified at any time t by giving the generalized coordinate q and the generalized momentum p. Thus the state of the system is specified by a point in phase space, whose elements are just the pairs (q, p). Classical mechanics asks how this point moves in phase space. Hamilton's canonical equations describe this motion and the two variables q and p are called canonical variables,

$$\dot{q} = \frac{\partial H}{\partial p},$$
$$\dot{p} = -\frac{\partial H}{\partial q}, \tag{4.1.1}$$

where $H \equiv H(q, p, t)$ is a given Hamiltonian function. An important expression is the Poisson bracket $\{S, R\}$ of the dynamical variable $S(q, p)$ with the dynamical variable $R(q, p)$,

$$\{S, R\} = \frac{\partial S}{\partial q}\frac{\partial R}{\partial p} - \frac{\partial S}{\partial p}\frac{\partial R}{\partial q}. \tag{4.1.2}$$

With respect to the correspondence with quantum cases it is allowed to introduce the classical commutator

$$[S, R] = i\hbar\{S, R\}, \tag{4.1.3}$$

where $\hbar$ (the Planck constant divided by 2π) does not properly belong to classical mechanics. As a particular case we mention the fundamental bracket $\{q, p\} = 1$. In this notation, the canonical equations (4.1.1) can be rewritten in the form

$$\dot{q} = \{q, H\},$$
$$\dot{p} = \{p, H\}, \tag{4.1.4}$$

and, more generally,

$$\dot{F} = \{F, H\}, \tag{4.1.5}$$

provided that the dynamical variable $F \equiv F(q, p)$ has no explicit dependence on t.

We will illustrate the particularly useful concept of the so-called action–angle variables by four examples. The action variable J is defined on each energy-level contour by the integral

$$J = \oint p\,dq, \qquad (4.1.6)$$

which is taken over a complete cycle of variation of q. To simplify the matter, we define the angle variable w so that w evolves extremely simply,

$$w = \nu t, \qquad (4.1.7)$$

where ν is the frequency of motion on the energy-level contour in the phase space, and so that w and J are connected with q, p by a canonical transformation, namely $\{w, J\} = 1$.

The first two examples concern vibration. In the case of a rectangular potential well of infinite depth we consider a bounded motion of a particle in the region between the walls $-a$, a. This motion is not treated regularly in classical mechanics, because the momentum is not defined at the turning points. We will see that this difficulty is not so serious. The generalized coordinate q and the momentum p are replaced here by a particular coordinate x and the momentum p_x. The appropriate Hamiltonian function reads

$$H = \frac{1}{2m}p_x^2, \qquad (4.1.8)$$

where m is the mass of the particle and the potential term is absent, because the motion is bounded and the infinite depth cannot be described easily. Applying formula (4.1.6), we obtain the action

$$\begin{aligned} J &= \int_{x=a,p_x<0}^{-a} p_x\,dx + \int_{x=-a,p_x>0}^{a} p_x\,dx \\ &= 4a|p_x| = 4a\sqrt{2mH}. \end{aligned} \qquad (4.1.9)$$

On assuming that the motion starts at

$$x(t)|_{t=0} = a \quad \text{with} \quad p_x(t)|_{t\to+0} < 0, \qquad (4.1.10)$$

we can describe it by

$$x \equiv x(t) = \frac{2a}{\pi}\mathrm{Sin}^{-1}[\cos(2\pi\nu t)], \qquad (4.1.11)$$

where

$$\nu = \frac{|p_x|}{4am}. \qquad (4.1.12)$$

Here Sin^{-1} stands for the arcsine function taking on the values in the interval $\left[-\frac{\pi}{2}, \frac{\pi}{2}\right]$. Substituting according to (4.1.7), we obtain that

$$x = \frac{2a}{\pi}\mathrm{Sin}^{-1}[\cos(2\pi w)]. \qquad (4.1.13)$$

As the function on the right-hand side is 1-periodic, we see that the angle variable can be related to x only by a multiple-valued transformation. Restricting ourselves to $w \in [0, 1)$, we may write

$$w = \theta(p_x) \frac{x + 3a}{4a} + \theta(-p_x) \frac{a - x}{4a}, \qquad (4.1.14)$$

resulting from the joint solution of (4.1.13) and

$$p_x = -\text{sgn}[\sin(2\pi w)] \frac{J}{4a}; \qquad (4.1.15)$$

here $\theta(p)$ is the unit-step function. Let us note that the initial condition (4.1.10) cannot be modified to comprise $p_x(t)|_{t \to +0} = 0$. The angle variable w is zero for $p_x(t)|_{t \to +0} = 0$.

For a suitable choice of the potential we encounter a harmonic oscillator. The generalized coordinate q and the momentum p become here again x and p_x. The respective Hamiltonian function reads

$$H = \frac{1}{2m} p_x^2 + \frac{1}{2b} x^2, \qquad (4.1.16)$$

where b is the elasticity constant. As the Hamiltonian is a constant of motion, we choose its value along with the conditions that $x(t)|_{t=0} > 0$ and the particle is at rest, which yields the initial condition

$$x(t)|_{t=0} = \sqrt{2bH}, \quad p_x(t)|_{t=0} = 0. \qquad (4.1.17)$$

From the formula (4.1.6), we get the action

$$
\begin{aligned}
J &= \int_{x=\sqrt{2bH}, p_x \leq 0}^{-\sqrt{2bH}} p_x \, dx + \int_{x=-\sqrt{2bH}, p_x \geq 0}^{\sqrt{2bH}} p_x \, dx \\
&= 2 \int_{-\sqrt{2bH}}^{\sqrt{2bH}} \sqrt{2mH - \frac{m}{b} x^2} \, dx \\
&= \frac{H}{\nu}, \qquad (4.1.18)
\end{aligned}
$$

where

$$\nu = \frac{1}{2\pi \sqrt{mb}}. \qquad (4.1.19)$$

The motion initially obeying (4.1.17) can be described by

$$x \equiv x(t) = \sqrt{2bH} \cos(2\pi \nu t). \qquad (4.1.20)$$

Substituting according to (4.1.7), we eliminate the time, and obtain

$$x = \sqrt{2bH}\,\cos(2\pi w) = \sqrt[4]{\frac{b}{m}}\sqrt{\frac{J}{\pi}}\,\cos(2\pi w). \tag{4.1.21}$$

Let us remark that the formulae (4.1.11) and (4.1.13) have been built after the analytical ones (4.1.20) and (4.1.21), respectively. The cosine function on the right-hand side of (4.1.21) is a typical 1-periodic function of the angle variable w. Restricting ourselves to $w \in [0,1)$, we may write

$$w = \begin{cases} \frac{1}{2\pi}\mathrm{Cos}^{-1}\left(\frac{x}{\sqrt{2bH}}\right) & \text{for } p_x = 0, \\[2mm] \theta(p_x)\left[1 - \frac{1}{2\pi}\mathrm{Cos}^{-1}\left(\frac{x}{\sqrt{2bH}}\right)\right] \\[2mm] \quad + \theta(-p_x)\frac{1}{2\pi}\mathrm{Cos}^{-1}\left(\frac{x}{\sqrt{2bH}}\right) & \text{for } p_x \neq 0, \end{cases} \tag{4.1.22}$$

where Cos^{-1} stands for the arccosine function taking on the values in the interval $[0,\pi]$. Recalling that

$$J = \pi\left(\sqrt{\frac{b}{m}}\,p_x^2 + \sqrt{\frac{m}{b}}\,x^2\right), \tag{4.1.23}$$

we understand that (4.1.22) and (4.1.23) result from the joint solution of (4.1.21) and

$$p_x = -\sqrt[4]{\frac{m}{b}}\sqrt{\frac{J}{\pi}}\,\sin(2\pi w). \tag{4.1.24}$$

Let us note that for $H = 0$ it is impossible to define the angle variable w.

There is a useful transformation, which leads to a very simple expression for the action and angle variables. Introducing the complex amplitude

$$\alpha = \frac{1}{\sqrt{2\hbar}}\left(\sqrt[4]{\frac{m}{b}}\,x + i\sqrt[4]{\frac{b}{m}}\,p_x\right), \tag{4.1.25}$$

we first observe that

$$[\alpha, \alpha^*] = 1, \tag{4.1.26}$$

where we used the classical commutator (4.1.3). Considering the inverse formulae to (4.1.25) and using (4.1.16) in (4.1.18), we obtain the Hamiltonian function

$$H = h\nu|\alpha|^2, \tag{4.1.27}$$

where $h = 2\pi\hbar$ is Planck's constant, and

$$J = h|\alpha|^2. \tag{4.1.28}$$

Although equation (4.1.22) can be rewritten as

$$
w = \begin{cases}
\frac{1}{2\pi}\mathrm{Cos}^{-1}\left(\frac{\mathrm{Re}\,\alpha}{|\alpha|}\right) & \text{for } \mathrm{Im}\,\alpha = 0, \\[2ex]
\theta(\mathrm{Im}\,\alpha)\left[1 - \frac{1}{2\pi}\mathrm{Cos}^{-1}\left(\frac{\mathrm{Re}\,\alpha}{|\alpha|}\right)\right] & \\
\quad +\theta(-\mathrm{Im}\,\alpha)\frac{1}{2\pi}\mathrm{Cos}^{-1}\left(\frac{\mathrm{Re}\,\alpha}{|\alpha|}\right) & \text{for } \mathrm{Im}\,\alpha \neq 0,
\end{cases}
\tag{4.1.29}
$$

it is useful to turn to the formulae (4.1.21) and (4.1.24) and to build an equation

$$
\alpha = \sqrt{\frac{J}{h}}\,\exp(-i2\pi w).
\tag{4.1.30}
$$

Introducing the polar decomposition

$$
\alpha = |\alpha|\exp(i\varphi),
\tag{4.1.31}
$$

we see that

$$
w = -\frac{\varphi}{2\pi}\ (\mathrm{mod}\ 1).
\tag{4.1.32}
$$

Denoting $\mathrm{Arg}_{-2\pi}\alpha$ the phase in (4.1.31), taking values in $(-2\pi, 0]$ (the value -2π will not be utilized), we can simplify dramatically the formula (4.1.29) to the form

$$
w = \begin{cases}
0 & \text{for } \alpha > 0 \text{ real,} \\
-\frac{1}{2\pi}\mathrm{Arg}_{-2\pi}\alpha & \text{otherwise.}
\end{cases}
\tag{4.1.33}
$$

More generally, we may require that the phase in (4.1.31) take on values in $(\theta_0, \theta_0+2\pi]$, where $\theta_0 = \mathrm{Arg}_{\theta_0}[\alpha(t)|_{t=0}]$.

In the case when the physical system under study is a plane (or axial) rotator, the generalized coordinate q and the momentum p become the rotation angle ϕ and the angular momentum L_z. The Hamiltonian function reads

$$
H = \frac{1}{2I}L_z^2,
\tag{4.1.34}
$$

where I is the moment of inertia. We assume the initial condition

$$
\phi|_{t=0} = 0, \quad L_z(t)|_{t=0} = \pm\sqrt{2IH}.
\tag{4.1.35}
$$

Applying the formula (4.1.6), we arrive at the action in the form

$$
\begin{aligned}
J_\pm &= \int_0^{\pm\pi} L_z\,d\phi + \int_{\pm\pi}^{\pm 2\pi} L_z\,d\phi \\
&= 2\pi|L_z| = 2\pi\sqrt{2IH}.
\end{aligned}
\tag{4.1.36}
$$

The rotation initially obeying (4.1.35) is described by

$$\phi \equiv \phi(t) = \pm 2\pi\nu t, \tag{4.1.37}$$

where (cf. (4.1.12))

$$\nu = \frac{|L_z|}{2\pi I}. \tag{4.1.38}$$

To eliminate the time, we substitute according to (4.1.7) and obtain

$$\phi = s2\pi w_s, \tag{4.1.39}$$

where

$$s = \text{sgn}(L_z). \tag{4.1.40}$$

We will use the notation J_s as if $s = \pm$. We see that the phase angle

$$w_s \equiv \text{sgn}(L_z)\frac{\phi}{2\pi} \ (\text{mod } 1). \tag{4.1.41}$$

Similarly, w_s means w_+ and w_-, although $s = \{-1, 0, 1\}$. Rather trivially, restricting $w_s \in [0, 1)$ and $\phi \in [0, 2\pi)$ or $\phi \in (-2\pi, 0]$ for $L_z > 0$ or $L_z < 0$, respectively, we observe that

$$
\begin{aligned}
w_+ &= \frac{\phi}{2\pi} &&\text{for } \phi \in [0, 2\pi), \\
w_- &= 0 &&\text{for } \phi = -2\pi, \\
w_- &= -\frac{\phi}{2\pi} &&\text{for } \phi \in (-2\pi, 0].
\end{aligned}
\tag{4.1.42}
$$

For this rotator the phase angle is closely related to the rotation angle. In fact, the direction of rotation may be anticlockwise, then the rotation angle grows or it may be clockwise, then the rotation angle falls. When the rotation angle changes by 2π, and the rotator has turned about in any direction, the phase angle increases by unity.

Historically, the spectroscopy of light emitted by atoms triggered the interest in the explanation of the unequal energy level spacing of emitors. The solution was the Bohr–Sommerfeld quantization rule, which postulated equal level spacing of action. Because in classical mechanics there is a relationship between action and energy, this meant also the quantization of energy. Interestingly, the validity of this quantization principle can be verified in modern quantum theory, although there is no accepted operator of action, except the case of the harmonic oscillator, where the classical proportion between the Hamiltonian function and action may be interpreted as the definition of the action operator. Equivalently, modern quantum theory neglects any representation in terms of phase angle resolution of the identity, except the investigations concentrating on the optical phase. Not being for this difficulty, the steady states of radiating systems could be characterized as states with especially simple phase-angle representations. Let us remark, that it relates to the period after

London (1926, 1927) who has been remembered in the course of the investigations into the quantum optical phase. The verification proceeds in the framework of the Wentzel–Kramers–Brillouin method.

To illustrate the Bohr–Sommerfeld quantization rule, we note that in the case of the vibration the rule prescribes the values of the action

$$J = \left(n + \frac{1}{2}\right)h, \quad n = 0, 1, \ldots, \tag{4.1.43}$$

leading to the condition

$$|p_x| = \left(n + \frac{1}{2}\right)\frac{h}{4a} \tag{4.1.44}$$

in the case of the rectangular potential well of the infinite depth and to the energy levels

$$H = \hbar\omega\left(n + \frac{1}{2}\right), \quad \omega = 2\pi\nu, \tag{4.1.45}$$

in the case of the harmonic oscillator. Whereas in the case of the rectangular potential well the quantization rule is not accurate enough, in the case of the harmonic oscillator and of suitable anharmonic ones it can be derived from the WKB approximation.

In the case of the axial rotator this rule indicates the values of the action

$$J = |N|h, \quad N = 0, \pm 1, \ldots, \tag{4.1.46}$$

which provide the values of the angular momentum

$$L_z = -N\hbar, \tag{4.1.47}$$

where we have chosen the minus sign to link the situation of the clockwise motion with positive values of N.

The above one-dimensional examples may characterize the situation in the Bohr–Sommerfeld quantization only superficially, while this quantization rules were applied to the Kepler orbits of electrons in the forcefield of the nucleus. With respect to the analogy with detection schemes, we address here a two-dimensional physical system. We assume that the system consists of two identical harmonic oscillators. The Hamiltonian function reads

$$H = H_x + H_y, \tag{4.1.48}$$

where

$$
\begin{aligned}
H_x &= \frac{1}{2m}p_x^2 + \frac{1}{2b}x^2, \\
H_y &= \frac{1}{2m}p_y^2 + \frac{1}{2b}y^2.
\end{aligned}
\tag{4.1.49}
$$

Introducing the complex amplitudes α_x, α_y,

$$\alpha_x = \frac{1}{\sqrt{2\hbar}}\left(\sqrt[4]{\frac{m}{b}}\,x + i\sqrt[4]{\frac{b}{m}}\,p_x\right),$$

$$\alpha_y = \frac{1}{\sqrt{2\hbar}}\left(\sqrt[4]{\frac{m}{b}}\,y + i\sqrt[4]{\frac{b}{m}}\,p_y\right), \tag{4.1.50}$$

we observe that

$$[\alpha_x, \alpha_x^*] = 1, \quad [\alpha_y, \alpha_y^*] = 1, \quad [\alpha_x, \alpha_y] = 0,$$
$$[\alpha_x^*, \alpha_y^*] = 0, \quad [\alpha_x, \alpha_y^*] = 0, \quad [\alpha_y, \alpha_x^*] = 0, \tag{4.1.51}$$

and we can rewrite the Hamiltonian function

$$H = h\nu|\alpha_x|^2 + h\nu|\alpha_y|^2. \tag{4.1.52}$$

Regarding the analogy with the detection schemes, we introduce also the complex amplitudes

$$\alpha_\mp = \frac{1}{\sqrt{2}}(\alpha_x \pm i\alpha_y). \tag{4.1.53}$$

Since this is the U(2) transformation, the relations in (4.1.51) hold also after the replacement of the subscripts x, y with $-$, $+$, respectively. The role of (4.1.53) can be illustrated by another analogy, namely, by the description of monochromatic light in the linear and circular polarization bases. The Hamiltonian function is then of the form

$$H = h\nu|\alpha_-|^2 + h\nu|\alpha_+|^2. \tag{4.1.54}$$

Taking into account that (cf. (4.1.28))

$$J_\mp = h|\alpha_\mp|^2, \tag{4.1.55}$$

we may perform the Bohr–Sommerfeld quantization in the two harmonic oscillators independently, and we state the conditions

$$|\alpha_\mp|^2 = n_\mp + \frac{1}{2}, \quad n_\mp = 0, 1, \ldots . \tag{4.1.56}$$

Introducing the polar coordinates,

$$\phi = \arg(x + iy),$$
$$r = \sqrt{x^2 + y^2}, \tag{4.1.57}$$

which extend to the whole phase space as

$$L_z = xp_y - yp_x,$$
$$p_r = \frac{xp_x + yp_y}{r}, \tag{4.1.58}$$

we may formulate the property

$$J_\phi + J_r = J_- + J_+,$$ (4.1.59)

where

$$J_\phi = \oint L_z \, d\phi,$$
$$J_r = \oint p_r \, dr.$$ (4.1.60)

Respecting the fact that

$$L_z = -\hbar(|\alpha_-|^2 - |\alpha_+|^2)$$ (4.1.61)

and the conservation of the angular momentum, we obtain by (4.1.59) that

$$J_r = 2\hbar \min(|\alpha_-|^2, |\alpha_+|^2).$$ (4.1.62)

Because the system of the "radial" oscillator cannot be separated from the plane "rotator", it is only a seeming paradox that (4.1.62) cannot obey the Bohr–Sommerfeld quantization condition,

$$J_r = 2\hbar \left[\min(n_-, n_+) + \frac{1}{2} \right],$$ (4.1.63)

while the angular momentum is quantized as

$$L_z = -\hbar(n_- - n_+).$$ (4.1.64)

It can easily be verified that the appropriate canonically conjugate variables, namely, the phase angles are

$$w_\phi = \theta(|\alpha_-|^2 - |\alpha_+|^2)w_- + \theta(|\alpha_+|^2 - |\alpha_-|^2)w_+,$$
$$w_r = \frac{1}{2}(w_- + w_+),$$ (4.1.65)

where (cf. (4.1.32))

$$w_\mp = -\frac{\varphi_\mp}{2\pi} \pmod{1},$$ (4.1.66)

with

$$\varphi_\mp = \arg(\alpha_\mp).$$ (4.1.67)

4.2 Geometric phase

Berry (1984) discovered that in the quantum adiabatic theorem the usual dynamical phase must be corrected by the geometric phase which can cause measurable interference effects after the parameters of the Hamiltonian have gone through a closed circuit in their space. The geometric phase is zero if the system has retraced the same arc to and back. The standard Born–Oppenheimer treatment of molecules could not provide the result of Longuet–Higgins and Herzberg, and of Stone [Berry (1984)] that nuclear wave function cannot be chosen to be single valued near the degeneracy. This problem was solved by Mead and Truhlar (1979), who showed that the choice of an external gauge potential can help. They call their result the molecular Aharonov–Bohm effect. Krakovsky and Birman (1995) have derived geometric potentials arising in the theory of coupling between slow and fast systems in connection with an elimination of the fast degrees of freedom. The geometric phase has been associated also with an incompressible flow issuing through a minimal surface [Martinez (1995)].

Similarly, the Foucault pendulum is interesting by the anholonomy, which occurs after a diurnal revolution. The anholonomy is the angle between the initial and final directions of swing and is equal to $2\pi \sin \lambda \equiv 2\pi \cos \theta$, where λ is the latitude and θ is the polar angle of the flagging pendulum when the centrifugal force is neglected. The anholonomy is zero for $\lambda = 0$, i. e., on the equator. At the poles the anholonomy is equal to a complete revolution for $\lambda = \pm\frac{\pi}{2}$. In general, the anholonomy is the spherical area of the zone between the equator and the parallel of the latitude we consider when the Earth radius R is set to be unity.

Let us consider an inertial frame x, y, z with the z axis oriented towards to the World Pole. We introduce the parametrization of the unit sphere, i. e.,

$$
\begin{aligned}
x &= \sin\theta \cos\varphi, \\
y &= \sin\theta \sin\varphi, \\
z &= \cos\theta, \qquad \theta \in (0,\pi), \quad \varphi \in [0, 2\pi),
\end{aligned}
\tag{4.2.1}
$$

where φ is the azimuthal angle. Let us denote, for a moment, $t_\theta \equiv \theta$, $t_\varphi \equiv \varphi$, and remember a formula for the Riemann connection ∇ whose coordinates for any surface are Christoffel's symbols Γ^l_{jk},

$$
\nabla_{\frac{\partial r}{\partial t_j}} \frac{\partial r}{\partial t_k} = \sum_l \Gamma^l_{jk} \frac{\partial r}{\partial t_l}, \quad j = \theta, \varphi, \quad k = \theta, \varphi, \quad l = \theta, \varphi.
\tag{4.2.2}
$$

For the sphere with the chosen parametrization, the Christoffel symbols read

$$
\Gamma^\varphi_{\theta\varphi} = \cot\theta, \quad \Gamma^\theta_{\theta\theta} = \Gamma^\varphi_{\theta\theta} = \Gamma^\theta_{\theta\varphi} = 0,
\tag{4.2.3}
$$

$$
\Gamma^\theta_{\varphi\theta} = \Gamma^\varphi_{\varphi\theta} = 0, \quad \Gamma^\theta_{\varphi\varphi} = -\sin\theta\cos\theta, \quad \Gamma^\varphi_{\varphi\theta} = \cot\theta.
\tag{4.2.4}
$$

The parallel transport of a contravariant vector with components A^θ, A^φ along the φ-coordinate line, $\theta = $ const, is described by the system of ordinary differential equations

$$
\begin{aligned}
\frac{dA^\theta}{d\varphi} &= -\Gamma^\theta_{\varphi\theta} A^\theta - \Gamma^\theta_{\varphi\varphi} A^\varphi, \\
\frac{dA^\varphi}{d\varphi} &= -\Gamma^\varphi_{\varphi\theta} A^\theta - \Gamma^\varphi_{\varphi\varphi} A^\varphi,
\end{aligned}
\tag{4.2.5}
$$

or

$$\frac{dA^\theta}{d\varphi} = \sin\theta\cos\theta\, A^\varphi,$$

$$\frac{dA^\varphi}{d\varphi} = -\cot\theta\, A^\theta. \qquad (4.2.6)$$

The coordinate vectors at (θ, φ) provide a useful observer frame. Because the coordinate vector $\frac{\partial \boldsymbol{r}}{\partial \varphi}$ is not normalized, we replace it by a normalized one and name Y the appropriate component

$$Y = \sin\theta\, A^\varphi. \qquad (4.2.7)$$

Y is the eastern component and A^θ is the southern component. The equations (4.2.6) simplify to the form

$$\frac{dA^\theta}{d\varphi} = \cos\theta\, Y,$$

$$\frac{dY}{d\varphi} = -\cos\theta\, A^\theta. \qquad (4.2.8)$$

If the vector with the components $A^\theta(0)$, $Y(0)$ is given at the point $(\theta, \varphi = 0)$, the parallel transported vector at (θ, φ) has the components

$$A^\theta(\varphi) = \cos(\varphi\cos\theta)A^\theta(0) + \sin(\varphi\cos\theta)Y(0),$$

$$Y(\varphi) = -\sin(\varphi\cos\theta)A^\theta(0) + \cos(\varphi\cos\theta)Y(0). \qquad (4.2.9)$$

For φ increasing with time as the Earth rotates, the simplified equations (4.2.8) describe the rotation from the East (via the South) to the West (cf. (4.2.9)) for the particular choice $A^\theta(0) = 0$, $Y(0) = 1$. As far as the pendulum swings linearly to a certain degree of approximation or its bob draws an ellipse, the equations (4.2.8) describe the evolution of the pendulum whose dynamical phase is independent of the rotation angle. When the pendulum draws a circle, the dynamical phase appears as a dynamical angle and the rotation angle contributes to it.

A time-dependent Hamiltonian may arise due to its dependence on parameters which are varied in time. As an example we present the Hamiltonian

$$\hat{H}(\boldsymbol{B}(t)) = \kappa\hbar\boldsymbol{B}(t)\cdot\hat{\boldsymbol{S}}, \qquad (4.2.10)$$

where $\boldsymbol{B}(t)$ is a magnetic field, κ is a constant involving the gyromagnetic ratio and $\hat{\boldsymbol{S}}$ is the vector spin operator. More generally, we consider the Hamiltonian $\hat{H}(\boldsymbol{R}(t))$, where $\boldsymbol{R}(t) = (R_1(t), R_2(t), \ldots, R_m(t))$ and $R_j(t)$, $j = 1, \ldots, m$, are environmental parameters, and we assume that the system was in a stationary state initially. If $\boldsymbol{R}(t)$ is slowly varied and, hence, $\hat{H}(\boldsymbol{R}(t))$ is slowly changed, it follows from the adiabatic theorem that the system will be in the eigenstate of $\hat{H}(\boldsymbol{R}(t))$ all the times. The state $|\psi(t)\rangle$ of the system evolves according to the Schrödinger equation

$$\frac{d}{dt}|\psi(t)\rangle = -\frac{i}{\hbar}\hat{H}(\boldsymbol{R}(t))|\psi(t)\rangle \tag{4.2.11}$$

and

$$|\psi(t)\rangle|_{t=0} = |n(\boldsymbol{R}(0))\rangle. \tag{4.2.12}$$

Here $|n(\boldsymbol{R}(0))\rangle$ is the stationary state that satisfies

$$\hat{H}(\boldsymbol{R}(0))|n(\boldsymbol{R}(0))\rangle = E_n(\boldsymbol{R}(0))|n(\boldsymbol{R}(0))\rangle, \tag{4.2.13}$$

with an energy $E_n(\boldsymbol{R}(0))$. We formulate the adiabatic approximation as $|\psi(t)\rangle \approx |\psi(t)\rangle_{\text{appr}}$, where

$$|\psi(t)\rangle_{\text{appr}} = \exp\left\{-\frac{i}{\hbar}\int_0^t E_n(\boldsymbol{R}(t'))\,dt'\right\}\exp[i\gamma_n(t)]|n(\boldsymbol{R}(t))\rangle. \tag{4.2.14}$$

Here the state $|\psi(t)\rangle_{\text{appr}}$ differs only by a phase factor from the eigenstate $|n(\boldsymbol{R}(t))\rangle$ that obeys

$$\hat{H}(\boldsymbol{R}(t))|n(\boldsymbol{R}(t))\rangle = E_n(\boldsymbol{R}(t))|n(\boldsymbol{R}(t))\rangle, \tag{4.2.15}$$

with an energy $E_n(\boldsymbol{R}(t))$. Compared with the operator $|n(\boldsymbol{R}(t))\rangle\langle n(\boldsymbol{R}(t))|$, the ket $|n(\boldsymbol{R}(t))\rangle$ has a phase property, which for present purposes is provisional. To define the definitive "right" state in the excursion between times 0 and t, we use the phase factor $\exp[i\gamma_n(t)]$ introduced by Berry (1984). It is assumed that $\gamma_n(0) = 0$. The dynamical phase factor in (4.2.14) reads as

$$\exp\left\{-\frac{i}{\hbar}\int_0^t E_n(\boldsymbol{R}(t'))\,dt'\right\}. \tag{4.2.16}$$

From the approximation

$$\frac{d}{dt}|\psi(t)\rangle_{\text{appr}} \approx -\frac{i}{\hbar}\hat{H}(\boldsymbol{R}(t))|\psi(t)\rangle_{\text{appr}}, \tag{4.2.17}$$

using the exact relation

$$\hat{H}(\boldsymbol{R}(t))|\psi(t)\rangle_{\text{appr}} = E_n(\boldsymbol{R}(t))|\psi(t)\rangle_{\text{appr}}, \tag{4.2.18}$$

we obtain that

$$\frac{d}{dt}\left\{\exp[i\gamma_n(t)]|n(\boldsymbol{R}(t))\rangle\right\} \approx 0. \tag{4.2.19}$$

The relation (4.2.19) cannot be exact, because this assumption should lead to a contradiction. Rather, this relation means that

$$\langle n(\boldsymbol{R}(t))|\frac{d}{dt}\left\{\exp[i\gamma_n(t)]|n(\boldsymbol{R}(t))\rangle\right\} = 0. \tag{4.2.20}$$

Hence, the phase correction obeys the differential equation

$$\frac{d}{dt}\gamma_n(t) = i\langle n(\mathbf{R}(t))|\mathrm{grad}|n(\mathbf{R}(t))\rangle \cdot \frac{d\mathbf{R}(t)}{dt} \tag{4.2.21}$$

and the geometric phase is $\gamma_n(T)$ provided that $\mathbf{R}(T) = \mathbf{R}(0)$. From this it follows that

$$|n(\mathbf{R}(T))\rangle\langle n(\mathbf{R}(T))| = |n(\mathbf{R}(0))\rangle\langle n(\mathbf{R}(0))|, \tag{4.2.22}$$

i. e., the system under study has returned to its original state.

From (4.2.21) it follows that

$$\gamma_n(t) = i\int_0^t \langle n(\mathbf{R}(t))|\mathrm{grad}|n(\mathbf{R}(t))\rangle \cdot \frac{d\mathbf{R}(t)}{dt}\, dt. \tag{4.2.23}$$

We introduce the notation $A = \mathbf{R}(0)$, $B = \mathbf{R}(t)$ to indicate the fact that the phase correction depends on the path, but the dynamics can be eliminated,

$$\gamma_n(AB) = i\int_{AB} \langle n(\mathbf{R})|\mathrm{grad}|n(\mathbf{R})\rangle \cdot d\mathbf{R}. \tag{4.2.24}$$

In the case of the geometric phase, we have an integral around the loop C,

$$\gamma_n(C) = i\int_C \langle n(\mathbf{R})|\mathrm{grad}|n(\mathbf{R})\rangle \cdot d\mathbf{R}. \tag{4.2.25}$$

Stokes' theorem applied to (4.2.25) provides

$$\gamma_n(C) = -\mathrm{Im}\left\{\int\int_S \mathrm{rot}\langle n(\mathbf{R})|\mathrm{grad}|n(\mathbf{R})\rangle \cdot d\mathbf{S}\right\}, \tag{4.2.26}$$

$$= -\mathrm{Im}\left\{\int\int_S \mathrm{grad}\langle n(\mathbf{R})| \times \mathrm{grad}|n(\mathbf{R})\rangle \cdot d\mathbf{S}\right\}, \tag{4.2.27}$$

where S is a surface bounded by C and $d\mathbf{S}$ is the area element in $\mathbf{R}$ space. The relation (4.2.27) can be rewritten as

$$\gamma_n(C) = -\int\int_S \mathbf{V}_n(\mathbf{R}) \cdot d\mathbf{S}, \tag{4.2.28}$$

where

$$\mathbf{V}_n(\mathbf{R}) = \mathrm{Im}\left\{\mathrm{grad}\langle n(\mathbf{R})| \times \mathrm{grad}|n(\mathbf{R})\rangle\right\}. \tag{4.2.29}$$

We will denote the corrected state as $|n(\mathbf{R}(t))\rangle'$,

$$|n(\mathbf{R}(t))\rangle' = \exp[i\gamma_n(t)]|n(\mathbf{R}(t))\rangle. \tag{4.2.30}$$

The new state is not single valued in a parameter domain which includes the loop C. The corresponding phase correction $\gamma_n'(t) \equiv 0$, which entails that $\gamma_n'(C)$ is not a geometric phase. On the contrary,

$$\exp[i\gamma_n(C)] = {}'\langle n(\boldsymbol{R}(0))|n(\boldsymbol{R}(T))\rangle'. \tag{4.2.31}$$

Instead of equation (4.2.21) we may consider the equation

$$\frac{d}{dt}|n(\boldsymbol{R}(t))\rangle' = i\hat{G}(\boldsymbol{R}(t)) \cdot \frac{d\boldsymbol{R}(t)}{dt}|n(\boldsymbol{R}(t))\rangle', \tag{4.2.32}$$

where the vector operator

$$\hat{G}(\boldsymbol{R}) = -i\sum_{\substack{n,m \\ n \neq m}} \frac{\langle m(\boldsymbol{R})|\mathrm{grad}[\hat{H}(\boldsymbol{R})]|n(\boldsymbol{R})\rangle}{E_n(\boldsymbol{R}) - E_m(\boldsymbol{R})}|m(\boldsymbol{R})\rangle\langle n(\boldsymbol{R})|. \tag{4.2.33}$$

In the formula (4.2.33) the primes may or need not be considered, because the terms do not depend on the phase factors.

For the Hamiltonian (4.2.10) Berry (1984) obtained

$$\boldsymbol{V}_n(\boldsymbol{B}) = n\frac{\boldsymbol{B}}{B^3}, \tag{4.2.34}$$

where n is the eigenvalue of the spin component along the magnetic field $\boldsymbol{B}$, and showed that

$$\gamma_n(C) = -in\Omega(C), \tag{4.2.35}$$

where $\Omega(C)$ is the solid angle that C subtends at $\boldsymbol{B} = \boldsymbol{0}$.

The restriction to adiabaticity has been lifted by Aharonov and Anandan (1987) by factoring out the time integral of the expectation of the Hamiltonian as the dynamical phase from the wave function. Since the state vector of a harmonic oscillator in a finite-dimensional Hilbert space changes sign only when the Hilbert space is of even dimension, Pati and Lawande (1995) have assumed this and calculated the cyclic geometric phase for this finite-dimensional Hilbert-space harmonic oscillator. Basic issues regarding the meaning of the adiabatic approximation and the implication of their answering to the understanding of the geometric phase have been addressed [Mostafazadeh (1997a, b)]. A precise definition of an adiabaticity parameter of a time-dependent Hamiltonian as a criterium for the applicability of the adiabatic approximation has been proposed there. Parmenter and Valentine (1996) have discussed the geometric phase using the de Broglie–Bohm pilot wave (causal) interpretation of quantum mechanics. This picture has also served as basis of an alternative interpretation for the quantum adiabatic approximation [Mostafazadeh (1997b)]. A general formula for the Berry phase and the corresponding Hannay angle [Berry (1985), Hannay (1985)] for an arbitrary Hamiltonian with equally spaced, nondegenerate eigenvalues has been obtained [Seshadri, Lakshmibala and Balakrishnan (1997)]. The situation when a geometric phase goes to zero in the limit of high frequency of measurement has been studied to illustrate that the repeated measurement on a quantum state not only arrests the quantum transition (the famous quantum Zeno effect), but also stops the development of the geometric phase [Pati and Lawande (1996)]. Joshi and Lawande (1992) and Joshi, Pati and Banerjee (1994) have investigated the Aharonov–Anandan phase and inferred that the measurement of the geometric phase in a dispersive fibre can be used as a method of discerning the statistical character of the field. The general relation between the nonadiabatic geometric phase and the photon statistics of the two-mode light field propagating through a Kerr-like medium fibre have been established [Zhu, Tang and Huang (1996)]. Among applications Tang (1995) has proposed a nonunitary transformation to approach the generalized Jaynes–Cummings model. For a special initial state, the Jaynes–Cummings model shows the property of cyclic evolution and exhibits the Aharonov–Anandan phase. Ji, Kim, Kim and Soh (1995) have derived exact wave functions for the eigenstates of the Lewis–Riesenfeld invariant [Lewis Jr. and Riesenfeld (1969)]. Furthermore, they have found the cyclic initial states and have presented the corresponding nonadiabatic Berry (i. e., Aharonov–Anandan) phase for a harmonic oscillator with τ-periodic Hamiltonian. Particularly, Ji and Kim (1995) have considered the oscillator alternating between two frequencies and found that even if the classical solutions are aperiodic there exist cyclic initial states which are τ-periodic. The nonadiabatic Berry phase occurs even for quantum systems with time-independent Hamiltonians and it is intimately connected with the nonstationarity of a quantum state [Zeng and Lei (1995)]. The study of wave-packet revivals has been extended to the case of Hamiltonians which are time independent through the adiabatic cycling of some parameters [Jarzynski (1995)]. The revival consists of dynamical phase effects and geometric phase effects.

Manifestations of the geometric phase factor for a photon were considered by Chiao and Wu (1986) and the experimental verification was performed by Tomita and Chiao (1986). They extended Berry's treatment to this particle which is a massless spin-1 boson. Its helicity $\hat{s} \cdot k$, where $\hat{s}$ is the vector spin operator and k is the direction of its propagation, can only be $+1$ or -1. Because the coupling between the photon and the magnetic field is not present, they replaced in the above example the direction of the magnetic field B by the direction of the propagation k. They used the helicity quantum number as an adiabatic invariant. Slow changes in the environment (in the index of refraction) will yield the same phase factor as above. They discussed the situations (i) when circularly polarized light propagates down a helically wound optical fibre, (ii) when linearly polarized light propagates down such a fibre, and (iii) when microwaves propagate down a helically wound circular waveguide. Berry (1986) suggested that changes could be accomplished with a transparent birefringent medium twisted along the axis of propagation. The geometry of a fibre constrains the direction k to trace out a loop C on the surface of a sphere in k space. Berry's argument leads to a geometric phase

$$\gamma(C) = -\sigma\Omega(C), \qquad (4.2.36)$$

where $\Omega(C)$ is the solid angle subtended by the loop C with respect to the origin $\mathbf{k} = \mathbf{0}$ and σ is the helicity quantum number. For a uniform helix the loop C is a circle and

$$\Omega(C) = 2\pi N(1 - \cos \theta), \qquad (4.2.37)$$

where N is the winding number of the helix and θ is the pitch angle of the helix. The phase factor is observable by interference. A circularly polarized laser beam can be injected into a single input optical fibre, which in turn feeds equal amounts of the light into two helically wound optical fibres each having N turns, but in contrary senses. The fibres are then coupled into a single output optical fibre, where interference occurs.

The case (ii) has been studied experimentally by Tomita and Chiao (1986). The initial state

$$|x\rangle = \frac{1}{\sqrt{2}}(|\sigma = +1\rangle + |\sigma = -1\rangle) \qquad (4.2.38)$$

has evolved into the state

$$|x'\rangle = \frac{1}{\sqrt{2}}\left[\exp(i\gamma_+)|\sigma = +1\rangle + \exp(i\gamma_-)|\sigma = -1\rangle\right]. \qquad (4.2.39)$$

Here $\gamma_\pm$ are Berry's phases for $\sigma = \pm 1$ and $\gamma_- = -\gamma_+$. Accordingly, $|\langle x|x'\rangle|^2 = \cos^2(\gamma_+)$. By Malus's law, this implies that the plane of polarization has been rotated by the angle γ_+. Frins and Dultz (1997) have reported a similar experiment that allows a direct observation of Berry's topological phase. Agarwal and Simon (1990) have not used the concept of "photon interference" and, distinguishing between the interference of particles and that of light beams, they have proved that the phase observed in the latter case would be the Hannay angle rather than the Berry phase.

Let us name also the interferometric experiment by Chiao, Antaramian, Ganga, Jiao, Wilkinson and Nathel (1988), where the case (i) has been accomplished although the two fibres were replaced with two nonplanar mirror configurations of opposite handedness. A demonstration of the existence of Berry's phase on the single-photon level has been provided by Kwiat and Chiao (1991). Berry (1987) also established the precise relation between earlier studies by Pancharatnam (1956, reprinted 1975) and the recently discovered phase change for slowly cycled quantum systems. Pancharatnam defined the phase difference between two beams in different states of polarization by considering their interference. He regarded the two beams as being "in phase" when the produced intensity was a maximum. In general, the phase difference was defined, even if the initial state of polarization had changed. It led to the observation that it was possible for a beam to go through two other polarization states without introducing any phase changes back to its original state and then to exhibit a phase shift. The magnitude of this phase shift (the Pancharatnam phase) was equal to half the solid angle subtended by the circuit at the centre of the Poincaré sphere. Several experiments have been described using interferometric techniques to measure this phase shift [Bhandari and Samuel (1988), Simon, Kimble and Sudarshan (1988), Chyba, Wang, Mandel and Simon (1988)]. Hence, there are two types of the geometric phases experimentally observed in optics: (i) The spin redirection phase predicted by Chiao and Wu (1986). (ii) The phase predicted and observed by Pancharatnam in 1956 and verified by Bhandari and Samuel (1988). With respect to both types of experimental observations there were discussions on the quantal or the classical nature of the geometric phase in optics [Tiwari (1992)]. He has outlined a tentative approach to understand this phenomenon based on the angular-momentum exchange between the light beam and the optical elements. De Vito and Levrero (1994) intended to distinguish the measurement of a closed circuit in momentum space [Tomita and Chiao (1986)] from Pancharatnam's "original" phase. They have accomplished a classical-theoretical analysis of the experiment described by Chyba, Wang, Mandel and Simon (1988). Hariharan, Larkin and Roy (1994) have presented quantitative measurements on the interference fringes obtained with white light in an interferometer using an achromatic phase shifter operating on the Pancharatnam phase. These observations confirm that the effects due to the introduction of a variable geometric phase, which is independent of the wavelength, differ significantly from those due to a change in the optical path difference. Since the transformations that produce and maintain squeezed states of the electromagnetic field form the Lorentz group $(SU(1,1))$, the use of degenerate parametric amplifier has been proposed for production of the Berry phases [Chiao and Jordan (1988)]. All ideas related to the geometric phase can be illustrated by using this nonlinear optical process [Cerveró and Lejarreta (1997)]. Observations of the Pancharatnam phase have been made by Hariharan, Roy, Robinson and O'Byrne (1993) at low light levels which ensured that the probability of more than one photon travelling through the interferometer was negligible. Exploiting the fact that the phase shift produced by a device which operates on the principle of the geometric phase, can be unbounded [Bhandari (1988)], Hariharan and Ciddor (1994) have shown how the Pancharatnam phase can be used in a simple optical system to produce a controllable phase shift that is independent of the wavelength. A demonstration of phase shifts due to the geometric phase or the sum of geometric and dynamical phases has been performed with twisted stack of polarizers or retarders [Berry and Klein (1996)]. An optical realization of the $SU(1,1)$ geometric phase is based on discrete transformations in an optical ring cavity containing partial reflectors [Benedek and Benedict (1997) and references therein].

The Berry–Pancharatnam topological phase can also be measured using two frequency down-converters aligned by the common idler beam [Grayson, Torgerson and Barbosa (1994)]. As shown by Klyshko (1989), a light field with n identically polarized photons per mode is expected to acquire n times the phase of a one-photon field. In the experiments of Brendel, Dultz and Martienssen (1995), the geometric phase has been measured by making a photon pair interfere with itself inside a Michelson interferometer. If both photons are in equal states of linear polarization, they have observed a doubling of the geometric phase compared to single photons; in the case of orthogonal states of linear polarization, the geometric phase is completely cancelled. Tiwari (1997) has compared the Pancharatnam phase in two-photon experiments to a pair of paths in the product manifold of two Poincaré spheres. In these experiments a geometric phase twice that of the classical value for parallel polarizations or zero for orthogonal polarizations is observed [Brendel, Dultz and Martienssen (1995)]. Appelt, Wäckerle and Mehring (1995) have reported on the geometric phases a spin-$\frac{1}{2}$ system accumulates in nonadiabatic figure-8 experiments. Experimental results for adiabatic figure-8 loops obtained by using a nuclear magnetic resonance gyroscope have been compared to those for a circle and a spherical triangle. A geometric phase factor has been derived for the split-beam experiment with a four-plate neutron interferometer as an example of cyclic evolutions [Hasegawa, Zawiski, Rauch and Joffe (1996)]. From rotation and linear translation operations, geometric and dynamical phases can be distinguished and separated by means of neutron interferometry [Wagle, Rakhecha, Summhammer, Badurek, Weinfurter, Allman, Kaiser, Hamacher, Jacobson and Werner (1997)], which provides a direct verification of Pauli anticommutation principle. The Berry–Pancharatnam phase can also be obtained for squeezed states [Mendaš (1997), Seshadri, Lakshmibala and Balakrishnan (1997)]. Another experimental demonstration of the Pancharatnam phase, which can be viewed as a demonstration of Berry's phase arising from quantum measurement, has been proposed [Hariharan, Ramachandran, Suresh and Samuel (1997)]. This paper points out that an analyzer which projects the beam onto a particular polarization state is a projection operator indeed. Anandan, Christian and Wanelik (1997) have provided a guide to the literature on the geometric angles and phases in classical and quantum physics.

4.3   Quantum phase problem

Although at the beginning of the quantum theory, whose fundamentals were laid down by Planck and Einstein, the discrete energy levels were more important than the wave–particle dualism, due attention to all the nonclassical phenomena has led Heisenberg and his contemporaries to their reinterpretation of classical mechanics and to the modern quantum theory. The rather philosophical principle of complementarity due to Bohr [Bohr (1949)] has been formalized in terms of canonically conjugate variables. Soon it became evident, that some physical properties exist, which cannot be derived through quantization of classical mechanics, for example the half-odd fermionic spin. Nevertheless, the canonical quantization is a most relevant procedure, although alternative procedures are available.

Considering that the operators of position coordinate and of the canonically conjugate momentum have resulted from a canonical quantization procedure, we may expect that the same procedure would yield the operators of action and angle. However, London (1926, 1927) showed that this objective could not be accomplished fully and provided a compromising solution. With quantum optics in mind, we have re-called the theory of harmonic oscillator in section 4.1 and we would no more speak of the action–angle variables. Instead we can speak of the dimensionless energy $|\alpha|^2$ and the phase φ as introduced in (4.1.28) and (4.1.32), respectively. In connection with this, we rewrite the Poisson bracket

$$\{w, J\} = 1 \tag{4.3.1}$$

in the form of a classical commutator (see (4.1.3))

$$[|\alpha|^2, \varphi] = i. \tag{4.3.2}$$

We will provide an indirect proof that the validity of (4.3.2) is related to the multiple-valued nature of the phase. The time-dependent phase variable exhibits this property, when it stays continuous as long as possible. On the contrary, the probabilistic and statistical treatment of the phase variable requires that it be single valued. The following consideration has its quantum counterpart. Let us assume that the single-valued phase variable φ_{sing} obeys (4.3.2) when substituted for φ. Introducing the coordinates

$$W = |\alpha|^2, \quad \theta = \arg \alpha, \tag{4.3.3}$$

we find that the classical commutator (4.3.2) is

$$[|\alpha|^2, \varphi] = i \frac{\partial \varphi}{\partial \theta}. \tag{4.3.4}$$

Assuming that θ_c is a continuity point of the single-valued phase variable, we perform the integration of equation (4.3.2) over the interval $[\theta_c, \theta_c + 2\pi)$ of θ and derive that $0 = \pm 2\pi i$, a contradiction. In fact, the multiple-valued phase increases or decreases by 2π according to the integration being performed in the anticlockwise or clockwise direction. It has been mentioned in the literature [Carruthers and Nieto (1968)] that the classical phase variable can be made single valued by allowing for a jump from $\theta_0 + 2\pi$ down to θ_0 in the counterclockwise direction and from θ_0 up to $\theta_0 + 2\pi$ in the clockwise direction. In this case the variable $\varphi_{\text{sing}} = \varphi_{\theta_0}$ has the classical commutator [Carruthers and Nieto (1968), Barnett and Pegg (1989)]

$$[|\alpha|^2, \varphi_{\theta_0}] = i[1 - 2\pi\delta(\theta - \theta_0)]. \tag{4.3.5}$$

The complexity of the commutator seems to indicate that also the dynamical prop-erties of the phase variable are complicated. Much worse, we obtain undesirable dynamical properties already from the assumption of the commutator (4.3.2). We assume the Hamiltonian function to be

$$H = \mp\hbar\varphi, \tag{4.3.6}$$

which leads to the equations of motion

$$\frac{d}{dt}|\alpha|^2 = \mp 1. \tag{4.3.7}$$

Although the choice of the plus sign in (4.3.6) does not lead to any difficulty when time passes, the process is irreversible and, in fact, the choice of the minus sign would lead to the violation of the condition $|\alpha|^2 \geq 0$ after a time long enough. Also this situation has its counterpart in quantum theory, where it is usually related to the lower boundedness of spectrum. The appropriate "dynamics" is a nonunitary discrete-time one and in the classical framework we have the example of a noncanonical transformation. Usually, i. e., in the framework of the quantum theory, it is emphasized that the unusual commutator (4.3.5) can be avoided using 2π-periodic functions of phase, especially $\cos \varphi$, $\sin \varphi$, etc. This cure cannot be used in the case of "irreversible" dynamics.

To be able to present the quantum plane rotator below as a "solution" of the quantum phase problem, we are going to provide here an analogy and a "solution" of the "canonical phase variable" problem. Although the classical plane rotator has its action–angle variables, it seems that no phase problem exists with this rotator. Let us recall that the action–angle variables are not defined globally, in fact, $(w_\pm, J_\pm)$ are defined in the half of the phase cylinder, where $\mathrm{sgn}(L_z) = \pm 1$. We see a similarity to the phase problem in the fact that, e. g., w_- may generate a shift of J_- to the negative values. Respecting the relation (4.1.36), we believe that a similar shift of L_z to the positive values, $L_z = -\frac{1}{2\pi}J_-$, would be allowed. Taking into account the relation (4.1.41) and assuming that $w_- = w$ for the harmonic oscillator, we obtain from (4.1.32) the relation

$$\phi = \varphi \pmod{2\pi} \tag{4.3.8}$$

connecting the plane rotator with the harmonic oscillator.

In quantum theory the harmonic oscillator is described by the Hamiltonian

$$\hat{H} = \frac{1}{2m}\hat{p}_x^2 + \frac{1}{2b}\hat{x}^2, \tag{4.3.9}$$

where m is the mass of a particle and b is the elasticity constant, $\hat{x}$ is the operator of position coordinate, $\hat{p}_x$ is the operator of appropriate conjugate momentum. Let us remember that the commutator,

$$[\hat{x}, \hat{p}_x] = i\hbar\hat{1}. \tag{4.3.10}$$

We consider the eigenstates of the position-coordinate operator $\hat{x}$,

$$\hat{x}|x\rangle = x|x\rangle, \quad x \in (-\infty, \infty), \tag{4.3.11}$$

and those of the conjugate momentum operator $\hat{p}_x$,

$$\hat{p}_x|p_x\rangle = p_x|p_x\rangle, \quad p_x \in (-\infty, \infty). \tag{4.3.12}$$

These eigenstates fulfill the relations of orthogonality

$$\langle x|x'\rangle = \delta(x - x'), \quad \langle p_x|p_x'\rangle = \delta(p_x - p_x') \tag{4.3.13}$$

and the conjugation relation

$$\langle x|p_x \rangle = \frac{1}{\sqrt{2\pi\hbar}} \exp\left(\frac{i}{\hbar}xp_x\right).$$ (4.3.14)

The fact that a one-dimensional harmonic oscillator exhibits equally-spaced energy levels motivated the definition of the annihilation and creation operators at the dawn of quantum theory. Here we introduce the annihilation operator

$$\hat{a}_x = \frac{1}{\sqrt{2\hbar}}\left(\sqrt[4]{\frac{m}{b}}\,\hat{x} + i\sqrt[4]{\frac{b}{m}}\,\hat{p}_x\right),$$ (4.3.15)

fulfilling the commutator relation

$$[\hat{a}_x, \hat{a}_x^\dagger] = \hat{1}$$ (4.3.16)

and we can rewrite the Hamiltonian (4.3.9) in the form

$$\hat{H} = \hbar\omega\left(\hat{a}_x^\dagger\hat{a}_x + \frac{1}{2}\hat{1}\right),$$ (4.3.17)

where

$$\omega = \frac{1}{\sqrt{mb}}.$$ (4.3.18)

We can consider the number of energy quanta for this oscillator

$$\hat{n}_x = \hat{a}_x^\dagger\hat{a}_x.$$ (4.3.19)

The basic single-mode states $|n_x\rangle$,

$$\hat{n}_x|n_x\rangle = n_x|n_x\rangle,$$ (4.3.20)

correspond to the oscillator energies $E_x = \hbar\omega(n_x + \frac{1}{2})$.

The usual Hilbert space $\mathcal{H}$ of the harmonic oscillator is the completion of the space Ψ consisting of the finite superpositions of the number states. Of course, $\Psi \subset \mathcal{H}$. The relations (4.3.13) indicate that the kets $|x\rangle$ and $|p_x\rangle$ do not belong to the Hilbert space $\mathcal{H}$, $\delta(0)$ being $+\infty$. Here we invoke the concept of the rigged Hilbert space or a Gelfand triplet [Böhm (1978), p. 33]. A useful notion are the elements $|\psi\rangle$ of the Hilbert space $\mathcal{H}$ whose position and momentum representations $\langle x|\psi\rangle$, $\langle p_x|\psi\rangle$ are elements of the space S of the test functions as introduced in the theory of generalized functions [Gelfand and Shilov (1964)] to define tempered distributions. The role of the number-state representation $\langle n|\psi\rangle$ can be seen in the definition of the physically preparable state [Vaccaro and Pegg (1990a)]. In a more mathematical setting, the completion of Ψ with respect to a τ_Φ-topology is the countably normed space Φ consisting of these states. As a motivation let us remember the space $S^\times$ consisting of the tempered distributions which are continuous linear functionals defined on S. We expect that the kets $|\psi\rangle^\times$, whose position and momentum representations $\langle x|\psi\rangle^\times$, $\langle p_x|\psi\rangle^\times$ belong to the space $S^\times$, are elements of a space $\Phi^\times$ of all τ_Φ-continuous linear functionals. The three spaces $\Phi \subset \mathcal{H} \subset \Phi^\times$ are called a Gelfand triplet. Any formalism may exploit the picture of the space $\Phi^\times$ as consisting of formal expansions in the Fock states. In addition to this, such a formalism should express these "improper" states through limiting procedures of the elements from $\mathcal{H}$, Φ, or even Ψ. In so a "rigged" formalism, sometimes a formal expansion in the Fock states is given a meaning of a convergent series. Let us remark that all the bras (dual vectors) $\langle\psi|$ need not be interpreted as the elements $^\times\langle\psi|$ of dual space of the continuous linear functionals (cf. [Böhm (1978)]). In our opinion the Hilbert space exists in two identical copies according to Dirac's picture and therefore also the Gelfand triplet exists in this bra and ket copies.

The famous quantum phase problem dates back to Dirac (1925, 1926a, b, c, 1927), who defined the phase operator implicitly by the relation

$$\hat{a} = \exp(i\hat{\varphi})\hat{n}^{\frac{1}{2}}, \quad \hat{n} = \hat{a}^\dagger\hat{a}, \tag{4.3.21}$$

where $\hat{a}$ is the annihilation operator and $\hat{n}$ is the number operator. Seeking to define the phase operator explicitly as

$$\hat{\varphi} = -i\ln\hat{u}, \tag{4.3.22}$$

where for the time being $\hat{u}$ is defined implicitly, so that

$$\hat{a} = \hat{u}\hat{n}^{\frac{1}{2}}, \tag{4.3.23}$$

we encounter several problems. These problems were not disregarded in the early criticism of this definition [London (1926, 1927), Louisell (1963), Susskind and Glogower (1964)].

Firstly the operator $\hat{u}$ is not defined by (4.3.23) uniquely, because it is not possible to divide both the sides of (4.3.23) by $\hat{n}^{\frac{1}{2}}$ carelessly. The operator $\hat{n}^{-\frac{1}{2}}$ does not exist. The pseudo-inverse of $\hat{n}$, $\hat{n}^-$, is given as $\hat{n}^- = \sum_{n=1}^\infty n^{-1}|n\rangle\langle n|$. We can multiply by $(\hat{n}^-)^{\frac{1}{2}}$. An analysis shows that this ambiguity affects only the matrix elements $\langle n|\hat{u}|0\rangle$, $n = 0, 1, 2, \dots$. Usually it is assumed that $\langle n|\hat{u}|0\rangle = 0$. Then

$$\hat{u} = \sum_{n=0}^\infty |n\rangle\langle n+1|. \tag{4.3.24}$$

Neglecting the difficulty, we, in fact, derive a correct commutation relation [Heitler (1954)],

$$[\exp(i\hat{\varphi}), \hat{n}] = \exp(i\hat{\varphi}), \tag{4.3.25}$$

which can be obeyed by a large family of operators [Lerner (1968)]. Expanding $\exp(i\hat{\varphi})$ on both sides of equation (4.3.25) and equating the terms of the same order, we obtain particularly for the lowest-order terms that $\hat{n}$ and $\hat{\varphi}$ are canonically conjugate variables [Lynch (1995)]

$$[\hat{n}, \hat{\varphi}] = i\hat{1}. \tag{4.3.26}$$

This at best imprecise equation leads to a number–phase uncertainty relation

$$\langle(\Delta\hat{n})^2\rangle\langle(\Delta\hat{\varphi})^2\rangle \geq \frac{1}{4}, \tag{4.3.27}$$

where

$$\begin{aligned}
\Delta\hat{n} &= \hat{n} - \langle\hat{n}\rangle\hat{1}, \\
\Delta\hat{\varphi} &= \hat{\varphi} - \langle\hat{\varphi}\rangle\hat{1}.
\end{aligned} \tag{4.3.28}$$

The phase operator $\hat{\varphi}$ is suspected to assume values from $-\infty$ to ∞. Concentrating only on the exact validity of equation (4.3.26), we take matrix elements of both sides of this equation in the number state basis $|n\rangle$ [Louisell (1963)],

$$\langle n|\hat{n}\hat{\varphi} - \hat{\varphi}\hat{n}|n'\rangle = i\langle n|n'\rangle. \tag{4.3.29}$$

Since the number states are eigenstates of the number operator $\hat{n}$, the relation (4.3.29) can be simplified

$$(n - n')\langle n|\hat{\varphi}|n'\rangle = i\delta_{nn'}. \tag{4.3.30}$$

The diagonal matrix elements have contradictory properties, $0 = i$ for $n = n'$. From these difficulties a lesson follows that we must begin with the properties of the operator $\hat{u}$, which has not the form $\exp(i\hat{\varphi})$, but which we cherish to denote as $\widehat{\exp}(i\varphi)$.

Let us remark that in the Pegg–Barnett formalism [Pegg and Barnett (1988)] (see section 4.7) the unsatisfactory relation (4.3.30) is replaced by the relation

$$(n - n')\langle n|\hat{\varphi}_{\theta_0 s}|n'\rangle = i\frac{(n - n' \pm 0)\pi}{s + 1}$$
$$\times \ \csc\left(\frac{(n - n' \pm 0)\pi}{s + 1}\right)\left\{\delta_{nn'} - \exp\left[i(n - n')\theta_{s+\frac{1}{2}}\right]\right\}, \tag{4.3.31}$$

where θ_0, $\theta_{s+\frac{1}{2}}$ are given in (4.7.36) and $(\pm 0)\csc(\pm 0) = 1$.

A formula for the general solution of equation (4.3.23) will be given below. Let us note that there does not exist a unitary solution to this problem, i. e., $\hat{u}^{\dagger}\hat{u} \neq \hat{1}$, although

$$\hat{u}\hat{u}^{\dagger} = \hat{1}. \tag{4.3.32}$$

To arrive at a good definition of the operator $\hat{u}$, it suffices to modify relation (4.3.23) to the form [Susskind and Glogower (1964)]

$$\hat{a} = (\hat{n} + \hat{1})^{\frac{1}{2}}\hat{u}. \tag{4.3.33}$$

The solution is just relation (4.3.24). Aiming to eliminate the square root from relations (4.3.23), (4.3.33), we derive the properties

$$\hat{n} + \hat{1} = \hat{u}\hat{n}\hat{u}^\dagger, \tag{4.3.34}$$

$$\hat{n} = \hat{u}^\dagger(\hat{n} + \hat{1})\hat{u} \tag{4.3.35}$$

characteristic of the operator $\hat{u}$.

Equation (4.3.23) has the general solution [Susskind and Glogower (1964)]

$$\hat{u}_{|\psi\rangle} = \sum_{n=0}^{\infty} |n\rangle\langle n+1| + |\psi\rangle\langle 0|, \tag{4.3.36}$$

where $|\psi\rangle$ is an arbitrary element of the underlying Hilbert space. This operator can be made unitary on a smaller Hilbert space by restricting the sum to $n = 0, ..., s - 1$ and putting $|\psi\rangle = \exp[i(s + 1)\theta_0]|s\rangle$, where s is the maximum number of photons allowed and θ_0 is an arbitrary real parameter [Barnett and Pegg (1989)], namely,

$$\hat{U}_{\theta_0, s} = \sum_{n=0}^{s-1} |n\rangle\langle n+1| + \exp[i(s + 1)\theta_0]|s\rangle\langle 0|. \tag{4.3.37}$$

The eigenkets of this unitary operator are s-phase states [Lerner, Huang and Walters (1970), Loudon (1973)]. The eigenvalues of this unitary operator are $\exp(i\theta_k)$, $k = 0, ..., s$, $\theta_0 < ... < \theta_s$, defined implicitly by the property $[\exp(i\theta_k)]^{s+1} = \exp[i(s+1)\theta_0]$ and by the property $\theta_s - \theta_0 < 2\pi$, θ_0 can be rather arbitrary. In this formalism the dimension of the Hilbert space is allowed to tend to infinity only after s-phase statistics have been calculated [Barnett and Pegg (1992)].

Another approach of constructing the phase operator is based on the enlarged Hilbert space [Newton (1980), Barnett and Pegg (1986), Lukš and Peřinová (1991)]. The property

$$\hat{u}^\dagger\hat{u} = \hat{1} - |0\rangle\langle 0| \tag{4.3.38}$$

suggests that the operator $\hat{u}$ is only one-sided unitary, because a number state with minimum number of photons (i. e., the vacuum state) exists. Therefore, it is convenient to consider number states with negative number of quanta as discussed in [Barnett and Pegg (1986)]. It is possible to compare the phase with the rotation angle and the photon number with the angular momentum in the bead-on-wire model, which can assume values of both signs. Considering the original mode interacting with an appropriate apparatus, we can interpret the states with negative number of photons as the vacuum state of the mode × the non-zero apparatus states, whereas we can view the states with nonnegative number of photons as the customary states of the mode × the zero apparatus state [Shapiro, Shepard and Wong (1990)], or formally $|N\rangle = |n = N\rangle|0\rangle_A$ for $N \geq 0$ and $|N\rangle = |0\rangle|n_A = -N\rangle_A$ for $N < 0$. In all cases the corresponding classical quantities are canonically conjugated. In contrast, quantum mechanics in a finite-dimensional Hilbert space suggested solutions [Santhanam

(1977a, b), Goldhirsch (1980)]. Similarly, in the usual theory of angular momentum, the irreducible representation spaces of the group SU(2) admit an introduction of the phase operator [Lévy-Leblond (1973), Santhanam (1976, 1977a)] and it has been interesting to investigate the harmonic oscillator group as the Inönü–Wigner contraction [Inönü and Wigner (1953)] of the SU(2) angular-momentum group [Vourdas (1990), Ellinas (1991a)].

The introduction of number states with negative number of photons leads to the definition of the phase operator

$$\hat{\phi} = -i \ln \hat{U}, \tag{4.3.39}$$

with a unitary operator

$$\hat{U} = \sum_{N=-\infty}^{\infty} |N\rangle\langle N+1|. \tag{4.3.40}$$

We will use also the notation $\hat{U} = \widehat{\exp}(i\phi)$ in other sections. Its eigenkets are

$$|\phi\rangle_e = \frac{1}{\sqrt{2\pi}} \sum_{N=-\infty}^{\infty} \exp(iN\phi)|N\rangle, \tag{4.3.41}$$

with the property

$$|\phi + 2\pi\rangle_e = |\phi\rangle_e, \tag{4.3.42}$$

It is known that these eigenkets have not finite norms. The resolution of the identity is

$$\hat{1}_e = \int_{\theta_0}^{\theta_0+2\pi} |\phi\rangle_e\ {}_e\langle\phi|\, d\phi \tag{4.3.43}$$

and the operator $\hat{U}$ can be expressed in the integral form

$$\hat{U} = \int_{\theta_0}^{\theta_0+2\pi} \exp(i\phi)|\phi\rangle_e\ {}_e\langle\phi|\, d\phi. \tag{4.3.44}$$

In the enlarged Hilbert space $\mathcal{H}_e$ it holds that

$$\exp(-i\tau\hat{N})|\phi\rangle_e = |\phi - \tau\rangle_e, \tag{4.3.45}$$

where

$$\hat{N} = \sum_{N=-\infty}^{\infty} N|N\rangle\langle N|. \tag{4.3.46}$$

In analogy to relations (4.3.34), (4.3.35) it is valid that

$$\hat{N} + \hat{1}_e = \hat{U}\hat{N}\hat{U}^\dagger, \tag{4.3.47}$$

$$\hat{N} = \hat{U}^\dagger(\hat{N} + \hat{1}_e)\hat{U}. \tag{4.3.48}$$

The orthogonality property reads

$$_e\langle\phi|\phi'\rangle_e = \sum_{k=-\infty}^{\infty} \delta(\phi - \phi' - 2k\pi). \qquad (4.3.49)$$

When we restrict the scalar product to $\phi, \phi' \in [\theta_0, \theta_0 + 2\pi)$ (a window), the sum in (4.3.49) can be reduced to the term $\delta(\phi - \phi')$. This simplification does not mean a complete analogy to the quantum mechanics, where for the momentum state $|p\rangle$ it holds that

$$\langle p|p'\rangle = \delta(p - p'), \qquad (4.3.50)$$

but the quantities p, p' are not restricted to finite intervals. The momentum state $|p\rangle$ is expressed by an analogue of the relation (4.3.41)

$$|p\rangle = \frac{1}{\sqrt{2\pi}} \int_{-\infty}^{\infty} \exp(ixp)|x\rangle\, dx. \qquad (4.3.51)$$

Although the drawback of the operator $\hat{u}$ not being unitary no more plagues $\hat{U}$, the definition (4.3.39) still needs a comment. We do not promise that the commutator relation (4.3.26) will be recovered as $[\hat{N}, \hat{\phi}] = i\hat{1}_e$, but compare an analysis of Loss and Mullen (1991). A purely mathematical difficulty consists in the fact that the natural logarithm of a complex argument is multiple valued and that a single-valued branch of this function must be chosen. As the single-valued branches of this function differ from one another by $2\pi i$, the classical phases differ from one another by 2π. In the case of the eigenvalues of unitary phase operator we consider the set of all complex units, for which there exists no single-valued branch of the logarithm. We try a slightly more general definition than the principal branch of the logarithm function [Lukš and Peřinová (1993a)]. We choose a point on the unit circle at which the single-valued branch $\ln_\kappa$ of the logarithm is allowed to have the discontinuity (jump) $i(\kappa + 2\pi)$, $i\kappa$. Any point on the unit circle can be denoted by ϕ modulo 2π with some real coordinate ϕ and accordingly, the point of jump is κ modulo 2π. With respect to this, the Hermitian phase operator is denoted as $\hat{\phi}_{\theta_0}$ for $\kappa = \theta_0$. The unitary Pegg–Barnett operator $\hat{U}_{\theta_0,s}$ defined by equation (4.3.37) leads to the Hermitian phase operator

$$\hat{\phi}_{\kappa,\theta_0,s} = -i\ln_\kappa(\hat{U}_{\theta_0,s}), \qquad (4.3.52)$$

where $\kappa = \theta_0 - \frac{\pi}{s+1}$.

To accomplish the definition ensuing from the substitution from (4.3.44) into (4.3.39), we must learn the properties of the 2π-periodic functions

$$\overline{\mathrm{id}}_\kappa(\phi) = -i\ln_\kappa[\exp(i\phi)], \quad \kappa \text{ real}. \qquad (4.3.53)$$

Such a function is undefined for $\phi = \kappa \pmod{2\pi}$,

$$\overline{\mathrm{id}}_\kappa(\phi) = \phi, \quad \phi \in (\kappa, \kappa + 2\pi). \qquad (4.3.54)$$

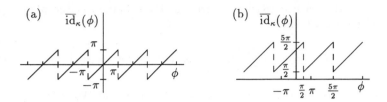

Figure 4.1: Saw-tooth (ratchet-like) function $\overline{\mathrm{id}}_\kappa(\phi) = \phi$; (a) $\overline{\mathrm{id}}_{-\pi}(\phi) = \phi$, $\phi \in (-\pi, \pi)$; (b) $\overline{\mathrm{id}}_{\frac{\pi}{2}}(\phi) = \phi$, $\phi \in (\frac{\pi}{2}, \frac{5\pi}{2})$.

According to the shape of its graph, it is called a saw-tooth or a ratchet-like function (cf. figure 4.1a, b).

Among other important properties let us name
(i)

$$\overline{\mathrm{id}}_\kappa(\phi) = \kappa + \pi + i \left\{ \ln[1 - \exp[-i(\phi - \kappa)]] - \ln[1 - \exp[i(\phi - \kappa)]] \right\}, \qquad (4.3.55)$$

(ii) the Fourier decomposition reads as

$$\overline{\mathrm{id}}_\kappa(\phi) = \kappa + \pi + i \sum_{k=1}^{\infty} \frac{1}{k} \exp[-ik(\phi - \kappa)] + i \sum_{k=-\infty}^{-1} \frac{1}{k} \exp[-ik(\phi - \kappa)]. \qquad (4.3.56)$$

According to the properties of Fourier series we may continue the function with

$$\overline{\mathrm{id}}_\kappa(\phi) = \kappa + \pi \ \text{for} \ \phi \equiv \kappa \ (\mathrm{mod}\ 2\pi). \qquad (4.3.57)$$

The right-hand side of (4.3.56) can be obtained also from the formula (4.3.55) invoking a formal use of Maclaurin series for the logarithm function. As a result of the above substitution, we obtain for $\kappa = \theta_0$,

$$\hat{\phi}_{\theta_0} = \int_{\theta_0}^{\theta_0 + 2\pi} \overline{\mathrm{id}}_{\theta_0}(\phi)|\phi\rangle_e\,{}_e\langle\phi|\,d\phi = \int_{\theta_0}^{\theta_0 + 2\pi} \phi|\phi\rangle_e\,{}_e\langle\phi|\,d\phi, \qquad (4.3.58)$$

and for $\kappa = \chi$

$$\hat{\phi}_\chi = \int_{\theta_0}^{\theta_0 + 2\pi} \overline{\mathrm{id}}_\chi(\phi)|\phi\rangle_e\,{}_e\langle\phi|\,d\phi = \overline{\mathrm{id}}_\chi(\hat{\phi}_{\theta_0}). \qquad (4.3.59)$$

Substituting from (4.3.56) into (4.3.58), we arrive at the operator decomposition

$$\begin{aligned}
\hat{\phi}_{\theta_0} &= (\theta_0 + \pi)\hat{1}_e + i \sum_{k=1}^{\infty} \frac{1}{k} \exp(-ik\theta_0)\hat{U}^k + i \sum_{k=-\infty}^{-1} \frac{1}{k} \exp(-ik\theta_0)\hat{U}^k \\
&= (\theta_0 + \pi)\hat{1}_e + i \sum_{l=1}^{\infty} \frac{1}{l} \left[\exp(-il\theta_0)\hat{U}^l - \exp(il\theta_0)\hat{U}^{-l} \right].
\end{aligned} \qquad (4.3.60)$$

Returning to the original space, we arrive at the definition of the phase operator [Lukš and Peřinová (1991)]

$$\hat{\varphi}_{\theta_0} = (\theta_0 + \pi)\hat{1} + i \sum_{k=1}^{\infty} \frac{1}{k} \left[\exp(-ik\theta_0)\hat{u}^k - \exp(ik\theta_0)\hat{u}^{\dagger k} \right], \qquad (4.3.61)$$

which is the Garrison–Wong phase operator [Garrison and Wong (1970)]. In fact, here we have introduced an arbitrary θ_0 [Damaskinsky and Yarunin (1978)]. Galindo (1984a, b) has suggested that the commutation relation (4.3.26) can be fulfilled for the phase operator $\hat{\phi}_{-\pi}$ when a suitable mathematical assumption is made. The phase operator should act only on an orthogonal complement of $\{c|\theta_0 = -\pi\rangle\}$, which is, of course, a subspace of $\Phi^\times$. Should $\{c|\theta_0 = -\pi\rangle\}^\perp$ be a subspace of Φ? Indications of the use of (4.3.55) can be found in [Popov and Yarunin (1973), Bergou and Englert (1991)]. In the extended Hilbert space, $\hat{U} = \exp(i\hat{\phi}_{\theta_0})$, any θ_0, whereas $\hat{u} \neq \exp(i\hat{\varphi}_{\theta_0})$, although such inequalities cannot be formulated easily.

The diagonal and off-diagonal number-state basis matrix elements of the phase operator (4.3.61) read

$$
\begin{aligned}
\langle n|\hat{\varphi}_{\theta_0}|n\rangle &= \theta_0 + \pi \\
\langle n'|\hat{\varphi}_{\theta_0}|n\rangle &= \frac{\exp[i(n'-n)\theta_0]}{i(n'-n)}, \quad n' \neq n.
\end{aligned}
\qquad (4.3.62)
$$

Unfortunately, it holds that $[\hat{\varphi}_{\theta_0}, \hat{\varphi}_\chi] \neq \hat{0}$ for $\theta_0 \neq \chi + 2k\pi$, k is a positive integer. This entails that the orthogonal system of the decomposition of $\hat{\varphi}_{\theta_0}$ differs from that of $\hat{\varphi}_\chi$. A phase operator indicated by Dirac (1927) and accepted by field theorists [Heitler (1954), Akhiezer and Berestetsky (1965)] could, at best, assume the form of a Garrison–Wong phase operator [Garrison and Wong (1970)].

The difficulties with the operator polar decomposition can be avoided when the phase operator is constructed using the resolution of the identity in terms of coherent states, which leads to the following definition

$$\hat{\phi}_{-\pi}^{(A)} = \frac{1}{2\pi i} \int \ln\left(\frac{\alpha}{\alpha^*}\right) |\alpha\rangle\langle\alpha| \, d^2\alpha \qquad (4.3.63)$$

given by Turski (1972).

In the system of harmonic oscillator there exists no phase operator. To understand this negative assertion more deeply, we observe that such a phase operator should provide an orthogonal system of eigenvectors (eigenkets) $|\varphi\rangle_i$ with the property

$$\exp(i\tau\hat{n})|\varphi\rangle_i = |\varphi + \tau\rangle_i. \qquad (4.3.64)$$

They may not be elements of the Hilbert space, but their scalar products with suitable elements of the Hilbert space exist. It is obvious that only such a phase operator can be considered to be well behaved. To be more persuasive, we mention further properties which can be expected by analogy with angular (rotation angle) states. The 2π-periodicity

$$|\varphi + 2\pi\rangle_i = |\varphi\rangle_i \qquad (4.3.65)$$

is assumed in the relation (4.3.64) and, perhaps, an open interval $(\theta_0, \theta_0 + 2\pi)$ of measured values could complete the picture. Let us denote

$$c_n^{(i)} = \langle n|\theta_0 + \pi\rangle_i \qquad (1.3.66)$$

and assume that

$$\lim_{n\to\infty} |c_n^{(i)}| = \frac{1}{\sqrt{2\pi}}. \qquad (4.3.67)$$

Let us introduce the continuous superposition

$$|\psi\rangle_i = \sqrt{\frac{3}{2\pi}} \int_{-\frac{\pi}{3}}^{\frac{\pi}{3}} |\theta_0 + \pi + \tau\rangle_i \, d\tau \qquad (4.3.68)$$

$$= \sqrt{\frac{6}{\pi}} \sum_{n=0}^{\infty} \frac{\sin\left(n\frac{\pi}{3}\right)}{n} c_n^{(i)}|n\rangle. \qquad (4.3.69)$$

From the assumption of orthogonality it follows that

$$_i\langle\psi| \exp(i\tau\hat{n})|\theta_0 + \pi\rangle_i = 0, \quad \tau \in \left(-\pi, -\frac{\pi}{3}\right) \cup \left(\frac{\pi}{3}, \pi\right). \qquad (4.3.70)$$

The requirement (4.3.70) can be rewritten as

$$\sqrt{\frac{6}{\pi}} \sum_{n=0}^{\infty} \frac{\sin\left(n\frac{\pi}{3}\right)}{n} |c_n^{(i)}|^2 \exp(i\tau n) = 0, \qquad (4.3.71)$$

but this cannot hold for all τ from a set of positive Lebesgue measure (e. g., [Rudin (1970)]). The lack of this orthogonal system does not preclude the possibility of constructing the phase operators based on the approximately orthogonal vectors

$$|\varphi\rangle = \frac{1}{\sqrt{2\pi}} \sum_{n=0}^{\infty} \exp(in\varphi)|n\rangle, \qquad (4.3.72)$$

which also have the property (4.3.64). One of the consequences of the nonexistence of the ideal orthogonal phase system is the incompatibility of cosine and sine operators. In spite of this, the discovery of quantum cosine and sine operators was an event [Nieto (1993)].

From the properties (4.3.32) and (4.3.38) it can be seen that the requirement of working with a unitary phase operator can be fulfilled algebraically by adopting a sort of the antinormal ordering of the operators $\hat{u}$, $\hat{u}^\dagger$ [Lukš and Peřinová (1991, 1993b), Brif and Ben-Aryeh (1994a), Vaccaro and Ben-Aryeh (1995)]. Of course, this leads to a classical–quantum correspondence. To each phase function $M(\varphi)$ there corresponds a phase operator

$$\hat{M} = \int_{\theta_0}^{\theta_0+2\pi} M(\varphi)|\varphi\rangle\langle\varphi|\,d\varphi, \tag{4.3.73}$$

which does not depend on θ_0 provided that $M(\varphi)$ is 2π-periodic. An example and an apparent exception is the phase operator according to (4.3.61), where the θ_0-dependence emerges, because either the function $M(\varphi) = \varphi$ is nonperiodic or the function $M(\varphi) = \overline{\mathrm{id}}_{\theta_0}(\varphi)$ is 2π-periodic, but θ_0-dependent. The Susskind–Glogower cosine and sine operators [Susskind and Glogower (1964)], which can be expressed as follows

$$\begin{aligned}
\widehat{\cos}\,\varphi &= \int_{\theta_0}^{\theta_0+2\pi} \cos\,\varphi|\varphi\rangle\langle\varphi|\,d\varphi, \\
\widehat{\sin}\,\varphi &= \int_{\theta_0}^{\theta_0+2\pi} \sin\,\varphi|\varphi\rangle\langle\varphi|\,d\varphi,
\end{aligned} \tag{4.3.74}$$

may serve as another example. The analogy with the antinormal ordering of the operators $\hat{a}$, $\hat{a}^\dagger$ is treated in section 4.5. The operators $\hat{M}$ can be called Toeplitz operators.

Rosenblum (1962) has been concerned with the spectral theory of the Toeplitz operators and has assumed that the functions $M(\varphi)$ satisfy the following two hypotheses: (i) $M(\varphi)$ is real, bounded below, and absolutely integrable on the circle, but it is not identically constant. (ii) For each real λ, the points φ, $M(\varphi) \leq \lambda$, form an arc of the circle up to some points of the measure zero. These hypotheses are satisfied by the Toeplitz operators $\widehat{\cos}\,\varphi$, $\widehat{\sin}\,\varphi$, $\hat{\varphi}_{\theta_0}$.

The quantum average of the operator (4.3.73)

$$\langle\hat{M}\rangle = \mathrm{Tr}\{\hat{\rho}\hat{M}\} = \int_{\theta_0}^{\theta_0+2\pi} \mathrm{Tr}\{\langle\varphi|\hat{\rho}|\varphi\rangle\}M(\varphi)\,d\varphi = \langle M(\varphi)\rangle_{\mathcal{A}}^{CS}, \tag{4.3.75}$$

where the superscript CS and the subscript "$\mathcal{A}$" indicate the antinormal ordering of the operators $\hat{u}$, $\hat{u}^\dagger$, can be reexpressed as

$$\langle\hat{M}\rangle = \mathrm{Tr}\{\hat{\hat{M}}\hat{\rho}\}, \tag{4.3.76}$$

where the superoperator $\hat{\hat{M}}$ is in a one-to-one correspondence with the operator $\hat{M}$. The system $\{|\varphi\rangle\}$ has the property similar to that of the orthogonal systems

$$[\hat{n}, \hat{M}] = i \int_{\theta_0}^{\theta_0+2\pi} \frac{\partial}{\partial\varphi}M(\varphi)|\varphi\rangle\langle\varphi|\,d\varphi. \tag{4.3.77}$$

Ban (1995a, b) has shown that there exists a unitary superoperator $\hat{\hat{u}}$ which displaces the lower and upper parallels to the diagonal of a matrix in the Fock basis from bottom up and from left to right, respectively. He has proved that the action of any analytic function $M(\hat{u})$ of the phase unitary superoperator on the identity operator leads to a Toeplitz operator. He has introduced a mapping from an operator to a superoperator. By using the mapping he has shown that the superoperator $\hat{M}$ obtained from a Toeplitz phase operator $\hat{M}$ is a function of the quantum phase operator $M(\hat{u})$.

Let us bear in mind that the Susskind–Glogower cosine and sine operators have the simplest possible matrix elements. According to [Paul (1974)] we can consider the cosine and sine operators

$$\hat{C}_{1P} = \frac{1}{\pi} \int \frac{\mathrm{Re}\,\alpha}{|\alpha|} \hat{\Delta}(\alpha, -1)\, d^2\alpha,$$

$$\hat{S}_{1P} = \frac{1}{\pi} \int \frac{\mathrm{Im}\,\alpha}{|\alpha|} \hat{\Delta}(\alpha, -1)\, d^2\alpha, \tag{4.3.78}$$

where $\hat{\Delta}(\alpha, -1)$ given in (3.4.15) is related to the antinormal ordering [Agarwal and Wolf (1970a, b)] Let us remark that Yao (1987) has modified the relation (4.3.78) to obtain the operators $\widehat{\cos}\,\varphi$, $\widehat{\sin}\,\varphi$. As expected, $\frac{1}{|\alpha|}$ has been replaced by a positive function of $|\alpha|$.

In analogy to the definition (4.3.78) we can consider the Weyl cosine and sine operators

$$\hat{C}_{1W} = \frac{1}{\pi} \int \frac{\mathrm{Re}\,\alpha}{|\alpha|} \hat{\Delta}(\alpha, 0)\, d^2\alpha,$$

$$\hat{S}_{1W} = \frac{1}{\pi} \int \frac{\mathrm{Im}\,\alpha}{|\alpha|} \hat{\Delta}(\alpha, 0)\, d^2\alpha, \tag{4.3.79}$$

where $\hat{\Delta}(\alpha, 0)$ defined in (3.4.14) is related to the symmetrical (Weyl) ordering [Lukš and Peřinová (1991)]. We note that the idea of the symmetrical ordering leads to the Wigner function for a phase state, namely, $\frac{1}{\pi}\langle\varphi|\hat{\Delta}(\alpha, 0)|\varphi\rangle$ [Herzog, Paul and Richter (1993)]. Another kind of symmetrization is encountered with the Lerner–Lynch cosine and sine operators [Lerner (1968), Lynch (1986)]

$$\widehat{\cos}_{LL}\,\varphi = \frac{1}{2}(\hat{u}_{LL} + \hat{u}_{LL}^\dagger), \quad \widehat{\sin}_{LL}\,\varphi = \frac{1}{2i}(\hat{u}_{LL} - \hat{u}_{LL}^\dagger), \tag{4.3.80}$$

where

$$\hat{u}_{LL} = \frac{1}{2}\left[\left(\hat{n} + \frac{1}{2}\hat{1}\right)^{-\frac{1}{2}} \hat{a} + \hat{a}\left(\hat{n} + \frac{1}{2}\hat{1}\right)^{-\frac{1}{2}}\right]. \tag{4.3.81}$$

This kind of explicit symmetrization can lead to many cosine and sine operators, because even the one-sided unitarity of exponential phase operators is not pursued. The remaining interesting question is whether the spectra of cosine and sine operators fill the interval $[-1, 1]$ [Lerner, Huang and Walters (1970), Ifantis (1971)].

It holds that

$$\langle n | \widehat{\cos} \, \varphi | n + 1 \rangle = \frac{1}{2}, \tag{4.3.82}$$

and according to Paul (1974)

$$\langle n | \hat{C}_{1\mathrm{P}} | n + 1 \rangle = \frac{1}{2\sqrt{n!(n+1)!}} \Gamma \left(n + \frac{3}{2} \right). \tag{4.3.83}$$

Because

$$\langle n | \hat{C}_{1\mathrm{W}} | n + 1 \rangle = \mathrm{Tr} \left\{ \hat{C}_{1\mathrm{W}} | n + 1 \rangle \langle n | \right\} \tag{4.3.84}$$

and

$$|n\rangle\langle m| = \frac{1}{\pi} \int l_{mn}(\alpha, 0) \hat{\Delta}(\alpha, 0) \, d^2\alpha, \tag{4.3.85}$$

where

$$\begin{aligned}
l_{nm}(\alpha, 0) &= (-1)^n \sqrt{\frac{n!}{m!}} \, (2\alpha)^{m-n} L_n^{m-n}(4|\alpha|^2)[2\exp(-2|\alpha|^2)] \\
&\quad \text{for } m \geq n, \\
l_{nm}(\alpha, 0) &= l_{mn}(\alpha^*, 0) \ \text{for } m \leq n,
\end{aligned} \tag{4.3.86}$$

and from the mapping theorem [Agarwal and Wolf (1970a, b)] it follows that

$$\begin{aligned}
\langle n | \hat{C}_{1\mathrm{W}} | n + 1 \rangle &= \frac{1}{\pi} \int \frac{\mathrm{Re}\,\alpha}{|\alpha|} l_{n+1,n}(\alpha, 0) \, d^2\alpha \\
&= \frac{1}{\pi} \int \frac{\mathrm{Re}\,\alpha}{|\alpha|} l_{n,n+1}(\alpha^*, 0) \, d^2\alpha \\
&= (-1)^n \sqrt{\frac{n+1}{2}} \sum_{k=0}^{n} \binom{n}{k} (-2)^k \frac{\Gamma\left(k + \frac{3}{2}\right)}{(k+1)!}.
\end{aligned} \tag{4.3.87}$$

The Laguerre polynomials in (4.3.86) are defined in (3.4.39). The summation in (4.3.87) can be carried out [Garraway and Knight (1992, 1993)]. Finally we find that

$$\langle n | \widehat{\cos}_{\mathrm{LL}} \, \varphi | n + 1 \rangle = \frac{1}{4} \left(\sqrt{\frac{n+1}{n+\frac{1}{2}}} + \sqrt{\frac{n+1}{n+\frac{3}{2}}} \right). \tag{4.3.88}$$

It holds that

$$\begin{aligned}
&\langle n | \hat{C}_{1\mathrm{W}} | n + 1 \rangle \geq \langle n | \widehat{\cos}_{\mathrm{LL}} \, \varphi | n + 1 \rangle \geq \langle n | \widehat{\cos} \, \varphi | n + 1 \rangle \\
&\geq \langle n | \hat{C}_{1\mathrm{P}} | n + 1 \rangle \ \text{ for } n \text{ even},
\end{aligned} \tag{4.3.89}$$

and

$$\begin{aligned}
&\langle n | \widehat{\cos}_{\mathrm{LL}} \, \varphi | n + 1 \rangle \geq \langle n | \widehat{\cos} \, \varphi | n + 1 \rangle \geq \langle n | \hat{C}_{1\mathrm{P}} | n + 1 \rangle \\
&\geq \langle n | \hat{C}_{1\mathrm{W}} | n + 1 \rangle \ \text{ for } n \text{ odd}.
\end{aligned} \tag{4.3.90}$$

Considering the trial states

$$|t_n\rangle = \frac{1}{\sqrt{2}}(|n\rangle + |n+1\rangle), \quad n = 0, 1, ..., \infty, \qquad (4.3.91)$$

with

$$\langle \widehat{\sin} \varphi \rangle = \langle \hat{S}_{1P} \rangle = \langle \hat{S}_{1W} \rangle = \langle \widehat{\sin}_{LL} \varphi \rangle = 0, \qquad (4.3.92)$$

where the brackets $\langle ... \rangle$ denote the averaging in the state $|t_n\rangle$, we obtain

$$\langle \widehat{\cos} \varphi \rangle = \langle n | \widehat{\cos} \varphi | n+1 \rangle, \quad \text{etc.} \qquad (4.3.93)$$

Relations (4.3.89), (4.3.90) can be verified in figure 4.2.

Superpositions of three number states are sufficient for the number–phase uncertainty relation (4.3.27) to be disproved using the Pegg–Barnett formalism [Lindner, Reiss, Wassiliadis and Freese (1996)]. Whereas the operators $\widehat{\cos} \varphi = \frac{1}{2}(\hat{u} + \hat{u}^\dagger)$ and $\widehat{\cos}_{LL} \varphi$ have originated due to a rescaling of the photon annihilation and creation operators with the photon-number operator, much more convenient way would be the rescaling of an appropriate quadrature operator with the averaged photon number. In this manner the "measured phase operators" have been devised [Barnett and Pegg (1986)],

$$\begin{aligned} \hat{C}_M &= k(\hat{a} + \hat{a}^\dagger), \\ \hat{S}_M &= -ik(\hat{a} - \hat{a}^\dagger), \end{aligned} \qquad (4.3.94)$$

where

$$k = \frac{1}{2\sqrt{\langle \hat{n} \rangle + \frac{1}{2}}}. \qquad (4.3.95)$$

Although these cosine and sine operators never became a formalism, cf. however [Lynch (1986), Vaccaro and Pegg (1994a)], there are several papers studying special states in this framework. An operationally determined preferred phase (set equal to zero) enables one to measure small phase fluctuations also using the operator $\hat{S}_M$ [Vaccaro and Pegg (1994a)].

The examples of cosine and sine operators differ from one another regarding the issue of the simultaneous measurement. From the viewpoint of the theory, any pair of these operators is incompatible. However, the Paul operators of cosine and sine are known to be simultaneously measurable in an enlarged system, i. e., by the homodyne detection technique. A similar property could be attributed to the Susskind–Glogower cosine and sine operators, but the questions of an enlarged system and the simultaneous measurability are not so clear.

The affections for cosine and sine of the phase were questioned by Lévy-Leblond (1976). He pointed out the analogy between the photon annihilation operator $\hat{a}$ and the Susskind–Glogower operator $\hat{u}$. Introducing the operators

$$\Delta \hat{u} = \hat{u} - \langle \hat{u} \rangle, \quad \Delta \hat{u}^\dagger = \hat{u}^\dagger - \langle \hat{u}^\dagger \rangle, \qquad (4.3.96)$$

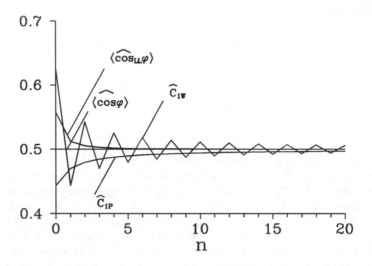

Figure 4.2: Average values $\langle \widehat{\cos} \, \varphi \rangle$, $\langle \widehat{\cos}_{LL}\varphi \rangle$, $\langle \hat{C}_{1P} \rangle$, and $\langle \hat{C}_{1W} \rangle$ computed in the trial state $|t_n\rangle$.

we note that

$$\langle \Delta \hat{u}^\dagger \Delta \hat{u} \rangle = \langle \hat{u}^\dagger \hat{u} \rangle - |\langle \hat{u} \rangle|^2, \tag{4.3.97}$$

which is the measure of the dispersion proposed by Lévy-Leblond (1976). In that time the role of the quadrature operators in quantum optics was not recognized and that is why the author was allowed to ask "who is afraid of non-Hermitian operators".

The difficulty that the cosine and sine phase operators are incompatible is evident when we take into account that their diagonal representations use distinct identity resolutions (cf. (4.5.31)–(4.5.34)). In fact, a suitable choice of the normalization allowed Susskind and Glogower (1964) to treat almost the same vectors as special phase states with two-valued phase. A very similar approach to the Garrison–Wong phase operator is in order. This operator has the diagonal representation

$$\hat{\varphi}_{\theta_0} = \int_{\theta_0}^{\theta_0+2\pi} \varphi_{\theta_0} |\varphi_{\theta_0}\rangle\langle\varphi_{\theta_0}| \, d\varphi_{\theta_0}, \tag{4.3.98}$$

where [Garrison and Wong (1970), Gantsog, Miranowicz and Tanaś (1992), Tanaś, Miranowicz and Gantsog (1996)]

$$|\varphi_{\theta_0}\rangle = \sum_{n=0}^{\infty} (\langle\varphi_{\theta_0}|n\rangle)^* |n\rangle. \tag{4.3.99}$$

The scalar products are given by the formulae

$$\langle \varphi_{\theta_0} | n \rangle = \sqrt{\frac{1}{\pi} \sin \left(\frac{\varphi_{\theta_0} - \theta_0}{2} \right)} f_n(\varphi_{\theta_0}), \qquad (4.3.100)$$

where, for $n \geq 1$,

$$f_n(\varphi_{\theta_0}) = - \sum_{m=0}^{n-1} \left(1 - \frac{m}{n} \right) g_{n-m}(\varphi_{\theta_0}) f_m(\varphi_{\theta_0}), \qquad (4.3.101)$$

$$g_n(\varphi_{\theta_0}) = \frac{1}{2\pi} \int_{\theta_0}^{\theta_0 + 2\pi} \ln \left| \varphi'_{\theta_0} - \varphi_{\theta_0} \right| e^{in\varphi'_{\theta_0}} d\varphi'_{\theta_0} - \frac{1}{2n} \left(e^{in\theta_0} + e^{in\varphi_{\theta_0}} \right), \qquad (4.3.102)$$

and

$$f_0(\varphi_{\theta_0}) = \exp[-g_0(\varphi_{\theta_0})], \qquad (4.3.103)$$

$$g_0(\varphi_{\theta_0}) = -\frac{1}{2} + \frac{1}{4\pi}[(2\pi + \theta_0 - \varphi_{\theta_0}) \ln(2\pi + \theta_0 - \varphi_{\theta_0}) + (\varphi_{\theta_0} - \theta_0) \ln(\varphi_{\theta_0} - \theta_0)]. \quad (4.3.104)$$

The eigenstates of the Garrison–Wong phase operator are orthogonal in the Dirac sense

$$\langle \varphi_{\theta_0} | \varphi'_{\theta_0} \rangle = \delta(\varphi_{\theta_0} - \varphi'_{\theta_0}). \qquad (4.3.105)$$

The Garrison–Wong phase operator has been addressed also by Popov and Yarunin (1973, 1992) and Damaskinsky and Yarunin (1978). A review of the quantum phase with a special attention to the Toeplitz operators, powers of the Susskind–Glogower exponential phase operators, the Garrison–Wong phase operator, and to the exponential phase and Hermitian phase operators symmetrically ordered in the photon annihilation and creation operators has been provided [Bergou and Englert (1991)] and the asymptotic properties of these operators have been studied for large-amplitude fields. In analogy to the cosine and sine phase states we must treat the eigenstates $|\varphi_{\theta_0}\rangle$ as two-valued phase states. One of the phase values is always θ_0. The nuisance of the phase θ_0 is reflected in the representation of the Fock states in the basis $\{|\varphi_{\theta_0}\rangle\}$. The distribution of the eigenvalue of φ_{θ_0} is not uniform as expected from the phase representation of the Fock states, instead it exhibits a concentration on $\theta_0 + \pi$ [Gantsog, Miranowicz and Tanaś (1992)]. This is related also to the simple result [Hennings, Smith and Dubin (1995a)]

$$\langle \varphi_{\theta_0} \rangle = \theta_0 + \pi,$$
$$\mathrm{var}\, \varphi_{\theta_0} = \frac{\pi^2}{3} - \sum_{k=n+1}^{\infty} \frac{1}{k^2}, \qquad (4.3.106)$$

which holds for the Fock state $|n\rangle$.

In the literature treating the phase problem and also in the publications whose authors are not aware of difficulties, the uncertainty principle for action and phase angle is frequently mentioned [Hilgevoord (1996)] and so is even the uncertainty principle for time and energy. An exact counterpart of the time–energy uncertainty relation is the relation between the displacement sufficient for resolution and the uncertainty in momentum [Yu (1996)]. The remarkable symmetries of the new quantum algebras suggest the idea that q-oscillators exist in Nature and that there will also exist a q-analogue quantum field which has such q-oscillators as its normal modes [MacFarlane (1989), Biederharn (1989), Sun and Fu (1989), Chaichian and Kulish (1990), Kulish and Damaskinsky (1990), Arik and Coon (1976), Chiu, Gray and Nelson (1992), Nelson (1993), Chaichian and Ellinas (1990)]. q-photon phase operators have been introduced in [Chiu, Gray and Nelson (1992)], q-photon phase difference operators have been studied [Yu (1993)]. The q-analogue coherent states $|z\rangle_q$ have been used not only to define a q-boson number distribution [Bužek (1991)], but also to investigate the q-analogue generalization of the phase (cosine and sine) operators of Susskind and Glogower and of the phase operator of Pegg and Barnett [Nelson and Fields (1995)]. Whereas the foregoing papers assumed that $q > 0$ and they only noted the possibility of $q = \exp(i\frac{2\pi}{s+1})$, Ellinas (1992) and Abe (1995) have used this assumption. The Pegg–Barnett phase operator formalism has been treated by Ellinas (1992) as a q-deformed theory. The asymptotic behaviour of the Pegg–Barnett formalism in large Hilbert-space dimension has been treated by Abe (1995) using the q-deformation theory.

4.4   Statistical treatment of angle and phase, measures of phase variability

The most of modern physics is based on quantum theory. For physicists quantum theory means to deal with operators representing the corresponding physical quantities. At a closer look one sees that the operators are connected with distributions. In this book phase distributions of "quantum origin" are treated in sections 4.6, 4.7, 4.10, 4.12, and 4.13.

We use, from time to time, the concept of a random variable, which seems to be very mathematical, but it may characterize a physical quantity. For instance, with a position coordinate, the notation $\langle X \rangle$ may be connected with a classical model and $\langle \hat{x} \rangle$ may occur in a quantum model of reality. The expectation value and moments of a random variable $X(\omega)$ are defined as

$$\langle X^k \rangle = \int_\Omega [X(\omega)]^k \, d\mathrm{Prob}(\omega), \tag{4.4.1}$$

where Prob is a probability measure on a σ-field $\mathcal{F}$ of subsets of Ω and Ω is the set of elementary random events. From the physical viewpoint the formula (4.4.1) is too much connected with the Kolmogorov axioms of the probability theory although a simple substitution $x = X(\omega)$ leads to an accepted formula

$$\langle X^k \rangle = \int_\mathbb{R} x^k \, d\mu_X(x), \tag{4.4.2}$$

where the Borel measure

$$\mu_X(E) = \text{Prob}(X^{-1}(E)), \tag{4.4.3}$$

with $E \in \mathcal{B}(\mathbb{R})$, $\mathcal{B}(\mathbb{R})$ is the σ-field of all Borel subsets of the real line. The variance of the random variable $X(\omega)$,

$$\text{var } X = \langle (\Delta X)^2 \rangle, \tag{4.4.4}$$

where $\Delta X = X - \langle X \rangle$, and it holds that

$$\text{var } X = \langle X^2 \rangle - \langle X \rangle^2 . \tag{4.4.5}$$

Modern physics does not adopt the term "random variable", but it still associates the statistical notions with physical quantities, which are now being represented by operators. In non-commutative measure theory, which is being developed because of the desire to investigate the mathematical foundations of quantum mechanics (see [Gudder (1979), Beltrametti and Cassinelli (1981), Varadarajan (1985), Pták and Pulmannová (1991)]), one replaces the notion of a Boolean algebra by the notion of an orthomodular lattice. In contrast, relation (4.4.1) assumes the representation of a Boolean algebraic structure with a σ-field of sets as usual in the probability theory. In the classical model the position coordinate is the random variable $X(\omega)$ and in quantum optics the position coordinate is represented by the operator $\hat{x}$. The expectation value and moments of this operator are defined as

$$\langle \hat{x}^k \rangle = \text{Tr}\{\hat{\rho}\hat{x}^k\}, \tag{4.4.6}$$

where $\hat{\rho}$ is the statistical operator. In the quantum theory of measurement, a measure on the Borel σ-field $\mathcal{B}(\mathbb{R})$ is introduced and, by analogy with (4.4.3), we denote it similarly,

$$\mu_{\hat{x}}(E) = \text{Tr}\{\hat{\rho}\hat{\Delta}_{\hat{x}}(E)\}, \tag{4.4.7}$$

where $\hat{\Delta}_{\hat{x}}(E)$ is a projection-valued measure with the property that $\hat{\Delta}_{\hat{x}}(\mathbb{R}) = \hat{1}$. Upon substituting $\mu_{\hat{x}}$ for μ_X into (4.4.2) and using the spectral decompositions (cf. section 3.1)

$$\int_{\mathbb{R}} x^k \hat{\Delta}_{\hat{x}}(dx) = \hat{x}^k \tag{4.4.8}$$

and the definition (4.4.6), we obtain the relation

$$\langle X^k \rangle = \langle \hat{x}^k \rangle. \tag{4.4.9}$$

That is why we have generalized the use of the left-hand side of (4.4.2) and that of random variables to the whole of this section. The variance of the operator $\hat{x}$,

$$\text{var } \hat{x} = \langle (\Delta\hat{x})^2 \rangle, \tag{4.4.10}$$

where $\Delta \hat{x} = \hat{x} - \langle \hat{x} \rangle \hat{1}$, and equivalently

$$\text{var } \hat{x} = \langle \hat{x}^2 \rangle - \langle \hat{x} \rangle^2. \tag{4.4.11}$$

The phase differs from the position coordinate in that, even in the case of classical phase, it is open to choose either a "right" phase distribution which is supported by an interval or a "convenient" phase distribution which is supported by a unit circle. As a consequence, many measures of phase variability have been devised. These have been invented either just for the purpose of quantum phase or as indicated above, they can be applied to the measurement of the quantum phase fluctuations.

4.4.1   One random phase variable and two independent random phase variables

The transformations to the polar coordinates (4.1.57) and similarly (4.1.55), (4.1.67) are not single valued. To achieve this, we modify them to the form

$$\phi_{\theta_0} = \text{Arg}_{\theta_0}(x + iy), \quad r = \sqrt{x^2 + y^2}, \tag{4.4.12}$$

$$\varphi_{\theta_0 \mp} = \text{Arg}_{\theta_0}(\alpha_\mp). \tag{4.4.13}$$

Nevertheless, they are not defined at the origin ($x = 0, y = 0$) and $\alpha_\mp = 0$, respectively. This detail may influence the transformation of joint distribution of the Cartesian coordinates x and y to the polar ones. It is plain, of course, except when the probability

$$\text{Prob}(X = 0, Y = 0) > 0, \tag{4.4.14}$$

where $\text{Prob}(A)$ is the probability of a random event A and X, Y are random variables assuming values x, y. Accordingly, the joint distribution of Φ and R must be renormalized. For a given distribution $f(x, y)$, the distribution $g(\phi, r)$ of angle and radius is determined according to the rule

$$g(\phi, r) = \frac{r_+ f(r \cos \phi, r \sin \phi)}{1 - \text{Prob}(X = 0, Y = 0)}, \quad r_+ = r \text{ for } r > 0, \ r_+ = 0 \text{ for } r < 0. \tag{4.4.15}$$

A technique of solving this problem for raw data is known as censoring in biometrics. Another method is connected with the idea that the phase for $r = 0$ should be randomized. In this case the distribution $g(\phi, r)$ of angle and radius is as follows

$$g(\phi, r) = \frac{1}{2\pi} \text{Prob}(X = 0, Y = 0) \delta(r) + r_+ f(r \cos \phi, r \sin \phi). \tag{4.4.16}$$

In quantum optics the quasidistributions of complex amplitudes are transformed according to (4.1.55), (4.1.67), and usually the problem of vacuum is not encountered. As a consequence of discretization, the positive probability (4.4.14) may be characteristic of the eight-port homodyne detection scheme as commented on in [Hradil and Bajer (1993)].

To illustrate the phase problem in classical and quantum theories, we have consulted a related branch of science. In the classical theory of statistics [Rao (1973), Frieden (1983), Lukš and Peřinová (1996)] there does not exist only one random variable (phase) Φ, but the whole set of random variables Φ_{θ_0} taking on values in the intervals $[\theta_0, \theta_0 + 2\pi)$. The random variables Φ_{θ_0} have to be consistent in the sense that each of their distributions $P_{\theta_0}(\varphi)$ is a single 2π-periodic function $P(\varphi)$ restricted to the appropriate interval,

$$P_{\theta_0}(\varphi) = \begin{cases} P(\varphi) & \text{for} \quad \varphi \in [\theta_0, \theta_0 + 2\pi), \\ 0 & \text{elsewhere.} \end{cases} \tag{4.4.17}$$

If one function $P_{\theta_0}(\varphi)$ is normalized,

$$\int_{\theta_0}^{\theta_0 + 2\pi} P(\varphi)\, d\varphi = 1, \quad \theta_0 \in \mathbb{R}, \tag{4.4.18}$$

then the others share this property and the function $P(\varphi)$ is normalized in this sense only, whereas

$$\int_{-\infty}^{\infty} P(\varphi)\, d\varphi = \infty. \tag{4.4.19}$$

With respect to what has been stated, the phase expectation $\langle \Phi \rangle$ and the phase variance $\langle (\Delta\Phi)^2 \rangle$ are, strictly speaking, undefined as long as the reference phase θ_0, and so the window $[\theta_0, \theta_0 + 2\pi)$, is left unspecified. One may consider the well-defined minimum variance [Judge (1963, 1964)]

$$V_{\mathrm{J}} = \min_{\theta_0} \langle (\Phi_{\theta_0} - \theta_0 - \pi)^2 \rangle, \tag{4.4.20}$$

as a measure of the phase fluctuation. This minimum is attained for some reference phases and the uniqueness case occurs whenever all these phases are congruent modulo 2π. In principle, the ambiguity of the optimum reference phase, even if identified modulo 2π, is possible. Sometimes, the condition

$$\langle \Phi_{\theta_0} \rangle = \theta_0 + \pi \tag{4.4.21}$$

is of importance. This condition could be obtained from the Fermat rule, as far as this rule is applicable to the problem (4.4.20). The relation (4.4.21) means that the phase variable arising through the choice of the phase window has the mean value in the centre of the window. The mean value of the phase computed in the window, for which also the phase variance is minimum, could be called a J-location of the phase distribution. Barnett and Pegg (1992) have proved that if the condition (4.4.21) is satisfied for all θ_0 then the phase distribution must be uniform.

In principle, the phases which are congruent modulo 2π can be identified. This property is respected by the periodicity of the functions like cosine and sine, etc. Technically, when the subscript θ_0 of the phase variable is omitted, this random variable may take on values in the factor group $\mathbb{R}/2\pi\mathbb{Z}$, where $\mathbb{R}$ means the set of real numbers and $\mathbb{Z}$ denotes the set of integers. In fact, we then sense the cosine and sine functions, in general, 2π-periodic functions, as functions of arguments from this factor group.

The circularity of the phase values is annoying not only with respect to a location of the phase distribution, but even regarding the phase fluctuation. When taken into account, the circularity may motivate a new measure of phase uncertainty. Let us consider first, for definiteness, the position x of a particle. For the sake of analysis, we introduce a random variable X. The uncertainty or fluctuation of this coordinate is measured as

$$\langle(\Delta X)^2\rangle = \int_{-\infty}^{\infty}(x - \langle x\rangle)^2 P(x)\,dx, \qquad (4.4.22)$$

where

$$\Delta X = X - \langle X\rangle, \quad \langle X\rangle = \int_{-\infty}^{\infty} xP(x)\,dx. \qquad (4.4.23)$$

The integral on the right-hand side in (4.4.22) can be rewritten in the form

$$\langle(\Delta X)^2\rangle = \frac{1}{2}\int_{-\infty}^{\infty}\int_{-\infty}^{\infty}(x - x')^2 P(x)P(x')\,dx\,dx'. \qquad (4.4.24)$$

This formal manipulation reflects the change from the idea of fluctuation to the meaning of dispersion. The dispersion is based on the quadratic moment of the distance between two random positions. More generally, we can consider a measure of separation $K(x, x')$ leading to a dispersion

$$V_K = \int_{-\infty}^{\infty}\int_{-\infty}^{\infty} K(x, x')P(x)P(x')\,dx\,dx'. \qquad (4.4.25)$$

Particularly,

$$K(x, x') = K_{\mathbb{R}}(x, x') = \frac{1}{2}(x - x')^2. \qquad (4.4.26)$$

Of course, $\langle(\Delta X)^2\rangle = V_{\mathbb{R}}$. In the case of phase variable, the analogue of (4.4.25) is

$$V_K = \int_{\theta_0}^{\theta_0+2\pi}\int_{\theta_0}^{\theta_0+2\pi} K(\varphi, \varphi')P(\varphi)P(\varphi')\,d\varphi\,d\varphi', \qquad (4.4.27)$$

with $K(\varphi, \varphi')$ 2π-periodic in both arguments. Motivated by the fact that

$$K_{\mathbb{R}}(x, x') = \frac{1}{2}\mu_{\mathbb{R}}^2(x, x'), \qquad (4.4.28)$$

where

$$\mu_{\mathbb{R}}(x, x') = |x - x'|, \qquad (4.4.29)$$

we will consider two cases of the separation kernel

$$K(\varphi, \varphi') = \frac{1}{2}\mu^2(\varphi, \varphi'), \qquad (4.4.30)$$

with $\mu(\varphi, \varphi')$ 2π-periodic in both arguments. On choosing the chord metric

$$\mu(\varphi, \varphi') = \mu_P(\varphi, \varphi') = 2\left|\sin\left(\frac{\varphi - \varphi'}{2}\right)\right| \tag{4.4.31}$$

and obtaining that

$$K_P(\varphi, \varphi') = 1 - \cos(\varphi - \varphi'), \tag{4.4.32}$$

we introduce the dispersion

$$V_P = 1 - \langle\cos\Phi\rangle^2 - \langle\sin\Phi\rangle^2 = 1 - |\langle\exp(i\Phi)\rangle|^2. \tag{4.4.33}$$

This measure would be transformed to reveal connections to other measures of phase fluctuations, e. g., [Holevo (1982)]

$$V_H = \frac{V_P}{1 - V_P} \tag{4.4.34}$$

and [Opatrný (1994)]

$$V_O = \text{Cos}^{-1}\left(\sqrt{1 - V_P}\right) = \text{Sin}^{-1}\left(\sqrt{V_P}\right). \tag{4.4.35}$$

From this

$$V_H = \tan^2(V_O). \tag{4.4.36}$$

We can approach the familiar Süssmann measure $\delta_n\Phi$ [Białynicki-Birula, Freyberger and Schleich (1993)] on considering the separation kernel

$$K_S(\varphi, \varphi') = 1 - \bar{\delta}(\varphi - \varphi'), \tag{4.4.37}$$

where the periodic $\bar{\delta}$-function is characterized by the property

$$\int_{-\infty}^{\infty} \bar{\delta}(\varphi) M(\varphi)\, d\varphi = \sum_{k=-\infty}^{\infty} M(2\pi k). \tag{4.4.38}$$

Inserting into the rule (4.4.27), we obtain

$$V_S = 1 - \int_{\theta_0}^{\theta_0 + 2\pi} P^2(\varphi) d\varphi. \tag{4.4.39}$$

For low phase fluctuations both the measures are asymptotically equal as is obvious from the relation

$$\delta_S\Phi = \frac{1}{1 - V_S}. \tag{4.4.40}$$

A new proposal of the measure of phase uncertainty may be based upon the separation kernel (4.4.30) resulting from the arc metric

$$\mu(\varphi, \varphi') = \mu_I(\varphi, \varphi') = 2\left|\text{Sin}^{-1}\left[\sin\left(\frac{\varphi - \varphi'}{2}\right)\right]\right|. \tag{4.4.41}$$

For brevity, we denote

$$\text{saw}(\varphi) = \text{Sin}^{-1}(\sin\varphi), \tag{4.4.42}$$

which is a 2π-periodic function

$$\text{saw}(\varphi) = \text{sgn}(\varphi)\min(|\varphi|, \pi - |\varphi|), \tag{4.4.43}$$

for $\varphi \in [-\pi, \pi)$. On inserting into the rule (4.4.27), we obtain the definition

$$V_{\text{I}} = 2 \int_{\theta_0}^{\theta_0 + 2\pi} \int_{\theta_0}^{\theta_0 + 2\pi} \text{saw}^2\left(\frac{\varphi - \varphi'}{2}\right) P(\varphi)P(\varphi')\, d\varphi\, d\varphi'. \tag{4.4.44}$$

It is possible to consider further measures of phase uncertainty. The measure

$$V_{\text{M}} = \min_{\varphi} \frac{1}{2\pi P^2(\varphi)} \tag{4.4.45}$$

has a similar form to the definition (4.4.20). In the literature these values are considered when commenting on the inverse-peak likelihood [Shapiro, Shepard and Wong (1989)] and here we have imprinted a formal similarity to the Süssmann measure.

Let us mention two measures of the form

$$\delta_L\Phi = \int_{\theta_0}^{\theta_0 + 2\pi} P(\varphi)f_L[P(\varphi)]\, d\varphi, \tag{4.4.46}$$

where the function $f_L[P(\varphi)]$ distinguishes the two cases. For

$$f_L(p) = f_{\text{S}}(p) = 1 - p, \tag{4.4.47}$$

we arrive easily at the relation (4.4.39). For

$$f_L(p) = f_{\text{E}}(p) = -\ln p, \tag{4.4.48}$$

we obtain the entropic measure

$$\delta_{\text{E}}\Phi = -\int_{\theta_0}^{\theta_0 + 2\pi} P(\varphi)\ln[P(\varphi)]\, d\varphi. \tag{4.4.49}$$

The comparison of the phase-uncertainty measures V_{J}, V_{M}, $\delta_{\text{S}}\Phi$, V_{P}, and $\delta_{\text{E}}\Phi$ has been performed in [Białynicki-Birula, Freyberger and Schleich (1993)].

It is easy to see that in the study of the phase all information on the phase distribution is contained in a characteristic sequence. This concept is defined as [Lukš and Peřinová (1996)]

$$\chi(s) = \langle \exp(is\Phi) \rangle, \quad s \in \mathbb{Z}. \tag{4.4.50}$$

Particularly, $\chi(0) = 1$. Assuming the probability density $P(\varphi)$, we can express the characteristic sequence as

$$\chi(s) = \int_{\theta_0}^{\theta_0 + 2\pi} \exp(is\varphi)P(\varphi)\, d\varphi, \quad s \in \mathbb{Z}. \tag{4.4.51}$$

At each point, where this density is continuous,

$$P(\varphi) = \frac{1}{2\pi} \sum_{s=-\infty}^{\infty} \exp(-is\varphi)\chi(s). \tag{4.4.52}$$

Let us note that the abstract factor group $\mathbb{R}/2\pi\mathbb{Z}$ is homeomorphic to the circle $S^1 - \{\alpha \subset \mathbb{C}, |\alpha|^2 = 1\}$. When respecting the circularity of phase, it is convenient to relate the integration to a positive number-valued measure ν. This measure is defined as a continuation of the length of the arcs on the circle S^1 to measures of more general sets. From the circle topology it is known that an open arc is determined by its end points and by some point that is known to be inside (outside) the arc. When the end points are $\exp(i\theta')$, $\exp(i\theta'')$, the equation for φ,

$$h(\varphi, \theta', \theta'') = \frac{1}{2}[\sin(\theta' - \varphi) + \sin(\varphi - \theta'') + \sin(\theta'' - \theta')] = 0, \tag{4.4.53}$$

has just two solutions $\varphi \equiv \theta'$, $\varphi \equiv \theta''$ (mod 2π); $h(\varphi, \theta', \theta'')$ means the signed triangular area. These end points determine just two arcs given by the inequalities for φ,

$$h(\varphi, \theta', \theta'') < 0, \quad h(\varphi, \theta', \theta'') > 0. \tag{4.4.54}$$

When a point $\exp(i\theta)$ is known to be outside an arc Δ, this arc is given by the equation

$$\text{sgn}[h(\varphi, \theta', \theta'')] = -\text{sgn}[h(\theta, \theta', \theta'')]. \tag{4.4.55}$$

On choosing the numbers θ', θ'' to satisfy the inequalities,

$$\theta \leq \theta' < \theta + 2\pi, \quad \theta \leq \theta'' < \theta + 2\pi, \tag{4.4.56}$$

we obtain that the length of the arc Δ

$$\nu(\Delta) = |\theta'' - \theta'|. \tag{4.4.57}$$

A possible continuation can be performed by the standard mathematical methods. For any set E contained in the unit circle S^1, we define its measure

$$\mu(E) = \text{Prob}(\{\omega; \exp[i\Phi(\omega)] \in E\}). \tag{4.4.58}$$

Assuming $\mu(\{\theta'\}) = \mu(\{\theta''\}) = 0$, we obtain a generalization of (4.4.52)

$$\frac{\mu(\Delta)}{\nu(\Delta)} = \frac{1}{2\pi} \sum_{s=-\infty}^{\infty} \text{Re}\{\exp(-is\theta)\chi(s)\}\text{sinc}[s\nu(\Delta)], \tag{4.4.59}$$

where $\theta \in \Delta$, $\theta \equiv \frac{1}{2}(\theta' + \theta'')$ (mod 2π), and on choosing the numbers θ', θ'' to satisfy the inequalities $\theta - \pi \leq \theta' < \theta + \pi$, $\theta - \pi \leq \theta'' < \theta + \pi$, we obtain the length of the arc Δ in the form (4.4.57). The function sinc x in (4.4.59) is defined as

$$\text{sinc}\, x = \begin{cases} \frac{\sin(\frac{x}{2})}{\frac{x}{2}} & \text{for } x \neq 0, \\ 1 & \text{for } x = 0. \end{cases} \tag{4.4.60}$$

The formula (4.4.59) can be written equivalently in the form

$$\frac{\mu(\Delta)}{\nu(\Delta)} = \frac{1}{2\pi} \lim_{S \to \infty} \sum_{s=-S}^{S} \exp(-is\theta)\chi(s)\text{sinc}[s\nu(\Delta)]. \tag{4.4.61}$$

Studying an ordinary random variable, we may use the ordinary characteristic function

$$\chi_{\theta_0}(s) = \langle \exp(is\Phi_{\theta_0}) \rangle, \quad s \in \mathbb{R}. \tag{4.4.62}$$

It is connected to the characteristic sequence by the interpolation formula

$$\chi_{\theta_0}(s) = \sum_{s'=-\infty}^{\infty} \chi(s') \exp[i(s-s')(\theta_0 + \pi)]\text{sinc}[2\pi(s-s')]. \tag{4.4.63}$$

Vice versa, the characteristic sequence is a restriction of any characteristic function to $\mathbb{Z}$,

$$\chi(s) = \chi_{\theta_0}(s), \quad s \in \mathbb{Z}. \tag{4.4.64}$$

Let us recall that the moments of Φ_{θ_0} can be derived from the characteristic function, especially the mean

$$\langle \Phi_{\theta_0} \rangle = \frac{1}{i}\frac{d}{ds}\chi_{\theta_0}(s)\Big|_{s=0} \tag{4.4.65}$$

and the second moment

$$\langle \Phi_{\theta_0}^2 \rangle = -\frac{d^2}{ds^2}\chi_{\theta_0}(s)\Big|_{s=0} \tag{4.4.66}$$

leading to the variance

$$\text{var}(\Phi_{\theta_0}) = \langle (\Delta\Phi_{\theta_0})^2 \rangle, \tag{4.4.67}$$

where

$$\Delta\Phi_{\theta_0} = \Phi_{\theta_0} - \langle \Phi_{\theta_0} \rangle, \tag{4.4.68}$$

so that

$$\text{var}(\Phi_{\theta_0}) = \langle \Phi_{\theta_0}^2 \rangle - \langle \Phi_{\theta_0} \rangle^2. \tag{4.4.69}$$

Given the random variable Φ which assumes the values in the quotient set $\mathbb{R}/2\pi\mathbb{Z}$, it is obvious that the random variable $k\Phi$, $k \in \mathbb{Z}$, takes on the values in the quotient set $\mathbb{R}/2\pi k\mathbb{Z}$. Using the natural surjection of $\mathbb{R}/2\pi k\mathbb{Z}$ onto $\mathbb{R}/2\pi\mathbb{Z}$, we get a random variable $\Phi^{(k)}$. According to the definition of the characteristic sequence (4.4.51) and observing that the random variable $\Phi^{(k)}$ has the density

$$P^{(k)}(\varphi) = \frac{1}{k}\sum_{j=0}^{k-1} P\left(\varphi - \frac{2\pi}{k}j\right), \tag{4.4.70}$$

we obtain that

$$\chi^{(k)}(s) = \chi(ks), \quad s \in \mathbb{Z}. \tag{4.4.71}$$

Using the characteristic sequence, we need a modification of the formula (4.4.65) for the mean. When $\chi(1) \neq 0$, we define the preferred phase

$$\text{Pref}_{\theta_0} \Phi = \text{Arg}_{\theta_0}[\chi(1)], \tag{4.4.72}$$

where $\text{Arg}_{\theta_0} z$ is the phase of z taking on the values in the interval $[\theta_0, \theta_0 + 2\pi)$. Recalling the multivaluedness from the analytic function theory, we obtain

$$\text{pref} \, \Phi = \arg[\chi(1)], \tag{4.4.73}$$

with $\arg z = \text{Im}(\ln z)$. Generally, the characteristics (4.4.72) and (4.4.65) differ. With the only exception of the random variable being distributed uniformly, there always exists a $\chi(k) \neq 0$ and we define the kth-order preferred phase

$$\text{Pref}_{\theta_0 j}^{[k]} \Phi = \frac{1}{k} \text{Arg}_{k\theta_0}[\chi(k)] + \frac{2\pi}{k} j, \quad j = 0, \ldots, k - 1, \tag{4.4.74}$$

where $\text{Arg}_{k\theta_0} z$ is the phase of z taking on the values in the interval $[k\theta_0, k\theta_0 + 2\pi)$, which is not unique, but takes on k different values, and

$$\text{pref}^{[k]} \Phi = \frac{1}{k} \arg[\chi(k)]. \tag{4.4.75}$$

These definitions can be applied to ℓ photon coherent states whose deteriorated phase properties they take into account [Peřinová, Lukš and Křepelka (1997a)].

The role of the variance (4.4.69) is taken over by the dispersion V_P (see (4.4.33)), which can be written as

$$D \, \Phi = 1 - |\chi(1)|^2. \tag{4.4.76}$$

In fact, $D \, \Phi = V_P$, but the new notation is explicit with respect to the random phase variable Φ. The role of the standard deviation $\sqrt{\text{var}(\Phi_{\theta_0})}$ is taken over by the measure V_O,

$$D \, \Phi = \sin^2(V_O). \tag{4.4.77}$$

By analogy to (4.4.76), we define a kth-order dispersion

$$D^{[k]} \, \Phi = \frac{1}{k^2} \left[1 - |\chi(k)|^2 \right], \tag{4.4.78}$$

where $[1 - |\chi(k)|^2]$ is equal to the generalized Paul–Bandilla phase-uncertainty measure introduced by Vaccaro and Orłowski (1995).

For comparison, let us formulate the definition of Collett measure of phase noise [Collett (1993a)]

$$\begin{aligned} \delta_C \, \Phi &= \langle \mu_P^2(\Phi, \text{pref} \, \Phi) \rangle \\ &= 2\langle 1 - \cos(\Phi - \text{pref} \, \Phi) \rangle. \end{aligned} \tag{4.4.79}$$

On using the first-order dispersion, Collett's measure can be written as

$$\delta_C \Phi = 2 \left(1 - \sqrt{1 - D\,\Phi}\right).$$
(4.4.80)

Using the Opatrný measure (4.4.35), we can rewrite (4.4.79) in the form

$$\delta_C \Phi = 2 \sin^2 \left(\frac{V_0}{2}\right).$$
(4.4.81)

In the case of two independent random phase variables Φ_1, Φ_2 with the respective probability densities $P_1(\varphi_1)$, $P_2(\varphi_2)$, the probability densities of their sum and difference $\Phi_\pm = \Phi_1 \pm \Phi_2$ in the sense of the quotient set $\mathbb{R}/2\pi\mathbb{Z}$ are given by the convolutions

$$P_\pm(\varphi) = \int_{\theta_0}^{\theta_0+2\pi} P_1(\varphi - \varphi_2)P_2(\pm\varphi_2)\,d\varphi_2.$$
(4.4.82)

The appropriate dispersions are connected by the relation (see [Opatrný (1994)] for the phase sum)

$$D(\Phi_\pm) = D(\Phi_1) + D(\Phi_2) - D(\Phi_1)D(\Phi_2).$$
(4.4.83)

A similar relation holds for the second-order dispersions

$$D^{[2]}(\Phi_\pm) = D^{[2]}(\Phi_1) + D^{[2]}(\Phi_2) - 4D^{[2]}(\Phi_1)D^{[2]}(\Phi_2).$$
(4.4.84)

The maximum dispersion of $D(\Phi_1) = 1$ (equivalently, $\mathrm{Pref}_{\theta_0}(\Phi_1)$ undefined) implies also $D(\Phi_\pm) = 1$ and $\mathrm{Pref}_{\theta_0}(\Phi_\pm)$ undefined.

The familiar theorem of the theory of probability on the characteristic function of the sum of independent random variables has its analogue for the characteristic sequences. It holds that

$$
\begin{aligned}
\chi_\pm(s) &= \langle\exp(is\Phi_\pm)\rangle = \langle\exp(is\Phi_1)\exp[is(\pm\Phi_2)]\rangle \\
&= \langle\exp(is\Phi_1)\rangle\langle\exp[is(\pm\Phi_2)]\rangle.
\end{aligned}
$$
(4.4.85)

According to the definition of the characteristic sequence (4.4.51) we obtain

$$\chi_+(s) = \chi_1(s)\chi_2(s), \quad \chi_-(s) = \chi_1(s)\chi_2^*(s),$$
(4.4.86)

where the asterisk denotes the complex conjugation.

The simple relationship between the characteristic functions enables one to establish that sum of N random phase variables of small dispersions $D_N = N^{-1}D_1$ can be approximated well by the wrapped Gaussian (normal) distribution of the phase in the limit of N tending to infinity, $D_\infty = 1 - \exp(-D_1)$ [Opatrný (1994)]. The previous proposition is an analogue of familiar central limit theorems of the probability theory. A similarly appealing is the idea of an analogue of the Tchebyshev inequality [Feller (1966)]

$$\mathrm{Prob}\{\mu_P(\Phi, \mathrm{pref}\,\Phi) \geq \epsilon\} \leq \frac{2}{\epsilon^2}D\Phi.$$
(4.4.87)

A version of (4.4.87) with the choice of metric $\mu_I(\varphi, \varphi')$ (4.4.41) can be found elsewhere [Opatrný (1994)].

In applications, we mostly encounter one- and two-peak distributions. We shall consider distributions on the circle [Rao (1973), Lukš and Peřinová (1991)].

(i) As a useful model we may consider uniform distribution of the angle

$$P(\varphi) = \frac{1}{2\pi}, \tag{4.4.88}$$

which describes a completely random angle. The characteristic sequence is of the form

$$\chi(s) = \delta_{s,0}. \tag{4.4.89}$$

There is no preferred value of angle (phase) and the phase dispersion is unity.

(ii) For the probability density

$$P(\varphi) = \frac{1}{2\pi I_0(\kappa)} \exp[\kappa \cos(\varphi - \beta)], \quad \beta \in [\theta_0, \theta_0 + 2\pi), \tag{4.4.90}$$

of the so-called von Mises distribution [von Mises (1931)], the characteristic sequence reads as

$$\chi(s) = \frac{I_s(\kappa)}{I_0(\kappa)} \exp(is\beta), \tag{4.4.91}$$

where $I_s(\kappa)$ is the modified Bessel function of order s. Here

$$\text{Pref}_{\theta_0}\, \Phi = \beta, \tag{4.4.92}$$

$$D\, \Phi = 1 - \left[\frac{I_1(\kappa)}{I_0(\kappa)}\right]^2. \tag{4.4.93}$$

(iii) For the probability density

$$P(\varphi) = \frac{1}{2\pi I_0(\kappa)} \exp\{\kappa \cos[2(\varphi - \beta)]\}, \quad \beta \in [\theta_0, \theta_0 + \pi), \tag{4.4.94}$$

the characteristic sequence reads as

$$\chi(s) = \begin{cases} \dfrac{I_{\frac{s}{2}}(\kappa)}{I_0(\kappa)} \exp(is\beta) & \text{for } s \text{ even,} \\ 0 & \text{for } s \text{ odd.} \end{cases} \tag{4.4.95}$$

In this case, $\text{Pref}_{\theta_0}\, \Phi$ is undefined, $D\, \Phi = 1$,

$$\text{Pref}^{[2]}_{\theta_0 j}\, \Phi = \beta + \pi j, \quad j = 0, 1, \tag{4.4.96}$$

and

$$D^{[2]}\, \Phi = \frac{1}{4}\left(1 - \left[\frac{I_1(\kappa)}{I_0(\kappa)}\right]^2\right). \tag{4.4.97}$$

(iv) If $P_{\mathbb{R}}(z)$ is any usual probability density, we introduce

$$P(\varphi) = \sum_{k=-\infty}^{\infty} P_{\mathbb{R}}(\varphi - 2\pi k), \tag{4.4.98}$$

which has all the properties of the phase distribution. For example, we choose

$$P_{\mathbb{R}}(z) = \frac{1}{\sqrt{2\pi}\sigma} \exp\left[-\frac{(z-m)^2}{2\sigma^2}\right], \tag{4.4.99}$$

where $\sigma > 0$, $m \in \mathbb{R}$. The probability density (4.4.98) yields the wrapped normal distribution in this case. According to a sampling theorem [Frieden (1983, 1991)] we observe that the appropriate characteristic sequence is of the form

$$\chi(s) = \exp\left(ism - \frac{1}{2}s^2\sigma^2\right). \tag{4.4.100}$$

The phase distribution given in (4.4.98) and (4.4.99) has the preferred phase

$$\mathrm{Pref}_{\theta_0}\Phi \equiv m \pmod{2\pi} \tag{4.4.101}$$

and the dispersion

$$D\Phi = 1 - \exp(-\sigma^2). \tag{4.4.102}$$

(v) For the probability density

$$P(\varphi) = \frac{1}{2\pi} \frac{1}{\cosh(2r) - \cos(\varphi - \beta)\sinh(2r)}, \quad \beta \in [\theta_0, \theta_0 + 2\pi), \tag{4.4.103}$$

where $r \geq 0$, the characteristic sequence reads as

$$\chi(s) = \exp(is\beta)(\tanh r)^{|s|}. \tag{4.4.104}$$

Here

$$\mathrm{Pref}_{\theta_0}\Phi = \beta, \tag{4.4.105}$$

$$D\Phi = \frac{1}{\cosh^2 r}. \tag{4.4.106}$$

From the form of the characteristic sequence it is obvious that the phase distribution (4.4.103) is the wrapped Cauchy probability distribution.

If we compare the probability density (4.4.94) with (4.4.90), we see that a two-peak distribution can become from a one-peak distribution by a mere replacement $\varphi - \beta \to 2(\varphi - \beta)$ in the simpler probability density. This formal substitution can be coupled with a transformation of the random phase variable, whose probability density exhibits one peak. This transformation can hardly be called otherwise than "halving" the values of the phase variable Φ. It is useful to note that the ambiguity of "halves" of Φ (mod 2π) is connected with the fifty-fifty splitting of the chance to result. Performing the replacement $\varphi - \beta \to 2(\varphi - \beta)$ in the probability density (4.4.103), we obtain the probability density

$$P(\varphi) = \frac{1}{2\pi} \frac{1}{\exp(2r)\sin^2(\varphi - \beta) + \exp(-2r)\cos^2(\varphi - \beta)}, \quad \beta \in [\theta_0, \theta_0 + 2\pi). \quad (4.4.107)$$

The characteristic sequence is a rarefied sequence (4.4.104),

$$\chi(s) = \begin{cases} \exp(i\frac{s}{2}\beta)(\tanh r)^{\left|\frac{s}{2}\right|} & \text{for } s \text{ even,} \\ 0 & \text{for } s \text{ odd.} \end{cases} \quad (4.4.108)$$

Similarly as in the case of the probability density (4.4.94), $\text{Pref}_{\theta_0} \Phi$ is undefined, $D\Phi = 1$, and the second-order preferred phase is given by (4.4.96), but

$$D^{[2]}\Phi = \frac{1}{4\cosh^2 r}. \quad (4.4.109)$$

4.4.2   Two correlated random phase variables

We consider a two-dimensional multiple-valued random phase vector (Φ_1, Φ_2) with which four random variables X_1, Y_1, X_2, Y_2 can be connected by the equations

$$X_j + iY_j = \exp(i\Phi_j), \quad j = 1, 2. \quad (4.4.110)$$

For $(\varphi_1, \varphi_2) \in (\mathbb{R}/2\pi\mathbb{Z})^2$, the mapping $(\exp(i\varphi_1), \exp(i\varphi_2))$ is an injection (a one-to-one function) of $(\mathbb{R}/2\pi\mathbb{Z})^2$ into $\mathbb{C}^2$. Its range is the unit torus $\mathbb{T}^2$. It is worth noting that quite similar equations

$$X_j + iY_j = \exp(iZ_j), \quad j = 1, 2, \quad (4.4.111)$$

where Z_j are usual random variables, may serve definitions of distributions on the unit torus $\mathbb{T}^2$. To this aim it is sufficient to pick Z_1, Z_2 obeying a (normal) Gaussian law. Now, these random variables are formally multiple valued according to their role in the relations (4.4.111). This will lead to summations in the formulae for the new probability densities and simple expressions cannot always be expected from this technique.

The study of two random phase variables seems to necessitate two variables instead of the four and the torus is mapped again into the plane. The simplest approach recently used in the literature on quantum optics [Barnett and Pegg (1990b)] consists in considering random phase variables $\Phi_{1\theta_{01}}$, $\Phi_{2\theta_{02}}$ and their joint probability density distributed on the square $Q = [\theta_{01}, \theta_{01} + 2\pi) \times [\theta_{02}, \theta_{02} + 2\pi)$. Nevertheless, the torus admits various "charts" even if polar angles are used solely. Some of them are mentioned below, but we adhere to an analysis of the simple chart $\text{Arg}_{\theta_{01}} \times \text{Arg}_{\theta_{02}}$ leading to the relation $\Phi_{j\theta_{0j}} = \text{Arg}_{\theta_{0j}}[\exp(i\Phi_j)]$, $j = 1, 2$. For simplicity, let us call the set $(2\pi\mathbb{Z})^2$ a lattice and denote it by L.

Let us note that $(\mathbb{R}/2\pi\mathbb{Z})^2 = \mathbb{R}^2/L$. Assuming that the random phase variables $\Phi_{1\theta_{01}}$, $\Phi_{2\theta_{02}}$ have a joint probability density $P_{\theta_{01}\theta_{02}}(\varphi_1,\varphi_2)$, we observe that $P_{\theta_{01}\theta_{02}}(\varphi_1,\varphi_2) = 0$ for (φ_1,φ_2) outside Q. Any doubly 2π-periodic function will be called L-periodic. The L-periodic continuation of the probability density from Q onto the whole $\mathbb{R}^2$ is useful and will be denoted by $P(\varphi_1,\varphi_2)$.

The L-periodicity may motivate the definition of a characteristic double sequence. We introduce this concept as

$$\chi(s_1,s_2) = \langle \exp[i(s_1\Phi_1 + s_2\Phi_2)]\rangle, \quad (s_1,s_2) \in \mathbb{Z}^2. \qquad (4.4.112)$$

Particularly, $\chi(0,0) = 1$. Supposing that the random phase vector (Φ_1,Φ_2) has an absolutely continuous distribution with the probability density $P(\varphi_1,\varphi_2)$ and choosing $\theta_{01}, \theta_{02} \in \mathbb{R}$, we can express the characteristic double sequence as

$$\chi(s_1,s_2) = \int_{\theta_{01}}^{\theta_{01}+2\pi} \int_{\theta_{02}}^{\theta_{02}+2\pi} \exp[i(s_1\varphi_1 + s_2\varphi_2)]P(\varphi_1,\varphi_2)\, d\varphi_1\, d\varphi_2. \qquad (4.4.113)$$

The inverse relation to (4.4.113) reads

$$P(\varphi_1,\varphi_2) = \frac{1}{4\pi^2} \sum_{s_1=-\infty}^{\infty} \sum_{s_2=-\infty}^{\infty} \exp[-i(s_1\varphi_1 + s_2\varphi_2)]\chi(s_1,s_2). \qquad (4.4.114)$$

Studying a pair of ordinary random variables, we may use the characteristic function

$$\chi_{\theta_{01}\theta_{02}}(s_1,s_2) = \langle \exp[i(s_1\Phi_{1\theta_{01}} + s_2\Phi_{2\theta_{02}})]\rangle, \quad (s_1,s_2) \in \mathbb{R}^2. \qquad (4.4.115)$$

It is related to the characteristic double sequence by the interpolation formula

$$\chi_{\theta_{01}\theta_{02}}(s_1,s_2) = \sum_{s_1'=-\infty}^{\infty} \sum_{s_2'=-\infty}^{\infty} \chi(s_1',s_2')\exp[i(s_1 - s_1')(\theta_{01} + \pi) + i(s_2 - s_2')(\theta_{02} + \pi)]$$

$$\times \operatorname{sinc}[2\pi(s_1 - s_1')]\operatorname{sinc}[2\pi(s_2 - s_2')]. \qquad (4.4.116)$$

The marginal characteristic functions for single random phase variables $\Phi_{1\theta_{01}}$ and $\Phi_{2\theta_{02}}$ are given by the formulae

$$\chi_{\theta_{01}}(s) = \chi_{\theta_{01}\theta_{02}}(s,0), \quad \chi_{\theta_{02}}(s) = \chi_{\theta_{01}\theta_{02}}(0,s), \quad s \in \mathbb{R}. \qquad (4.4.117)$$

Let us remember that the mixed second moment of $\Phi_{1\theta_{01}}$ and $\Phi_{2\theta_{02}}$ is derived as

$$\langle \Phi_{1\theta_{01}}\Phi_{2\theta_{02}}\rangle = -\frac{\partial^2}{\partial s_1 \partial s_2}\chi_{\theta_{01}\theta_{02}}(s_1,s_2)\bigg|_{s_1=0,s_2=0} \qquad (4.4.118)$$

and is used in the computation of the covariance

$$\operatorname{cov}(\Phi_{1\theta_{01}}, \Phi_{2\theta_{02}}) = \langle \Delta\Phi_{1\theta_{01}}\Delta\Phi_{2\theta_{02}}\rangle = \langle \Phi_{1\theta_{01}}\Phi_{2\theta_{02}}\rangle - \langle \Phi_{1\theta_{01}}\rangle\langle \Phi_{2\theta_{02}}\rangle. \qquad (4.4.119)$$

The mathematical approach to modelling the physical quantities would require the use of the lattice

$$L' = \{(2\pi(k_1 + k_2), 2\pi(k_1 - k_2)); k_1, k_2 \in \mathbb{Z}\}$$

$$= \{(2\pi k_+, 2\pi k_-); k_+, k_- \in 2\mathbb{Z} \text{ or } k_+, k_- \in 2\mathbb{Z} + 1\}. \quad (4.4.120)$$

Here $2\mathbb{Z} = \{2k; k \in \mathbb{Z}\}$ and $2\mathbb{Z} + 1 = \{2k + 1; k \in \mathbb{Z}\}$. The pair of multiple-valued random variables $\Phi_\pm$ has not an L-periodic probability density, but an L'-periodic one. This manifests itself in the characteristic double sequence of this pair of multiple-valued random variables, which is given by the formula

$$\chi'_{+-}(s_+, s_-) = \chi(s_+ + s_-, s_+ - s_-) \text{ for } s_+, s_- \in \mathbb{Z} \text{ or } s_+, s_- \in \mathbb{Z} + \frac{1}{2}, \quad (4.4.121)$$

where $\mathbb{Z} + \frac{1}{2} = \{k + \frac{1}{2}; k \in \mathbb{Z}\}$, and which provides the L'-periodic probability density as

$$P'_{+-}(\varphi_+, \varphi_-) = \frac{1}{8\pi^2} \sum_{s_+ \in \mathbb{Z}} \sum_{s_- \in \mathbb{Z}} \exp[-i(s_+\varphi_+ + s_-\varphi_-)]\chi'_{+-}(s_+, s_-)$$

$$+ \frac{1}{8\pi^2} \sum_{s_+ \in \mathbb{Z} \mid \frac{1}{2}} \sum_{s_- \in \mathbb{Z} \mid \frac{1}{2}} \exp[-i(s_+\varphi_+ + s_-\varphi_-)]\chi'_{+-}(s_+, s_-). \quad (4.4.122)$$

As multiple-valued random phase variables cannot have an L'-periodic probability density, the literature treated the problem of the recovery of the L-periodicity with a procedure. We reapproach this problem via characteristic double sequence. Quite simply

$$\chi_{+-}(s_+, s_-) = \chi'_{+-}(s_+, s_-), \quad s_+, s_- \in \mathbb{Z}. \quad (4.4.123)$$

Let us emphasize that the variables s_+, s_- in (4.4.123) are restricted to the domain of the definition in (4.4.112).

Returning to the formula (4.4.121), let us remark that the marginal characteristic functions $\chi_+(s)$ and $\chi_-(s)$ for single multiple-valued random phase variables Φ_+ and Φ_-, respectively, can be obtained as usual by substituting $s_- = 0 \in \mathbb{Z}$ and $s_+ = 0 \in \mathbb{Z}$, respectively. So the situation applies with $s_+, s_- \in \mathbb{Z}$ and

$$\chi_+(s) = \chi'_{+-}(s, 0), \quad \chi_-(s) = \chi'_{+-}(0, s), \quad s \in \mathbb{Z}. \quad (4.4.124)$$

By comparison with (4.4.123) it is obvious that L'-periodic probability density provides 2π-periodic marginals directly as well as after an intermediate computation of the L-periodic probability density in the formalism of characteristic (double and simple) sequences.

If the multiple-valued random phase variables Φ_1, Φ_2 have a joint probability density $P(\varphi_1, \varphi_2)$, we obtain the L-periodic probability density for the multiple-valued phase sum and multiple-valued phase difference

$$P_{+-}(\varphi_+, \varphi_-) = \frac{1}{4\pi^2} \sum_{s_+=-\infty}^{\infty} \sum_{s_-=-\infty}^{\infty} \exp[-i(s_+\varphi_+ + s_-\varphi_-)]\chi_{+-}(s_+, s_-). \quad (4.4.125)$$

Using (4.4.113), (4.4.121), and (4.4.123), we may rewrite (4.4.125) as

$$P_{+-}(\varphi_+, \varphi_-) = \int\int_Q \bar{\delta}(\varphi_1 + \varphi_2 - \varphi_+)\bar{\delta}(\varphi_1 - \varphi_2 - \varphi_-)P(\varphi_1, \varphi_2)\, d\varphi_1\, d\varphi_2, \quad (4.4.126)$$

where $\bar{\delta}(\varphi)$ is the 2π-periodic Dirac delta function and $\bar{\delta}(\varphi_1)\bar{\delta}(\varphi_2)$ is the L-periodic Dirac delta function. Using the definition (cf. (4.4.38))

$$\bar{\delta}(\varphi) = \sum_{k=-\infty}^{\infty} \delta(\varphi - 2\pi k), \qquad (4.4.127)$$

we obtain that the L-periodic probability density restricted to a square Q_{+-} is given by the formulae

$$P_{+-}(\varphi_+, \varphi_-) = \frac{1}{2}\left[P_{\theta_{01},\theta_{02}}\left(\frac{\varphi_+ + \varphi_-}{2}, \frac{\varphi_+ - \varphi_-}{2}\right) \right.$$
$$\left. + P_{\theta_{01},\theta_{02}}\left(\frac{\varphi_+ + \varphi_-}{2} - \pi, \frac{\varphi_+ - \varphi_-}{2} - \pi\right) \right], \quad (4.4.128)$$

for $\varphi_+ \geq \theta_{0+} + \pi$, $|\varphi_- - \theta_{0-} - \pi| < \varphi_+ - \theta_{0+} - \pi$;

$$P_{+-}(\varphi_+, \varphi_-) = \frac{1}{2}\left[P_{\theta_{01},\theta_{02}}\left(\frac{\varphi_+ + \varphi_-}{2}, \frac{\varphi_+ - \varphi_-}{2}\right) \right.$$
$$\left. + P_{\theta_{01},\theta_{02}}\left(\frac{\varphi_+ + \varphi_-}{2} - \pi, \frac{\varphi_+ - \varphi_-}{2} + \pi\right) \right], \quad (4.4.129)$$

for $\varphi_- \geq \theta_{0-} + \pi$, $|\varphi_+ - \theta_{0+} - \pi| < \varphi_- - \theta_{0-} - \pi$;

$$P_{+-}(\varphi_+, \varphi_-) = \frac{1}{2}\left[P_{\theta_{01},\theta_{02}}\left(\frac{\varphi_+ + \varphi_-}{2}, \frac{\varphi_+ - \varphi_-}{2}\right) \right.$$
$$\left. + P_{\theta_{01},\theta_{02}}\left(\frac{\varphi_+ + \varphi_-}{2} + \pi, \frac{\varphi_+ - \varphi_-}{2} + \pi\right) \right], \quad (4.4.130)$$

for $\varphi_+ \leq \theta_{0+} + \pi$, $|\varphi_- - \theta_{0-} - \pi| < |\varphi_+ - \theta_{0+} - \pi|$;

$$P_{+-}(\varphi_+, \varphi_-) = \frac{1}{2}\left[P_{\theta_{01},\theta_{02}}\left(\frac{\varphi_+ + \varphi_-}{2}, \frac{\varphi_+ - \varphi_-}{2}\right) \right.$$
$$\left. + P_{\theta_{01},\theta_{02}}\left(\frac{\varphi_+ + \varphi_-}{2} + \pi, \frac{\varphi_+ - \varphi_-}{2} - \pi\right) \right], \quad (4.4.131)$$

for $\varphi_- \leq \theta_{0-} + \pi$, $|\varphi_+ - \theta_{0+} - \pi| < |\varphi_- - \theta_{0-} - \pi|$. Here

$$Q_{+-} = [\theta_{0+}, \theta_{0+} + 2\pi) \times [\theta_{0-}, \theta_{0-} + 2\pi), \qquad (4.4.132)$$

with

$$\theta_{0\pm} = (\theta_{01} + \pi) \pm (\theta_{02} + \pi) - \pi = \theta_{01} \pm (\theta_{02} + \pi). \qquad (4.4.133)$$

From (4.4.128)–(4.4.131) it is obvious that two points of Q' are cast into a single one of Q.

After a casting (surjection) procedure, the formula $\langle \Phi_{1\theta_{01}} \pm \Phi_{2\theta_{02}} \rangle = \langle \Phi_{1\theta_{01}} \rangle \pm \langle \Phi_{2\theta_{02}} \rangle$ applies no more. A loophole in notation may obscure understanding of a failure of the twin formulae

$$\mathrm{cov}(\Psi_{1\theta_{01}}, \Phi_{2\theta_{02}}) = \pm \frac{1}{2} [\mathrm{var}(\Phi_{1\theta_{01}} \perp \Phi_{2\theta_{02}}) - \mathrm{var}(\Phi_{1\theta_{01}}) \quad \mathrm{var}(\Phi_{2\theta_{02}})]. \tag{4.4.134}$$

Recalling the results of Barnett and Pegg (1990b), we encounter the situation, where the left-hand side of (4.4.134) depends on a parameter r, but the right-hand side for the phase difference does not depend on r.

The dependence between the random phase variables $\Phi_{1\theta_{01}}$, $\Phi_{2\theta_{02}}$ can be assessed by the normalized covariance (the correlation coefficient)

$$\mathrm{cor}(\Phi_{1\theta_{01}}, \Phi_{2\theta_{02}}) = \frac{\mathrm{cov}(\Phi_{1\theta_{01}}, \Phi_{2\theta_{02}})}{\sqrt{\mathrm{var}(\Phi_{1\theta_{01}})\mathrm{var}(\Phi_{2\theta_{02}})}}, \tag{4.4.135}$$

which can assume values of both signs in the interval $[-1, 1]$. Actually this characteristic measures the dependence between the multiple valued random phase variables Φ_1, Φ_2, but with a flavour of ambiguity. We assume that a specific concept for the correlation between Φ_1 and Φ_2 should be introduced. To this end we apply the group correlation coefficient [Andĕl (1978), p. 307] between random vectors (X_1, Y_1) and (X_2, Y_2) given in (4.4.110) using a complexification. This quantity is given as

$$\rho^2_{\exp(i\Phi_1), \exp(i\Phi_2)} = 1 - \frac{|\mathbf{W}|}{|\mathbf{V}_{11}||\mathbf{V}_{22}|} \geq 0, \tag{4.4.136}$$

where $|\mathbf{A}|$ means the determinant of a matrix $\mathbf{A}$, and the matrix $\mathbf{W}$

$$\mathbf{W} = \begin{pmatrix} \mathbf{V}_{11} & \mathbf{V}_{12} \\ \mathbf{V}_{21} & \mathbf{V}_{22} \end{pmatrix}, \tag{4.4.137}$$

with

$$\mathbf{V}_{jj} = \begin{pmatrix} \mathrm{cov}(e^{i\Phi_j}, e^{-i\Phi_j}) & \mathrm{var}(e^{i\Phi_j}) \\ \mathrm{var}(e^{-i\Phi_j}) & \mathrm{cov}(e^{i\Phi_j}, e^{-i\Phi_j}) \end{pmatrix}, \quad j = 1, 2, \tag{4.4.138}$$

$$\mathbf{V}_{12} = \begin{pmatrix} \mathrm{cov}(e^{i\Phi_1}, e^{-i\Phi_2}) & \mathrm{cov}(e^{i\Phi_1}, e^{i\Phi_2}) \\ \mathrm{cov}(e^{-i\Phi_1}, e^{-i\Phi_2}) & \mathrm{cov}(e^{-i\Phi_1}, e^{i\Phi_2}) \end{pmatrix}, \tag{4.4.139}$$

$$\mathbf{V}_{21} = \mathbf{V}_{12}^\dagger. \tag{4.4.140}$$

The complexification is made without introducing complex conjugation in the variance ((4.4.69)) and the covariance ((4.4.119)). Motivated by the fact that the coefficient (4.4.136) cannot take on negative values and by the observation that the diagonal elements on the right-hand side of (4.4.138)

$$\mathrm{cov}(e^{i\Phi_j}, e^{-i\Phi_j}) = \langle \Delta e^{i\Phi_j} \Delta e^{-i\Phi_j} \rangle = D(\Phi_j), \quad j = 1, 2, \tag{4.4.141}$$

we introduce the codispersions of the random phase variables Φ_1, Φ_2 and of Φ_1, $-\Phi_2$,

$$
\begin{aligned}
\text{cod}(\Phi_1, \Phi_2) &= \text{cov}(e^{i\Phi_1}, e^{-i\Phi_2}), \\
\text{cod}(\Phi_1, -\Phi_2) &= \text{cov}(e^{i\Phi_1}, e^{i\Phi_2}).
\end{aligned}
\tag{4.4.142}
$$

Let us observe that: (i) We seeked for the signs ± 1, but we have obtained complex units, because the quantities (4.4.142) can be imaginary. (ii) We expected only a "covariance" of Φ_1 and Φ_2, but we have obtained also the codispersion of Φ_1 and $-\Phi_2$. With respect to the point (i) we remark that the codispersion can be analyzed using the familiar polar decomposition of complex numbers and the squared modulus can be compared with (4.4.136). As to (ii) we know that the relation between the complex random variables $\exp(i\Phi_2)$ and $\exp(-i\Phi_2)$ is nonlinear, i. e., $\exp(-i\Phi_2) = [\exp(i\Phi_2)]^{-1}$. In our opinion, no analogue of the twin formulae (4.4.134) giving a unified measure of correlation in terms of the phase sum and the phase difference is known yet. Exploiting this observation for single phases, we invent also the characteristics

$$
\text{cod}(\Phi_j, -\Phi_j) = \text{cov}(e^{i\Phi_j}, e^{i\Phi_j}) = \text{var}(e^{i\Phi_j}), \quad j = 1, 2.
\tag{4.4.143}
$$

To take into account that the random variables (X_1, Y_1), (X_2, Y_2) have the distribution concentrated on the torus $\mathbb{T}^2$, we consider the conditional distribution of the multiple-valued random phase variable Φ_1 given a value φ_2 of the random phase variable Φ_2 and that of Φ_2 given a value φ_1 of Φ_1. Whenever $P_2(\varphi_2) \neq 0$, $P_1(\varphi_1) \neq 0$, we introduce the conditional probability densities

$$
P(\varphi_1|\varphi_2) = \frac{P(\varphi_1, \varphi_2)}{P_2(\varphi_2)}, \quad P(\varphi_2|\varphi_1) = \frac{P(\varphi_1, \varphi_2)}{P_1(\varphi_1)},
\tag{4.4.144}
$$

respectively. Applying the relation (4.4.51), we obtain the conditional characteristic sequences $\chi_1(s|\varphi_2)$, $\chi_2(s|\varphi_1)$, but the definition (4.4.50) is modified to the forms

$$
\begin{aligned}
\chi_1(s|\varphi_2) &= E(\exp(is\Phi_1)|\varphi_2), \\
\chi_2(s|\varphi_1) &= E(\exp(is\Phi_2)|\varphi_1),
\end{aligned}
\tag{4.4.145}
$$

where E means the conditional expectation,

$$
\begin{aligned}
E(\exp(is\Phi_1)|\varphi_2) &= \int_{\theta_{01}}^{\theta_{01}+2\pi} \exp(is\varphi_1) P(\varphi_1|\varphi_2)\, d\varphi_1, \\
E(\exp(is\Phi_2)|\varphi_1) &= \int_{\theta_{02}}^{\theta_{02}+2\pi} \exp(is\varphi_2) P(\varphi_2|\varphi_1)\, d\varphi_2.
\end{aligned}
\tag{4.4.146}
$$

As a further characteristic of correlation, we may recommend the conditional preferred phases $\text{pref}(\Phi_1|\varphi_2)$, $\text{pref}(\Phi_2|\varphi_1)$ provided that these quantities are defined for all φ_2, φ_1, respectively. With the conditional preferred phases the connected lines $l_{1|2}$ and $l_{2|1}$ on the torus $\mathbb{T}^2$ consisting of the points $(\exp[\text{pref}(\Phi_1|\varphi_2)], \exp(i\varphi_2))$ for all

$\varphi_2 \in [\theta_{02}, \theta_{02}+2\pi)$ and $(\exp(i\varphi_1), \exp[\mathrm{pref}(\Phi_2|\varphi_1)])$ for all $\varphi_1 \in [\theta_{01}, \theta_{01}+2\pi)$, respectively, have been associated. In a situation, to chart the connected line $l_{1|2}$ on the torus $\mathbb{T}^2$, we choose a reference point with the coordinates $\varphi_{1\theta_{01}} = \mathrm{Pref}_{\theta_{01}}(\Phi_1|\varphi_2 = \theta_{02})$, $\varphi_{2\theta_{02}} = \theta_{02}$ in $\mathbb{R}^2$. For $\varphi_{2\theta_{02}} \neq \theta_{02}$, we pick up the values from $\mathrm{pref}(\Phi_1|\varphi_2)$, which will form an image of the line $l_{1|2}$ in $\mathbb{R}^2$ going through the reference point. This procedure is known in the optical literature as the phase unwrapping. An image of the line $l_{2|1}$ may be obtained by selecting the reference point with the coordinates $\varphi_{1\theta_{01}} = \theta_{01}$, $\varphi_{2\theta_{02}} = \mathrm{Pref}_{\theta_{02}}(\Phi_2|\varphi_1 = \theta_{01})$ in $\mathbb{R}^2$ and continuing the curve in $\mathbb{R}^2$ with suitable points from $\mathrm{pref}(\Phi_2|\varphi_1)$. The situations of better or worse fit between the curves $l_{1|2}$ and $l_{2|1}$ may be interpreted as a stronger or weaker correlation between the multiple-valued random phase variables Φ_1 and Φ_2. A prospective use can be associated also with the conditional kth-order preferred phases.

The covariance of complex random variables $\exp(i\Phi_1)$, $\exp(-i\Phi_1)$ can be decomposed in the form

$$\mathrm{cov}\left(e^{i\Phi_1}, e^{-i\Phi_1}\right)$$

$$= E_2\mathrm{cov}_1\left(e^{i\Phi_1}, e^{-i\Phi_1}|\Phi_2\right) + \mathrm{cov}_2\left(E_1(e^{i\Phi_1}|\Phi_2), E_1(e^{-i\Phi_1}|\Phi_2)\right). \qquad (4.4.147)$$

Let us remember that the conditional expectations $E(e^{i\Phi_1}|\varphi_2)$, $E(e^{-i\Phi_1}|\varphi_2)$ and the conditional covariance $\mathrm{cov}(e^{i\Phi_1}, e^{-i\Phi_1}|\varphi_2)$ are understandable as related to the conditional probability density $P(\varphi_1|\varphi_2)$ defined in (1.1.141). Of course, these quantities are functions of the value φ_2 of the random phase variable Φ_2 and when they are measurable functions we may substitute for φ_2 the appropriate random phase variable Φ_2. In this case we use the subscript notation $E_1(e^{i\Phi_1}|\Phi_2)$, $E_1(e^{-i\Phi_1}|\Phi_2)$, and $\mathrm{cov}_1\left(e^{i\Phi_1}, e^{-i\Phi_1}|\Phi_2\right)$ for the random quantities resulting from this substitution. Subscript 2 is used for the expectation of cov_1 and for the expectations and covariances of $E_1(e^{i\Phi_1}|\Phi_2)$, $E_1(e^{-i\Phi_1}|\Phi_2)$. A dispersion and the conditional dispersion can be introduced in (4.4.147) according to the formulae (4.4.141) and to

$$D_1(\Phi_1|\Phi_2) = \mathrm{cov}_1\left(e^{i\Phi_1}, e^{-i\Phi_1}|\Phi_2\right). \qquad (4.4.148)$$

As

$$\mathrm{cov}_2\left(E_1(e^{i\Phi_1}|\Phi_2), E_1(e^{-i\Phi_1}|\Phi_2)\right)$$

$$= \mathrm{var}_2\left(E_1(\cos \Phi_1|\Phi_2)\right) + \mathrm{var}_2\left(E_1(\sin \Phi_1|\Phi_2)\right) \geq 0, \qquad (4.4.149)$$

it holds that

$$E_2D_1(\Phi_1|\Phi_2) \leq D(\Phi_1). \qquad (4.4.150)$$

The difference (4.4.149) between the sides of (4.4.150) is a measure of uncertainty for the prediction of the values of Φ_1 based on the knowledge of Φ_2. Similar considerations are well known for the entropy of the distribution of Φ_1, which enters the right-hand side of an analogue of (4.4.150), whereas the conditional entropy enters the left-hand side. The right-hand side as reduced by the conditional entropy is called the transinformation [Frieden (1983, 1991)].

As an application, we will consider the two-dimensional wrapped normal distribution and the phase distribution of the two-mode squeezed vacuum state.

(i) The technique motivated by the relation (4.4.111) can be formulated, without introducing Z_j, in terms of the formula

$$P(\varphi_1, \varphi_2) = \sum_{k_1=-\infty}^{\infty} \sum_{k_2=-\infty}^{\infty} P_{\mathbb{R}^2}(\varphi_1 - 2\pi k_1, \varphi_2 - 2\pi k_2), \qquad (4.4.151)$$

where $P_{\mathbb{R}^2}(z_1, z_2)$ is a usual probability density, e. g.,

$$P_{\mathbb{R}^2}(z_1, z_2) = \frac{1}{2\pi \sigma_1 \sigma_2 \sqrt{1 - \rho^2}}$$

$$\times \exp\left\{\frac{-1}{2(1-\rho^2)}\left[\frac{(z_1 - m_1)^2}{\sigma_1^2} + \frac{(z_2 - m_2)^2}{\sigma_2^2} - \frac{2\rho(z_1 - m_1)(z_2 - m_2)}{\sigma_1 \sigma_2}\right]\right\},$$
$$(4.4.152)$$

with $\sigma_j > 0$, $j = 1, 2$, $|\rho| < 1$, $m_j \in \mathbb{R}$, $j = 1, 2$. According to a sampling theorem [Frieden (1991)], we observe that an equivalent expression consists of (4.4.114), with

$$\chi(s_1, s_2) = \exp\left[i(s_1 m_1 + s_2 m_2) - \frac{1}{2}(s_1^2 \sigma_1^2 + s_2^2 \sigma_2^2 + 2s_1 s_2 \sigma_1 \sigma_2 \rho)\right]. \qquad (4.4.153)$$

For $\rho = 0$ the probability density (4.4.152) is a product of two probability densities and this reappears in (4.4.151) as the independence of the random phase variables Φ_1, Φ_2.

(ii) Quantum-derived phase distribution was considered in an analysis of the two-mode squeezed vacuum state [Barnett and Pegg (1990b)]. The L-periodic joint probability density of random phase variables is here

$$P(\varphi_1, \varphi_2) = \frac{1}{4\pi^2} \frac{1}{\cosh(2r) - \cos(\varphi_1 + \varphi_2 - \xi)\sinh(2r)}, \qquad (4.4.154)$$

where r and ξ are parameters of squeezing, $r \geq 0$, and ξ is a phase. The characteristic double sequence (4.4.112) reads as

$$\chi(s_1, s_2) = \delta_{s_1, s_2} \exp(i\xi s_1)(\tanh r)^{s_1}, \quad (s_1, s_2) \in \mathbb{Z}^2. \qquad (4.4.155)$$

Considering the L'-periodic probability density of the multiple-valued phase sum and the multiple-valued phase difference Φ_+, Φ_-, we resort to the characteristic double sequence according to (4.4.121),

$$\chi'_{+-}(s_+, s_-) = \delta_{2s_-, 0} \exp(i\xi s_+)(\tanh r)^{s_+} \qquad (4.4.156)$$

for $s_+, s_- \in \mathbb{Z}$ or $s_+, s_- \in \mathbb{Z} + \frac{1}{2}$. Adopting the L-periodic probability density, we merely ignore s_+, s_- from $\mathbb{Z} + \frac{1}{2}$ and (4.4.123) holds. From (4.4.125) we obtain that

$$P_{+-}(\varphi_+, \varphi_-) = \frac{1}{4\pi^2} \frac{1}{\cosh(2r) - \cos(\varphi_+ - \xi)\sinh(2r)}. \qquad (4.4.157)$$

Applying the specific concept of the group correlation coefficient, we obtain that

$$\rho^2_{\exp(i\Phi_1),\exp(i\Phi_2)} = 1 - \frac{1}{(\cosh r)^4} \ . \tag{4.4.158}$$

Analyzing the calculations, we may rewrite (4.4.158) as

$$\rho^2_{\exp(i\Phi_1),\exp(i\Phi_2)} = 1 - \left[1 - |\text{cod}(\Phi_1, -\Phi_2)|^2 \right]^2 . \tag{4.4.159}$$

Let us note that the marginal phase probability densities are uniform, i. e.,

$$P_1(\varphi_1) = P_2(\varphi_2) = \frac{1}{2\pi} \tag{4.4.160}$$

and the appropriate dispersions are maximum

$$D(\Phi_1) = D(\Phi_2) = 1. \tag{4.4.161}$$

The codispersions (4.4.142)

$$\text{cod}(\Phi_1, \Phi_2) = 0, \quad \text{cod}(\Phi_1, -\Phi_2) = \exp(i\xi)\tanh r. \tag{4.4.162}$$

These values have a possible interpretation that the multiple-valued random phase variables Φ_1 and Φ_2 are uncorrelated, but the multiple-valued random phase variables Φ_1 and $-\Phi_2$ are more correlated for greater $\tanh r$ or simply for greater r. The phase parameter ξ could be helpful for choosing the double window Q. According to (4.4.144) the conditional probability densities

$$P(\varphi_1|\varphi_2) = P(\varphi_2|\varphi_1) = \frac{1}{2\pi} \frac{1}{\cosh(2r) - \cos(\varphi_1 + \varphi_2 - \xi)\sinh(2r)} \ . \tag{4.4.163}$$

The conditional characteristic sequences (4.4.145) become

$$\chi_1(s|\varphi_2) = \exp[is(\xi - \varphi_2)](\tanh r)^{|s|}, \tag{4.4.164}$$

$$\chi_2(s|\varphi_1) = \exp[is(\xi - \varphi_1)](\tanh r)^{|s|}. \tag{4.4.165}$$

The conditional preferred phases follow from (4.4.105) and (4.4.163)

$$\begin{aligned} \xi - \varphi_2 &\in \text{pref}(\Phi_1|\varphi_2), \\ \xi - \varphi_1 &\in \text{pref}(\Phi_2|\varphi_1). \end{aligned} \tag{4.4.166}$$

Here the graphs of these dependences are determined as

$$\begin{aligned} l_{1|2} &= \left\{ (e^{i(\xi-\varphi_2)}, e^{i\varphi_2}); \varphi_2 \in [\theta_{02}, \theta_{02} + 2\pi) \right\}, \\ l_{2|1} &= \left\{ (e^{i\varphi_1}, e^{i(\xi-\varphi_1)}); \varphi_1 \in [\theta_{01}, \theta_{01} + 2\pi) \right\} \end{aligned} \tag{4.4.167}$$

providing the same curve (a helix on torus, which for a specific closedness of torus is a topological circle). This identity does not ensure the minimum conditional dispersions,

$$D(\Phi_1|\varphi_2) = D(\Phi_2|\varphi_1) = \frac{1}{\cosh^2 r} \ . \tag{4.4.168}$$

Hence,

$$E_2 D_1(\Phi_1|\Phi_2) = E_1 D_2(\Phi_2|\Phi_1) = \frac{1}{\cosh^2 r} \tag{4.4.169}$$

and in the relation (4.4.150) the equality is not attained.

The assumption that the random variables are outcomes of a quantum measurement and the appropriate replacement of the random variables by operators has been discussed in [Lukš and Peřinová (1996)].

4.5   Orderings of exponential phase operators

In this section we intend to illustrate the similarities and differences between the antinormal ordering of the annihilation and creation operators and that of the exponential phase operators.

4.5.1   Analogy

The Susskind–Glogower cosine and sine operators can also be defined as

$$\widehat{\cos}\,\varphi = \mathrm{Re}[\widehat{\exp}(i\varphi)], \quad \widehat{\sin}\,\varphi = \mathrm{Im}[\widehat{\exp}(i\varphi)], \tag{4.5.1}$$

where $\mathrm{Re}\,\hat{M} = \frac{1}{2}(\hat{M} + \hat{M}^\dagger)$, $\mathrm{Im}\,\hat{M} = -\frac{i}{2}(\hat{M} - \hat{M}^\dagger)$, $\hat{M}$ being an arbitrary operator, and one of the exponential phase operators

$$\widehat{\exp}(i\varphi) = \hat{u} = (\hat{a}\hat{a}^\dagger)^{-\frac{1}{2}}\hat{a}, \tag{4.5.2}$$

from (4.3.24). Our study is based on observations that the commutator with the other exponential phase operator

$$\widehat{\exp}(-i\varphi) = \hat{u}^\dagger, \tag{4.5.3}$$

$$[\widehat{\exp}(i\varphi), \widehat{\exp}(-i\varphi)] = |0\rangle\langle 0|, \tag{4.5.4}$$

is a projection operator similarly as $[\hat{a}, \hat{a}^\dagger] = \hat{1}$ and that the commutator

$$[\widehat{\cos}\,\varphi, \widehat{\sin}\,\varphi] = \frac{i}{2}|0\rangle\langle 0| \tag{4.5.5}$$

is similar to $[\mathrm{Re}\,\hat{a}, \mathrm{Im}\,\hat{a}] = \frac{i}{2}\hat{1}$.

In addition to the properties of the Susskind–Glogower phase operators which are expounded by the commutators, there are some which are related to associative algebra. Let us name [Carruthers and Nieto (1968)]

$$(\widehat{\cos}\,\varphi)^2 + (\widehat{\sin}\,\varphi)^2 = \hat{1} - \frac{1}{2}|0\rangle\langle 0|. \tag{4.5.6}$$

This relation is suspicious at first sight. In fact, in the measurement theory we can assign, to the Hermitian operator on the left- or right-hand side of (4.5.6), a random variable taking only on the values $\frac{1}{2}$ and 1. In a state of the physical system, where the photon-number distribution is known to be $p(n)$, the sum of squared trigonometric operators takes on the value $\frac{1}{2}$ with the probability $p(0)$ and the value 1 with the probability $1 - p(0)$. It is the possibility that the random variable assumes the value $\frac{1}{2}$ that provokes. Although there is no need of ordering cosine and sine operators on the left-hand side of (4.5.6), we see that using (4.5.1), we sum the expressions

$$
\begin{aligned}
(\widehat{\cos}\,\varphi)^2 &= \frac{1}{4}\left[\widehat{\exp}(i2\varphi) + \widehat{\exp}(i\varphi)\widehat{\exp}(-i\varphi)\right.\\
&\quad \left. + \widehat{\exp}(-i\varphi)\widehat{\exp}(i\varphi) + \widehat{\exp}(-i2\varphi)\right], \tag{4.5.7}\\
&= \frac{1}{2}\left[\hat{1} - \frac{1}{2}|0\rangle\langle 0| + \widehat{\cos}(2\varphi)\right], \tag{4.5.8}
\end{aligned}
$$

$$
\begin{aligned}
(\widehat{\sin}\,\varphi)^2 &= -\frac{1}{4}\left[\widehat{\exp}(i2\varphi) - \widehat{\exp}(i\varphi)\widehat{\exp}(-i\varphi)\right.\\
&\quad \left. \widehat{\exp}(i\varphi)\widehat{\exp}(i\varphi) + \widehat{\exp}(-i2\varphi)\right], \tag{4.5.9}\\
&= \frac{1}{2}\left[\hat{1} - \frac{1}{2}|0\rangle\langle 0| - \widehat{\cos}(2\varphi)\right]. \tag{4.5.10}
\end{aligned}
$$

Here the relations (4.5.7) and (4.5.9) comprise the operators

$$\widehat{\exp}(ik\varphi) = \begin{cases} \hat{u}^k & \text{for } k \geq 0, \\ \hat{u}^{\dagger|k|} & \text{for } k < 0 \end{cases} \tag{4.5.11}$$

and are ordered symmetrically in the exponential phase operators. But, if we replace the product $\widehat{\exp}(-i\varphi)\widehat{\exp}(i\varphi)$ with $\widehat{\exp}(i\varphi)\widehat{\exp}(-i\varphi)$, this is what we name the anti-normal ordering of the operators $\widehat{\exp}(i\varphi)$, $\widehat{\exp}(-i\varphi)$, and denote it appropriately by the subscript $\mathcal{A}$ and the superscript CS, we obtain the operators

$$\left[(\widehat{\cos}\,\varphi)^2\right]_{\mathcal{A}}^{CS} = \frac{1}{2}[\hat{1} + \widehat{\cos}(2\varphi)], \quad [(\widehat{\sin}\,\varphi)^2]_{\mathcal{A}}^{CS} = \frac{1}{2}[\hat{1} - \widehat{\cos}(2\varphi)]. \tag{4.5.12}$$

On summing, we get

$$\left[(\widehat{\cos}\,\varphi)^2 + (\widehat{\sin}\,\varphi)^2\right]_{\mathcal{A}}^{CS} = \hat{1}. \tag{4.5.13}$$

A similar use of the normal ordering leads to

$$\left[(\widehat{\cos}\,\varphi)^2 + (\widehat{\sin}\,\varphi)^2\right]_{\mathcal{N}}^{CS} = \hat{1} - |0\rangle\langle 0|. \tag{4.5.14}$$

In the measurement theory, we assign to the Hermitian operator of (4.5.14) a random variable taking on the values 0 and 1. The possibility that the random variable assumes the value 0 is undesirable. On the contrary, to the identity operator in (4.5.13), we assign the deterministic value 1 [Lukš and Peřinová (1991)].

To complete the analogy, we return to the photon annihilation and creation operators, which have the properties

$$\left[(\operatorname{Re}\hat{a})^2 + (\operatorname{Im}\hat{a}^2)\right]_{\mathcal{N}}^{QP} = \hat{a}^\dagger\hat{a}, \tag{4.5.15}$$

where the superscript QP means the usual phase-space approach and $\mathcal{N}$ means the normal ordering of the operators $\hat{a}, \hat{a}^\dagger$,

$$(\operatorname{Re}\hat{a})^2 + (\operatorname{Im}\hat{a})^2 = \hat{a}^\dagger\hat{a} + \frac{1}{2}\hat{1}, \tag{4.5.16}$$

and

$$\left[(\operatorname{Re}\hat{a})^2 + (\operatorname{Im}\hat{a})^2\right]_{\mathcal{A}}^{QP} = \hat{a}^\dagger\hat{a} + \hat{1}, \tag{4.5.17}$$

where $\mathcal{A}$ means the antinormal ordering of the operators $\hat{a}, \hat{a}^\dagger$. Not only are the results (4.5.15), (4.5.16), (4.5.17) presented in the increasing order, but also the relations (4.5.14), (4.5.6), (4.5.13), respectively, exhibit the growth of the right-hand sides. In more detail, using the order relation of Hermitian operators [Davies (1976)], the analogy can be illustrated by the relations

$$\left[(\operatorname{Re}\hat{a})^2\right]_{\mathcal{N}}^{QP} \leq (\operatorname{Re}\hat{a})^2 \leq \left[(\operatorname{Re}\hat{a})^2\right]_{\mathcal{A}}^{QP}, \tag{4.5.18}$$

$$\left[(\widehat{\cos}\,\varphi)^2\right]_{\mathcal{N}}^{CS} \leq (\widehat{\cos}\,\varphi)^2 \leq \left[(\widehat{\cos}\,\varphi)^2\right]_{\mathcal{A}}^{CS}, \tag{4.5.19}$$

which can be derived also for the operators $\operatorname{Im}\hat{a}$ and $\widehat{\sin}\,\varphi$.

Introducing the operators

$$\Delta\widehat{\cos}\,\varphi = \widehat{\cos}\,\varphi - \langle\widehat{\cos}\,\varphi\rangle, \quad \Delta\widehat{\sin}\,\varphi = \widehat{\sin}\,\varphi - \langle\widehat{\sin}\,\varphi\rangle \tag{4.5.20}$$

and taking into account the relations for the operator $\widehat{\cos}\,\varphi$

$$\begin{aligned}
\langle(\Delta\widehat{\cos}\,\varphi)^2\rangle_{\mathcal{N}}^{CS} &= \langle(\widehat{\cos}\,\varphi)^2\rangle_{\mathcal{N}}^{CS} - \langle\widehat{\cos}\,\varphi\rangle^2, \\
\langle(\Delta\widehat{\cos}\,\varphi)^2\rangle &= \langle(\widehat{\cos}\,\varphi)^2\rangle - \langle\widehat{\cos}\,\varphi\rangle^2, \\
\langle(\Delta\widehat{\cos}\,\varphi)^2\rangle_{\mathcal{A}}^{CS} &= \langle(\widehat{\cos}\,\varphi)^2\rangle_{\mathcal{A}}^{CS} - \langle\widehat{\cos}\,\varphi\rangle^2
\end{aligned} \tag{4.5.21}$$

and the similar relations for the operator $\widehat{\sin}\,\varphi$, we find that

$$\langle(\Delta\widehat{\cos}\,\varphi)^2\rangle_{\mathcal{N}}^{CS} + \langle(\Delta\widehat{\sin}\,\varphi)^2\rangle_{\mathcal{N}}^{CS} = \langle\hat{1} - |0\rangle\langle 0|\rangle - |\langle\widehat{\exp}(i\varphi)\rangle|^2, \tag{4.5.22}$$

Standard	$\sin^m \varphi \cos^n \varphi \to \widehat{\cos}^n \varphi \, \widehat{\sin}^m \varphi$
Antistandard	$\sin^m \varphi \cos^n \varphi \to \widehat{\sin}^m \varphi \, \widehat{\cos}^n \varphi$
Normal	$\exp(-im\varphi)\exp(in\varphi) \to \widehat{\exp}(-im\varphi)\widehat{\exp}(in\varphi)$
Antinormal	$\exp(-im\varphi)\exp(in\varphi) \to \widehat{\exp}(in\varphi)\widehat{\exp}(-im\varphi)$

Table 4.1: Rules of association between the trigonometric and exponential functions of the variable φ and functions of the operators $\widehat{\exp}(i\varphi)$, $\widehat{\exp}(-i\varphi)$, $\widehat{\cos}\,\varphi$, $\widehat{\sin}\,\varphi$.

$$\langle(\Delta\widehat{\cos}\,\varphi)^2\rangle + \langle(\Delta\widehat{\sin}\,\varphi)^2\rangle \;=\; \langle\hat{1} - \frac{1}{2}|0\rangle\langle 0|\rangle - |\langle\widehat{\exp}(i\varphi)\rangle|^2, \quad (4.5.23)$$

$$\langle(\Delta\widehat{\cos}\,\varphi)^2\rangle_{\mathcal{A}}^{CS} + \langle(\Delta\widehat{\sin}\,\varphi)^2\rangle_{\mathcal{A}}^{CS} \;=\; 1 - |\langle\widehat{\exp}(i\varphi)\rangle|^2. \qquad (4.5.24)$$

Taking into account that

$$\langle\Delta\widehat{\exp}(-i\varphi)\Delta\widehat{\exp}(i\varphi)\rangle = \langle(\Delta\widehat{\cos}\,\varphi)^2\rangle_{\mathcal{N}}^{CS} + \langle(\Delta\widehat{\sin}\,\varphi)^2\rangle_{\mathcal{N}}^{CS}, \qquad (4.5.25)$$

we can see that the formula (4.5.22) is the original proposal of Lévy-Leblond (1976), cf. [Lynch (1995)]. The relation (4.5.24) is most satisfactory [Lukš and Peřinová (1991)]. The use of the antinormal ordering leads to the usual real phase (mod 2π) and avoids introducing a complex phase. The formulae (4.5.22) and (4.5.23) for the variances indicate that the adoption of the normal and symmetrical orderings is equivalent to the introduction of a complex quantum phase. We can deduce that this complex phase should have a positive imaginary part.

The analogy between the operators $\hat{a}$, $\hat{a}^\dagger$, $\mathrm{Re}\,\hat{a}$, $\mathrm{Im}\,\hat{a}$ on the one hand and the operators $\widehat{\exp}(i\varphi)$, $\widehat{\exp}(-i\varphi)$, $\widehat{\cos}\,\varphi$, $\widehat{\sin}\,\varphi$ on the other hand is so far going that the terms the standard and antistandard orderings for the operators $\widehat{\cos}\,\varphi$, $\widehat{\sin}\,\varphi$ and the terms the normal and antinormal orderings for the operators $\widehat{\exp}(i\varphi)$, $\widehat{\exp}(-i\varphi)$ are natural. These orderings are given by the rules of assignment according to table 4.1. In relation to the symmetrical or Weyl ordering the four phase operators behave equally well. Nevertheless, just in this case the Pythagorean theorem, as applied to the unit circle, is not obeyed. In the only case of the antinormal ordering, results are desirable. An observation of the Weyl ordering of the Susskind–Glogower cosine and sine operators has been made, when q-deformation has been considered [Kimler IV and Nelson (1996)].

Let us note that in the formulae (4.5.13), (4.5.14), (4.5.19) last two lines of table 4.1 have been applied in the form

$$\widehat{\exp}(-im\varphi)\widehat{\exp}(in\varphi) = [\widehat{\exp}(-im\varphi)\widehat{\exp}(in\varphi)]_{\mathcal{N}}^{CS}, \qquad (4.5.26)$$

$$\widehat{\exp}(in\varphi)\widehat{\exp}(-im\varphi) = [\widehat{\exp}(in\varphi)\widehat{\exp}(-im\varphi)]_{\mathcal{A}}^{CS}. \qquad (4.5.27)$$

As the change affects the term to the left of the arrow, becoming the content of the bracket $[\;\;]^{CS}$, so deeply, let us understand that the meaning of these formal

Standard	$S^m C^n \to \widehat{\cos}^n \varphi\, \widehat{\sin}^m \varphi$
Antistandard	$S^m C^n \to \widehat{\sin}^m \varphi\, \widehat{\cos}^n \varphi$
Normal	$(C - iS)^m (C + iS)^n \to \widehat{\exp}(-im\varphi)\widehat{\exp}(in\varphi)$
Antinormal	$(C - iS)^m (C + iS)^n \to \widehat{\exp}(in\varphi)\widehat{\exp}(-im\varphi)$

Table 4.2: Rules of association between functions of the variables C, S and functions of the operators $\widehat{\exp}(i\varphi)$, $\widehat{\exp}(-i\varphi)$, $\widehat{\cos}\,\varphi$, $\widehat{\sin}\,\varphi$.

expressions can be best conveyed by the modification $\widehat{\exp}(-im\varphi^*)\widehat{\exp}(in\varphi)$ taking into account that the phase can be complex. In this manner, we respect that the theories of ordering start with functions [Moyal (1949), Agarwal and Wolf (1970a)], even if the formalisms of quantum field theory and solid state physics deal only with operators (cf. our bracket with operators inside). In this spirit, $\cos\varphi$ and $\sin\varphi$ should be modified to become

$$C \equiv C(\varphi) = \frac{1}{2}\left[\exp(i\varphi) + \exp(-i\varphi^*)\right],$$

$$S \equiv S(\varphi) = \frac{1}{2i}\left[\exp(i\varphi) - \exp(-i\varphi^*)\right]. \qquad (4.5.28)$$

The inverse transformation to (4.5.28) reads as

$$\varphi = -i\ln(C + iS). \qquad (4.5.29)$$

It is more advantageous to work with new variables C and S than with the complex phase φ. The corrected rules of assignment are given in table 4.2.

In this place we want to comment on the symmetrical ordering of any two operators, which has been expounded in section 3.1 ((3.1.44)). It is necessary, because the symmetrical ordering is missing in table 4.1. We can take the operator $\exp(i\theta\hat{A}+i\tau\hat{B})$ for the generator of the symmetrically ordered products of n operators $\hat{A}$ and m operators $\hat{B}$, $n, m = 0, 1, \ldots, \infty$. Here $\hat{A} = \widehat{\cos}\,\varphi$, $\hat{B} = \widehat{\sin}\,\varphi$. We may also remove the requirement of the Hermiticity of operators $\hat{A}$, $\hat{B}$ and interpret the operator

$$\exp(i\theta\widehat{\cos}\,\varphi + i\tau\widehat{\sin}\,\varphi) = \exp\left[\frac{(\tau + i\theta)}{2}\widehat{\exp}(i\varphi) + \frac{(-\tau + i\theta)}{2}\widehat{\exp}(-i\varphi)\right], \qquad (4.5.30)$$

as the generating operator of the symmetrically ordered products of n operators $\widehat{\exp}(i\varphi)$ and m operators $\widehat{\exp}(-i\varphi)$. Treatment of a different kind of ordering than the symmetrical one is more or less connected to the role of $\hat{A} + i\hat{B}$ as a physically meaningful operator. The standard and antistandard orderings from table 4.1 are easily generalized to operators $\hat{A}$, $\hat{B}$. In fact, this is reported in [Scully and Cohen (1987)]. The principle is clear, but the use of the words standard and antistandard may be awkward or even impossible.

The cosine and sine operators are incompatible, because their diagonal representations use distinct identity resolutions, i. e.,

$$\widehat{\cos}\,\varphi = \int_{-1}^{1} C|C\rangle\langle C|\,dC, \tag{4.5.31}$$

where [Peřinová and Lukš (1996)]

$$|C\rangle = \frac{1}{i\sqrt{1-C^2}}[\exp(i\varphi)|\varphi\rangle - \exp(-i\varphi)|-\varphi\rangle], \quad \varphi = \mathrm{Cos}^{-1}\,C, \tag{4.5.32}$$

and

$$\widehat{\sin}\,\varphi = \int_{-1}^{1} S|S\rangle\langle S|\,dS, \tag{4.5.33}$$

where

$$|S\rangle = i^{\hat{n}}|C = S\rangle, \tag{4.5.34}$$

with

$$i^{\hat{n}} = \exp\left(i\frac{\pi}{2}\hat{n}\right). \tag{4.5.35}$$

Whereas $|C\rangle$ is the eigenstate of the cosine operator appropriate to the eigenvalue C, $|C = S\rangle$ means the eigenstate of the cosine operator appropriate to the eigenvalue S. In comparison with [Carruthers and Nieto (1968)] we have assumed simplified formulae (4.5.31), (4.5.33) and obtained the slightly more complicated formula (4.5.32). Although the phase states (4.3.72), which enter the explicit formulae for $|C\rangle$ and $|S\rangle$, are not orthogonal, the eigenstates of cosine and sine operators are orthogonal in a Dirac sense,

$$\begin{aligned} \langle C|C'\rangle &= \delta(C - C'), \\ \langle S|S'\rangle &= \delta(S - S'). \end{aligned} \tag{4.5.36}$$

The eigenstates of the cosine operator are also given by the expansion

$$|C\rangle = \begin{cases} \sqrt{\frac{2}{\pi}}\sqrt[4]{1-C^2}\sum_{n=0}^{\infty} U_n(C)|n\rangle & \text{for } C \in [-1,1], \\ 0 & \text{for } C \notin [-1,1], \end{cases} \tag{4.5.37}$$

where $U_n(C)$ are the Chebyshev polynomials of the second kind [Mathai (1993)]. In fact, no eigenstates can be the null vectors, but we add them having in mind the spectral density measure.

The phase states $|\varphi\rangle$ have the property (cf. (4.6.33)),

$$\langle\varphi|\varphi'\rangle = \frac{1}{4\pi} + \frac{1}{2}\delta(\varphi - \varphi') - i\frac{1}{4\pi}P_\varphi\cot\left(\frac{\varphi - \varphi'}{2}\right), \tag{4.5.38}$$

where $\mathcal{P}_\varphi$ (principal value) denotes a generalized function of φ, which is the limit of the indicated function, but replaced by zero in a symmetrical neighbourhood of $\varphi \equiv \varphi' \pmod{2\pi}$. From this we derive the following formula

$$\langle C|S \rangle = \sqrt[4]{1-C^2}\sqrt[4]{1-S^2}\Big[-\frac{1}{\pi}\mathcal{P}_C\frac{1}{C^2+S^2-1}$$
$$+i\,\mathrm{sgn}(CS)\delta(C^2+S^2-1)\Big], \tag{4.5.39}$$

where $\mathcal{P}_C$ relates to the singularities $C = \pm\sqrt{1-S^2}$. The "unphysical features," in fact quantum oscillations, of the probability distribution of the Susskind–Glogower cosine operator for number states [D'Ariano and Paris (1993)] can be compared with the shape of the probability distribution of the quadrature operator for these states. The peculiar quantum statistics are avoided adopting a joint measurement of the two quadratures similarly as the canonical phase probability distribution defines satisfactorily a joint distribution of quantum cosine and sine.

4.5.2   Quasidistributions for quantum cosine and sine

We introduce the quasidistributions of eigenvalues of the cosine and sine phase operators using the method of quantum characteristic function which was outlined in [Scully and Cohen (1987)]. The method of characteristic function can be extended to other orderings. We introduce generating functions as operators

$$\hat{D}^{CS}_{\mathrm{st}}\left(-\frac{\tau}{2}+i\frac{\theta}{2}\right) = \exp(i\theta\,\widehat{\cos}\,\varphi)\exp(i\tau\,\widehat{\sin}\,\varphi), \tag{4.5.40}$$

$$\hat{D}^{CS}_{\mathrm{antist}}\left(-\frac{\tau}{2}+i\frac{\theta}{2}\right) = \exp(i\tau\,\widehat{\sin}\,\varphi)\exp(i\theta\,\widehat{\cos}\,\varphi), \tag{4.5.41}$$

$$\hat{D}^{CS}_{\mathcal{N}}\left(-\frac{\tau}{2}+i\frac{\theta}{2}\right) = \exp\left[\frac{-\tau+i\theta}{2}\widehat{\exp}(-i\varphi)\right]$$
$$\times\exp\left[\frac{\tau+i\theta}{2}\widehat{\exp}(i\varphi)\right], \tag{4.5.42}$$

$$\hat{D}^{CS}_{\mathcal{A}}\left(-\frac{\tau}{2}+i\frac{\theta}{2}\right) = \exp\left[\frac{\tau+i\theta}{2}\widehat{\exp}(i\varphi)\right]$$
$$\times\exp\left[\frac{-\tau+i\theta}{2}\widehat{\exp}(-i\varphi)\right], \tag{4.5.43}$$

$$\hat{D}^{CS}_{\mathcal{S}}\left(-\frac{\tau}{2}+i\frac{\theta}{2}\right) = \exp(i\theta\,\widehat{\cos}\,\varphi+i\tau\,\widehat{\sin}\,\varphi), \tag{4.5.44}$$

where the subscripts st, antist, $\mathcal{N}$, $\mathcal{A}$, $\mathcal{S}$ denote the standard, antistandard, normal, antinormal, and symmetrical orderings, respectively. The generating function (4.5.44) of symmetrically ordered monomials in the operators $\widehat{\exp}(i\varphi)$, $\widehat{\exp}(-i\varphi)$ admits a dynamical interpretation. Sukumar (1989b) has adopted this approach and produced new states from the vacuum state using this unitary operator. He has described also the production of states from any given number state $|m\rangle$. The coefficients of the number-state representations are just the matrix elements of our operator generating function (4.5.44). The occurrence of the Bessel functions is reminiscent of the phase optimized states of Bandilla, Paul and Ritze (1991) (see section 4.9). Sukumar (1989b) has advertised that a suitable interaction could suppress a specific number-state component.

Quantum characteristic functions are defined by the relation

$$C_{\text{ord}}^{CS}\left(-\frac{\tau}{2}+i\frac{\theta}{2}\right) = \text{Tr}\left\{\hat{\rho}\hat{D}_{\text{ord}}^{CS}\left(-\frac{\tau}{2}+i\frac{\theta}{2}\right)\right\}, \tag{4.5.45}$$

where ord = st, antist, $\mathcal{N}$, $\mathcal{A}$, $\mathcal{S}$ and $\hat{\rho}$ is the statistical operator to be represented by the quasidistributions. The respective quasidistributions can be obtained according to the relation

$$\Phi_{\text{ord}}^{CS}(C+iS) = \mathcal{F}^{-1}\left[C_{\text{ord}}^{CS}\left(-\frac{\tau}{2}+i\frac{\theta}{2}\right)\right](C+iS), \tag{4.5.46}$$

where the inverse Fourier transformation

$$\mathcal{F}^{-1}\left[C_{\text{ord}}^{CS}\left(-\frac{\tau}{2}+i\frac{\theta}{2}\right)\right](C+iS)$$

$$= \frac{1}{4\pi^2}\int\int \exp(-i\theta C - i\tau S)C_{\text{ord}}^{CS}\left(-\frac{\tau}{2}+i\frac{\theta}{2}\right) d\theta\, d\tau. \tag{4.5.47}$$

For ord = st, antist, $\mathcal{N}$, $\mathcal{A}$, $\mathcal{S}$ we introduce the operator densities

$$\hat{\Phi}_{\text{ord}}^{CS}(C+iS) = \mathcal{F}^{-1}\left[\hat{D}_{\text{ord}}^{CS}\left(-\frac{\tau}{2}+i\frac{\theta}{2}\right)\right](C+iS), \tag{4.5.48}$$

and we arrive at

$$\hat{\Phi}_{\text{st}}^{CS}(C+iS) = \langle C|S\rangle|C\rangle\langle S|, \tag{4.5.49}$$

$$\hat{\Phi}_{\text{antist}}^{CS}(C+iS) = \langle S|C\rangle|S\rangle\langle C|, \tag{4.5.50}$$

$$\hat{\Phi}_{\mathcal{N}}^{CS}(C+iS) = \frac{1}{\sqrt{1-C^2-S^2}}\sum_{n=0}^{\infty}\sum_{m=0}^{\infty}\frac{1}{n!m!}\left[-\frac{\partial}{\partial(C-iS)}\right]^n$$

$$\times\left[-\frac{\partial}{\partial(C+iS)}\right]^m \delta(C+iS)|n\rangle\langle m|, \tag{4.5.51}$$

$$\hat{\Phi}_{\mathcal{A}}^{CS}(C+iS) = 2\delta(C^2+S^2-1)|\varphi=\arg(C+iS)\rangle\langle\varphi=\arg(C+iS)|, \tag{4.5.52}$$

$$\hat{\Phi}_{\mathcal{S}}^{CS}(C+iS) = \frac{1}{\pi}\sum_{n=0}^{\infty}(n+1)\sum_{m=0}^{n}(C+iS)^{n-m}R_m^{(1,n-m)}(C^2+S^2)|n\rangle\langle m|$$

$$+\frac{1}{\pi}\sum_{m=0}^{\infty}(m+1)\sum_{n=0}^{m-1}(C-iS)^{m-n}R_n^{(1,m-n)}(C^2+S^2)|n\rangle\langle m|$$

$$\text{for } C^2+S^2 < 1, \tag{4.5.53}$$

$$\hat{\Phi}_{\mathcal{S}}^{CS}(C+iS) = \hat{0} \text{ for } C^2+S^2 > 1, \tag{4.5.54}$$

where $R_m^{(\alpha,\beta)}(x)$ are the shifted Jacobi polynomials [Mathai (1993)]

$$R_m^{(\alpha,\beta)}(x) = \frac{(-1)^m}{m!}(1-x)^{-\alpha}x^{-\beta}\frac{d^m}{dx^m}\left[(1-x)^{\alpha+m}x^{\beta+m}\right]. \tag{4.5.55}$$

Let us note that the operator densities (4.5.49) and (4.5.50) can be derived almost immediately without the generators (4.5.40) and (4.5.41) and the corresponding method can be formulated for two and more noncommuting operators [Barut (1957)].

The operator densities (4.5.51), (4.5.52) in terms of the variables $\rho = \exp(-\text{Im}\,\varphi)$ and $\text{Re}\,\varphi$ read

$$\hat{\Phi}_{\mathcal{N}}^{CS}\left(\rho e^{i\text{Re}\,\varphi}\right) = \frac{1}{\pi\rho\sqrt{1-\rho^2}}\sum_{n=0}^{\infty}\sum_{m=0}^{\infty}\frac{1}{(n+m)!}e^{i(n-m)\text{Re}\,\varphi}$$

$$\times\left(-\frac{\partial}{\partial\rho}\right)^{n+m}\delta(\rho)|n\rangle\langle m|, \tag{4.5.56}$$

$$\hat{\Phi}_{\mathcal{A}}^{CS}\left(\rho e^{i\text{Re}\,\varphi}\right) = 2\delta(\rho^2-1)|\text{Re}\,\varphi\rangle\langle\text{Re}\,\varphi|. \tag{4.5.57}$$

Still for ord = st, antist, $\mathcal{N}$, $\mathcal{A}$, $\mathcal{S}$, an application of (4.5.46) to (4.5.47) leads to the rule

$$\Phi_{\text{ord}}^{CS}(C+iS) = \text{Tr}\left\{\hat{\rho}\hat{\Phi}_{\text{ord}}^{CS}(C+iS)\right\}. \tag{4.5.58}$$

The case of normal ordering is difficult, because the operator density is an operator-valued generalized function. The same formal property in the case of antinormal ordering is not so strong, because the generalized function can be separated in (4.5.57). Above all, the Dirac delta function present there expresses the very pleasant fact that the variables C and S have the unit circle property. In other words, the problem of complex phase does not arise for the antinormal ordering.

On substituting the concrete operator densities (4.5.49)–(4.5.54) into (4.5.58), we obtain the appropriate quasidistributions

$$\Phi_{\text{st}}^{CS}(C+iS) = \langle S|\hat{\rho}|C\rangle\langle C|S\rangle, \tag{4.5.59}$$

$$\Phi_{\text{antist}}^{CS}(C+iS) = \langle C|\hat{\rho}|S\rangle\langle S|C\rangle, \tag{4.5.60}$$

$$\Phi_{\mathcal{N}}^{CS}(C+iS) = \frac{1}{\sqrt{1-C^2-S^2}}\sum_{n=0}^{\infty}\sum_{m=0}^{\infty}\rho_{mn}\frac{1}{n!m!}$$

$$\times\left[-\frac{\partial}{\partial(C-iS)}\right]^n\left[-\frac{\partial}{\partial(C+iS)}\right]^m\delta(C+iS)$$

$$= \frac{1}{\pi\rho\sqrt{1-\rho^2}}\sum_{n=0}^{\infty}\sum_{m=0}^{\infty}\rho_{mn}\frac{1}{(n+m)!}e^{i(n-m)\mathrm{Re}\,\varphi}$$

$$\times\left(-\frac{\partial}{\partial\rho}\right)^{n+m}\delta(\rho), \tag{4.5.61}$$

$$\Phi_{\mathcal{A}}^{CS}(C+iS) = 2P\left(\arg(C+iS)\right)\delta(C^2+S^2-1),$$

$$= 2P(\mathrm{Re}\,\psi)\delta(\rho^2-1), \tag{4.5.62}$$

$$\Phi_{\mathcal{S}}^{CS}(C+iS) = \frac{1}{\pi}\sum_{n=0}^{\infty}(n+1)\sum_{m=0}^{n}\rho_{mn}(C+iS)^{n-m}R_m^{(1,n-m)}(C^2+S^2)$$

$$+\frac{1}{\pi}\sum_{m=0}^{\infty}(m+1)\sum_{n=0}^{m-1}\rho_{mn}(C-iS)^{m-n}R_n^{(1,m-n)}(C^2+S^2)$$

$$\text{for } C^2+S^2<1, \tag{4.5.63}$$

$$\Phi_{\mathcal{S}}^{CS}(C+iS) = 0 \text{ for } C^2+S^2>1. \tag{4.5.64}$$

The quasidistributions related to the standard and antistandard orderings of the cosine and sine operators are products of a regular function and a generalized function. An analogous property can be seen in the quasidistribution related to the antinormal ordering of the exponential phase operators. Moreover, we can observe that the decomposition into usual and generalized functions coincides with the separation between real and imaginary parts of the complex phase. The generalized function then defines that the imaginary part of the phase vanishes and so the usual function introduces the "antinormal" distribution $P(\varphi)$ of the resulting real phase.

Another role of the operator densities (4.5.49)–(4.5.54) can be seen in a generation of operators according to a given complex-phase space function $M(C+iS)$,

$$\hat{M}_{\mathrm{ord}} = \int_{-\infty}^{\infty}\int_{-\infty}^{\infty}M(C+iS)\hat{\Phi}_{\mathrm{ord}}^{CS}(C+iS)\,dC\,dS. \tag{4.5.65}$$

The reconstruction of the original statistical operator can or cannot be accomplished according to the relation

$$\hat{\rho} = \int\int\Phi_{\mathrm{ord}}^{CS}(C+iS)\hat{\Delta}_{\mathrm{ord}}^{CS}(C+iS)\,dC\,dS, \tag{4.5.66}$$

where the operators to be mixed

$$\hat{\Delta}_{\mathrm{st}}^{CS}(C+iS) = -\pi\frac{C^2+S^2-1}{\sqrt[4]{1-C^2}\sqrt[4]{1-S^2}}|S\rangle\langle C|, \tag{4.5.67}$$

$$\hat{\Delta}_{\mathrm{antist}}^{CS}(C+iS) = -\pi\frac{C^2+S^2-1}{\sqrt[4]{1-C^2}\sqrt[4]{1-S^2}}|C\rangle\langle S|, \tag{4.5.68}$$

$$\hat{\Delta}_{\mathcal{N}}^{CS}(C+iS) = |C+iS\rangle\langle C+iS|, \tag{4.5.69}$$

with $|C+iS\rangle$ a coherent phase state,

$$\hat{\Delta}_{\mathcal{S}}^{CS}(C+iS) = \sum_{n=0}^{\infty}\sum_{m=0}^{\infty}u_{nm}(C^2+S^2)\langle n|\hat{\Phi}_{\mathcal{S}}^{CS}(C+iS)|m\rangle|n\rangle\langle m|, \tag{4.5.70}$$

with

$$u_{nm}(C^2 + S^2) = \pi \frac{(n+m+1)}{(n+1)(m+1)}(1 - C^2 - S^2).   \tag{4.5.71}$$

In the case of antinormal ordering the scheme (4.5.66) cannot be embodied, because the state of the physical system cannot be determined completely by the mere phase properties.

It is obvious that the relation between operator density $\hat{\Phi}_{\mathrm{ord}}^{CS}(C+iS)$ and the operator $\hat{\Delta}_{\mathrm{ord}}^{CS}(C+iS)$ (if any) is not very simple, there is a difference from an archetypal distribution of quadratures, where the relationship is established by a simple multiplication by π and by the Hermitian conjugation. For example, the usual quadrature Wigner function can be defined as

$$\Phi_S^{QP}(\alpha) = \mathrm{Tr}\left\{\hat{\rho}\hat{\Phi}_S^{QP}(\alpha)\right\},   \tag{4.5.72}$$

where the Wigner operator (cf. (3.4.14) for $s = 0$)

$$\hat{\Phi}_S^{QP}(\alpha) = \frac{2}{\pi}(-1)^{(\hat{a}^\dagger - \alpha^* \hat{1})(\hat{a} - \alpha \hat{1})}.   \tag{4.5.73}$$

Recalling the theory in section 3.4, we observe that for $s = 0$ by the inversion of (4.5.72)

$$\hat{\rho} = \int \Phi_S^{QP}(\alpha)\hat{\Delta}_S^{QP}(\alpha)\, d^2\alpha,   \tag{4.5.74}$$

where the operators to be mixed

$$\hat{\Delta}_S^{QP}(\alpha) = \pi\hat{\Phi}_S^{QP}(\alpha).   \tag{4.5.75}$$

Arbitrarily small thermalization of (4.5.75) leads to an operator whose trace exists and is unity.

Contrary to the usual presentation, we expound the mapping of operators whose expectations are to be taken onto functions, i. e., an inverse relation to (4.5.65). Given an arbitrary operator $\hat{M}$, its c-number equivalent is defined as

$$M_{\mathrm{ord}}(C + iS) = \mathrm{Tr}\left\{\hat{\Delta}_{\mathrm{ord}}^{CS}(C + iS)\hat{M}\right\}, \quad \mathrm{ord} \neq \mathcal{A}.   \tag{4.5.76}$$

Again, the simplicity of the usual presentation is due to the proportionalities like (4.5.75), whereas we had to distinguish the relation (4.5.76) from (4.5.58).

4.6   Enlargement of the Hilbert space of a harmonic oscillator

Susskind and Glogower (1964) and Carruthers and Nieto (1968) knew that part of quantum phase problem would reappear as the quantum angle problem. Difficulties with the latter problem are so minute that any reduction of the quantum phase problem to the angle problem can be considered a solution of the phase problem. Since the angular momentum, which is canonically conjugate to the rotation angle, takes on values of both signs, there has been a trend of enlarging the Hilbert space of the harmonic oscillator with negative number states [Barnett and Pegg (1986)]. Newton has achieved the enlargement by supplementing the harmonic oscillator with spin [Newton (1980)].

4.6.1   Number-product vacuum states

We will adhere to the approach of Shapiro and Shepard (1991) appropriate to quantum optics. From this a thread leads to the formalism of Ban [Ban (1993), Shapiro (1993)]. Shapiro and Shepard (1991) derived the number-product vacuum states as elements of the Hilbert space $\mathcal{H}_e$,

$$|n_-, 0\rangle, \quad |0, n_+\rangle, \tag{4.6.1}$$

which belong to the Hilbert spaces $\mathcal{H}_{-0} \subset \mathcal{H}_{-+}$, $\mathcal{H}_{0+} \subset \mathcal{H}_{-+}$, respectively, where $\mathcal{H}_{-+} = \mathcal{H}_- \otimes \mathcal{H}_+$, whereas Ban (1993) considered the relative number states (see section 4.13). We can introduce the relative number states $|n, m\rangle\rangle$ using formally asymmetrical identity resolutions

$$\hat{1}_e = \sum_{n_-=0}^{\infty} |n = n_-, 0\rangle\rangle\langle n_-, 0| + \sum_{n_+=1}^{\infty} |n = -n_+, 0\rangle\rangle\langle 0, n_+| = \hat{e}_{-,0} + \hat{e}'_{+,0}, \tag{4.6.2}$$

$$\hat{1}_e = \sum_{n=0}^{\infty} |n_- = n, 0\rangle\langle\langle n, 0| + \sum_{n=-\infty}^{-1} |0, n_+ = -n\rangle\langle\langle n, 0| = \hat{p}_{-,0} + \hat{p}'_{+,0}. \tag{4.6.3}$$

The assumption of photon vacuum expressed in (4.6.1) as $n_+ = 0$ and $n_- = 0$, respectively, enables one to reduce some two-mode concepts to the single-mode ones. From the relations (4.6.2) and (4.6.3) it is obvious that

$$\mathcal{H}_e = \mathcal{H}_{-0} \oplus \mathcal{H}'_{0+}, \tag{4.6.4}$$

where $\mathcal{H}'_{0+} = \mathcal{H}_{0+} \cap \mathcal{H}_{00}^{\perp}$, with $\mathcal{H}_{00} = \mathcal{H}_{-0} \cap \mathcal{H}_{0+}$.

A similar, but simpler solution has been presented by Newton (1980) and it has been generalized in the study of formal properties of operator polar decomposition by Stenholm (1993). Reflecting somewhat the 2×2 matrix notation of these authors and remembering that the reduced exponential phase operators

$$\hat{u}_{-,0} = \sum_{n_-=0}^{\infty} |n_-, 0\rangle\langle n_- + 1, 0|, \tag{4.6.5}$$

$$\hat{u}_{+,0} = \sum_{n_+=0}^{\infty} |0, n_+\rangle\langle 0, n_+ + 1|, \tag{4.6.6}$$

we introduce the Shapiro–Shepard unitary exponential phase operator

$$\widehat{\exp}(i\phi_{-+,0}) = \hat{e}_{-,0}\hat{u}_{-,0}\hat{p}_{-,0} + \hat{e}'_{+,0}|0,1\rangle\langle 0,0|\hat{p}_{-,0} + \hat{e}'_{+,0}\hat{u}^{\dagger}_{+,0}\hat{p}'_{+,0}, \tag{4.6.7}$$

or

$$\widehat{\exp}(i\phi_{-+,0}) = \sum_{n=-\infty}^{\infty} |n,0\rangle\rangle\langle\langle n+1,0|. \tag{4.6.8}$$

The usual notation $\hat{Y} \equiv \hat{Y}_0$ [Shapiro (1993)] is modified here and the subscript zero indicates the assumption of number-product vacuum states. Let us note that here and in what follows the mirror convention for bras is not used. It can easily be verified that the phase state in the Newton expansion

$$|\phi,0\rangle\rangle = \hat{e}_{-,0}|\phi,0\rangle + \hat{e}'_{+,0}|0,-\phi\rangle, \tag{4.6.9}$$

where

$$|\phi,0\rangle = \frac{1}{\sqrt{2\pi}} \sum_{n_-=0}^{\infty} \exp(in_-\phi)|n_-,0\rangle,$$

$$|0,\phi\rangle = \frac{1}{\sqrt{2\pi}} \sum_{n_+=0}^{\infty} \exp(in_+\phi)|0,n_+\rangle. \tag{4.6.10}$$

So simply

$$|\phi,0\rangle\rangle = \frac{1}{\sqrt{2\pi}} \sum_{n=-\infty}^{\infty} \exp(in\phi)|n,0\rangle\rangle, \tag{4.6.11}$$

in the Shapiro–Shepard formalism. To a 2π-periodic function $M(\phi)$ of an observed phase the operator $\widehat{M}_0$ can be assigned in the Shapiro–Shepard system,

$$\widehat{M}_0 = \int_{\theta_0}^{\theta_0+2\pi} M(\phi)|\phi,0\rangle\rangle\langle\langle\phi,0|\,d\phi. \tag{4.6.12}$$

Applying the Newton expansion, we get

$$\begin{aligned}\widehat{M}_0 &= \hat{e}_{-,0}\widehat{M}_{-,0}\hat{p}_{-,0} + \hat{e}_{-,0}\widehat{M}_{-,+,0}\hat{p}'_{+,0} \\ &\quad + \hat{e}'_{+,0}\widehat{M}_{+,-,0}\hat{p}_{-,0} + \hat{e}'_{+,0}\widehat{M}^-_{+,0}\hat{p}'_{+,0},\end{aligned} \tag{4.6.13}$$

where the Toeplitz operators

$$\widehat{M}_{-,0} = \int_{\theta_0}^{\theta_0+2\pi} M(\varphi)|\varphi,0\rangle\langle\varphi,0|\,d\varphi,$$

$$\widehat{M}^-_{+,0} = \int_{\theta_0}^{\theta_0+2\pi} M(-\varphi)|0,\varphi\rangle\langle 0,\varphi|\,d\varphi, \tag{4.6.14}$$

and the cross operator terms

$$\widehat{M}_{-,+,0} = \int_{\theta_0}^{\theta_0+2\pi} M(\varphi)|\varphi,0\rangle\langle 0,-\varphi|\,d\varphi,$$

$$\widehat{M}_{+,-,0} = \int_{\theta_0}^{\theta_0+2\pi} M(\varphi)|0,-\varphi\rangle\langle\varphi,0|\,d\varphi. \tag{4.6.15}$$

The above elaborated notation is ready for use in situations when the harmonic oscillator system is treated together with a quantum rotator.

4.6.2   States of a plane rotator

It is worthwhile to distinguish two mechanical models, a plane rotator (such as a bead on a circular wire) [Susskind and Glogower (1964)] and a massive particle in the xy plane [Lukš, Peřinová and Křepelka (1994a)](see section 4.13). In the former case a rotation angle suffices to determine the position and in the latter case it does together with a radius. Thus there are two operators of a rotation angle. Considering a rotation about the z axis in the case of the plane rotator, the operator of the angular momentum is of importance. In quantum theory, the (orbital) angular momentum takes on only the integer multiples of the Planck constant divided by 2π, $\hbar = \frac{h}{2\pi}$. In other words, the operator of the angular momentum $\hat{L}_z$ fulfills the relation

$$\hat{L}_z = -\hbar \hat{N}, \tag{4.6.16}$$

where $\hat{N}$ is an operator with the eigenvalues $N \in \mathbb{Z}$ and the eigenvectors $|N\rangle$, i. e.,

$$\hat{N} = \sum_{N=-\infty}^{\infty} N |N\rangle \langle N|. \tag{4.6.17}$$

The exponential operator of the rotation angle reads

$$\widehat{\exp}(i\phi) = \sum_{N=-\infty}^{\infty} |N\rangle\langle N+1|. \tag{4.6.18}$$

The case of the massive particle is more complicated, because in quantum theory the state of the motion can be circular, elliptic, and linear. Nevertheless, we may assume that at most the circular motion is present. So, we also dispose of the notation (4.6.1)

$$|N\rangle = \begin{cases} |n_- = N, 0\rangle & \text{for } N \geq 0, \\[2mm] |0, n_+ = -N\rangle & \text{for } N \leq 0, \end{cases} \tag{4.6.19}$$

where $-$ and $+$ mean the clockwise and anticlockwise rotations, respectively. As a consequence, we interpret the exponential angular operator as the exponential phase operator for the clockwise rotation. Annihilation operators are introduced,

$$\hat{a}_- = \sum_{N=0}^{\infty} \sqrt{N+1} |N\rangle\langle N+1|, \tag{4.6.20}$$

$$\hat{a}_+ = \sum_{N=-\infty}^{-1} \sqrt{-N} |N+1\rangle\langle N|. \tag{4.6.21}$$

Introducing the Susskind–Glogower exponential phase operators for the clockwise and anticlockwise motion (cf. (4.6.5), (4.6.6)),

$$\widehat{\exp}(i\varphi_-) = \sum_{N=0}^{\infty} |N\rangle\langle N+1|, \tag{4.6.22}$$

$$\widehat{\exp}(i\varphi_+) = \sum_{N=-\infty}^{-1} |N+1\rangle\langle N|, \tag{4.6.23}$$

we observe that the exponential operator of the rotation angle for the plane rotator (cf. (4.6.7))

$$\widehat{\exp}(i\phi) = \widehat{\exp}(i\varphi_-) + [\widehat{\exp}(i\varphi_+)]^\dagger. \tag{4.6.24}$$

In terms of the states (4.6.19) the rotation angle states can be expanded just as done in (4.3.41). The phase states of the clockwise and anticlockwise rotations are given in the form

$$|\varphi_- = \phi\rangle = \hat{1}_- |\phi\rangle_e, \quad |\varphi_+ = \phi\rangle = \hat{1}_+ |-\phi\rangle_e, \tag{4.6.25}$$

respectively, where

$$\hat{1}_- = \sum_{N=0}^{\infty} |N\rangle\langle N|, \quad \hat{1}_+ = \sum_{N=-\infty}^{0} |N\rangle\langle N|. \tag{4.6.26}$$

Moreover, the relative number operator

$$\hat{N} = \hat{n}_- - \hat{n}_+, \tag{4.6.27}$$

where the number operators

$$\hat{n}_- = \hat{a}_-^\dagger \hat{a}_-, \quad \hat{n}_+ = \hat{a}_+^\dagger \hat{a}_+. \tag{4.6.28}$$

Believing that concepts of the quantum theory of rotation angles need regularization, Barnett and Pegg (1990a) have accomplished a truncation of the Hilbert space $\mathcal{H}_e$.

4.6.3   The original and enlarged Hilbert spaces

After the enlarged Hilbert space $\mathcal{H}_e$ has been introduced in section 4.3, the identity operator of the original Hilbert space $\mathcal{H} \equiv \mathcal{H}_-$ becomes interesting as the projection operator onto the space, whereas the enlarged Hilbert space has its extended identity operator (4.3.43). Using the resolution of the original identity

$$\hat{1} = \int_{\theta_0}^{\theta_0+2\pi} |\varphi\rangle\langle\varphi|\, d\varphi, \tag{4.6.29}$$

we can obtain the integral expression for any ket of $\mathcal{H}$

$$|\psi\rangle = \int_{\theta_0}^{\theta_0+2\pi} \psi(\varphi)|\varphi\rangle\, d\varphi. \tag{4.6.30}$$

Here we have introduced a phase representation

$$\psi(\varphi) = \langle\varphi|\psi\rangle. \tag{4.6.31}$$

This representation can be obtained also as an angular representation in the space of plane rotator,

$$\psi(\varphi) = {}_e\langle\phi = \varphi|\psi\rangle. \tag{4.6.32}$$

In contrast to the angle states $|\phi\rangle_e$ being orthogonal, it holds that

$$
\begin{aligned}
\langle\varphi|\varphi'\rangle &= \frac{1}{2\pi}\sum_{n=0}^{\infty}\exp[-in(\varphi-\varphi')] \\
&= \frac{1}{2}\delta(\varphi-\varphi') + \frac{1}{4\pi}\left\{1 - i\mathcal{P}_\varphi\cot\left[\frac{1}{2}(\varphi-\varphi')\right]\right\} \\
&= \frac{1}{2}\delta(\varphi-\varphi') + \frac{1}{2\pi}\mathcal{P}_\varphi\left\{1 - \exp[-i(\varphi-\varphi')]\right\}^{-1}.
\end{aligned}
\tag{4.6.33}
$$

The phase states are mutually dependent, viz. [Susskind and Glogower (1964)]

$$
\int_{\theta_0}^{\theta_0+2\pi} \exp(ik\varphi)|\varphi\rangle\, d\varphi = 0, \quad k = 1, 2, ..., \infty.
\tag{4.6.34}
$$

In comparison with the enlarged Hilbert space, the phase states of which comprise vector components with negative number of photons, this can be interpreted as the absence of such a vector component in the phase space of the original Hilbert space. It is well known that the system of coherent states has a similar property [Peřina (1972, 1991)]. The overcompleteness of the phase states $|\varphi\rangle$ means that the ket $|\psi\rangle$ may be expanded with many weighting factors other than $\langle\varphi|\psi\rangle$ ((4.6.31)), but only $\langle\varphi|\psi\rangle$ has projections only on the negative integer exponentials [Susskind and Glogower (1964)].

These phase representations are the so-called boundary functions in the Hardy space $H_2(D)$ (D is the unit disk) [Duran (1970)] of the functions $f(\alpha^*)$,

$$
f(\alpha^*) = \frac{1}{\sqrt{2\pi}}\sum_{n=0}^{\infty} c_n\alpha^{*n}.
\tag{4.6.35}
$$

The functions (4.6.35) are also z transforms of the signals c_n [Oppenheim and Schafer (1975)]. Putting $\alpha^* = \exp(-i\varphi)$ and

$$
c_n = \langle n|\psi\rangle
\tag{4.6.36}
$$

in (4.6.35), we obtain the phase representation

$$
\psi(\varphi) = \frac{1}{\sqrt{2\pi}}\sum_{n=0}^{\infty} c_n\exp(-in\varphi).
\tag{4.6.37}
$$

From the identity resolution it follows that the overlap of two kets has the simple form

$$
\langle\psi|\psi'\rangle = \int_{\theta_0}^{\theta_0+2\pi} \psi^*(\varphi)\psi'(\varphi)\, d\varphi,
\tag{4.6.38}
$$

where

$$
\psi'(\varphi) = \langle\varphi|\psi'\rangle.
\tag{4.6.39}
$$

Another property

$$\int_{\theta_0}^{\theta_0+2\pi} \psi(\varphi) \exp(iN\varphi)\, d\varphi = 0, \quad N = -1, -2, ..., -\infty, \tag{4.6.40}$$

is pointed out in [Garrison and Wong (1970)].

Using repeatedly the resolution of identity (4.6.29), we get an integral expression for the statistical operator

$$\hat{\rho} = \int_{\theta_0}^{\theta_0+2\pi} \int_{\theta_0}^{\theta_0+2\pi} \rho(\varphi, \varphi') |\varphi\rangle \langle \varphi'|\, d\varphi\, d\varphi', \tag{4.6.41}$$

where the matrix elements in the phase state basis,

$$\rho(\varphi, \varphi') = \langle \varphi | \hat{\rho} | \varphi' \rangle, \tag{4.6.42}$$
$$= {}_e\langle \phi = \varphi | \hat{\rho} | \phi = \varphi' \rangle_e. \tag{4.6.43}$$

More generally, we could consider these matrix elements for other operators, but one must be cautious, because the phase and angular states go beyond both the original and enlarged Hilbert spaces. The utility of the phase state basis $|\varphi\rangle$ to represent various quantum operators and states is reflected in the literature [Susskind and Glogower (1964), Lukš and Peřinová (1991, 1993b), Lukš, Peřinová and Křepelka (1992a), Brif and Ben-Aryeh (1994b)].

The number and phase operators on the enlarged Hilbert space $\mathcal{H}_e$ have the commutator

$$\left[\hat{N}, \hat{\phi}_{\theta_0}\right] = i(\hat{1}_e - 2\pi |\theta_0\rangle_{ee}\langle\theta_0|), \tag{4.6.44}$$

with $|\theta_0\rangle_e$ defined in (4.3.41). From this the uncertainty relation for these operators in a state with the statistical operator $\hat{\rho}_e$ acting on the space $\mathcal{H}_e$ follows

$$\langle (\Delta \hat{N})^2 \rangle_e \langle (\Delta \hat{\phi}_{\theta_0})^2 \rangle_e \geq \frac{1}{4}\left[1 - 2\pi P_e(\theta_0)\right]^2, \tag{4.6.45}$$

where the subscript e refers to the state $\hat{\rho}_e$ and

$$P_e(\theta_0) = {}_e\langle\theta_0|\hat{\rho}_e|\theta_0\rangle_e. \tag{4.6.46}$$

Particularly, for the states described by the statistical operator $\hat{\rho}$ acting on the original Hilbert space $\mathcal{H}$ the number–phase uncertainty relation simplifies

$$\langle (\Delta \hat{n})^2 \rangle \langle (\Delta \hat{\phi}_{\theta_0})^2 \rangle \geq \frac{1}{4}\left[1 - 2\pi P(\theta_0)\right]^2, \tag{4.6.47}$$

where

$$P(\theta_0) = \langle\theta_0|\hat{\rho}|\theta_0\rangle. \tag{4.6.48}$$

The absence of the subscript e refers to the state with the operator $\hat{\rho}$, the operator $\hat{N}$ can be replaced by the operator $\hat{n}$, but the operator $\hat{\phi}_{\theta_0}$ must still be used. Whereas our derivation is based on the usual scheme (3.5.21), an operator-free derivation of the uncertainty relation (4.6.47) for pure states is also possible [Judge (1964), Shapiro, Shepard and Wong (1990)]. A rotationally symmetrical modification of the uncertainty relation (4.6.45) for the plane rotator has been derived by Judge (1964) and considered for the quantum harmonic oscillator by Fain (1967).

In the following we will show to what extent we can apply the method of characteristic function in the case of the photon-number and phase-function operators. At present we know that the method of the characteristic function used by other authors [Mukunda (1979), Bizarro (1994)] can lead to the Wigner function for rotation angle and orbital angular momentum without half-odd numbers of orbital angular-momentum quanta. Their results are relevant for the quantum phase problem. Whereas Mukunda (1979) shortens 4π-domain of the characteristic function without further explanation, the detailed analysis of Bizarro (1994) reveals that the 4π-domain would lead to an introduction of the half-odd values of the angular momentum, which are carefully excluded from the system of the rigid (plane) rotator.

We define the operator generating function for the symmetrically ordered products of number and phase operators [Lukš and Peřinová (1991)]

$$\hat{G}_{n\varphi_{\theta_0}}(\mu,\lambda) = \hat{1}\hat{G}_{N\phi_{\theta_0}}(\mu,\lambda)\hat{1}, \qquad (4.6.49)$$

where

$$\hat{G}_{N\phi_{\theta_0}}(\mu,\lambda) = \exp\left[i(\mu\hat{N} + \lambda\hat{\phi}_{\theta_0})\right] \qquad (4.6.50)$$

and μ, λ are real numbers. In fact, the appropriate operator generating function is defined in the enlarged Hilbert space and projected onto the original one with the original identity operator $\hat{1}$. Although the number and phase operators on the enlarged Hilbert space have the quantum Poisson bracket not equal to the identity operator (cf. (4.6.44)), it holds at least that

$$\left[\hat{N}, \hat{\phi}_{\theta_0}\right] \approx i\hat{1}_e, \qquad (4.6.51)$$

which is a good approximation for well-chosen quantum states. From the Baker–Hausdorff identity (3.2.4) we derive the group commutation relation

$$\exp(\hat{A})\exp(\hat{B})\exp(-\hat{A})\exp(-\hat{B}) = \exp\left([\hat{A},\hat{B}]\right). \qquad (4.6.52)$$

We observe that an analogous group commutation relation holds

$$\exp\left(i\lambda\hat{\phi}_{\theta_0}\right)\exp(i\mu\hat{N})\exp\left(-i\lambda\hat{\phi}_{\theta_0}\right)\exp(-i\mu\hat{N}) = \exp(i\mu\lambda\hat{1}_e) \text{ for } \lambda \text{ integer.}$$
$$(4.6.53)$$

In the following, we restrict ourselves to λ integer, because on this assumption the operator $\exp\left(i\lambda\hat{\phi}_{\theta_0}\right)$ is a displacement operator for the angular-momentum representation. As the group commutator on the left-hand side of (4.6.53) is a c-number, the operator generating function can be approximated,

$$\hat{G}_{N\phi_{\theta_0}}(\mu,\lambda) \approx \hat{G}_{N\phi}^{\text{appr}}(\mu,\lambda), \qquad (4.6.54)$$

where

$$\hat{G}_{N\phi}^{\text{appr}}(\mu, \lambda) = \exp(i\mu\hat{N}) \exp\left(i\lambda\hat{\phi}_{\theta_0}\right) \exp(i\mu\lambda) \qquad (4.6.55)$$

$$= \exp\left(i\lambda\hat{\phi}_{\theta_0}\right) \exp(i\mu\hat{N}) \exp(-i\mu\lambda). \qquad (4.6.56)$$

The right-hand side in (4.6.56) is ordered so that the operator $\hat{\phi}_{\theta_0}$ is to the left from $\hat{N}$ (the antistandard ordering), and the right-hand side in (4.6.55) comprise the operator $\hat{N}$ to the left from $\hat{\phi}_{\theta_0}$ (the standard ordering). As it holds that

$$\hat{1} \exp(i\lambda\hat{\phi}_{\theta_0})\hat{1} = \begin{cases} \hat{u}^{\lambda}, & \lambda \geq 0, \\ \hat{u}^{\dagger|\lambda|}, & \lambda \leq 0, \end{cases} \qquad (4.6.57)$$

we rewrite (4.6.55), (4.6.56) in the respective ways

$$\hat{1}\hat{G}_{N\phi_{\theta_0}}(\mu, \lambda)\hat{1} \approx \begin{cases} \hat{u}^{\lambda} \exp(i\mu\hat{n}) \exp\left(-i\frac{\mu\lambda}{2}\right), & \lambda \geq 0, \\ \hat{u}^{\dagger|\lambda|} \exp(i\mu\hat{n}) \exp\left(-i\frac{\mu\lambda}{2}\right), & \lambda \leq 0, \end{cases} \qquad (4.6.58)$$

$$\hat{1}\hat{G}_{N\phi_{\theta_0}}(\mu, \lambda)\hat{1} \approx \begin{cases} \exp(i\mu\hat{n})\hat{u}^{\lambda} \exp\left(i\frac{\mu\lambda}{2}\right), & \lambda \geq 0, \\ \exp(i\mu\hat{n})\hat{u}^{\dagger|\lambda|} \exp\left(i\frac{\mu\lambda}{2}\right), & \lambda \leq 0. \end{cases} \qquad (4.6.59)$$

Using the expansions in the number state basis

$$\hat{u}^{\lambda} = \sum_{n=0}^{\infty} |n\rangle\langle n + \lambda|, \quad \lambda \geq 0,$$

$$\hat{u}^{\dagger|\lambda|} = \sum_{n=0}^{\infty} |n - \lambda\rangle\langle n|, \quad \lambda \leq 0, \qquad (4.6.60)$$

and respecting the sign of λ, we get for $\lambda \geq 0$

$$\hat{u}^{\lambda} \exp(i\mu\hat{n}) = \sum_{n=0}^{\infty} \exp[i\mu(n + \lambda)]|n\rangle\langle n + \lambda|, \qquad (4.6.61)$$

$$\exp(i\mu\hat{n})\hat{u}^{\lambda} = \sum_{n=0}^{\infty} \exp(i\mu n)|n\rangle\langle n + \lambda|, \qquad (4.6.62)$$

and hence

$$\hat{u}^{\lambda} \exp(i\mu\hat{n}) \exp\left(-i\frac{\mu\lambda}{2}\right) = \exp(i\mu\hat{n})\hat{u}^{\lambda} \exp\left(i\frac{\mu\lambda}{2}\right)$$

$$= \sum_{n=0}^{\infty} \exp\left[i\mu\left(n + \frac{\lambda}{2}\right)\right] |n\rangle\langle n + \lambda|. \qquad (4.6.63)$$

Similarly, the relations (4.6.61), (4.6.62), and (4.6.63) have the counterparts for $\lambda \leq 0$

$$\hat{u}^{\dagger|\lambda|}\exp(i\mu\hat{n}) = \sum_{n=0}^{\infty}\exp(i\mu n)|n - \lambda\rangle\langle n|, \tag{4.6.64}$$

$$\exp(i\mu\hat{n})\hat{u}^{\dagger|\lambda|} = \sum_{n=0}^{\infty}\exp[i\mu(n - \lambda)]|n - \lambda\rangle\langle n|, \tag{4.6.65}$$

and

$$\hat{u}^{\dagger|\lambda|}\exp(i\mu\hat{n})\exp\left(-i\frac{\mu\lambda}{2}\right) = \exp(i\mu\hat{n})\hat{u}^{\dagger|\lambda|}\exp\left(i\frac{\mu\lambda}{2}\right)$$

$$= \sum_{n=0}^{\infty}\exp\left[i\mu\left(n - \frac{\lambda}{2}\right)\right]|n - \lambda\rangle\langle n|, \tag{4.6.66}$$

respectively. By a compromising change of notation we arrive at the identity of the last expressions in (4.6.63) and (4.6.66). The multiplicative constant $\exp\left(i\frac{\mu\lambda}{2}\right)$ on the right-hand side in (4.6.59), which has the standard ordering, as well as the multiplicative constant $\exp\left(-i\frac{\mu\lambda}{2}\right)$ on the right-hand side in (4.6.58), which has the antistandard ordering, transform the corresponding operator to an approximately symmetrical ordering. This analogy allows us to approximate the operator generating function $\hat{G}_{n\varphi_{\theta_0}}(\mu, \lambda)$ in the form

$$\hat{G}_{n\varphi}^{\text{appr}}(\mu, \lambda) = \sum_{n=\frac{\lambda}{2}}^{\infty}\exp(i\mu n)\left|n - \frac{\lambda}{2}\right\rangle\left\langle n + \frac{\lambda}{2}\right|; \tag{4.6.67}$$

the subscript λ is an integer and μ is a real number (modulo 4π) or, equivalently, a number from the interval $[-2\pi, 2\pi)$. The standard or the antistandard ordering by itself leads to a complex-valued quasidistribution. But it is possible to use the standard ordering for $\lambda > 0$ and the antistandard ordering for $\lambda < 0$, as Ban (1995b) has done and the result is a real-valued quasidistribution.

To appreciate the role of the approximations (4.6.58) and (4.6.59), we observe the properties

$$\hat{G}_{n\varphi}^{\text{appr}}(\mu, \lambda) = \exp\left(-i\frac{\mu\lambda}{2}\right)\hat{u}^{\lambda}\exp(i\mu\hat{n})$$

$$= \exp\left(i\frac{\mu\lambda}{2}\right)\exp(i\mu\hat{n})\hat{u}^{\lambda}, \quad \lambda \geq 0, \tag{4.6.68}$$

$$\hat{G}_{n\varphi}^{\text{appr}}(\mu, \lambda) = \exp\left(-i\frac{\mu\lambda}{2}\right)\hat{u}^{\dagger|\lambda|}\exp(i\mu\hat{n})$$

$$= \exp\left(i\frac{\mu\lambda}{2}\right)\exp(i\mu\hat{n})\hat{u}^{\dagger|\lambda|}, \quad \lambda \leq 0. \tag{4.6.69}$$

From (4.6.60) it is evident that

$$
\hat{G}_{n\varphi}^{\text{appr}}(\mu, \lambda) = \begin{cases} \exp\left(-i\frac{\mu}{2\lambda}\hat{n}^2\right) \hat{u}^\lambda \exp\left(i\frac{\mu}{2\lambda}\hat{n}^2\right), & \lambda > 0, \\ \exp\left(-i\frac{\mu}{2\lambda}\hat{n}^2\right) \hat{u}^{\dagger|\lambda|} \exp\left(i\frac{\mu}{2\lambda}\hat{n}^2\right), & \lambda < 0, \end{cases}
\tag{4.6.70}
$$

or simply, because $\hat{G}_{n\varphi}^{\text{appr}}(0, \lambda) = \hat{u}^\lambda, \hat{u}^{\dagger|\lambda|}$ for $\lambda > 0$, $\lambda < 0$, respectively,

$$
\hat{G}_{n\varphi}^{\text{appr}}(\mu, \lambda) = \exp\left(-i\frac{\mu}{2\lambda}\hat{n}^2\right) \hat{G}_{n\varphi}^{\text{appr}}(0, \lambda) \exp\left(i\frac{\mu}{2\lambda}\hat{n}^2\right).
\tag{4.6.71}
$$

Equations (4.6.70) mean the evolution in the Heisenberg picture of the phase operators $\hat{u}^\lambda, \hat{u}^{\dagger|\lambda|}$, respectively, after the time $t = -\frac{\mu}{2\lambda\kappa}$ when we concentrate on a nonlinear oscillator described by the interaction Hamiltonian

$$
\hat{H}_{\text{int}} = \hbar\kappa(\hat{a}^\dagger\hat{a})^2,
\tag{4.6.72}
$$

where κ is a real constant for the intensity dependence [Peřinová and Lukš (1994) and references therein].

It is possible to prove that

$$
\hat{G}_{N\phi_{\theta_0}}(\mu, \lambda) = \exp\left(i\frac{\lambda}{2\mu}\hat{\phi}_{\theta_0}^2\right) \exp(i\mu\hat{N}) \exp\left(-i\frac{\lambda}{2\mu}\hat{\phi}_{\theta_0}^2\right).
\tag{4.6.73}
$$

This relation is exact and it has a surprisingly similar form to the relation

$$
\exp\left[i(\mu\hat{q} + \lambda\hat{p})\right] = \exp\left(i\frac{\lambda}{2\mu}\hat{p}^2\right) \exp(i\mu\hat{q}) \exp\left(-i\frac{\lambda}{2\mu}\hat{p}^2\right),
\tag{4.6.74}
$$

where $\hat{q}$ and $\hat{p}$ are the generalized canonically conjugate operators. The quantum characteristic function $C_{n\varphi}^{\text{appr}}(\mu, \lambda)$ reads

$$
C_{n\varphi}^{\text{appr}}(\mu, \lambda) = \text{Tr}\left\{\hat{\rho}\hat{G}_{n\varphi}^{\text{appr}}(\mu, \lambda)\right\}.
\tag{4.6.75}
$$

Now we can use the method of characteristic function to define the Wigner function $\overline{\Phi}_{n\varphi}^{\text{appr}}(n, \varphi)$ for number and phase,

$$
\overline{\Phi}_{n\varphi}^{\text{appr}}(n, \varphi) = \frac{1}{8\pi^2} \sum_{\lambda=-\infty}^{\infty} \int_{-2\pi}^{2\pi} C_{n\varphi}^{\text{appr}}(\mu, \lambda) \exp\left[-i(\mu n + \lambda\varphi)\right] d\mu.
\tag{4.6.76}
$$

The variable n is half a nonnegative integer and φ is a real number (modulo 2π). The multiplicative factor $\frac{1}{8\pi^2}$ differs from $\frac{1}{4\pi^2}$ in the usual Wigner function for the operators $\hat{q}$, $\hat{p}$ (see section 3.1) due to the fact that the normalization condition for $\overline{\Phi}_{n\varphi}^{\text{appr}}(n, \varphi)$ is chosen as follows

$$
\sideset{}{'}\sum_{n} \int_{\theta_0}^{\theta_0+2\pi} \overline{\Phi}_{n\varphi}^{\text{appr}}(n, \varphi) \, d\varphi = 1
\tag{4.6.77}
$$

instead of a possible normalization condition

$$\sum_n{}' \int_{\theta_0}^{\theta_0+2\pi} \overline{\Phi}'_{n\varphi}(n,\varphi) \, \frac{d\varphi}{2} = 1 \qquad (4.6.78)$$

for $\overline{\Phi}'_{n\varphi}(n,\varphi) = 2\overline{\Phi}^{\text{appr}}_{n\varphi}(n,\varphi)$ with the properties closer to those of a quasidistribution. We can define a screw operator,

$$\hat{D}_{n\varphi}(\overline{n},\overline{\varphi}) = \hat{G}_{n\varphi}(\overline{\varphi},-\overline{n}). \qquad (4.6.79)$$

In spite of this terminology, this operator has a restricted dynamics and only its continuation to the physical system of rotator is interesting [Lukš, Peřinová and Křepelka (1992b)]. This leads to a modification of the definition (4.6.75) in the form

$$\begin{aligned} C^{\text{appr}\,D}_{n\varphi}(\overline{n},\overline{\varphi}) &= \text{Tr}\left\{\hat{\rho}\hat{D}_{n\varphi}(\overline{n},\overline{\varphi})\right\} \\ &= C^{\text{appr}\,G}_{n\varphi}(\overline{\varphi},-\overline{n}) \end{aligned} \qquad (4.6.80)$$

and to the appropriate modification of (4.6.76)

$$\overline{\Phi}_{n\varphi}(n,\varphi) = \frac{1}{8\pi^2} \sum_{\overline{n}=-\infty}^{\infty} \int_{-2\pi}^{2\pi} C^{\text{appr}\,D}_{n\varphi}(\overline{n},\overline{\varphi}) \exp\left[i(\overline{n}\varphi - n\overline{\varphi})\right] d\overline{\varphi}. \qquad (4.6.81)$$

We will use the matrix elements method (MEM) to define the Wigner function for angular momentum and angle or, in other words, the extended Wigner function for number and phase [Lukš and Peřinová (1993b)]

$$\begin{aligned} \overline{\Phi}^{\text{MEM}}_{N\phi}(N,\phi) &= \frac{1}{2\pi} \sum_{N-k=-\infty}^{\infty} \rho_{N-k,N+k} \exp(i2k\phi) \\ &= \text{Tr}\left\{\hat{\rho}\hat{\overline{\Phi}}^{\text{MEM}}_{N\phi}(N,\phi)\right\}, \end{aligned} \qquad (4.6.82)$$

where

$$\hat{\overline{\Phi}}^{\text{MEM}}_{N\phi}(N,\phi) = \frac{1}{2\pi} \sum_{N-k=-\infty}^{\infty} \exp(-i2k\phi)|N-k\rangle\langle N+k|; \qquad (4.6.83)$$

$\hat{\rho}$ denotes a statistical operator from $\mathcal{L}(\mathcal{H}_e)$, a space of bounded linear operators acting on $\mathcal{H}_e$. We assume that N takes on half-integer values $N = 0, \pm\frac{1}{2}, \pm1, ..., \pm\infty$, lest $N-k$ and $N+k$ should be of the same parity only, and note that the value of k is half-odd when N is half-odd.

Although the Wigner function for angular momentum and angle, which can assume negative values, is a quasidistribution, it holds that

$$\int_{\theta_0}^{\theta_0+2\pi} \overline{\Phi}^{\text{MEM}}_{N\phi}(N,\phi) \, d\phi = p(N) \quad \text{for } N \text{ integer}, \qquad (4.6.84)$$

where

$$p(N) = \rho_{NN}. \tag{4.6.85}$$

Similarly it is valid that

$$\sum_{N=-\infty}^{\infty}{}' \overline{\Phi}_{N\phi}^{\mathrm{MEM}}(N, \phi) = P(\phi), \tag{4.6.86}$$

where

$$P(\phi) = {}_e\langle\phi|\hat{\rho}|\phi\rangle_e; \tag{4.6.87}$$

let us remember that $\sum'$ here and in the following means that the summation step is $\frac{1}{2}$.

Let us note that we can rewrite the formula (4.6.83) in the form

$$\hat{\overline{\Phi}}_{N\phi}^{\mathrm{MEM}}(N, \phi) = \frac{1}{2\pi}\int_{\theta_0}^{\theta_0+2\pi} \exp(-i2N\phi')|\phi + \phi'\rangle_{ee}\langle\phi - \phi'|\,d\phi'. \tag{4.6.88}$$

The system of operators $\hat{\overline{\Phi}}_{N\phi}^{\mathrm{MEM}}(N, \phi)$ is overcomplete in the space $\mathcal{L}(\mathcal{H}_e)$. As a consequence, the quasidistribution possesses the following properties

$$\overline{\Phi}_{N\phi}^{\mathrm{MEM}}(N, \phi) = \begin{cases} \overline{\Phi}_{N\phi}^{\mathrm{MEM}}(N, \phi + \pi) & \text{for } N \text{ integer,} \\ \\ -\overline{\Phi}_{N\phi}^{\mathrm{MEM}}(N, \phi + \pi) & \text{for } N \text{ half-odd.} \end{cases} \tag{4.6.89}$$

Taking into account the relations (4.6.89), we obtain in addition to (4.6.84)

$$\int_{\theta_0}^{\theta_0+2\pi} \overline{\Phi}_{N\phi}^{\mathrm{MEM}}(N, \phi)\,d\phi = 0 \quad \text{for } N \text{ half-odd.} \tag{4.6.90}$$

It can be proved that the inverse formula to (4.6.82) reads

$$\hat{\rho} = \sum_{N=-\infty}^{\infty}{}' \int_{\theta_0}^{\theta_0+2\pi} \overline{\Phi}_{N\phi}^{\mathrm{MEM}}(N, \phi)\hat{\Delta}_{N\phi}^{\mathrm{MEM}}(N, \phi)\,d\phi, \tag{4.6.91}$$

where

$$\hat{\Delta}_{N\phi}^{\mathrm{MEM}}(N, \phi) = 2\pi\hat{\overline{\Phi}}_{N\phi}^{\mathrm{MEM}}(N, \phi). \tag{4.6.92}$$

The Wigner function $\overline{\Phi}_{N\phi}^{\mathrm{MEM}}(N, \phi)$ is related to a mapping of operators. The quantum average $\langle\hat{M}\rangle$ of an operator $\hat{M}$ can be expressed in the form

$$\mathrm{Tr}\left\{\hat{\rho}\hat{M}\right\} = \sum_{N=-\infty}^{\infty}{}' \int_{\theta_0}^{\theta_0+2\pi} \overline{\Phi}_{N\phi}^{\mathrm{MEM}}(N, \phi)M(N, \phi)\,d\phi, \tag{4.6.93}$$

where

$$M(N, \phi) = \text{Tr}\left\{\hat{\Delta}_{N\phi}^{\text{MEM}}(N, \phi)\hat{M}\right\}$$

$$= \sum_{N-k=-\infty}^{\infty} \langle N - k|\hat{M}|N + k\rangle \exp(-i2k\phi). \tag{4.6.94}$$

For $\hat{M} = \hat{1}$ it holds that $M(N, \phi) \equiv 1$ for N integer, $M(N, \phi) = 0$ for N half-odd.

In quantum optical problems the statistical operators $\hat{\rho}$ possess zero matrix elements $\rho_{N'N} = 0$ for $N' < 0$ and/or $N < 0$. The same is valid with respect to the operators $\hat{M}$ except the operators of phase. The phase operators can be related naturally to the angle operators and these virtual phase operators have some nonzero matrix elements $M_{NN'}$ for $N < 0$ and/or $N' < 0$.

With respect to the projection from the space $\mathcal{L}(\mathcal{H}_e)$ onto $\mathcal{L}(\mathcal{H})$ we define the Wigner function for number and phase as follows

$$\overline{\Phi}_{n\varphi}^{\text{MEM}}(n, \varphi) = \frac{1}{2\pi} \sum_{k=-n}^{n} \rho_{n-k,n+k} \exp(i2k\varphi) \tag{4.6.95}$$

$$= \text{Tr}\left\{\hat{\rho}\hat{\overline{\Phi}}_{n\varphi}^{\text{MEM}}(n, \varphi)\right\}, \tag{4.6.96}$$

where

$$\hat{\overline{\Phi}}_{n\varphi}^{\text{MEM}}(n, \varphi) = \frac{1}{2\pi} \sum_{k=-n}^{n} \exp(-i2k\varphi)|n - k\rangle\langle n + k|. \tag{4.6.97}$$

We assume that n takes on half-integer values, $n = 0, \frac{1}{2}, 1, ..., \infty$. We note that the value of k in the sum is half-odd when n is half-odd. Evaluating the formula (4.6.76) in the number state basis, we find that

$$\hat{\overline{\Phi}}_{n\varphi}^{\text{appr}}(n, \varphi) = \hat{\overline{\Phi}}_{n\varphi}^{\text{MEM}}(n, \varphi). \tag{4.6.98}$$

The formulae analogous to (4.6.84), (4.6.85) and (4.6.86), (4.6.87) read

$$\int_{\theta_0}^{\theta_0+2\pi} \overline{\Phi}_{n\varphi}^{\text{MEM}}(n, \varphi)\, d\varphi = p(n) \text{ for } n \geq 0 \text{ and an integer}, \tag{4.6.99}$$

where

$$p(n) = \rho_{nn}, \tag{4.6.100}$$

and

$$\sum_{n=0}^{\infty}{}' \overline{\Phi}_{n\varphi}^{\text{MEM}}(n, \varphi) = P(\varphi), \tag{4.6.101}$$

where

$$P(\varphi) = \langle\varphi|\hat{\rho}|\varphi\rangle. \tag{4.6.102}$$

After the projection the quasidistribution (4.6.82) retains the properties

$$
\overline{\Phi}_{n\varphi}^{\text{MEM}}(n,\varphi) =
\begin{cases}
\overline{\Phi}_{n\varphi}^{\text{MEM}}(n,\varphi+\pi) & \text{for } n \text{ integer,} \\[2ex]
-\overline{\Phi}_{n\varphi}^{\text{MEM}}(n,\varphi+\pi) & \text{for } n \text{ half-odd.}
\end{cases}
\tag{4.6.103}
$$

Let us note that

$$
\int_{\theta_0}^{\theta_0+2\pi} \overline{\Phi}_{n\varphi}^{\text{MEM}}(n,\varphi)\, d\varphi = 0 \ \text{ for } n \text{ half-odd.}
\tag{4.6.104}
$$

It can be proved that the inverse formula to (4.6.96) is of the form

$$
\hat{\rho} = \sum_{n=0}^{\infty}{}' \int_{\theta_0}^{\theta_0+2\pi} \overline{\Phi}_{n\varphi}^{\text{MEM}}(n,\varphi)\hat{\Delta}_{n\varphi}^{\text{MEM}}(n,\varphi)\, d\varphi,
\tag{4.6.105}
$$

where

$$
\hat{\Delta}_{n\varphi}^{\text{MEM}}(n,\varphi) = 2\pi \hat{\overline{\Phi}}_{n\varphi}^{\text{MEM}}(n,\varphi).
\tag{4.6.106}
$$

The Wigner function (4.6.96) is useful for the so-called mapping theorem

$$
\text{Tr}\left\{\hat{\rho}\hat{M}\right\} = \sum_{n=0}^{\infty}{}' \int_{\theta_0}^{\theta_0+2\pi} \overline{\Phi}_{n\varphi}^{\text{MEM}}(n,\varphi)m(n,\varphi)\, d\varphi,
\tag{4.6.107}
$$

where

$$
\begin{aligned}
m(n,\varphi) &= \text{Tr}\left\{\hat{\Delta}_{n\varphi}^{\text{MEM}}(n,\varphi)\hat{M}\right\} \\
&= \sum_{k=-n}^{n} \langle n-k|\hat{M}|n+k\rangle \exp(i2k\varphi).
\end{aligned}
\tag{4.6.108}
$$

Let us remark that Ban (1995b) has invented an alternative matrix elements method leading to a particular quasidistribution for number and phase.

Taking into account the dynamics by introducing the notation

$$
C_{n\varphi}^{\text{appr}\,G}(\mu,\lambda;t) = \text{Tr}\left\{\hat{\rho}\hat{U}^{\dagger}(t)\hat{G}_{n\varphi}^{\text{appr}}(\mu,\lambda)\hat{U}(t)\right\},
\tag{4.6.109}
$$

where $\hat{U}(t)$ is a unitary operator of the time evolution, we can choose

$$
\hat{U}(t) = \exp\left(-i\kappa t \hat{n}^2\right).
\tag{4.6.110}
$$

By repeated use of the formula (4.6.71), we obtain

$$
C_{n\varphi}^{\text{appr}\,G}(\mu,\lambda;t) = \text{Tr}\left\{\hat{\rho}\hat{G}_{n\varphi}^{\text{appr}}(\mu - 2\lambda\kappa t,\lambda)\right\}, \quad \lambda \neq 0.
\tag{4.6.111}
$$

Using (4.6.109) for $t=0$, we arrive at the relation

$$
C_{n\varphi}^{\text{appr}\,G}(\mu,\lambda;t) = C_{n\varphi}^{\text{appr}\,G}(\mu - 2\lambda\kappa t,\lambda;0),
\tag{4.6.112}
$$

which is obvious for $\lambda = 0$. The tomography of a quantum state due to Leonhardt (1996) is based on the assumption $\mu = 0$. The characteristic function of the initial state can be reconstructed almost fully in terms of characteristic sequences of quantum phase in the evolved states (cf. section 4.4),

$$
\begin{aligned}
C_{n\varphi}^{\text{appr}\,G}(-2\lambda\kappa t, \lambda; 0) &= C_{n\varphi}^{\text{appr}\,G}(0, \lambda; t) \\
&= \chi(\lambda; t), \quad \lambda \neq 0.
\end{aligned}
\tag{4.6.113}
$$

Of course,

$$
C_{n\varphi}^{\text{appr}\,G}(\mu, 0; 0) = \langle \exp[i\mu\hat{n}(0)] \rangle .
\tag{4.6.114}
$$

Let us note that

$$
\begin{aligned}
C_{n\varphi}^{\text{appr}\,D}(\overline{n}, 2\overline{n}\kappa t, \lambda; 0) &= C_{n\varphi}^{\text{appr}\,D}(\overline{n}, 0; t) \\
&= \chi(-\overline{n}; t) = \chi^*(\overline{n}; t), \quad \overline{n} \neq 0,
\end{aligned}
\tag{4.6.115}
$$

$$
C_{n\varphi}^{\text{appr}\,D}(0, \overline{\varphi}; 0) = \langle \exp[i\overline{\varphi}\hat{n}(0)] \rangle .
\tag{4.6.116}
$$

Using the screw operator, we can describe the Kerr evolution as follows

$$
C_{n\varphi}^{\text{appr}\,D}(\overline{n}, \overline{\varphi}; t) = C_{n\varphi}^{\text{appr}\,D}(\overline{n}, \overline{\varphi} + 2\overline{n}\kappa t; 0).
\tag{4.6.117}
$$

A closer connection with Leonhardt's theory will be established in section 4.7.

4.7 Restricted Hilbert space of a harmonic oscillator

It is not so simple to disprove the existence of a suitably behaved phase operator, but let us see section 4.3 for it. The situation is characterized by saying that "there have been difficulties in finding" such an operator [Pegg and Barnett (1988)]. The distrust of the Susskind–Glogower phase states (4.3.72), or

$$
|\varphi\rangle = \lim_{s \to \infty} |\varphi; s\rangle,
\tag{4.7.1}
$$

where unnormalized vectors

$$
|\varphi; s\rangle = \frac{1}{\sqrt{2\pi}} \sum_{n=0}^{s} \exp(in\varphi)|n\rangle,
\tag{4.7.2}
$$

could be provoked by the fact that the limit does not exist in the Hilbert space of the harmonic oscillator, but it must be defined in a more general topological space [Böhm (1978)]. Unfortunately, no solution can be obtained by the mere normalization of the vectors $|\varphi; s\rangle$. The limit [Loudon (1973)]

$$
|\theta\rangle = \lim_{s \to \infty} |\theta; s\rangle,
\tag{4.7.3}
$$

where an s-phase state [Lerner, Huang and Walters (1970)]

$$|\theta; \mathbf{s}\rangle = \frac{1}{\sqrt{\mathbf{s}+1}} \sum_{n=0}^{\mathbf{s}} \exp(in\theta)|n\rangle, \qquad (4.7.4)$$

does not exist at all. Similar blind alleys were anticipated by the mathematicians who developed nonstandard calculus (for the basics and applications, see [Albeverio, Fenstad, Høegh-Krohn and Lindstrøm (1986)]). The nonstandard calculus is still too much an unusual formalism, nevertheless, its nonstandard integers can be perceived not only as infinites, but also as divergent sequences of standard integers. This idea has been touched on by Vaccaro and Bonner (1995) and modified by Vaccaro (1995). Loudon (1973) focused on the evolution of the electric field for a phase state. Vaccaro, Barnett and Pegg (1992) have investigated this evolution and concluded that the well-defined times when the mean electric field changes sign suggest that a single-mode field could be used as a quantum clock. Admitting that some of the ideas incorporated in and arising out of his paper have been considered by a number of other authors, Pegg (1991) described the entanglement between a clock and another physical system as a characteristic property of the total reality vector.

Although the "phase state" (4.7.3) does not exist in standard mathematical physics, it can motivate the following remarkable characteristic of the operator $\hat{u}$ given in (4.3.24),

$$\begin{aligned} \langle\theta|\hat{u}^{\dagger}\hat{u}|\theta\rangle &= \lim_{\mathbf{s}\to\infty} \langle\theta; \mathbf{s}|\hat{u}^{\dagger}\hat{u}|\theta; \mathbf{s}\rangle \\ &= \lim_{\mathbf{s}\to\infty} \langle\theta; \mathbf{s}|\hat{u}\hat{u}^{\dagger}|\theta; \mathbf{s}\rangle = \langle\theta|\hat{u}\hat{u}^{\dagger}|\theta\rangle = 1. \end{aligned} \qquad (4.7.5)$$

Moreover, this nonexisting phase state has a phase distribution

$$|\langle\varphi'|\theta\rangle|^2 = \lim_{\mathbf{s}\to\infty} |\langle\varphi'|\theta; \mathbf{s}\rangle|^2 = \delta(\varphi' - \theta). \qquad (4.7.6)$$

Pegg and Barnett (1988) observed that in the theory of optical oscillator it is convenient to restrict oneself to the pre-Hilbert space, $\Psi_{\mathbf{s}}$, spanned by the lowest $(\mathbf{s}+1)$ number states, which coincides with the appropriate Hilbert space $\mathcal{H}_{\mathbf{s}} = \Psi_{\mathbf{s}}$. In this finite-dimensional Hilbert space they considered the truncated creation operators $\hat{a}_{\mathbf{s}}^{\dagger}$ defined by the property

$$\hat{a}_{\mathbf{s}}^{\dagger}|n\rangle = \begin{cases} \sqrt{n+1}|n+1\rangle & \text{for } n < \mathbf{s}, \\ 0 & \text{for } n \geq \mathbf{s}, \end{cases} \qquad (4.7.7)$$

where $\mathbf{s}$ is the maximum photon number. Of course, the truncated annihilation operator obeys the relation

$$\hat{a}_{\mathbf{s}} = \left(\hat{a}_{\mathbf{s}}^{\dagger}\right)^{\dagger}. \qquad (4.7.8)$$

The truncated version of Dirac's problem (4.3.23) reads [Lukš and Peřinová (1997b)]

$$\hat{a}_{\mathbf{s}} = \hat{U}_{\mathbf{s}}\hat{n}^{\frac{1}{2}}, \qquad (4.7.9)$$

where $\hat{U}_s$ is a unitary operator to be determined. The solution $\hat{U}_s$ is characterized by the property

$$\hat{U}_s|n\rangle = \begin{cases} |n-1\rangle & \text{for } n = 1,\ldots,s, \\ U_{s0}|s\rangle & \text{for } n = 0, \end{cases} \tag{4.7.10}$$

where U_{s0} is a complex unit. From the cyclic property (4.7.10) it is obvious that the inverse of the operator $\hat{U}_s$ exists and it holds that

$$\hat{U}_s^{-1}|n\rangle = \begin{cases} |n+1\rangle & \text{for } n = 0,\ldots,s-1, \\ U_{s0}^*|0\rangle & \text{for } n = s. \end{cases} \tag{4.7.11}$$

From this property and according to the relation $\hat{U}_s\hat{U}_s^\dagger = \hat{1}$ we obtain that

$$\hat{U}_s^\dagger = \hat{U}_s^{-1}, \tag{4.7.12}$$

the required unitarity. The notion of an index or an index theorem provides a tool for an analysis of the representation of linear operators such as $\hat{a}$ and $\hat{a}^\dagger$ [Fujikawa (1995)]. A unit index results for the original Hilbert space, where Hermitian phase operator cannot be consistently defined, while the vanishing index results for the truncated Hilbert space. The truncation Ψ_s of the Hilbert space $\mathcal{H}$ of the harmonic oscillator enables one to use the elementary characteristic polynomial of the operator $\hat{U}_s$,

$$\begin{vmatrix} -\lambda & 1 & 0 & \ldots & 0 & 0 & 0 \\ 0 & -\lambda & 1 & \ldots & 0 & 0 & 0 \\ \vdots & \vdots & \vdots & \ddots & \vdots & \vdots & \vdots \\ 0 & 0 & 0 & \ldots & 0 & -\lambda & 1 \\ U_{s0} & 0 & 0 & \ldots & 0 & 0 & -\lambda \end{vmatrix} = (-1)^{s+1}\left(\lambda^{s+1} - U_{s0}\right). \tag{4.7.13}$$

The determinant on the left-hand side in (4.7.13) has been written in the number state basis to obtain the right-hand side, which is invariant with respect to a choice of basis. The eigenvalues of this operator are $(s+1)$th roots of the complex unit U_{s0}. They are situated at $(s+1)$ vertices of a regular polygon inscribed in the unit circle. The appropriate polar angles can be denoted by θ_m,

$$\theta_m = \theta_0 + \frac{2\pi m}{s+1}, \quad m = 0,\ldots,s. \tag{4.7.14}$$

To take into account this situation, we shall use an extra subscript, $\hat{U}_s \equiv \hat{U}_{\theta_0,s}$.

Pegg and Barnett [Pegg and Barnett (1988, 1989), Barnett and Pegg (1989)] introduced phase states in the space $\mathcal{H}_s$,

$$|\theta_m;s\rangle = \frac{1}{\sqrt{s+1}}\sum_{n=0}^{s} \exp(in\theta_m)|n\rangle, \tag{4.7.15}$$

where θ_m is given in (4.7.14), with θ_0 being a selected phase. These phase states are the eigenstates of the unitary operator $\hat{U}_s$ for $U_{s0} = \exp[i(s+1)\theta_0]$, they are orthogonal and complete in $\mathcal{H}_s$. Although not complete in the space $\mathcal{H}$, they satisfy a physical notion of completeness in the large-s limit. The unitary operator $\hat{F}_s$, which assigns to each number state $|n\rangle$, $n = 0, 1, ..., s$, the phase state $|\theta_n; s\rangle$, has the singular decomposition (cf. [Ellinas (1991b)], where $\theta_0 = 0$)

$$\hat{F}_s = \sum_{n=0}^{s} |\theta_n; s\rangle\langle n|. \tag{4.7.16}$$

Multiplying the truncated annihilation operator $\hat{a}_s$ from the left by this unitary operator and from the right by its Hermitian conjugate, we mimic the annihilation operator of the phase quanta $\hat{p}_s$ [Bužek, Wilson-Gordon, Knight and Lai (1992)],

$$\hat{p}_s = \sum_{m=0}^{s} \sqrt{\theta_m} \, |\theta_{m-1}; s\rangle\langle\theta_m; s|. \tag{4.7.17}$$

If θ_0 is taken to be zero,

$$\hat{p}_s = \sqrt{\frac{2\pi}{s+1}} \, \hat{F}_s \hat{a}_s \hat{F}_s^\dagger \tag{4.7.18}$$

and all the new concepts can be characterized using the unitary operator $\hat{F}_s$. The idea of duality can be traced back to Vourdas (1990). He has introduced the phase (ladder or step) operators $\hat{\theta}_+$ and $\hat{\theta}_-$ which play a dual role in relation to the angular-momentum ladder operators $\hat{J}_+$ and $\hat{J}_-$.

With the s-phase states a probability distribution in the state described by the statistical operator $\hat{\rho}$ can be associated,

$$\Phi_{\varphi s}(\theta_m) = \langle\theta_m; s|\hat{\rho}|\theta_m; s\rangle, \quad m = 0, \ldots, s. \tag{4.7.19}$$

A direct means for measuring the s-phase distribution which involves synthesizing the projection onto an s-phase state has been proposed [Barnett and Pegg (1996)]. The ingredient is a reciprocal-binomial state, which is coherently mixed with the field on a 50:50 beamsplitter. The projection operator arises also from the output condition that s counts are registered in one detector and no counts in the other. The idea of Barnett and Pegg (1996) can be used also for the Q (Φ_A) function measurement [Baseia, Moussa and Bagnato (1997)].

From the technical point of view it is interesting that θ_m and $|\theta_m; s\rangle$ are defined by (4.7.14) and (4.7.15), respectively, for all m and that the states are periodic,

$$|\theta_{m+s+1}; s\rangle = |\theta_m; s\rangle. \tag{4.7.20}$$

Moreover, it holds that

$$\exp(i\overline{\theta}_{\overline{m}}\hat{n})|\theta_m; s\rangle = |\theta_{m+\overline{m}}; s\rangle, \tag{4.7.21}$$

where

$$\overline{\theta}_{\overline{m}} = \frac{2\pi\overline{m}}{s+1}, \quad \overline{m} \text{ integral}. \tag{4.7.22}$$

In the formula (4.7.21) the property (4.7.20) is used, which expresses also the physical property of circularity of s-phase states.

Equally important from the technical point of view is that the circularity appears in the number states, which should be called accordingly s-number states. Of course, the number states are

$$|n\rangle = \frac{1}{\sqrt{s+1}} \sum_{m=0}^{s} \exp(-in\theta_m)|\theta_m; s\rangle, \quad n = 0, \quad , s. \tag{4.7.23}$$

Let us define the s-number states

$$|n; s\rangle = \frac{1}{\sqrt{s+1}} \sum_{m=0}^{s} \exp(-in\theta_m)|\theta_m; s\rangle, \tag{4.7.24}$$

for all n integral.

By comparison

$$|n; s\rangle = |n\rangle, \quad n = 0, \ldots, s. \tag{4.7.25}$$

Moreover,

$$|n + s + 1; s\rangle = \exp[-i(s+1)\theta_0]|n; s\rangle, \tag{4.7.26}$$

for all n integral.

Using the states (4.7.24) (cf. [Ellinas (1991b)], where $\theta_0 = 0$), we present the Pegg Barnett unitary exponential phase operator defined on $\mathcal{H}_s$,

$$\begin{aligned} U_{\theta_0, s} &= \sum_{n=0}^{s} |n; s\rangle\langle n+1; s| - \sum_{n=0}^{s-1} |n\rangle\langle n \mid 1| + |s; s\rangle\langle s+1; s| \\ &= \sum_{n=0}^{s-1} |n\rangle\langle n+1| + \exp[i(s+1)\theta_0]|s\rangle\langle 0|, \end{aligned} \tag{4.7.27}$$

in exact agreement with [Barnett and Pegg (1989), Lukš and Peřinová (1993a)]. An alternative derivation of the Pegg–Barnett phase operator has been based on the quantized nonlinear nonbijective canonical transformation [Luis and Sánchez-Soto (1993a)]. According to (4.7.24), the s-number states are not periodic, they are skew periodic. But the s-circularity of s-number states is somewhat unphysical.

With the introduction of s-number states we will consider not only the coefficients

$$c_{ns} = \langle n|\psi_s\rangle, \quad n = 0, \ldots, s, \tag{4.7.28}$$

for the ket $|\psi_s\rangle$ of $\mathcal{H}_s$, but also the s-number representation

$$\bar{c}_{ns} = \langle n; s|\psi_s\rangle. \tag{4.7.29}$$

Obviously, the representation (4.7.29) has the properties

$$\begin{aligned} \bar{c}_{ns} &= c_{ns}, \quad n = 0, \ldots, s, \\ \bar{c}_{n+s+1; s} &= \exp[i(s+1)\theta_0]\bar{c}_{ns}, \quad n \text{ integral}. \end{aligned} \tag{4.7.30}$$

In quantum optics also the following notions and properties could be useful. For averaging operators, the matrix elements in the number state basis for the state operator $\hat{\rho}_s$ of $\mathcal{H}_s$ are of use,

$$\rho_{mns} = \langle m|\hat{\rho}_s|n\rangle, \quad m,n = 0,\ldots,s. \tag{4.7.31}$$

In analogy, the s-number state basis matrix elements are

$$\overline{\rho}_{mns} = \langle m; s|\hat{\rho}_s|n; s\rangle, \quad m,n \text{ integral}, \tag{4.7.32}$$

and they have the properties

$$\overline{\rho}_{mns} = \rho_{mns} \text{ for } m,n = 0,\ldots,s, \tag{4.7.33}$$

and

$$\begin{aligned}\overline{\rho}_{m+s+1,n,s} &= \exp[i(s+1)\theta_0]\overline{\rho}_{mns}, \\ \overline{\rho}_{m,n+s+1,s} &= \exp[-i(s+1)\theta_0]\overline{\rho}_{mns} \text{ for } m,n \text{ integral.}\end{aligned} \tag{4.7.34}$$

An attempt to suppress the s-circularity of s-number states has been made [Lukš and Peřinová (1990)] to reflect the overcompleteness and the non-orthogonality of the phase states in the space $\mathcal{H}$ by considering the s-phase states

$$|\theta_m; s\rangle = \frac{1}{\sqrt{s+1}} \sum_{n=0}^{s} \exp(in\theta_m)|n\rangle, \quad m = 0, \frac{1}{2}, 1, \ldots, s + \frac{1}{2}, \tag{4.7.35}$$

where

$$\begin{aligned}\theta_m &= \theta_0 + \frac{2\pi m}{s+1}, \quad m = 0,\ldots,s, \\ \theta_m &= \theta_{\frac{1}{2}} + \frac{2\pi(m - \frac{1}{2})}{s+1}, \quad m = \frac{1}{2},\ldots,s + \frac{1}{2},\end{aligned} \tag{4.7.36}$$

with

$$\theta_{\frac{1}{2}} = \theta_0 + \frac{\pi}{s+1}. \tag{4.7.37}$$

The circularity of s-phase states respects an enlargement of possibilities, the set of $0,\ldots,s$ to all integers and that of $\frac{1}{2},\ldots,s + \frac{1}{2}$ to all half-odd numbers. A simple discretization of (4.6.33) reads

$$\langle\theta_m; s|\theta_{m'}; s\rangle = \overline{\delta}_{mm'} + \frac{1}{s+1}\frac{1-(-1)^{2(m-m')}}{1-\exp[\pm 0 - i(\theta_m - \theta_{m'})]}, \tag{4.7.38}$$

where the correction ± 0 helps define $\frac{0}{\pm 0} = 0$ for $m = m' \pmod{(s+1)}$ and

$$\overline{\delta}_{mm'} = \sum_{k=-\infty}^{\infty} \delta_{m,m'+k(s+1)}, \tag{4.7.39}$$

with the Kronecker delta function $\delta_{mm'}$ generalized to half-odd subscripts.

Similarly as the phase states (4.3.72) form the identity resolution (4.6.29), the s-phase states (4.7.35) form an identity resolution,

$$\hat{1}_s = \frac{1}{2} \sum_{m=0}^{s+\frac{1}{2}}{}' |\theta_m; s\rangle\langle\theta_m; s|, \tag{4.7.40}$$

where the prime here and in the following means that the summation step is $\frac{1}{2}$ and the s-identity operator

$$\hat{1}_s = \sum_{n=0}^{s} |n\rangle\langle n|. \tag{4.7.41}$$

Similarly as the Pegg–Barnett formalism, the s-phase states (4.7.35)–(4.7.37) provide a sound basis for quantum statistical computations, but they are overcomplete and non-orthogonal.

In quantum theory of measurement the identity resolution is a specific property of the probability-operator measure. Also, a construction of operators is provided related to this measure [Helstrom (1976), Hall (1991)]. The overcomplete system (4.7.35) can be used as an illustration of the probability-operator measure. According to previous work [Lukš and Peřinová (1993a, 1990)], operators can be constructed, which are means lying between the Pegg–Barnett s-phase operators, with selected phases θ_0 and $\theta_{\frac{1}{2}}$. Concretely, the choice of the phase θ_0 leads to

$$\hat{M}_{\theta_0,s} = \sum_{m=0}^{s} M(\theta_m)|\theta_m; s\rangle\langle\theta_m; s| \tag{4.7.42}$$

corresponding to the phase function $M(\varphi)$. From the same function it is possible to generate another diagonalized operator

$$\hat{M}_{\theta_{\frac{1}{2}},s} = \sum_{m=\frac{1}{2}}^{s+\frac{1}{2}} M(\theta_m)|\theta_m; s\rangle\langle\theta_m; s|. \tag{4.7.43}$$

The mean between the operators (4.7.42) and (4.7.43),

$$\hat{m}_{\theta_0,s} = \frac{1}{2}\left(\hat{M}_{\theta_0,s} + \hat{M}_{\theta_{\frac{1}{2}},s}\right), \tag{4.7.44}$$

still corresponds to the function $M(\varphi)$,

$$\hat{m}_{\theta_0,s} = \frac{1}{2} \sum_{m=0}^{s+\frac{1}{2}}{}' M(\theta_m)|\theta_m; s\rangle\langle\theta_m; s|. \tag{4.7.45}$$

For the choice $M(\varphi) = \exp(i\varphi)$ we obtain the exponential s-phase operator

$$\hat{u}_s = \frac{1}{2}(\hat{U}_{\theta_0,s} + \hat{U}_{\theta_{\frac{1}{2}},s}), \tag{4.7.46}$$

or

$$\hat{u}_{\mathbf{s}} = \sum_{n=0}^{\mathbf{s}-1} |n\rangle\langle n+1|. \tag{4.7.47}$$

The prescription (4.7.42) provides the Hermitian s-phase operator

$$\hat{\phi}_{\theta_0,\mathbf{s}} = \sum_{m=0}^{\mathbf{s}} \theta_m |\theta_m; \mathbf{s}\rangle\langle\theta_m; \mathbf{s}| \tag{4.7.48}$$

$$= \theta_{\frac{\mathbf{s}}{2}}\hat{1}_{\mathbf{s}} + \sum_{l=0}^{\mathbf{s}} \left[\frac{i\pi}{\mathbf{s}+1} \frac{\exp\left(-il\theta_{-\frac{1}{2}}\right)}{\sin\left(\frac{\pi l}{\mathbf{s}+1}\right)} \hat{u}_{\mathbf{s}}^l + \text{c. c.} \right]. \tag{4.7.49}$$

It is not easy to guess a counter-poise (4.7.43) to (4.7.42) to carry out the construction of the phase operator in the formalism of the overcomplete states. The solution is to consider the partial identity function $\overline{\mathrm{id}}_\theta(\varphi)$ defined in (4.3.54) and (4.3.57), which is 2π-periodic

$$\overline{\mathrm{id}}_\theta(\varphi + 2\pi) = \overline{\mathrm{id}}_\theta(\varphi) \quad \text{for all} \quad \varphi \text{ real.} \tag{4.7.50}$$

In this case the original definition (4.7.48) can be written in the form

$$\hat{\phi}_{\theta_0,\mathbf{s}} = \sum_{m=0}^{\mathbf{s}} \overline{\mathrm{id}}_{\theta_{-\frac{1}{2}}}(\theta_m)|\theta_m; \mathbf{s}\rangle\langle\theta_m; \mathbf{s}| \tag{4.7.51}$$

and the counter-poise (4.7.43) can be denoted by [Lukš and Peřinová (1993a, 1997c)]

$$\hat{\phi}_{\theta_{-\frac{1}{2}},\theta_{\frac{1}{2}},\mathbf{s}} = \sum_{m=\frac{1}{2}}^{\mathbf{s}+\frac{1}{2}} \overline{\mathrm{id}}_{\theta_{-\frac{1}{2}}}(\theta_m)|\theta_m; \mathbf{s}\rangle\langle\theta_m; \mathbf{s}|; \tag{4.7.52}$$

here the summation is extended to the half-odd numbers only. Let us note that in (4.7.51) it holds that $\overline{\mathrm{id}}_{\theta_{-\frac{1}{2}}}(\theta_m) = \theta_m$ for $m = 0,\ldots,\mathbf{s}$, and similarly for $m = \frac{1}{2},\ldots,\mathbf{s}-\frac{1}{2}$, whereas

$$\overline{\mathrm{id}}_{\theta_{-\frac{1}{2}}}(\theta_{\mathbf{s}+\frac{1}{2}}) = \theta_{\mathbf{s}+\frac{1}{2}} - \pi. \tag{4.7.53}$$

Popov and Yarunin (1973) have introduced essentially the s-operator $\hat{\phi}_{0,\mathbf{s}}$ prior to Pegg and Barnett (1988) and have given its matrix elements in the form of a sum, which reflects also the decomposition

$$\hat{\phi}_{0,\mathbf{s}} = -\pi|0; \mathbf{s}\rangle\langle 0; \mathbf{s}| + \hat{\phi}_{0,0,\mathbf{s}}, \tag{4.7.54}$$

where the ket $|0; \mathbf{s}\rangle = |\theta = 0; \mathbf{s}\rangle$. According to Cibils, Cuche, Marvulle and Wreszinski (1991) there is a connection between the general Pegg–Barnett formalism and a system developed for large-amplitude fields [Białynicki-Birula and Białynicka-Birula (1976), Białynicki-Birula (1977, 1980)]. Tsui and Reid (1992) adhere to the alternative derivation of the Pegg–Barnett formalism starting with a unitary s-phase operator. They have presented a generalization to K degrees of freedom. They have contributed to the analysis of the phase-uncertainty measurements of Gerhardt, Büchler and Liftin (1974). Nieto (1977) analyzed these measurements, but his results exhibited a bias for small mean photon numbers due to the symmetrical ordering of the Susskind–Glogower exponential phase operators. The analysis according to the Pegg–Barnett formalism is equivalent to the use of the antinormal ordering of these exponential phase operators.

For the development of s-phase formalism, the definition of a physically preparable state is of importance [Barnett and Pegg (1989)]. Any state $|\psi\rangle$ is physically preparable if

$$\langle\psi|\hat{n}^p|\psi\rangle < \infty \quad \text{for any integer } p \geq 0. \qquad (4.7.55)$$

This state can be called physical in brief [Vaccaro and Pegg (1993)]. The state $|\psi\rangle$ can be approximated by its restriction $|\psi_s\rangle$, s large enough,

$$|\psi_s\rangle = \frac{1}{\sqrt{\langle\psi|\hat{1}_s|\psi\rangle}}\hat{1}_s|\psi\rangle. \qquad (4.7.56)$$

Here we generalize this property to a mixed state described by the statistical operator $\hat{\rho}$ and such that

$$\text{Tr}\{\hat{\rho}\hat{n}^p\} < \infty \quad \text{for any integer } p \geq 0. \qquad (4.7.57)$$

Also this statistical operator $\hat{\rho}$ can be approximated by its restriction

$$\hat{\rho}_s = \frac{1}{\text{Tr}\{\hat{1}_s\hat{\rho}\hat{1}_s\}}\hat{1}_s\hat{\rho}\hat{1}_s. \qquad (4.7.58)$$

On considering the $\mathcal{H}_s$-restriction of a bounded operator $\hat{M}$,

$$\hat{M}_s = \hat{1}_s\hat{M}\hat{1}_s, \qquad (4.7.59)$$

it can be derived that [Vaccaro and Pegg (1993)]

$$\text{Tr}\{\hat{\rho}\hat{M}\} = \lim_{s\to\infty} \text{Tr}\{\hat{\rho}_s\hat{M}_s\}. \qquad (4.7.60)$$

Although many important operators (the number operator $\hat{n}$, the photon annihilation (creation) operator $\hat{a}$ ($\hat{a}^\dagger$), ...) have interesting $\mathcal{H}_s$-restrictions, with respect to the quantum phase, the $\mathcal{H}_s$-restrictions have not been either accepted ($\hat{u}_s = \hat{1}_s\hat{u}\hat{1}_s$, the exponential phase operator) or introduced (the Garrison–Wong phase operator). We have remembered above that in the Pegg-Barnett formalism the unitary exponential phase operator is introduced, $\hat{U}_{\theta_0,s} \neq \hat{u}_s$, as well as the Hermitian s-phase operator $\hat{\phi}_{\theta_0,s}$ is not a restriction of the Garrison–Wong phase operator. Moreover, quantum optics has been enriched by the procedure $\lim_{s\to\infty} \text{Tr}\{\hat{\rho}_s\hat{M}_s\}$, where $\hat{M}_s = \hat{U}_{\theta_0,s}$, $\hat{\phi}_{\theta_0,s}, \ldots$ The fact that Pegg and Barnett provided "nice" concepts as, e. g., the Hermitian s-phase operator for s finite, but they defined the quantum phase only in the infinite-s limit, has been criticized repeatedly, e. g., by Ma and Rhodes (1991). See also the appropriate explanation by Pegg and Barnett (1991).

The relation of canonical conjugation may be formulated in two ways in quantum theory. The first possibility is the trust that the canonical quantization makes all the classically conjugate quantities quantum-mechanically canonically conjugate, i. e., that this procedure ensures the appropriate commutation relation. The other approach supposes the existence of the orthogonal eigenstates of the conjugate observables, or at least eigenstates which provide the resolution of identity. In this approach the characteristic property of conjugation or complementarity [Sánchez-Ruiz (1994) and references therein] is the maximum uncertainty of the other observable, whenever the first one takes on a definite value. As usual, the conditional probabilities are squared moduli of the scalar products and their phase factors express this definite value.

On using the discrete Fourier transformation, the s-phase vectors (4.7.15) have been defined [Pegg and Barnett (1988), Lukš and Peřinová (1990)] and an s-phase observable has also been defined. With respect to the commutator, calculations of the matrix elements have been performed as well as the appropriate limiting procedures.

As the optical quadratures have the commutator very similar to that of generalized coordinate and momentum, the problem of s-quadratures may be formulated. The generalized coordinates and momenta have been treated by Pegg, Vaccaro and Barnett (1990). Following this approach, we could assume and introduce s-quadrature states. We denote the assumed s-quadrature states belonging to $\mathcal{H}_\mathbf{s}$,

$$|\mathrm{Re}_n; \mathbf{s}\rangle, \tag{4.7.61}$$

where

$$\mathrm{Re}_n = \mathrm{Re}_0 + \frac{\sqrt{\pi}n}{\sqrt{\mathbf{s}+1}}, \quad n = -\left[\frac{\mathbf{s}+1}{2}\right], \ldots, \left[\frac{\mathbf{s}}{2}\right]. \tag{4.7.62}$$

Using the states (4.7.61), we could approximate the quadrature operator $\mathrm{Re}\,\hat{a}$ which has the eigenstates $|\mathrm{Re}\,\alpha\rangle$. Let us emphasize that we cannot study the eigenstates of the s-quadrature operator $\mathrm{Re}(\hat{a}_\mathbf{s})$, because it has not the equally spaced eigenvalues Re_n [Figurny, Orłowski and Wódkiewicz (1993)]. It is desirable that the function $|\langle \mathrm{Re}\,\alpha|\mathrm{Re}_n; \mathbf{s}\rangle|^2$ of the argument α has a pronounced peak at Re_n. These functions express strong correlation between s-quadratures and the continuous ones.

Introducing the conjugate s-quadrature states

$$|\mathrm{Im}_m; \mathbf{s}\rangle = \frac{1}{\sqrt{\mathbf{s}+1}} \sum_{n=-\left[\frac{\mathbf{s}+1}{2}\right]}^{\left[\frac{\mathbf{s}}{2}\right]} \exp(i2\mathrm{Re}_n\mathrm{Im}_m)|\mathrm{Re}_n; \mathbf{s}\rangle, \tag{4.7.63}$$

where

$$\mathrm{Im}_m = \mathrm{Im}_0 + \frac{\sqrt{\pi}m}{\sqrt{\mathbf{s}+1}}, \quad m = -\left[\frac{\mathbf{s}+1}{2}\right], \ldots, \left[\frac{\mathbf{s}}{2}\right], \tag{4.7.64}$$

we care that the function $|\langle \mathrm{Im}\,\alpha | \mathrm{Im}_m; s \rangle|^2$ of the argument α is concentrated around Im_m. We cannot study the eigenstates of the s-quadrature operator $\mathrm{Im}(\hat{a}_s)$, because it has not the eigenvalues Im_m and its eigenstates are not canonically conjugate with those of $\mathrm{Re}(\hat{a}_s)$. In contrast, our ideal s-quadrature states have this property,

$$\langle \mathrm{Re}_n; s | \mathrm{Im}_m; s \rangle = \frac{1}{\sqrt{s+1}} \exp(i2\mathrm{Re}_n\mathrm{Im}_m). \tag{4.7.65}$$

In fact, we assume that the expansion of the states (4.7.61) reads

$$|\mathrm{Re}_m; s \rangle = \sum_{n=0}^{s} \langle n | \mathrm{Re}_m; s \rangle |n\rangle, \tag{4.7.66}$$

but the coefficients $\langle n | \mathrm{Re}_m; s \rangle$ are not known yet. From the natural requirement that

$$|\mathrm{Im}_m; s \rangle = \sum_{n=0}^{s} i^n \langle n | \mathrm{Re}_m; s \rangle |n\rangle, \tag{4.7.67}$$

when $\mathrm{Re}_0 = \mathrm{Im}_0$, and from (4.7.63), we derive the property that the arithmetic vectors

$$\mathbf{h}_n = \left\{ \langle n | \mathrm{Re}_m; s \rangle : m = - \left[\frac{s+1}{2} \right], \ldots, \left[\frac{s}{2} \right] \right\}, \quad n = 0, \ldots, s, \tag{4.7.68}$$

are the eigenvectors of the discrete Fourier transformation with the eigenvalues i^n. However, this property is not sufficient for the computation of the vectors $\mathbf{h}_n$, since the eigenspaces appropriate to the eigenvalues $1, -1, i, -i$, have the dimensions $\left[\frac{s+5}{4} \right], \left[\frac{s+3}{4} \right], \left[\frac{s+2}{4} \right], \left[\frac{s}{4} \right]$, respectively [Mehta (1987)]. Nevertheless, it is tempting to construct $\langle n | \mathrm{Re}_m; s \rangle$ as

$$\langle n | \mathrm{Re}_m; s \rangle \propto \sum_{p=-\infty}^{\infty} \langle n | \mathrm{Re}_{p(s+1)+m}; s \rangle, \tag{4.7.69}$$

but it operates only for s even when the range of the subscripts in (4.7.62) is of the form $m = -\frac{s}{2}, \ldots, \frac{s}{2}$. For s odd this prescription would cause a difficulty [Mehta (1987)]. Also this construction must eventually be rejected even though on other grounds than the eigenstates of s-quadrature operators. This construction does not yield exact orthogonal states.

The commutator complicates in the Pegg–Barnett formalism. Whereas the truncated number operator $\hat{n}_s$ and the exponential s-phase operator (4.7.46) have the commutator

$$[\hat{n}_s, \hat{u}_s^l] = -l\hat{u}_s^l, \tag{4.7.70}$$

the truncated number operator $\hat{n}_s$ and the unitary exponential phase operator (4.7.27) have only the more complicated property

$$[\hat{n}_s, \hat{U}_{\theta_0,s}^l] = -l\hat{U}_{\theta_0,s}^l + (s+1)\exp[i(s+1)\theta_0]\hat{u}_s^{\dagger s+1-l}, \quad 0 \le l \le s. \tag{4.7.71}$$

Using (4.7.70) and the Hermitian conjugate, we can easily derive from the relation (4.7.49) that

$$[\hat{n}_{\mathbf{s}}, \hat{\phi}_{\theta_0,\mathbf{s}}] = i \sum_{l=1}^{\mathbf{s}} \left[-\frac{\pi l}{\mathbf{s}+1} \frac{\exp\left(-il\theta_{-\frac{1}{2}}\right)}{\sin\left(\frac{\pi l}{\mathbf{s}+1}\right)} \hat{u}_{\mathbf{s}}^l + \text{c. c.} \right]. \tag{4.7.72}$$

Although the Heisenberg commutation relation is not achieved in (4.7.72), the Weyl group commutation relation emerges in the form [Weyl (1932), section 4.14, Ellinas (1991b)]

$$(\hat{U}_{\theta_0,\mathbf{s}}^{\dagger})^{\overline{n}} \exp(i\overline{\theta}_{\overline{m}}\hat{n}_{\mathbf{s}}) = \exp(-i\overline{n}\overline{\theta}_{\overline{m}}) \exp(i\overline{\theta}_{\overline{m}}\hat{n}_{\mathbf{s}})(\hat{U}_{\theta_0,\mathbf{s}}^{\dagger})^{\overline{n}}, \tag{4.7.73}$$

where $\overline{\theta}_{\overline{m}}$ is given in (4.7.22) and $\overline{n}$ is an integer.

Let us consider, quite generally, the operators

$$\hat{m}_{\mathbf{s}} = \sum_{l=0}^{\mathbf{s}} (\tilde{m}_{\mathbf{s},-l}\hat{u}_{\mathbf{s}}^l + \tilde{m}_{\mathbf{s},-l+\mathbf{s}+1}\hat{u}_{\mathbf{s}}^{\dagger \mathbf{s}+1-l}). \tag{4.7.74}$$

The matrix representations of these operators in the number state basis are just the Toeplitz matrices [Voevodin and Tyrtyshnikov (1987)]. If it holds that

$$\tilde{m}_{\mathbf{s},-l} = \tilde{m}_{\mathbf{s},\mathbf{s}+1-l} \tag{4.7.75}$$

for all $l = 1, 2, \ldots, \mathbf{s}$, the matrix

$$(\langle n'|\hat{m}_{\mathbf{s}}|n\rangle) \tag{4.7.76}$$

is called a circulant. If it is valid that

$$\tilde{m}_{\mathbf{s},-l} = -\tilde{m}_{\mathbf{s},\mathbf{s}+1-l}, \quad l = 1, 2, \ldots, \mathbf{s}, \tag{4.7.77}$$

the matrix (4.7.76) is called a skew circulant. If

$$\tilde{m}_{\mathbf{s},-l} = v\tilde{m}_{\mathbf{s},\mathbf{s}+1-l}, \quad l = 1, 2, \ldots, \mathbf{s}, \tag{4.7.78}$$

where v is a complex number, $|v| = 1$, we call the matrix (4.7.76) a v-skew circulant.

Every 2π-periodic piecewise smooth function $M(\varphi)$ may be expanded in a Fourier series

$$M(\varphi) = \sum_{l=-\infty}^{\infty} \tilde{M}_{-l} \exp(il\varphi). \tag{4.7.79}$$

Restricting ourselves to the choice $\varphi = \theta_0, \theta_1, \ldots, \theta_{\mathbf{s}}$ following Pegg and Barnett, we use the finite discrete Fourier expansion

$$M(\varphi) = \sum_{l=0}^{\mathbf{s}} \tilde{M}_{\theta_0,\mathbf{s},-l} \exp(il\varphi), \tag{4.7.80}$$

where

$$
\begin{aligned}
\tilde{M}_{\theta_0,s,-l} &= \frac{1}{s+1} \sum_{m=0}^{s} M(\theta_m) \exp(-il\theta_m) \\
&= \sum_{l'=-\infty}^{\infty} \tilde{M}_{-l'} \exp[i(l'-l)\theta_0] \sum_{k=-\infty}^{\infty} \delta_{l'-l,k(s+1)} \\
&= \sum_{k=-\infty}^{\infty} \exp[ik(s+1)\theta_0] \tilde{M}_{-l-k(s+1)}.
\end{aligned} \tag{4.7.81}
$$

Taking into account, alternatively, the choice $\varphi = \theta_0, \theta_{\frac{1}{2}}, \ldots, \theta_{s+\frac{1}{2}}$, we get the finite discrete Fourier expansion

$$
M(\varphi) = \sum_{l=-(s+1)}^{s} \tilde{m}_{\theta_0,s,-l} \exp(il\varphi), \tag{4.7.82}
$$

where

$$
\tilde{m}_{\theta_0,s,-l} = \sum_{k=-\infty}^{\infty} \exp[i2k(s+1)\theta_0] \tilde{M}_{-l-2k(s+1)}. \tag{4.7.83}
$$

It is evident that

$$
\tilde{M}_{\theta_0,s,-l} = \tilde{m}_{\theta_0,s,-l} + \exp[-i(s+1)\theta_0] \tilde{m}_{\theta_0,s,-l+s+1} \tag{4.7.84}
$$

and

$$
\tilde{M}_{\theta_{\frac{1}{2}},s,-l} = \tilde{m}_{\theta_0,s,-l} - \exp[-i(s+1)\theta_0] \tilde{m}_{\theta_0,s,-l+s+1}. \tag{4.7.85}
$$

The following relations are illustrative [Lukš and Peřinová (1993a)],

$$
\hat{U}^l_{\theta_0,s} = \hat{u}^l_s + \exp[i(s+1)\theta_0] \hat{u}^{\dagger s+1-l}_s, \tag{4.7.86}
$$

$$
\hat{U}^l_{\theta_{\frac{1}{2}},s} = \hat{u}^l_s - \exp[i(s+1)\theta_0] \hat{u}^{\dagger s+1-l}_s, \quad 0 \le l \le s. \tag{4.7.87}
$$

The appropriate matrix representations are v-skew circulants for $v = \pm \exp[-i(s+1)\theta_0]$. Since

$$
\hat{M}_{\theta_0,s} = \sum_{l=0}^{s} \tilde{M}_{\theta_0,s,-l} \hat{U}^l_{\theta_0,s}, \tag{4.7.88}
$$

$$
\hat{M}_{\theta_{\frac{1}{2}},s} = \sum_{l=0}^{s} \tilde{M}_{\theta_{\frac{1}{2}},s,-l} \hat{U}^l_{\theta_{\frac{1}{2}},s}, \tag{4.7.89}
$$

the formulae (4.7.42) and (4.7.43) provide v-skew circulants with respect to the number state basis with $v = \pm \exp[-i(s+1)\theta_0]$ and it holds that

$$
\hat{m}_{\theta_0,s} = \sum_{l=0}^{s} \left(\tilde{m}_{\theta_0,s,-l} \hat{u}^l_s + \tilde{m}_{\theta_0,s,-l+s+1} \hat{u}^{\dagger s+1-l}_s \right). \tag{4.7.90}
$$

The appropriate matrix representation is a Toeplitz matrix. As

$$\hat{u}_s^{\dagger s+1-l} = \hat{0}, \quad l = 0, \tag{4.7.91}$$

the formula (4.7.90) forgets the coefficient $\tilde{m}_{\theta_0,s,s+1}$.

We compare the Wigner function for number and phase presented in section 4.6 with the Wigner function for finite number and discrete phase [Vaccaro and Pegg (1990b)]. The Wigner function for number and phase does not exhibit any new properties of circularity, but introduces half-odd numbers, whereas its analogue for finite number and s-phase is plagued by the circularities, although it avoids the unusual half-odd numbers. In [Lukš a Peřinová (1994)] both the generalization to half-odd numbers and the perseverance of circularities have been addressed. The approach has been new in that it uses the concept of characteristic function, which plays the central role in the fundamental exposition of the Glauber quasidistributions. An axiomatic approach to the derivation of the Wigner function in finite dimensional systems has been presented in [Luis and Peřina (1998)].

The unitary exponential phase operator (4.7.27) has the property

$$\hat{U}_{\theta_0,s}^{\bar{n}+s+1} = \exp[i(s+1)\theta_0]\hat{U}_{\theta_0,s}^{\bar{n}}, \quad \bar{n} \text{ integral}. \tag{4.7.92}$$

We can define the s-screw operator on $\mathcal{H}_s$,

$$\hat{D}_{n\varphi,\theta_0,s}(\bar{n},\bar{\theta}_{\bar{m}}) = \begin{cases} \exp(\frac{1}{2}i\bar{n}\bar{\theta}_{\bar{m}})(\hat{U}_{\theta_0,s}^\dagger)^{\bar{n}}\exp(i\bar{\theta}_{\bar{m}}\hat{n}_s), \\[2mm] \exp(-\frac{1}{2}i\bar{n}\bar{\theta}_{\bar{m}})\exp(i\bar{\theta}_{\bar{m}}\hat{n}_s)(\hat{U}_{\theta_0,s}^\dagger)^{\bar{n}}. \end{cases} \tag{4.7.93}$$

So defined an operator depends on θ_0 $\left(\bmod \frac{2\pi}{s+1}\right)$ and we will omit the subscript θ_0 in what follows. This operator has the skew-periodicity properties

$$\hat{D}_{n\varphi s}(\bar{n}+2s+2,\bar{\theta}_{\bar{m}}) = \exp[-i(2s+2)\theta_0]\hat{D}_{n\varphi s}(\bar{n},\bar{\theta}_{\bar{m}}), \tag{4.7.94}$$

$$\hat{D}_{n\varphi s}(\bar{n},\bar{\theta}_{\bar{m}+2s+2}) = \hat{D}_{n\varphi s}(\bar{n},\bar{\theta}_{\bar{m}}). \tag{4.7.95}$$

The s-screw operator has many other interesting properties in common with the displacement operator underlying the Glauber exposition of normal and antinormal characteristic functions and of the appropriate quasidistributions [Ellinas (1991b)]. The operator (4.7.93) can be expressed in the s-number state basis of $\mathcal{H}_s$

$$\hat{D}_{n\varphi s}(\bar{n},\bar{\theta}_{\bar{m}}) = \sum_{n=-\frac{\bar{n}}{2}}^{-\frac{\bar{n}}{2}+s} \exp(i\bar{\theta}_{\bar{m}}n)\left|n+\frac{\bar{n}}{2};s\right\rangle\left\langle n-\frac{\bar{n}}{2};s\right|. \tag{4.7.96}$$

Let us note that the technical problem with the summation domain on the right-hand side is made easier, because an operator is summed, which depends on n periodically. The appropriate s-phase expansion for the s-screw operator (4.7.93) is

$$\hat{D}_{n\varphi s}(\bar{n},\bar{\theta}_{\bar{m}}) = \sum_{m=-\frac{\bar{m}}{2}}^{-\frac{\bar{m}}{2}+s} \exp(-i\bar{n}\theta_m)\left|\theta_{m+\frac{\bar{m}}{2}};s\right\rangle\left\langle\theta_{m-\frac{\bar{m}}{2}};s\right|. \tag{4.7.97}$$

Like the displacement operator, the s-screw operator can be applied in defining quantum characteristic function for the state of system. Generally, if an operator $\hat{M}_s$ acting on the Hilbert space $\mathcal{H}_s$ is averaged, we can define its quantum Fourier transformation

$$\tilde{M}_{n\varphi s}(\bar{n}, \bar{\theta}_{\overline{m}}) = \frac{1}{s+1}\text{Tr}\{\hat{M}_s \hat{D}_{n\varphi s}(-\bar{n}, -\bar{\theta}_{\overline{m}})\}. \qquad (4.7.98)$$

The quantum characteristic function for a state described by $\hat{\rho}_s$ of $\mathcal{H}_s$ is defined as

$$C_{n\varphi s}(\bar{n}, \bar{\theta}_{\overline{m}}) = \text{Tr}\{\hat{\rho}_s \hat{D}_{n\varphi s}(\bar{n}, \bar{\theta}_{\overline{m}})\}. \qquad (4.7.99)$$

From the normalization property $\text{Tr}\{\hat{\rho}_s\} = 1$ it follows that

$$C_{n\varphi s}(0,0) = 1. \qquad (4.7.100)$$

In addition to the periodicity properties, which easily follow from the relations (4.7.94) and (4.7.95), the quantum characteristic function fulfills the following conditional periodicities

$$C_{n\varphi s}(\bar{n} + s + 1, \bar{\theta}_{\overline{m}}) = (-1)^{\overline{m}}\exp[-i(s+1)\theta_0]C_{n\varphi s}(\bar{n}, \bar{\theta}_{\overline{m}}), \qquad (4.7.101)$$
$$C_{n\varphi s}(\bar{n}, \bar{\theta}_{\overline{m}+s+1}) = (-1)^{\bar{n}}C_{n\varphi s}(\bar{n}, \bar{\theta}_{\overline{m}}). \qquad (4.7.102)$$

Letting $\bar{n} = 0$, we obtain the characteristic function of photon number

$$C_{ns}(\bar{\theta}_{\overline{m}}) = C_{n\varphi s}(0, \bar{\theta}_{\overline{m}}) = \text{Tr}\{\hat{\rho}_s \exp(i\bar{\theta}_{\overline{m}}\hat{n}_s)\}. \qquad (4.7.103)$$

For $\overline{m} = 0$, we get the characteristic function of the s-phase

$$C_{\varphi s}(\bar{n}) = C_{n\varphi s}(\bar{n}, 0) = \text{Tr}\{\hat{\rho}_s (\hat{U}_{\theta_0,s}^\dagger)^{\bar{n}}\}. \qquad (4.7.104)$$

In these cases the period is just $(s+1)$ not $(2s+2)$.

From the physical point of view it is important that the quantum expectation has a quasiclassical expression

$$\text{Tr}\{\hat{\rho}_s \hat{M}_s\} = \frac{1}{4}\sum_{\bar{n}=0}^{2s+1}\sum_{\overline{m}=0}^{2s+1} C_{n\varphi s}(\bar{n}, \bar{\theta}_{\overline{m}})\tilde{M}_{n\varphi s}(\bar{n}, \bar{\theta}_{\overline{m}}), \qquad (4.7.105)$$

which can be proved using the expansions

$$C_{n\varphi s}(\bar{n}, \bar{\theta}_{\overline{m}}) = \sum_{n=-\frac{\bar{n}}{2}}^{-\frac{\bar{n}}{2}+s} \exp(i\bar{\theta}_{\overline{m}}n)\left\langle n - \frac{\bar{n}}{2}; s\middle|\hat{\rho}_s\middle|n + \frac{\bar{n}}{2}; s\right\rangle, \qquad (4.7.106)$$

$$\tilde{M}_{n\varphi s}(\bar{n}, \bar{\theta}_{\overline{m}}) = \frac{1}{s+1}\sum_{n=\frac{\bar{n}}{2}}^{\frac{\bar{n}}{2}+s} \exp(-i\bar{\theta}_{\overline{m}}n)\left\langle n + \frac{\bar{n}}{2}; s\middle|\hat{M}_s\middle|n - \frac{\bar{n}}{2}; s\right\rangle. \qquad (4.7.107)$$

We arrive at the Wigner function for finite number and s-phase by interpreting the quantum characteristic function as a classical one, i. e., using the transformation of a characteristic function to a quasidistribution,

$$\Phi_{n\varphi s}(n, \theta_m) = \frac{1}{2(s+1)^2} \sum_{\overline{n}=0}^{2s+1} \sum_{\overline{m}=0}^{2s+1} \exp[i(\overline{n}\theta_m - \overline{\theta}_{\overline{m}}n)] C_{n\varphi s}(\overline{n}, \overline{\theta}_{\overline{m}}); \qquad (4.7.108)$$

n, m are integers and half-odd numbers.

This quasidistribution can be used for the computation of photon-number distribution and s-phase distribution,

$$p_s(n) = \frac{1}{2} \sideset{}{'}\sum_{m=0}^{s+\frac{1}{2}} \Phi_{n\varphi s}(n, \theta_m), \qquad (4.7.109)$$

$$\Phi_{\varphi s}(\theta_m) = \frac{1}{2} \sideset{}{'}\sum_{n=0}^{s+\frac{1}{2}} \Phi_{n\varphi s}(n, \theta_m). \qquad (4.7.110)$$

Let us observe that $p_s(n) = 0$ for n half-odd and, moreover, $\Phi_{\varphi s}(\theta_m) = 0$ for m half-odd.

In terms of operators the Wigner function reads

$$\Phi_{n\varphi s}(n, \theta_m) = \frac{1}{s+1} \text{Tr}\{\hat{\rho}_s \hat{\Delta}_{n\varphi s}(n, \theta_m)\}, \qquad (4.7.111)$$

where the basis operator

$$\hat{\Delta}_{n\varphi s}(n, \theta_m) = \frac{1}{2s+2} \sum_{\overline{n}=0}^{2s+1} \sum_{\overline{m}=0}^{2s+1} \exp[i(\overline{n}\theta_m - \overline{\theta}_{\overline{m}}n)] \hat{D}_{n\varphi s}(\overline{n}, \overline{\theta}_{\overline{m}}). \qquad (4.7.112)$$

The basis operator has the periodicity properties

$$\hat{\Delta}_{n\varphi s}(n + s + 1, \theta_m) = \hat{\Delta}_{n\varphi s}(n, \theta_m), \qquad (4.7.113)$$

$$\hat{\Delta}_{n\varphi s}(n, \theta_{m+s+1}) = \hat{\Delta}_{n\varphi s}(n, \theta_m). \qquad (4.7.114)$$

Viewed physically, much more interesting property is that for s even

$$\text{Tr}\{\hat{\Delta}_{n\varphi s}(n, \theta_m)\} = \begin{cases} 1 & \text{for } n \text{ integral,} \\ (-1)^{2m} & \text{for } n \text{ half-odd.} \end{cases} \qquad (4.7.115)$$

To the contrary, the case of s odd is pathological, because

$$\text{Tr}\{\hat{\Delta}_{n\varphi s}(n, \theta_m)\} = \begin{cases} 1 + (-1)^{2m} & \text{for } n \text{ integral,} \\ 0 & \text{for } n \text{ half-odd.} \end{cases} \qquad (4.7.116)$$

Essentially, the paper [Vaccaro and Pegg (1990b)] has used the basis operator in the even-s case, but the use of half-odd numbers for labelling quasiprobabilities has not been approved.

The Wigner function for number and s-phase has periodicity properties, which easily follow from the relations (4.7.113) and (4.7.114), and obeys the linear dependence relations

$$\Phi_{n\varphi s}\left(n + \frac{s+1}{2}, \theta_m\right) = (-1)^{2m}\Phi_{n\varphi s}(n, \theta_m), \qquad (4.7.117)$$

$$\Phi_{n\varphi s}\left(n, \theta_{m+\frac{s+1}{2}}\right) = (-1)^{2n}\Phi_{n\varphi s}(n, \theta_m). \qquad (4.7.118)$$

The inverse discrete Fourier transformation is used to obtain an image of the averaged operator as the function

$$M_{n\varphi s}(n, \theta_m) = \sum_{\bar{n}=0}^{2s+1}\sum_{\bar{m}=0}^{2s+1} \exp[i(-\bar{n}\theta_m + \bar{\theta}_{\bar{m}}n)]\tilde{M}_{n\varphi s}(\bar{n}, \bar{\theta}_{\bar{m}}), \qquad (4.7.119)$$

which leads to the relation

$$M_{n\varphi s}(n, \theta_m) = \text{Tr}\{\hat{\Delta}_{n\varphi s}(n, \theta_m)\hat{M}_s\}. \qquad (4.7.120)$$

The Wigner function and this image enter the expression for the quantum average

$$\text{Tr}\{\hat{\rho}_s\hat{M}_s\} = \frac{1}{4}\sum_{n=0}^{s+\frac{1}{2}}\!\!'\sum_{m=0}^{s+\frac{1}{2}}\!\!' \Phi_{n\varphi s}(n, \theta_m)M_{n\varphi s}(n, \theta_m). \qquad (4.7.121)$$

For s even, this formula can be simplified

$$\text{Tr}\{\hat{\rho}_s\hat{M}_s\} = \sum_{n=0}^{s}\sum_{m=0}^{s} \Phi_{n\varphi s}(n, \theta_m)M_{n\varphi s}(n, \theta_m). \qquad (4.7.122)$$

Let us emphasize here that the Vaccaro–Pegg formalism for the Wigner function uses only n, m integral.

Here, we have the expansions

$$\Phi_{n\varphi s}(n, \theta_m) = \frac{1}{s+1} \sum_{\substack{k=0 \\ k=n(\text{mod}\,1)}}^{s+\frac{1}{2}} \exp(-i2k\theta_m)\langle n+k; s|\hat{\rho}_s|n-k; s\rangle, \qquad (4.7.123)$$

$$M_{n\varphi s}(n, \theta_m) = \sum_{\substack{k=0 \\ k=n(\text{mod}\,1)}}^{s+\frac{1}{2}} \exp(-i2k\theta_m)\langle n+k; s|\hat{M}_s|n-k; s\rangle. \qquad (4.7.124)$$

The fact that the formula (4.7.121) cannot be reduced for s odd is an argument in favour of the half-odd values of photon number and of subscripts. For further analysis of this dichotomy we express the state $\hat{\rho}_s$ of the optical system as

$$\hat{\rho}_s = \frac{1}{4}\sum_{n=0}^{s+\frac{1}{2}}\!\!'\sum_{m=0}^{s+\frac{1}{2}}\!\!' \Phi_{n\varphi s}(n, \theta_m)\hat{\Delta}_{n\varphi s}(n, \theta_m). \qquad (4.7.125)$$

The s-phase representation of the basis operator reads

$$\hat{\Delta}_{n\varphi s}(n, \theta_m) = \sum_{\substack{k=0 \\ k=m(\bmod 1)}}^{s+\frac{1}{2}} \exp(-i2kn\bar{\theta}_1)|\theta_{m+k}; s\rangle\langle\theta_{m-k}; s|. \tag{4.7.126}$$

Using its s-number representation

$$\hat{\Delta}_{n\varphi s}(n, \theta_m) = \sum_{\substack{k=0 \\ k=n(\bmod 1)}}^{s+\frac{1}{2}} \exp(-i2k\theta_m)|n - k; s\rangle\langle n + k; s|, \tag{4.7.127}$$

we obtain that

$$\rho_{mns} = \frac{1}{4} \left\{ \sum_{m'=0}^{s+\frac{1}{2}}{}' \Phi_{n\varphi s}\left(\frac{m+n}{2}, \theta_{m'}\right) \exp[i(m - n)\theta_{m'}] \right.$$

$$\left. + \sum_{m'=0}^{s+\frac{1}{2}}{}' \Phi_{n\varphi s}\left(\frac{m+n+s+1}{2}, \theta_{m'}\right) \exp[i(m - n)\theta_{m'}] \right\}. \tag{4.7.128}$$

For s even this formula can be simplified to the form

$$\rho_{mns} = \sum_{m'=0}^{s} \Phi_{n\varphi s}\left(\left[\frac{m+n}{2}\right]_{s+1}, \theta_{m'}\right) \exp[i(m - n)\theta_{m'}], \tag{4.7.129}$$

where

$$\left[\frac{m+n}{2}\right]_{s+1} = \begin{cases} \frac{m+n}{2} & \text{for} \quad m + n \quad \text{even}, \\ \frac{m+n+s+1}{2} & \text{for} \quad m + n \quad \text{odd}. \end{cases} \tag{4.7.130}$$

This proves that for s even the Vaccaro–Pegg Wigner function can work without half-odd numbers [Vaccaro and Pegg (1990b)]. Nevertheless, for s odd the reduction of the expression (4.7.128) is not possible and it comprises integral first arguments (photon numbers) for $(m + n)$ even only, while for $(m + n)$ odd our formalism has to use half-odd first variables.

Considering the characteristic function for s even appropriate to the Wigner function due to Vaccaro and Pegg

$$C_{n\varphi s}^{\text{VP}}(\bar{n}, \bar{\theta_m}) = \sum_{n=0}^{s} \sum_{m=0}^{s} \exp[i(-\bar{n}\theta_m + \bar{\theta_m}n)]\Phi_{n\varphi s}(n, \theta_m), \tag{4.7.131}$$

we easily obtain that the quantum characteristic function $C_{n\varphi s}(\bar{n}, \bar{\theta_m})$ from (4.7.99) has the property

$$C_{n\varphi s}(\bar{n}, \bar{\theta_m}) = \frac{1}{2}\left[1 + (-1)^{\bar{n}} + (-1)^{\bar{m}} - (-1)^{\bar{n}+\bar{m}}\right] C_{n\varphi s}^{\text{VP}}(\bar{n}, \bar{\theta_m}). \tag{4.7.132}$$

The relationship (4.7.132) can simply be rewritten

$$
C_{n\varphi s}(\overline{n}, \overline{\theta_m}) = \begin{cases} C_{n\varphi s}^{\mathrm{VP}}(\overline{n}, \overline{\theta_m}) & \text{for } \overline{n}, \overline{m} \text{ even,} \\[2mm] -C_{n\varphi s}^{\mathrm{VP}}(\overline{n}, \overline{\theta_m}) & \text{for } \overline{n}, \overline{m} \text{ odd,} \\[2mm] C_{n\varphi s}^{\mathrm{VP}}(\overline{n}, \overline{\theta_m}) & \text{otherwise.} \end{cases} \tag{4.7.133}
$$

With respect to finite discrete systems like atoms or spins, there is an interest in transcriptions of the continuous position-momentum Wigner formalism for discrete quantum mechanics. Wootters (1987) pioneered a Wigner formalism for finite systems with, however, a number-theoretic flavour. A similar treatment has been provided by Galetti and De Toledo Piza (1988). Cohendet, Combe, Sirugue and Sirugue-Collin (1988) considered the Wigner function for odd-dimensional systems.

With regard to the twofold role, which is ascribed to the limit $s \to \infty$ in the Pegg–Barnett system, we introduce a notion concerning the operator convergence. For the sake of brevity we will use the notation $\hat{A}_s \overset{m}{\to} \hat{A}$ if it holds that [Lukš and Peřinová (1993a)]

$$
\lim_{s \to \infty} \langle n' | \hat{A}_s | n \rangle = \langle n' | \hat{A} | n \rangle \tag{4.7.134}
$$

for all number states $|n'\rangle$, $|n\rangle$. In the Pegg–Barnett formalism the m-limit has a restricted use, but it is contained in their work.

First of all, the matrix elements of the Pegg–Barnett exponential s-phase operators converge to the matrix elements of the corresponding Susskind–Glogower exponential phase operators, i. e., $\hat{U}_{\theta_0,s}^l \overset{m}{\to} \hat{u}^l$, $\hat{U}_{\theta_0,s}^{\dagger l} \overset{m}{\to} \hat{u}^{\dagger l}$. The same is valid for the exponential s-phase operators based on the finite overcomplete system of vectors, namely, $\hat{u}_s^l \overset{m}{\to} \hat{u}^l$, $\hat{u}_s^{\dagger l} \overset{m}{\to} \hat{u}^{\dagger l}$. It can be proved as follows. We may assume that $0 \leq l \leq s$. It holds that

$$
\langle n' | \hat{u}_s^l | n \rangle = \delta_{n'+l,n} \quad (0 \leq n \leq s), \tag{4.7.135}
$$

where $(P) = 1$ when P is valid and $(P) = 0$ otherwise. For $s \to \infty$ $(0 \leq n \leq s) \to (0 \leq n)$ and therefore

$$
\langle n' | \hat{u}_s^l | n \rangle \to \langle n' | \hat{u}^l | n \rangle. \tag{4.7.136}
$$

In contrast, cf. (4.7.86),

$$
\langle n' | \hat{u}_s^{\dagger s+1-l} | n \rangle = \delta_{n'+l,n+s+1}(0 \leq n' \leq s). \tag{4.7.137}
$$

As for $s \to \infty$ the condition $0 \leq n' \leq s$ becomes $0 < n'$ unlike the Kronecker delta function variables, so

$$
\langle n' | \hat{u}_s^{\dagger s+1-l} | n \rangle \to 0. \tag{4.7.138}
$$

We will use the more usual notation $\hat{A}_s \overset{w}{\to} \hat{A}$ (the weak limit), if

$$
\lim_{s \to \infty} \langle \psi' | \hat{A}_s | \psi \rangle = \langle \psi' | \hat{A} | \psi \rangle \tag{4.7.139}
$$

for all states $|\psi'\rangle$, $|\psi\rangle$ belonging to the Hilbert space $\mathcal{H}$. It is obvious that

$$\hat{u}_{\mathrm{s}}^l \xrightarrow{w} \hat{u}^l \tag{4.7.140}$$

and

$$\hat{u}_{\mathrm{s}}^{\dagger\mathrm{s}+1-l} \xrightarrow{w} \hat{0}. \tag{4.7.141}$$

From this

$$\hat{U}_{\theta_0,\mathrm{s}}^l \xrightarrow{w} \hat{u}^l. \tag{4.7.142}$$

The well-known property (4.3.32) of the Susskind–Glogower exponential phase operator $\hat{u}$ may be compared with the appropriate property of the Pegg–Barnett unitary s-phase operators

$$\hat{U}_{\theta_0,\mathrm{s}}\hat{U}_{\theta_0,\mathrm{s}}^{\dagger} \xrightarrow{w} \hat{u}\hat{u}^{\dagger}. \tag{4.7.143}$$

Moreover, it is valid that

$$\hat{u}_{\mathrm{s}}\hat{u}_{\mathrm{s}}^{\dagger} \xrightarrow{w} \hat{u}\hat{u}^{\dagger}. \tag{4.7.144}$$

Similarly, it holds that

$$\hat{u}_{\mathrm{s}}^{\dagger}\hat{u}_{\mathrm{s}} \xrightarrow{w} \hat{u}^{\dagger}\hat{u}, \tag{4.7.145}$$

but

$$\hat{U}_{\theta_0,\mathrm{s}}^{\dagger}\hat{U}_{\theta_0,\mathrm{s}} = \hat{1}_{\mathrm{s}} \tag{4.7.146}$$

and the operator on the right-hand side of (4.7.145) takes the form (4.3.38). The Pegg–Barnett system arrives at the unitarity of the operator $\hat{U}_{\theta_0,\mathrm{s}}$, whereas the Susskind–Glogower approach does not.

Generally, if $\hat{M}_{\theta_0,\mathrm{s}}$ and $\hat{m}_{\theta_0,\mathrm{s}}$ are given by the formulae (4.7.42) and (4.7.44), respectively, then it holds that

$$\hat{M}_{\theta_0,\mathrm{s}} \xrightarrow{m} \hat{M}, \quad \hat{m}_{\theta_0,\mathrm{s}} \xrightarrow{m} \hat{M}, \tag{4.7.147}$$

where the operator $\hat{M}$ is given in (4.3.73). It can easily be computed that

$$\langle n'|\hat{M}_{\theta_0,\mathrm{s}}|n\rangle = \tilde{M}_{\theta_0,\mathrm{s},-(n-n')}. \tag{4.7.148}$$

These elements form a Toeplitz matrix. As to the limiting procedures (4.7.147), we observe that the series expansion for $\tilde{M}_{\theta_0,0,-(n-n')}$ is decimated for s tending to the infinity except the term of $k = 0$ with respect to the formula (4.7.81), so $\lim_{\mathrm{s}\to\infty} \tilde{M}_{\theta_0,\mathrm{s},-(n-n')} = \tilde{M}_{-(n-n')}$. In other words $\langle n'|\hat{M}_{\theta_0,\mathrm{s}}|n\rangle \to \langle n'|\hat{M}|n\rangle$. The matrix elements of the operator $\hat{m}_{\theta_0,\mathrm{s}}$,

$$\langle n'|\hat{m}_{\theta_0,\mathrm{s}}|n\rangle = \tilde{m}_{\theta_0,\mathrm{s},-(n-n')}. \tag{4.7.149}$$

Comparing (4.7.83) with (4.7.81), we observe that the series expansion for $\tilde{m}_{\theta_0,\mathrm{s},-(n-n')}$ is selected from that for $\tilde{M}_{\theta_0,\mathrm{s},-(n-n')}$ respecting the term of $k = 0$. As a consequence, $\lim_{\mathrm{s}\to\infty} \tilde{m}_{\theta_0,\mathrm{s},-(n-n')} = \tilde{M}_{-(n-n')}$ and again $\langle n'|\hat{m}_{\theta_0,\mathrm{s}}|n\rangle \xrightarrow{m} \langle n'|\hat{M}|n\rangle$. Using a suitable definition of an antinormally ordered product of Toeplitz operators, Vaccaro and Ben-Aryeh (1995) have proved that (let us note that the product of weak limits is not necessarily the weak limit of a sequence of products of operators) the weak limit of a product of Pegg–Barnett v-skew circulant operators is the antinormally ordered product of the respective weak limits.

Let us remark that in the integral (4.3.73) the quantity θ_0 can be replaced by an arbitrary number with regard to the 2π-periodicity of the function $M(\varphi)$. It is well-known that a positive operator-valued measure $\hat{F}$ is assigned to the overcomplete system (4.3.72) [Holevo (1982), Hall (1991)]

$$\hat{F}(\Delta) = \int_\Delta |\varphi\rangle\langle\varphi| \, d\varphi, \tag{4.7.150}$$

where Δ is a measurable subset of the phase window. An orthogonal or projection-valued measure $\hat{E}_{\theta_0,s}$ is associated with the orthogonal system $|\theta_m; s\rangle$ given in (4.7.15), $m = 0, 1, \ldots, s$,

$$\hat{E}_{\theta_0,s}(\Delta) = \sum_{\theta_m \in \Delta} |\theta_m; s\rangle\langle\theta_m; s|. \tag{4.7.151}$$

The positive operator-valued measure $\hat{F}$ can be obtained in the sense of the convergence of operator measures, i. e., $\hat{E}_{\theta_0,s} \to \hat{F}$ for $s \to \infty$ [Hall (1991)]. Analogously, a positive operator-valued measure $\hat{F}_{\theta_0,s}$ can be attached to the overcomplete system $|\theta_m; s\rangle$, $m - \frac{1}{2}, \frac{3}{2}, \ldots, s + \frac{1}{2}$, and it can be found that $\hat{F}_{\theta_0,s} \to \hat{F}$ for $s \to \infty$.

It can be proved that at a continuity point φ of the probability density $P(\varphi)$,

$$P(\varphi) = \lim_{s \to \infty} \frac{(s+1)}{2\pi} \Phi_{\varphi s}(\theta_{m(\varphi,s)}), \tag{4.7.152}$$

with the approximations $\theta_{m(\varphi,s)}$ of the argument

$$\varphi = \lim_{s \to \infty} \theta_{m(\varphi,s)}, \tag{4.7.153}$$

where $\{m(\varphi, s)\}$ is a sequence of numbers from $0, 1, \ldots, \infty$. An ingredient of the proof could be the uniform convergence of the sequence $\{\frac{(s+1)}{2\pi}\Phi_{\varphi s}(\varphi)\}$ to the probability density $P(\varphi)$. This occurs for pure states [Pegg and Barnett (1997)]. In a more complicated case of the Wigner functions, we have a refinement of the previous statement

$$\Phi_{n\varphi}(n, \varphi) = \lim_{s \to \infty} \frac{(s+1)}{2\pi} \Phi_{n\varphi s}(n, \theta_{m(\varphi,s)}), \tag{4.7.154}$$

whenever (4.7.153) holds.

Analyzing the property of circularity, we recall that ((4.7.86))

$$\hat{U}^l_{\theta_0,s} = \begin{cases} \hat{u}^l_s + \exp[i(s+1)\theta_0]\hat{u}^{\dagger s+1-l}_s & \text{for } 0 \leq l \leq s, \\[2mm] \hat{u}^{\dagger|l|}_s + \exp[-i(s+1)\theta_0]\hat{u}^{s+1-|l|}_s & \text{for } -s \leq l \leq 0. \end{cases} \tag{4.7.155}$$

Upon taking the $\mathcal{H}_s$-restriction of the operator $\hat{D}_{n\varphi}(\overline{n}, \overline{\theta}_{\overline{m}})$ introduced in (4.6.79) as follows

$$\hat{d}_{n\varphi s}(\overline{n}, \overline{\theta}_{\overline{m}}) = \sum_{n=0}^{s-|\overline{n}|} \exp\left[i\overline{\theta}_{\overline{m}}\left(n + \frac{|\overline{n}|}{2}\right)\right] |n\rangle\langle n + |\overline{n}||$$
$$\text{for } -s \le \overline{n} \le 0, \tag{4.7.156}$$

$$\hat{d}_{n\varphi s}(\overline{n}, \overline{\theta}_{\overline{m}}) = \sum_{n=0}^{s-\overline{n}} \exp\left[i\overline{\theta}_{\overline{m}}\left(n + \frac{\overline{n}}{2}\right)\right] |n + \overline{n}\rangle\langle n|$$
$$\text{for } 0 \le \overline{n} \le s, \tag{4.7.157}$$

we can write the decompositions of the s-screw operator (4.7.93),

$$\hat{D}_{n\varphi s}(\overline{n}, \overline{\theta}_{\overline{m}}) = \hat{d}_{n\varphi s}(\overline{n}, \overline{\theta}_{\overline{m}}) + \exp[i(s+1)\theta_0]\hat{d}_{n\varphi s}(\overline{n} + s + 1, \overline{\theta}_{\overline{m}})$$
$$\text{for } -s \le \overline{n} \le 0, \tag{4.7.158}$$

$$\hat{D}_{n\varphi s}(\overline{n}, \overline{\theta}_{\overline{m}}) = \hat{d}_{n\varphi s}(\overline{n}, \overline{\theta}_{\overline{m}}) + \exp[-i(s+1)\theta_0]\hat{d}_{n\varphi s}(\overline{n} - s - 1, \overline{\theta}_{\overline{m}})$$
$$\text{for } 0 \le \overline{n} \le s. \tag{4.7.159}$$

Let us note that

$$\hat{d}_{n\varphi s}(\overline{n}, \overline{\varphi}) = \exp\left(i\frac{\overline{\varphi}}{2\overline{n}}\hat{n}^2\right)\hat{d}_{n\varphi s}(\overline{n}, 0)\exp\left(-i\frac{\overline{\varphi}}{2\overline{n}}\hat{n}^2\right) \tag{4.7.160}$$

as a simple consequence of the identity

$$\hat{D}(\overline{n}, \overline{\varphi}) = \exp\left(i\frac{\overline{\varphi}}{2\overline{n}}\hat{n}^2\right)\hat{D}(\overline{n}, 0)\exp\left(-i\frac{\overline{\varphi}}{2\overline{n}}\hat{n}^2\right). \tag{4.7.161}$$

Now we will derive the modification of the rule (4.6.71). It holds that

$$\hat{U}^\dagger(t)\hat{D}_{n\varphi s}(\overline{n}, 0)\hat{U}(t) = \hat{d}_{n\varphi s}(\overline{n}, 2\overline{n}\kappa t) + \exp[i(s+1)\theta_0]$$
$$\times \hat{d}_{n\varphi s}(\overline{n} + s + 1, 2(\overline{n} + s + 1)\kappa t)$$
$$\text{for } -s \le \overline{n} \le 0, \tag{4.7.162}$$

$$\hat{U}^\dagger(t)\hat{D}_{n\varphi s}(\overline{n}, 0)\hat{U}(t) = \hat{d}_{n\varphi s}(\overline{n}, 2\overline{n}\kappa t) + \exp[-i(s+1)\theta_0]$$
$$\times \hat{d}_{n\varphi s}(\overline{n} - s - 1, 2(\overline{n} - s - 1)\kappa t)$$
$$\text{for } 0 \le \overline{n} \le s. \tag{4.7.163}$$

Assuming that $\kappa t \equiv 0 \left(\text{mod } \frac{2\pi}{s+1}\right)$, we obtain that

$$\hat{D}_{n\varphi s}(\overline{n}, 2\overline{n}\kappa t) = \hat{U}^\dagger(t)\hat{D}_{n\varphi s}(\overline{n}, 0)\hat{U}(t). \tag{4.7.164}$$

Here $t = \frac{2\pi\tau}{\kappa(s+1)}$, τ integer.

It is useful to modify (4.6.115) to the form

$$C_{n\varphi s}\left(\overline{n}, \frac{4\pi\overline{n}\tau}{s+1}; 0\right) = C_{n\varphi s}\left(\overline{n}, 0; \frac{2\pi\tau}{\kappa(s+1)}\right). \tag{4.7.165}$$

It is instructive to ask whether to any $\overline{m}$ there exists a value of τ, for which

$$\frac{4\pi\overline{n}\tau}{s+1} \equiv \overline{\theta}_{\overline{m}} \pmod{4\pi}. \tag{4.7.166}$$

It is equivalent to solving an equation for τ

$$2\overline{n}\tau \equiv \overline{m} \pmod{2(s+1)}. \tag{4.7.167}$$

This equation can be solved for a prime modulus by division. Complete residue systems modulo a prime form a field, i. e., a set of numbers for which addition, subtraction, multiplication, and division (except by zero) are defined and for which the usual commutative, associative, and distributive laws apply [Schröder (1990)]. If we now assume that $(s+1)$ is a prime and we, moreover, restrict ourselves to $\overline{m}$ even, we observe that the equation

$$\overline{n}\tau \equiv \frac{\overline{m}}{2} \pmod{(s+1)} \tag{4.7.168}$$

has a unique solution for $\overline{n} \not\equiv 0 \pmod{(s+1)}$. In fact, the quantum characteristic function is to be determined for $\overline{n} = 0 \pmod{(s+1)}$ according to (4.7.103) in this case. On the contrary, the relation (4.7.102) indicates that the restriction to $\overline{m}$ even does not affect the generality. This is a tomography of the quantum state in the Hilbert space $\mathcal{H}_s$. The theory of discrete Wigner functions and of discrete quantum state tomography has been presented in general terms [Leonhardt (1995, 1996)]. It is worth noting that

$$\hat{D}_{n\varphi s}\left(\overline{n}, \overline{\theta}_{\overline{m}(\overline{\varphi}, s)}\right) \overset{w}{\to} \hat{D}_{n\varphi}\left(\overline{n}, \overline{\varphi}\right), \tag{4.7.169}$$

with the approximations $\overline{\theta}_{\overline{m}(\overline{\varphi}, s)}$ of the reciprocal photon number

$$\overline{\varphi} = \lim_{s\to\infty} \overline{\theta}_{\overline{m}(\overline{\varphi}, s)}. \tag{4.7.170}$$

A discrete Q function can be defined with the Wódkiewicz concept of propensities, i. e., as a discrete convolution of two Wigner functions based on Wootters's formalism [Opatrný, Bužek, Bajer and Drobný (1995)].

In order to prepare predetermined field states, the conditional measurement approach has been suggested [Sherman and Kurizki (1992), Sherman, Kurizki and Kadyshevitch (1992), Garraway, Sherman, Moya-Cessa, Knight and Kurizki (1994), Harel, Kurizki, McIver and Coutsias (1996)]. This approach has been enriched by a recipe for constructing an arbitrary superposition of Fock states [Vogel, Akulin and Schleich (1993)].

4.8   Quantum phase for multimode fields

In this section we outline the solution of the quantum phase problem for the multimode fields postponing the multimode quantum phase based on the phase-space description to section 4.10.

4.8.1   Multimode formalisms of quantum phase

In the quantum description of J monochromatic modes of light the Hilbert space $\mathcal{H}$ is the tensor product of the Hilbert spaces $\mathcal{H}_j$, $j = 1, 2, \ldots, J$,

$$\mathcal{H} = \bigotimes_{j=1}^{J} \mathcal{H}_j. \tag{4.8.1}$$

Accordingly, the number states have J components,

$$|\{n_j\}\rangle = \bigotimes_{j=1}^{J} |n_j\rangle_j. \tag{4.8.2}$$

The same notation is used in defining phase states

$$|\{\varphi_j\}\rangle = \bigotimes_{j=1}^{J} |\varphi_j\rangle_j. \tag{4.8.3}$$

It is useful to recall that any single-mode operator $\hat{M}$ can become a multimode operator $\hat{M}_j$ acting on the jth mode of the J-mode system,

$$\hat{M}_j = \hat{1}^{\otimes(j-1)} \otimes \hat{M} \otimes \hat{1}^{\otimes(J-j)}, \tag{4.8.4}$$

where $\hat{1}^{\otimes(j-1)}$ and $\hat{1}^{\otimes(J-j)}$ are $(j-1)$- and $(J-j)$-mode identity operators, respectively. Of course,

$$\begin{aligned} \hat{1}^{\otimes(j-1)} &\equiv \hat{1}_1 \otimes \ldots \otimes \hat{1}_{j-1}, \\ \hat{1}^{\otimes(J-j)} &\equiv \hat{1}_{j+1} \otimes \ldots \otimes \hat{1}_J, \end{aligned} \tag{4.8.5}$$

whereas $\hat{M}$ on the right-hand side in (4.8.4) is not provided with the subscript j, because this simple notation is reserved for the multimode operator.

From the properties of the tensor product it follows that the exponential phase operators for the jth mode

$$\begin{aligned} \widehat{\exp}(i\varphi_j) &= (\hat{n}_j + \hat{1})^{-\frac{1}{2}} \hat{a}_j, \tag{4.8.6} \\ \widehat{\exp}(-i\varphi_j) &= [\widehat{\exp}(i\varphi_j)]^\dagger, \tag{4.8.7} \end{aligned}$$

and the trigonometrical (cosine and sine) operators for the jth mode

$$\widehat{\cos}(\varphi_j) = \frac{1}{2}[\widehat{\exp}(i\varphi_j) + \widehat{\exp}(-i\varphi_j)], \tag{4.8.8}$$

$$\widehat{\sin}(\varphi_j) = \frac{1}{2i}[\widehat{\exp}(i\varphi_j) - \widehat{\exp}(-i\varphi_j)], \tag{4.8.9}$$

$j = 1, \ldots, J$, can also be constructed according to the prescription (4.8.4).

Denoting the single-mode operator-valued density for the phase measurement

$$\hat{\Gamma}(\varphi) = |\varphi\rangle\langle\varphi|, \tag{4.8.10}$$

we use the same tool to define the operator-valued densities $\hat{\Gamma}_j(\varphi_j)$, $j = 1, \ldots, J$. We introduce also the multimode operator-valued density

$$\hat{\Gamma}(\{\varphi_j\}) = \prod_{j=1}^{J} \hat{\Gamma}_j(\varphi_j). \tag{4.8.11}$$

This notation helps unify the exposition with that in section 4.10, whereas here it also holds that

$$\hat{\Gamma}(\{\varphi_j\}) = |\{\varphi_j\}\rangle\langle\{\varphi_j\}|, \tag{4.8.12}$$

which follows easily from the properties of the Kronecker products. We have two expressions for the probability density,

$$P(\{\varphi_j\}) = \langle\{\varphi_j\}|\hat{\rho}|\{\varphi_j\}\rangle \tag{4.8.13}$$

$$= \mathrm{Tr}\left\{\hat{\rho}\hat{\Gamma}(\{\varphi_j\})\right\}. \tag{4.8.14}$$

The marginal phase probability densities of separate modes can easily be obtained,

$$P_j(\varphi_j) = \mathrm{Tr}\left\{\hat{\rho}\hat{\Gamma}_j(\varphi_j)\right\}. \tag{4.8.15}$$

We easily generalize the Wigner function for number and phase given in (4.6.95) to the multimode case. To simplify the presentation, we use the prescription (4.8.4) to introduce the operators $\hat{\Delta}_{n_j\varphi_j}^{\mathrm{MEM}}(n_j, \varphi_j)$ related to modes $j = 1, \ldots, J$ as tensor products of the identity operators and the single-mode operator $\hat{\Delta}_{n\varphi}^{\mathrm{MEM}}(n, \varphi)$ given in (4.6.106) and (4.6.97). The appropriate multimode Wigner function for numbers and phases reads

$$\overline{\Phi}_{\{n_j\varphi_j\}}^{\mathrm{MEM}}(\{n_j, \varphi_j\}) = \frac{1}{(2\pi)^J}\mathrm{Tr}\left\{\hat{\rho}\hat{\Delta}_{\{n_j\varphi_j\}}^{\mathrm{MEM}}(\{n_j, \varphi_j\})\right\}, \tag{4.8.16}$$

where

$$\hat{\Delta}_{\{n_j\varphi_j\}}^{\mathrm{MEM}}(\{n_j, \varphi_j\}) = \prod_{j=1}^{J} \hat{\Delta}_{n_j\varphi_j}^{\mathrm{MEM}}(n_j, \varphi_j). \tag{4.8.17}$$

This generalization for $J = 2$ has been used for the derivation of a quasidistribution of the photon-number difference and the phase difference in [Luis and Peřina (1998)].

To expound how the phase problem is jointly "solved" in the J monochromatic modes of light, we assume at first that in these modes the formalism of enlargement has been adopted. The Hilbert spaces $\mathcal{H}_j$ are treated as imbedded in the enlarged Hilbert spaces $\mathcal{H}_{je}$, $\mathcal{H}_j \subset \mathcal{H}_{je}$, $j = 1, \ldots, J$. The product Hilbert space

$$\mathcal{H}_e = \bigotimes_{j=1}^{J} \mathcal{H}_{je} \tag{4.8.18}$$

can be interpreted also as the Hilbert space of J plane rotators. The angular-momentum states have J components,

$$|\{N_j\}\rangle = \bigotimes_{j=1}^{J} |N_j\rangle_j. \tag{4.8.19}$$

The angular-momentum component L_{jz} is appropriate to the jth plane rotator,

$$L_{jz} = -\hbar N_j. \tag{4.8.20}$$

Besides this interpretation, it is also possible to treat $\mathcal{H}_e$ as part of the Hilbert space of $2J$ light modes and $\mathcal{H}_{je}$ as parts of the Hilbert spaces $\mathcal{H}_{j-0} \otimes \mathcal{H}_{j0+}$, quite as Shapiro and Shepard (1991) considered these (cf. section 4.6). In the physical system of the plane rotators we consider rotation-angle states as e-phase states of the clockwise rotations and interpret them as e-phase states of light modes. These are

$$|\{\phi_j\}\rangle_e = \bigotimes_{j=1}^{J} |\phi_j\rangle_{je}. \tag{4.8.21}$$

Any single-rotator operator $\hat{M}_e$ becomes an operator $\hat{M}_{je}$ acting on the jth Hilbert space of the J-rotator system according to the following relation,

$$\hat{M}_{je} = \hat{1}_e^{\otimes(j-1)} \otimes \hat{M}_e \otimes \hat{1}_e^{\otimes(J-j)}, \tag{4.8.22}$$

where

$$\hat{1}_e^{\otimes(j-1)} = \hat{1}_{1e} \otimes \ldots \otimes \hat{1}_{(j-1)e}, \quad \hat{1}_e^{\otimes(J-j)} = \hat{1}_{(j+1)e} \otimes \ldots \otimes \hat{1}_{Je}. \tag{4.8.23}$$

Setting $\hat{M}_e = \widehat{\exp}(i\phi)$, $\widehat{\exp}(-i\phi)$, $\widehat{\cos}\,\phi$, $\widehat{\sin}\,\phi$ in (4.8.22), we obtain the definitions of the operators $\hat{M}_{je} = \widehat{\exp}(i\phi_j)$, $\widehat{\exp}(-i\phi_j)$, $\widehat{\cos}(\phi_j)$, $\widehat{\sin}(\phi_j)$.

Denoting the single-rotator operator-valued density for the rotation-angle measurement by

$$\hat{\Gamma}_e(\phi) = |\phi\rangle_{e\,e}\langle\phi| \tag{4.8.24}$$

and setting $\hat{M}_e = \hat{\Gamma}_e(\phi)$ in (4.8.22), we get $\hat{M}_{je} \equiv \hat{\Gamma}_{je}(\phi)$. Introducing the operator-valued density for J plane rotators,

$$\hat{\Gamma}_e(\{\phi_j\}) = \prod_{j=1}^{J} \hat{\Gamma}_{je}(\phi_j) \tag{4.8.25}$$

$$= |\{\phi_j\}\rangle_e \, _e\langle\{\phi_j\}|, \tag{4.8.26}$$

we have two expressions for the probability density

$$P(\{\phi_j\}) = \, _e\langle\{\phi_j\}|\hat{\rho}|\{\phi_j\}\rangle_e \tag{4.8.27}$$

$$= \mathrm{Tr}\left\{\hat{\rho}\hat{\Gamma}_e(\{\phi_j\})\right\}. \tag{4.8.28}$$

We may suppose that the statistical operator $\hat{\rho}$ describes no anticlockwise rotation in any plane rotator, i. e., the state $\hat{\rho}$ can be interpreted as the state of a multimode optical system. In the same manner, we provide $\hat{\Delta}_{N\phi}^{\mathrm{MEM}}(N, \phi)$ ((4.6.92), (4.6.83)) with subscript j to obtain $\hat{\Delta}_{N_j\phi_j}^{\mathrm{MEM}}(N_j, \phi_j)$. The appropriate multidimensional Wigner function for numbers (angular momenta) and phases reads

$$\hat{\overline{\Phi}}_{\{N,\phi_j\}}^{\mathrm{MEM}}(\{N_j\}, \{\phi_j\}) = \frac{1}{(2\pi)^J} \mathrm{Tr}\left\{\hat{\rho}\hat{\Delta}_{\{N,\phi_j\}}^{\mathrm{MEM}}(\{N_j, \phi_j\})\right\}, \tag{4.8.29}$$

where

$$\hat{\Delta}_{\{N,\phi_j\}}^{\mathrm{MEM}}(\{N_j, \phi_j\}) = \prod_{j=1}^{J} \hat{\Delta}_{N_j\phi_j}^{\mathrm{MEM}}(N_j, \phi_j). \tag{4.8.30}$$

Now, we expound the multimode Pegg–Barnett formalism. The Hilbert spaces are truncated to yield the Hilbert spaces $\mathcal{H}_{js_j}$, $\mathcal{H}_{js_j} \subset \mathcal{H}_j$, $j = 1, \ldots, J$. The product Hilbert space

$$\mathcal{H}_{\{s_j\}} = \bigotimes_{j=1}^{J} \mathcal{H}_{js_j} \tag{4.8.31}$$

is part of the original Hilbert space $\mathcal{H}$. The finite-dimensional Hilbert space $\mathcal{H}_{\{s_j\}}$ has a finite orthogonal basis of $\{s_j\}$-phase states,

$$|\{\theta_{m,j}\}; \{s_j\}\rangle = \bigotimes_{j=1}^{J} |\theta_{m,j}; s_j\rangle, \tag{4.8.32}$$

where

$$\theta_{m,j} = \theta_{0j} + \frac{2\pi m_j}{s_j + 1}, \quad m_j = 0, \ldots, s_j, \tag{4.8.33}$$

and θ_{0j} are given. We assume that the operators $\hat{M}_j = \hat{U}_{j,\theta_{0j},s_j}$, $\hat{U}_{j,\theta_{0j},s_j}^\dagger$ are defined according to (4.8.4) with $\hat{M} = \hat{U}_{\theta_{0j},s_j}$, $\hat{U}_{\theta_{0j},s_j}^\dagger$, respectively.

We denote the s-phase state statistical operator for single mode by

$$\hat{\gamma}_s(\theta_m) = |\theta_m; s\rangle\langle\theta_m; s| \tag{4.8.34}$$

and setting $\hat{M} = \hat{\gamma}_{\mathbf{s}_j}(\theta_{mjj})$ in (4.8.4), we define $\hat{M}_j = \hat{\gamma}_{j\mathbf{s}_j}(\theta_{mjj})$. We can introduce the multimode probability-operator measure

$$
\hat{\gamma}_{\{\mathbf{s}_j\}}(\{\theta_{mjj}\}) = \prod_{j=1}^{J} \hat{\gamma}_{j\mathbf{s}_j}(\theta_{mjj}) \tag{4.8.35}
$$

$$
= |\{\theta_{mjj}\}; \{\mathbf{s}_j\}\rangle\langle\{\theta_{mjj}\}; \{\mathbf{s}_j\}|. \tag{4.8.36}
$$

Let us note that the representation of the operator (4.8.35) in the number state basis does not comprise any number states $|n_{j'}\rangle_{j'}$ with $n_{j'} > s_{j'}$, whereas for convenience of the definition the operators $\hat{\gamma}_{j\mathbf{s}_j}(\theta_{mjj})$ do not obey this rule except $j' = j$.

In the multimode case, the definition of the physically preparable state $|\psi\rangle$ can be based on the property

$$
\langle\psi| \prod_{j=1}^{J} \hat{n}_j^{p_j} |\psi\rangle < \infty \quad \text{for any } \{p_j\}, \tag{4.8.37}
$$

where $p_j \geq 0$ are integers. Its generalization to the mixed state described by the statistical operator $\hat{\rho}$ reads

$$
\text{Tr}\left\{\hat{\rho} \prod_{j=1}^{J} \hat{n}_j^{p_j}\right\} < \infty \quad \text{for any } \{p_j\}. \tag{4.8.38}
$$

The phase properties of the physical state described by the statistical operator $\hat{\rho}$ can be, at least in the limit $\mathbf{s}_1 \to \infty, \ldots, \mathbf{s}_J \to \infty$, described by the probabilities

$$
P_{\{\mathbf{s}_j\}}(\{\theta_{mjj}\}) = \langle\{\theta_{mjj}\}; \{\mathbf{s}_j\}|\hat{\rho}|\{\theta_{mjj}\}; \{\mathbf{s}_j\}\rangle \tag{4.8.39}
$$

$$
= \text{Tr}\left\{\hat{\rho}\hat{\gamma}_{\{\mathbf{s}_j\}}(\{\theta_{mjj}\})\right\}. \tag{4.8.40}
$$

Just as done with the operators $\hat{\gamma}_{\mathbf{s}}(\theta_m)$, $\hat{\gamma}_{\mathbf{s}_j}(\theta_{mjj})$, and $\hat{\gamma}_{j\mathbf{s}_j}(\theta_{mjj})$, we attach the subscript j to $\hat{\Delta}_{n\varphi\mathbf{s}}(n, \theta_m)$ ((4.7.112)) to obtain $\hat{\Delta}_{jn_j\varphi_j\mathbf{s}_j}(n_j, \theta_{mjj})$. The appropriate multimode Wigner function for numbers and phases reads

$$
\Phi_{\{n_j\varphi_j\}\{\mathbf{s}_j\}}(\{n_j\}, \{\theta_{mjj}\}) = \prod_{j=1}^{J} \frac{1}{\mathbf{s}_j + 1} \text{Tr}\left\{\hat{\rho}\hat{\Delta}_{\{n_j\varphi_j\}\{\mathbf{s}_j\}}(\{n_j\}, \{\theta_{mjj}\})\right\}, \tag{4.8.41}
$$

where

$$
\hat{\Delta}_{\{n_j\varphi_j\}\{\mathbf{s}_j\}}(\{n_j\}, \{\theta_{mjj}\}) = \prod_{j=1}^{J} \hat{\Delta}_{jn_j\varphi_j\mathbf{s}_j}(n_j, \theta_{mjj}). \tag{4.8.42}
$$

4.8.2   Grey is the phase sum and the phase difference is green

In the applications of quantum theory, the sum and difference of phases of distinct modes are treated frequently. A good picture of phase-sum and phase-difference operators can be based on representations of the groups SU(2) and SU(1,1). The original use of the group $SU(2)$ is connected to the quantum-mechanical angular momentum [Wigner (1959)]. We have mentioned the effort to introduce the azimuthal angle operators for the angular momentum in section 4.3 oriented to unitary angle operators. In contrast, Vaglica and Vetri (1984) have introduced nonunitary exponential angle operators reminiscent of Susskind and Glogower (1964). The operators for the spherical angles of the quantum angular-momentum vector have been discussed by Nienhuis and van Enk (1993). Apart from pure mathematical reasons, the interest in these groups can be accounted for by the use of these concepts in the description of linear and nonlinear interferometers [Yurke, McCall and Klauder (1986)]. More generally, the group $SU(2)$ is suitable for the description of passive lossless linear quantum-optical devices with two input and output light modes. On the contrary, the group $SU(1,1)$ is convenient for the description of active lossless nonlinear devices with classical pumping, e. g., four-wave mixing under the parametric approximation.

For the analysis of an $SU(2)$ interferometer it is instructive to consider the Hermitian operators

$$
\begin{aligned}
\hat{J}_1 &= \frac{1}{2}(\hat{a}_1^\dagger \hat{a}_2 + \hat{a}_2^\dagger \hat{a}_1), \\
\hat{J}_2 &= \frac{1}{2i}(\hat{a}_1^\dagger \hat{a}_2 - \hat{a}_2^\dagger \hat{a}_1), \\
\hat{J}_3 &= \frac{1}{2}(\hat{a}_1^\dagger \hat{a}_1 - \hat{a}_2^\dagger \hat{a}_2).
\end{aligned}
\tag{4.8.43}
$$

This notation mirrors the Jordan–Schwinger representation of a quantum angular-momentum vector [Jordan (1935), Schwinger (1965)] (see section 3.6). These operators form the two-mode boson realization of the Lie algebra su(2),

$$
\begin{aligned}
[\hat{J}_1, \hat{J}_2] &= i\hat{J}_3, \\
[\hat{J}_2, \hat{J}_3] &= i\hat{J}_1, \\
[\hat{J}_3, \hat{J}_1] &= i\hat{J}_2.
\end{aligned}
\tag{4.8.44}
$$

The commutators (4.8.44) allow us to interpret the operators $\hat{J}_1$, $\hat{J}_2$, $\hat{J}_3$ as the usual "elliptic" angular-momentum components. The raising and lowering operators $\hat{J}_+$ and $\hat{J}_-$, respectively, have been introduced in (3.6.1) and expressed in terms of boson annihilation and creation operators in (3.6.8). The Casimir operator for any unitary irreducible representation of the group $SU(2)$ is a constant,

$$
J_{\text{irr}}^2 = J_{1\text{irr}}^2 + J_{2\text{irr}}^2 + J_{3\text{irr}}^2 = j(j+1)1_{\text{irr}},
\tag{4.8.45}
$$

and a representation of this group is determined by a single number j that acquires discrete positive values $j = \frac{1}{2}, 1, \frac{3}{2}, 2, \dots$. By using the operators given in (4.8.43), we get

$$
\hat{J}^2 = \hat{J}_1^2 + \hat{J}_2^2 + \hat{J}_3^2 = \frac{\hat{n}_s}{2}\left(\frac{\hat{n}_s}{2} + \hat{1}\right),
\tag{4.8.46}
$$

where

$$\hat{n}_s = \hat{a}_1^\dagger \hat{a}_1 + \hat{a}_2^\dagger \hat{a}_2 \tag{4.8.47}$$

is the operator of total number of photons entering the interferometer. We see that total photon number n is an SU(2) invariant related to the index j via $j = \frac{n}{2}$. The representation Hilbert space $\mathcal{H}_n^{\text{SU}(2)}$ is spanned by the complete orthonormal basis $|j, m\rangle$ $(m = -j, -j+1, ..., j-1, j)$ that can be expressed in terms of the Fock states of two modes,

$$|j, m\rangle = |j + m\rangle_1 |j - m\rangle_2. \tag{4.8.48}$$

Similarly, a two-mode system, which behaves according to the group SU(1,1), is described conveniently by the Hermitian operators,

$$\begin{aligned}
\hat{K}_1 &= \frac{1}{2}(\hat{a}_1^\dagger \hat{a}_2^\dagger + \hat{a}_1 \hat{a}_2), \\
\hat{K}_2 &= \frac{1}{2i}(\hat{a}_1^\dagger \hat{a}_2^\dagger - \hat{a}_1 \hat{a}_2), \\
\hat{K}_3 &= \frac{1}{2}(\hat{a}_1^\dagger \hat{a}_1 + \hat{a}_2 \hat{a}_2^\dagger).
\end{aligned} \tag{4.8.49}$$

These operators form the two-mode boson realization of the Lie algebra su(1,1),

$$\begin{aligned}
[\hat{K}_1, \hat{K}_2] &= -i\hat{K}_3, \\
[\hat{K}_2, \hat{K}_3] &= i\hat{K}_1, \\
[\hat{K}_3, \hat{K}_1] &= i\hat{K}_2.
\end{aligned} \tag{4.8.50}$$

According to the commutators (4.8.50) we interpret the operators $\hat{K}_1$, $\hat{K}_2$, $\hat{K}_3$ as the hyperbolic angular-momentum components. It is also useful to introduce the raising and lowering operators

$$\begin{aligned}
\hat{K}_+ &= \hat{K}_1 + i\hat{K}_2 = \hat{a}_1^\dagger \hat{a}_2^\dagger, \\
\hat{K}_- &= \hat{K}_1 - i\hat{K}_2 = \hat{a}_1 \hat{a}_2.
\end{aligned} \tag{4.8.51}$$

The Casimir operator for any unitary irreducible representation is a constant,

$$K_{\text{irr}}^2 = K_{3\text{irr}}^2 - K_{1\text{irr}}^2 - K_{2\text{irr}}^2 = k(k-1)1_{\text{irr}}. \tag{4.8.52}$$

Thus, a representation of the group SU(1,1) is determined by a single number k that is called the Bargmann index. For the discrete-series representations [Bargmann (1947)], the Bargmann index acquires discrete values $k = \frac{1}{2}, 1, \frac{3}{2}, 2, ...$ By using the operators given in (4.8.49), we get

$$\hat{K}^2 = \hat{K}_3^2 - \hat{K}_1^2 - \hat{K}_2^2 = \frac{1}{4}(\hat{n}_d^2 - \hat{1}), \tag{4.8.53}$$

where

$$\hat{n}_d = \hat{a}_1^\dagger \hat{a}_1 - \hat{a}_2^\dagger \hat{a}_2 \tag{4.8.54}$$

is the photon-number difference operator between the modes. We see that the photon-number difference n_d is an SU(1,1) invariant related to the Bargmann index k via $k = \frac{1}{2}(n_d + 1)$. The representation Hilbert space $\mathcal{H}_{n_d}^{SU(1,1)}$ is spanned by the complete orthonormal basis $|k, n_2\rangle$ $(n_2 = 0, 1, 2, ...)$ that can be expressed in terms of the Fock states of two modes,

$$|k, n_2\rangle = |n_2 + 2k - 1\rangle_1 |n_2\rangle_2. \tag{4.8.55}$$

Having noted the importance of the sum and difference of phases, we may also remark that a quite innocent transformation to the sum or difference of phases can incur a loss of information. In fact, if we shift both phases by the same angle π (mod 2π), we cannot arrive at a different sum and difference of phases. This fact has no character of an absolute law and is likely to be connected with our preference of 2π-periodic functions of sum and difference of phases. The joint distribution of original phases is not "transformed", but "cast" onto the joint distribution of the sum and difference of phases [Barnett and Pegg (1990b), Luis and Sánchez-Soto (1996), Lukš and Peřinová (1996)].

The definitions of sine and cosine operators of the sum and difference of phases depend on the formalism adopted. As we reserve a space to the Susskind–Glogower exponential phase operators, let us remark, that there are two ways how to define the exponential phase operators of the phase sum and the phase difference. Firstly, we can multiply the exponential phase operator of one mode by the exponential phase operator of the other mode or by its Hermitian conjugate. Equivalently, we can perform the polar decomposition of the operators $\hat{K}_-$ and $\hat{J}_-$, given in (4.8.51) and (3.6.1), respectively. The cosine and sine operators will be the Hermitian and "anti-Hermitian" parts of these exponential phase operators. Their analogy with the operators $\hat{K}_1, -\hat{K}_2, \hat{J}_1, -\hat{J}_2$ is so far reaching that there is no problem to find that the commutators between the cosine operator of the phase sum and the cosine operator of the phase difference, etc., do not vanish.

The group SU(1,1) is only locally compact and the appropriate theory of representations is more complicated than in the case of the group SU(2). In spite of this fact just the group-theoretic approach helps understand that the phase-sum operators will not have simpler properties than the single-mode cosine and sine operators. The literature devoted to the abstract theory of the phase sum (à la Dirac) is not copious [Agarwal (1993)]. Gerry (1988) has paid attention to the phase operators for SU(1,1), but he has restricted the realization to the single-mode case, where we encounter a doubled phase instead of the phase sum.

On the contrary, the group SU(2) is compact and also its representation in terms of two harmonic oscillators is decomposable into the irreducible finite-dimensional representations. The cosine and sine operators of the phase difference were studied by Susskind and Glogower [Susskind and Glogower (1964)].

In any investigation of optical systems, the conservation of photon number is of importance. If $\hat{A}$ is any operator and $\hat{n}_s$ is the total photon-number operator given in (4.8.47) and if they commute,

$$[\hat{A}, \hat{n}_s] = \hat{0}, \tag{4.8.56}$$

then $\hat{A}$ can be decomposed,

$$\hat{A} = \sum_{n=0}^{\infty} \hat{A}_n, \tag{4.8.57}$$

using $\hat{A}_n$ of the form

$$\hat{A}_n = \sum_{j=0}^{n} \sum_{k=0}^{n} A_{jk}^{(n)} |j, n - j\rangle \langle k, n - k|, \tag{4.8.58}$$

where

$$A_{jk}^{(n)} = \langle j, n - j | \hat{A} | k, n - k \rangle. \tag{4.8.59}$$

Particularly, the identity operator has the property (4.8.56) and it decomposes in terms of the identity operators acting on the Hilbert spaces $\mathcal{H}_n^{\text{SU}(2)}$ of the irreducible representations of the group SU(2),

$$\hat{1}_n^{\text{SU}(2)} = \sum_{j=0}^{n} |j, n - j\rangle \langle j, n - j|. \tag{4.8.60}$$

This enables us to write concisely,

$$\hat{A}_n = \hat{1}_n^{\text{SU}(2)} \hat{A} \hat{1}_n^{\text{SU}(2)}. \tag{4.8.61}$$

The Susskind–Glogower exponential phase-difference operator

$$\widehat{\exp}[i(\varphi_1 - \varphi_2)] = \widehat{\exp}(i\varphi_1)\widehat{\exp}(-i\varphi_2) \tag{4.8.62}$$

has the property (4.8.56) and it can be decomposed as

$$\widehat{\exp}[i(\varphi_1 - \varphi_2)] = \sum_{n=0}^{\infty} \left\{ \widehat{\exp}[i(\varphi_1 - \varphi_2)] \right\}_n. \tag{4.8.63}$$

Susskind and Glogower (1964) (see also [Carruthers and Nieto (1968)]) studied the operators

$$[\widehat{\cos}(\varphi_1 - \varphi_2)]_n = \frac{1}{2} \left(\left\{ \widehat{\exp}[i(\varphi_1 - \varphi_2)] \right\}_n + \left\{ \widehat{\exp}[-i(\varphi_1 - \varphi_2)] \right\}_n \right), \tag{4.8.64}$$

$$[\widehat{\sin}(\varphi_1 - \varphi_2)]_n = \frac{1}{2i} \left(\left\{ \widehat{\exp}[i(\varphi_1 - \varphi_2)] \right\}_n - \left\{ \widehat{\exp}[-i(\varphi_1 - \varphi_2)] \right\}_n \right), \tag{4.8.65}$$

where

$$\left\{ \widehat{\exp}[-i(\varphi_1 - \varphi_2)] \right\}_n = \left\{ \widehat{\exp}[i(\varphi_1 - \varphi_2)] \right\}_n^\dagger. \tag{4.8.66}$$

The commutator of these operators reads

$$\left[[\widehat{\cos}(\varphi_1 - \varphi_2)]_n , [\widehat{\sin}(\varphi_1 - \varphi_2)]_n\right] = \frac{i}{2}\left(|0,n\rangle\langle 0,n| - |n,0\rangle\langle n,0|\right). \tag{4.8.67}$$

The characteristic polynomial of the cosine operator (4.8.64)

$$\begin{vmatrix} -\lambda & \frac{1}{2} & \cdots & 0 & 0 \\ \frac{1}{2} & -\lambda & \cdots & 0 & 0 \\ \vdots & \vdots & \vdots & \ddots & \vdots \\ 0 & 0 & \cdots & -\lambda & \frac{1}{2} \\ 0 & 0 & \cdots & \frac{1}{2} & -\lambda \end{vmatrix} = \left(-\frac{1}{2}\right)^{n+1} U_{n+1}(\lambda), \tag{4.8.68}$$

where the Chebyshev polynomial of the second kind,

$$U_n(x) = \frac{\sin[(n+1)\mathrm{Cos}^{-1}x]}{\sin(\mathrm{Cos}^{-1}x)}. \tag{4.8.69}$$

The eigenvalues are distinct,

$$\lambda = C_{n,k} \equiv \cos(\theta_{n,k}), \quad k = \frac{1}{2}, 1, \ldots, \frac{n+1}{2}, \tag{4.8.70}$$

where

$$\theta_{n,k} = \frac{2\pi k}{n+2}, \quad k = \frac{1}{2}, 1, \ldots, \frac{n+1}{2}. \tag{4.8.71}$$

From the viewpoint of pure mathematics it is interesting that to any proper fraction $\frac{\theta}{2\pi} \in Q \cap \left(0, \frac{1}{2}\right)$, where Q is the set of rational numbers, there corresponds an eigenvalue of the cosine operator of the phase difference,

$$C = \cos \theta. \tag{4.8.72}$$

Although the eigenvalues form a dense set, they do not form a continuum, but only a countable set, which is expressed by our subscripts. Distinct symbols $\theta_{n,k}$, $\theta_{n',k'}$ may denote the same phase difference and on this condition the eigenvalue C,

$$C = \cos(\theta_{n,k}) = \cos(\theta_{n',k'}), \tag{4.8.73}$$

is degenerate. As all eigenvalues are degenerate, for instance, $\theta_{n,k} = \theta_{2n+2,2k}$, we speak of eigenspaces spanned by the simultaneous eigenvectors of the cosine operator and of the operator, which determines this basis in the eigenspace. But this auxiliary operator is known to be the operator of total photon number. In accordance with this we can write the eigenvectors

$$|C_{n,k}, n\rangle = \sqrt{\frac{2}{n+2}}\sqrt{1 - C_{n,k}^2} \sum_{j=0}^{n} U_j(C_{n,k})|j, n-j\rangle. \tag{4.8.74}$$

The solution to the eigenvalue problem for the sine operator of the phase difference can be advised by the fact that the appropriate eigenvalues form the same dense countable set as in the case of the cosine operator. This can be written as

$$\text{Spec}[\widehat{\sin}(\varphi_1 - \varphi_2)] = \text{Spec}[\widehat{\cos}(\varphi_1 - \varphi_2)], \qquad (4.8.75)$$

where Spec stands for spectrum. The simultaneous eigenvectors of the sine operator of the phase difference and of the operator of total photon number can be characterized by the property,

$$|S, n\rangle = i^{\hat{n}_1}|C = S, n\rangle, \qquad (4.8.76)$$

where $S \in \text{Spec}[\widehat{\sin}(\varphi_1 - \varphi_2)]$.

In the case of two modes the construction of a "well-behaved" phase-difference operator can be conceived in at least two ways. First, we may adopt the enlargement formalism or the truncation formalism, which are known to define the absolute phase operators in the two separate modes. In this case the distributions of the cosine and sine of the phase difference will have probability densities and will be supported by a continuum. The second way is motivated by the polar decomposition of the lowering operator $\hat{J}_-$ as has been mentioned above. One polar decomposition can define the Susskind–Glogower exponential operator of the phase difference,

$$\hat{J}_- = (\hat{n}_1 + \hat{1})^{\frac{1}{2}}\widehat{\exp}[i(\varphi_1 - \varphi_2)](\hat{n}_2 + \hat{1})^{\frac{1}{2}}. \qquad (4.8.77)$$

Another polar decomposition admits not only the Susskind–Glogower exponential phase operator, but also a unitary operator of the phase difference [Luis and Sánchez-Soto (1993b)],

$$\hat{J}_- = \hat{U}_{\text{d}}\hat{n}_1^{\frac{1}{2}}(\hat{n}_2 + \hat{1})^{\frac{1}{2}}, \qquad (4.8.78)$$

where $\hat{U}_{\text{d}}$ is a unitary operator to be determined. Equation (4.8.78) can be rewritten as

$$\hat{a}_1\widehat{\exp}(-i\varphi_2) = \hat{U}_{\text{d}}\hat{n}_1^{\frac{1}{2}}. \qquad (4.8.79)$$

Using the restrictions to $\mathcal{H}_n^{\text{SU}(2)}$ and the result of Lévy-Leblond (1973, 1976), we obtain

$$\hat{U}_{\text{d}} = \widehat{\exp}[i(\varphi_1 - \varphi_2)] + \sum_{n=0}^{\infty} |\psi_n\rangle\langle 0, n|, \qquad (4.8.80)$$

where

$$|\psi_n\rangle = U_{n00n}|n, 0\rangle, \qquad (4.8.81)$$

with U_{n00n} complex units. If the transposition of the modes is to lead to the inverse or Hermitian conjugate of the considered operator, then U_{n00n} should be real. Also this formalism leads to the unusual intrinsicly quantum property that the distributions of the cosine and sine of the phase difference are singular discrete and have no usual probability density. The Luis–Sánchez-Soto phase-difference operator reappears in the Stokes parameters of the electromagnetic field [Sánchez-Soto and Luis (1994)].

The Stokes operators can be defined as $\hat{S}_0 = \hat{n}_s$, $\hat{S}_1 = 2\hat{J}_1$, $\hat{S}_2 = 2\hat{J}_2$, $\hat{S}_3 = 2\hat{J}_3$, when $\hat{a}_1$ and $\hat{a}_2$ are the photon annihilation operators of appropriate polarization components. The opposition between $\hat{U}_d$ and $\hat{U}_{1,\theta_{01},s_1}\hat{U}^\dagger_{2,\theta_{02},s_2}$ provoked a discussion [Pegg and Vaccaro (1995), Luis and Sánchez-Soto (1995)]. The relationship between the azimuthal-angle operators in the system of angular momentum and the phase-difference operator associated with the two oscillators introduced à la Schwinger has been discussed [Kar and Bhaumik (1995)]. The situation of two light modes can be changed deeply and an operator representing the relative phase between a single mode of the electromagnetic field and the atomic dipole in the Jaynes–Cummings model can be introduced, $\hat{a}_2 \to \hat{a}$, $\hat{a}_1 \to \hat{\sigma}_-$, where $\hat{\sigma}_-$ is the Pauli lowering atomic operator [Luis and Sánchez-Soto (1997a)]. In that paper the time evolution of the characteristics of the relative phase has been analyzed. Luis and Sánchez-Soto (1997b) have compared their solution of the relative quantum phase with two positive operator-valued measures for field and dipole phases. Their solution is based on the polar decomposition of the operator $\hat{\sigma}_-\hat{a}^\dagger$. Yu (1997) has rediscovered the Luis–Sánchez-Soto phase-difference operator [Luis and Sánchez-Soto (1993b)] and discussed it thoroughly with respect to the nonclassical properties of sums of quantum phase differences in a more-than-two-mode field.

4.9   Number–phase extremal states

In quantum theory, we encounter physical quantities that bear names of the classical ones, but they have a different meaning. This was called a reinterpretation by Heisenberg (1925). The new meaning of the quantities manifests itself in a new role of measurement. Not all pairs of the quantities are compatible, i. e., the measurement of one observable may affect the accuracy of the measurement of the other. This phenomenon is described by the uncertainty relations. It is marvellous that, in spite of the prominent physical role of the uncertainty relations, their mathematical proof rests upon the mere Schwartz inequality (see section 3.5).

The quantum phase problem is formulated for the quantum optical oscillator. Both early and recent work on this problem comprises also the study of the rotation angle of the quantal plane rotator. Do not let us forget the attention paid to the phase difference and phase sum of two quantum optical oscillators. Of many quantities, which can be considered in physical systems, some are basic, for instance, the quadrature components in the radiation modes in quantum optics and the angular momentum and the rotation angle in the plane rotator. The quantum theory of these systems arises in the quantum interpretation of these quantities as operators and of the Poisson brackets as the commutators. Usually, we can discern whether an examined classical observable is basic or not. In the latter case, it must be possible to express the investigated observable in terms of the basic ones. This possibility is yielded also by the quantum theory if we accept the necessity of ordering quantum operators in some cases. The photon-number operator can be simply expressed in terms of the basic quadrature operators in the system of optical oscillator and a similar "number operator" is basic in the system of the plane rotator. To the contrary, there is no (Hermitian) phase operator in the system of optical oscillator, but, fortunately, a similar "phase operator" is basic again in the system of the plane rotator.

The eigenvalue problem (3.5.24) directly leads to intelligent or minimum-uncertainty states. The eigenvalue problem (3.5.19) for the Hermitian operator can be solved as the variational problem for $\langle \hat{A} \rangle$ with a subsidiary condition that $\langle \hat{B} \rangle$ is known or the variational problem for $\langle \hat{B} \rangle$ with a condition that $\langle \hat{A} \rangle$ (in many cases, the mean photon number) is known. In the problems we consider in the one-mode case (and also in the cases of the plane rotator and of the truncated Hilbert space), the eigenvalues are nondegenerate. They are not multiple even in the two-mode case, but this is achieved by simultaneously considering an eigenspace (belonging to an eigenvalue) of the operator of photon-number sum (of photon-number difference).

We shall focus on the quantum phase of the optical oscillator. We shall obtain special states for this physical system not only directly, but also indirectly using the Hilbert space of the plane rotator and the truncated Hilbert space of optical oscillator.

4.9.1 Special states of the optical oscillator

The original problem treated in the harmonic oscillator is that of position-coordinate–momentum minimum-uncertainty states. It can be modified for the optical oscillator on considering canonically conjugate quadrature operators

$$\text{(i)} \quad \hat{A} = \hat{Q}, \ \hat{B} = \hat{P}, \ [\hat{Q}, \hat{P}] = 2i\hat{1}, \tag{4.9.1}$$

which are to be substituted into (3.5.24). If $\gamma > 0$, this eigenvalue problem defines the ordinary squeezed states (cf. section 3.5).

For comparison, we may treat the eigenvalue problem, which can be obtained by setting

$$\text{(ii)} \quad \hat{A} = \widehat{\cos}\,\varphi, \ \hat{B} = \widehat{\sin}\,\varphi, \ [\widehat{\cos}\,\varphi, \widehat{\sin}\,\varphi] = \frac{i}{2}|0\rangle\langle 0|, \tag{4.9.2}$$

in (3.5.24). Contrary to the case of the usual squeezed state, we must assume that

$$|\text{Re}\,\lambda|^2 + \frac{1}{\gamma^2}|\text{Im}\,\lambda|^2 < 1. \tag{4.9.3}$$

The solutions of this problem for $\gamma = 1$ are the Holstein–Primakoff generalized SU(1,1) coherent states with the Bargmann index $\frac{1}{2}$ [Bužek (1989a, b)] or coherent phase states [Shapiro and Shepard (1991), Vourdas (1993)], $\lambda = \zeta$, $|\psi\rangle = |\zeta\rangle$,

$$|\zeta\rangle = \sqrt{1 - |\zeta|^2} \sum_{n=0}^{\infty} \zeta^n |n\rangle, \ |\zeta| < 1, \tag{4.9.4}$$

and for $\gamma \geq 0$, the solutions to the problem (3.5.24), (4.9.2) are superpositions of two coherent phase states $|\zeta_1\rangle$, $|\zeta_2\rangle$ generally, where [Sukumar (1993), Lukš and Peřinová (1997b)]

$$\zeta_{1,2} = \frac{\lambda \pm \sqrt{\lambda^2 - 1 - \gamma^2}}{1 + \gamma}. \tag{4.9.5}$$

The exceptions are $\lambda = \pm\sqrt{1 - \gamma^2}$, which are the foci of the ellipse described by (4.9.3) with the equality sign. The solutions are called "squeezed phase states" by Shapiro and Shepard (1991). A proposal how the coherent phase states can be ideally achieved has been made by D'Ariano (1992). For $\gamma = -1$, dual coherent phase states should be defined. Although it is not possible using usual techniques, dual eigenkets of the Susskind–Glogower phase operator have been introduced by Fan and Xiao (1996).

The solution of the eigenvalue problem (3.5.24) for the case

$$\text{(iii)} \quad \hat{A} = \hat{n}, \quad \hat{B} = \hat{P}, \quad [\hat{n}, \hat{P}] = i\hat{Q}, \tag{4.9.6}$$

is of slightly different character [Hradil (1991)]. The appropriate eigenvalues are

$$\lambda = M + \gamma^2, \quad M = 0, 1, \ldots, \infty. \tag{4.9.7}$$

From this it follows that $\langle \hat{P} \rangle = 0$. The eigenkets are given as

$$\begin{aligned}
|\psi\rangle &= \frac{\exp\left(-\frac{\gamma^2}{2}\right)}{\sqrt{L_M^0(-4\gamma^2)}} \left\{ \sum_{n=0}^{M} \sqrt{\frac{n!}{M!}} \gamma^{M-n} L_n^{M-n}(-\gamma^2)|n\rangle \right. \\
&\quad \left. + \sum_{n-M+1}^{\infty} \sqrt{\frac{M!}{n!}} \gamma^{n-M} L_M^{n-M}(-\gamma^2)|n\rangle \right\}.
\end{aligned} \tag{4.9.8}$$

Contrary to cases (i) and (ii), where $\gamma > 0$, here γ is a real number. Nevertheless, the case $\gamma > 0$ seems to be more "canonical" than the case $\gamma < 0$, which can be reduced to the former by rotation by the phase shift π. This leads to the conclusion that a generalization with the operator $\hat{P}$ replaced by the operator $\hat{P}(\tau)$ is possible. The canonical phase distribution $P(\varphi)$ of this intelligent state has one peak and it holds that $\text{Pref}_{\theta_0}\varphi = \text{Arg}_{\theta_0}\gamma$. The quantum statistical properties of this state follow from a generalization of this eigenvalue problem using the displacement and the rotation of the phase space [Peřinová, Lukš and Křepelka (1994), Lukš, Peřinová and Křepelka (1994b)]. The number–quadrature-P intelligent states exhibit sub-Poissonian behaviour (number squeezing), which is significant for near-number states and is almost negligible for the asymptotic coherent state. As for the quadrature variances, the momentum-like quadrature exhibits no squeezing, whereas the position-like quadrature exhibits squeezing for a suitable value of the parameter γ. The phase narrowing develops with the parameter γ increasing, but its amount is no greater than that in the coherent state. Each of number–quadrature intelligent states can be obtained from a two-mode squeezed coherent state after reduction of the idler-mode component to a number state [Watanabe and Yamamoto (1988), Agarwal (1990), Peřinová, Lukš and Křepelka (1996a, b)].

The photon-number and phase properties of the number–quadrature-P intelligent state change after a displacement, but a suitable displacement leads to the appealing eigenvalue problem,

$$(\hat{n} - \gamma\hat{a}^\dagger)|\psi\rangle = \lambda|\psi\rangle. \tag{4.9.9}$$

The eigenvalues are $\lambda = M$ and the appropriate eigenvectors read

$$|\psi\rangle = \frac{\exp\left(-\frac{\gamma^2}{2}\right)}{\sqrt{M!L_M^0(-\gamma^2)}} \sum_{n=M}^{\infty} \frac{\sqrt{n!}}{(n-M)!}\gamma^{n-M}|n\rangle. \tag{4.9.10}$$

These states are known as the photon-added coherent states [Agarwal and Tara (1991)],

$$|\psi\rangle = \frac{1}{\sqrt{M!L_M^0(-\gamma^2)}}\hat{a}^{\dagger M}|\gamma\rangle. \tag{4.9.11}$$

The above mentioned states belong at least formally to the noise minimum states of Hradil (1990a).

Substituting

$$\text{(iv)} \quad \hat{A} = \hat{n}, \;\; \hat{B} = -\hat{Q}, \;\; [\hat{n}, \hat{Q}] = -i\hat{P}, \tag{4.9.12}$$

into the scheme (3.5.19), we obtain an eigenvalue problem leading to displaced number states. The eigenvalues are

$$\lambda = M - \gamma^2 \tag{4.9.13}$$

and the appropriate eigenvectors

$$|\psi\rangle = \hat{D}(\gamma)|M\rangle. \tag{4.9.14}$$

The scheme (3.5.19) leads to special states with $\langle[\hat{A}, \hat{B}]\rangle = 0$ or $\langle\psi|\hat{P}|\psi\rangle = 0$. Of course, they are not intelligent states. Moreover,

$$\begin{aligned}
\langle\psi|\hat{n}|\psi\rangle &= M + \gamma^2, \;\; \langle\psi|\hat{Q}|\psi\rangle = 2\gamma, \\
\langle\psi|(\Delta\hat{n})^2|\psi\rangle &= \gamma^2(2M+1), \;\; \langle\psi|(\Delta\hat{Q})^2|\psi\rangle = 2M+1, \\
\text{cov}(\hat{n}, \hat{Q}) &= \gamma(2M+1).
\end{aligned} \tag{4.9.15}$$

A suitably defined quantum correlation characterizes the desired extremal property $\text{cor}(\hat{n}, \hat{Q}) = \text{sgn}\,\gamma$.

Replacing the quadrature operator $\hat{P}$ in the scheme (4.9.6) by the Susskind–Glogower sine operator and substituting

$$\text{(v)} \quad \hat{A} = \hat{n}, \;\; \hat{B} = \widehat{\sin}\,\varphi, \;\; [\hat{n}, \widehat{\sin}\,\varphi] = i\widehat{\cos}\,\varphi, \tag{4.9.16}$$

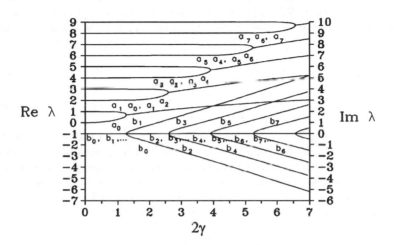

Figure 4.3: Eigenvalues for a superposition of the operators of the photon number and the sine of phase in the dependence on γ (the real parts of eigenvalues (curves $a_0, a_1, \ldots$), the imaginary parts of eigenvalues (curves $b_0, b_1, \ldots$)).

into (3.5.24), we arrive at the eigenvalue problem of Jackiw (1968). The eigenvalues λ are given implicitly as the roots of the equation

$$I_{-1-\lambda}(\gamma) = 0, \qquad (4.9.17)$$

with the modified Bessel function $I_\nu(\gamma)$. Although we have no explicit formula for the roots of (4.9.17), we present the relation

$$\lambda = \lambda_M(\gamma), \quad M = 0, 1, \ldots, \infty, \qquad (4.9.18)$$

in analogy with the formula (4.9.7). Whereas Jackiw (1968) concentrated on real roots of equation (4.9.18), which we could denote $\lambda_{2k}(\gamma)$, $\lambda_{2k+1}(\gamma)$, whenever $2k < \lambda < 2k + 1$, complex roots have been considered in [Lukš, Peřinová and Křepelka (1992a)].

Computing the (real) roots of equation (4.9.17) by means of numerical analysis and plotting them in the dependence on γ, we obtain a plot consisting of an infinite number of disjoint ovals stacked symmetrically above one another, as can be seen in figure 4.3, where we restricted ourselves to $\gamma \geq 0$. Numbering the ovals with $k \in \mathbb{N}$ from bottom up, we find that the kth oval is contained in the strip $\{(\gamma, \lambda); \gamma \in \mathbb{R}, \lambda \in [2k, 2k + 1]\}$. The projection of the kth oval onto the γ axis is the interval $[-\gamma_k, \gamma_k]$. From the picture it is likely that

$$0 < \gamma_0 < \gamma_1 < \ldots . \qquad (4.9.19)$$

Taking into account also complex roots of equation (4.9.17), we associate with the kth oval the two roots, $\lambda_{2k}(\gamma)$, $\lambda_{2k+1}(\gamma)$. For γ and k such that the two roots are real, the mean value $\frac{1}{2}[\lambda_{2k}(\gamma) + \lambda_{2k+1}(\gamma)]$ can be observed in the picture and when the roots are complex, the common value, $\text{Re}[\lambda_{2k}(\gamma)] = \text{Re}[\lambda_{2k+1}(\gamma)]$, is plotted as a natural continuation of the mean value. The imaginary parts $\pm\text{Im}[\lambda_{2k}(\gamma)] = \mp\text{Im}[\lambda_{2k+1}(\gamma)]$ are also plotted.

The appropriate eigenvector is

$$|\psi\rangle = \left[\sum_{l=0}^{\infty} |I_{l-\lambda_M(\gamma)}(\gamma)|^2\right]^{-\frac{1}{2}} \sum_{n=0}^{\infty} I_{n-\lambda_M(\gamma)}(\gamma)|n\rangle. \tag{4.9.20}$$

The phase properties of this number–sine intelligent state have been studied thoroughly in [Lukš, Peřinová and Křepelka (1992a)]. The photon-number and phase characteristics, such as the phase distribution, the phase dispersion, the number–sine uncertainty ratio, the uncertainty product comprising the phase dispersion, and the expectation value of the sine operator have been investigated in the dependence on γ. For example, the numerical results illustrate the uncertainty relation

$$M \geq \frac{1}{4}, \tag{4.9.21}$$

where the number–phase uncertainty product

$$M = \left[\langle(\Delta\hat{n})^2\rangle + \frac{1}{4}\right]\left[1 - |\langle\widehat{\exp}(i\varphi)\rangle|^2\right]. \tag{4.9.22}$$

With respect to the analogy between the quadrature operator $\hat{P}$ and the sine operator $\widehat{\sin}\,\varphi$, a similarity between the quantum statistics of the number–quadrature-P intelligent states and the number–sine intelligent states for the eigenvalue λ real can be observed. Particularly, there is a possibility to call the state of Jackiw a crescent state. Let us note that $\text{Im}\,\lambda = 0$ here, whenever $\langle\widehat{\sin}\,\varphi\rangle = 0$. The eigenvalues $M + \gamma^2$ and $\lambda_M(\gamma)$ behave similarly for $M = 0, 2, 4, \ldots$. In the case of complex imaginary $\lambda_M(\gamma)$ the dissimilarity is due mainly to the peak in the phase distribution, which occurs not for $\varphi = 0 \pmod{2\pi}$, but rather for $\varphi \approx \mp\frac{\pi}{2}$ for M even, odd, respectively. Thus, the number–sine intelligent state can be studied as a perturbed number state. Yamamoto, Imoto and Machida (1986) have shown numerically that the Jackiw state exhibits a clear crescent shape and used this state as a standard for physical states produced in a Kerr nonlinear interferometer. Assuming that an interference on a beamsplitter reduces to a displacement, Kitagawa and Yamamoto (1986) have determined the optimum reference signal and also optimum interaction length to obtain the optimum value of the Fano factor. Approximate number–sine intelligent states and near-number states via Kerr coupling and quadrature measurement have been investigated by Kitagawa, Imoto and Yamamoto (1987).

The use of the phase representation for the solution of the eigenvalue problem according to (4.9.16) is not impressive, but it is convincing for the eigenvalue problem $(\hat{n} - \gamma\widehat{\exp}(-i\varphi))|\psi\rangle = \lambda|\psi\rangle$ [Brif and Ben-Aryeh (1994b)]. The Barut–Girardello states [Barut and Girardello (1971)] $|\frac{1}{2}, \zeta\rangle$ can be considered in the connection with the quantum phase [Brif (1995)]. The appropriate modification of the problem and systematic investigation of statistical and phase properties and number–phase uncertainty relations have been presented there. Rotationally symmetrical number–phase uncertainty relations are very attractive, but no appropriate intelligent states exist except number states. The relations (4.9.21) and (4.9.22) can be rewritten in the form

$$\left[\langle(\Delta\hat{n})^2\rangle + \frac{1}{4}\right] D\varphi \geq \frac{1}{4}, \qquad (4.9.23)$$

where the phase dispersion (cf. section 4.4)

$$D\varphi = \langle\Delta\widehat{\exp}(i\varphi)\Delta\widehat{\exp}(-i\varphi)\rangle_A^{CS}, \qquad (4.9.24)$$

with $\Delta\widehat{\exp}(\pm i\varphi) = \widehat{\exp}(\pm i\varphi) - \langle\widehat{\exp}(\pm i\varphi)\rangle$. This can be related to the formula $D\varphi = 1 - |\langle\widehat{\exp}(i\varphi)\rangle|^2$ by Bandilla and Paul (1969) and to the solution of Newton (1980) (cf. also (4.9.38), (4.9.39) below). An alternative procedure leads to the form of Holevo (1982)

$$\langle(\Delta\hat{n})^2\rangle\frac{D\varphi}{1 - D\varphi} \geq \frac{1}{4}. \qquad (4.9.25)$$

Originally, this derivation leads to the relation (cf. [Carruthers and Nieto (1965)])

$$\langle(\Delta\hat{n})^2\rangle\frac{(D\varphi)_S^{CS}}{1 - D\varphi} \geq \frac{1}{4}, \qquad (4.9.26)$$

where

$$(D\varphi)_S^{CS} = \langle(\Delta\widehat{\sin}\varphi)^2\rangle + \langle(\Delta\widehat{\cos}\varphi)^2\rangle. \qquad (4.9.27)$$

It is useful to interpret (3.5.19) after dividing by γ as the variational problem $\langle\hat{B}\rangle = \min$, with a subsidiary condition that $\langle\hat{A}\rangle$ is known. Substituting

$$(vi) \quad \hat{A} = \hat{n}, \quad \hat{B} = -|\varphi = 0\rangle\langle\varphi = 0| \qquad (4.9.28)$$

into (3.5.19), we get the Shapiro–Shepard–Wong problem. In this original formulation, the problem has no solution, because it leads to logical contradictions related to the infinite Hilbert space. In [Shapiro, Shepard and Wong (1989)] a solution is achieved after adopting the finite-dimensional Hilbert space approach.

The choice (4.9.28) means to maximize the peak of the phase distribution at $\varphi = 0$ and comes from the quantum estimation theory, as well as the problem to maximize $\langle\widehat{\cos}\varphi\rangle$. Now we substitute

$$(vii) \quad \hat{A} = \hat{n}, \quad \hat{B} = -\widehat{\cos}\varphi, \quad [\hat{n}, \widehat{\cos}\varphi] = -i\widehat{\sin}\varphi \qquad (4.9.29)$$

into (3.5.19) and we arrive at the eigenvalue problem according to Bandilla, Paul and Ritze (1991). The eigenvalues λ are given implicitly as the roots of the equation

$$J_{-1-\lambda}(\gamma) = 0, \qquad (4.9.30)$$

with the Bessel function $J_\nu(\gamma)$. The appropriate eigenvector reads

$$|\psi\rangle = \left\{ \sum_{l=0}^{\infty} \left[J_{l-\lambda_M(\gamma)}(\gamma) \right]^2 \right\}^{-\frac{1}{2}} \sum_{n=0}^{\infty} J_{n-\lambda_M(\gamma)}(\gamma)|n\rangle, \qquad (4.9.31)$$

where $\lambda_M(\gamma)$, which has the property $\lambda_M(0) = M$, is a "perturbed" eigenvalue M of the number operator $\hat{n}$. By symmetry, it holds that $\langle \sin \varphi \rangle = 0$. As this condition can be fulfilled without loss of generality, the variational problem is the problem of minimizing the phase dispersion under the condition that the mean photon number is known and $\langle \widehat{\sin} \varphi \rangle = 0$. The eigenvalue problem (3.5.19) with (4.9.29) is the stationarity (Fermat) condition. A comparative study of possible problems can be found in [Białynicki–Birula, Freyberger and Schleich (1993)], cf. also the problem examined in [Daeubler, Miller, Risken and Schoendorff (1993)]. Let us note that this eigenvalue problem was motivated by the paper of Summy and Pegg (1990). The substitution of the ordinary squeezed states for the phase optimized states has been discussed by Bandilla and Ritze (1994). The dispersions of the phase probability distributions have been determined based on the Pegg–Barnett formalism, the Wigner function, and the Q function for fixed mean photon numbers in dependence on the squeezing parameter. In this situation, the phase optimized state is characterized only by a plateau corresponding to the Pegg–Barnett phase uncertainty. The asymptotic formulae of Collett (1993b) for the optimum squeezing parameter and the attained Pegg–Barnett and Wigner dispersions have been rederived. Leaving out the condition on the mean photon number, we obtain the problem treated by Brunet (1964) and Harms and Lorigny (1964). Being aware that improper elements of the Hilbert space $\mathcal{H}$ (cf. (4.5.32), (4.5.34)) would not be useful, they have considered the finite-dimensional Hilbert space.

The variational formulation can be based also on the subsidiary constraints that $\langle \hat{n} \rangle = \bar{n}$ and $\langle (\Delta \hat{n})^2 \rangle$ are known. The variational technique leads to the choice

$$(viii) \quad \hat{A} = (\hat{n} - \bar{n}'\hat{1})^2, \quad \hat{B} = -\widehat{\cos} \varphi, \qquad (4.9.32)$$

in (3.5.19). In fact, this variational formulation is beyond the scheme (3.5.19), which can be seen also from the dependence of the operator $\hat{A}$ chosen on a parameter $\bar{n}'$. The replacement $\gamma \to 2\gamma^2$ is made. This problem has been formulated in [Lukš, Peřinová and Křepelka (1992b)] and solved in [Opatrný (1995)].

The optimization with respect to other measures of phase variability is more complicated, because they need not be quadratic functionals in $|\psi\rangle$. Therefore, it is impossible to define the operator $\hat{B}$ independent of $|\psi\rangle$. With this proviso in mind, we shall formulate the variational problem for the Süssmann measure and substitute

$$(ix) \quad \hat{A} = \hat{n}, \quad \hat{B} = -\frac{(|\psi\rangle\langle\psi|)_{\text{Toep}}}{\langle\psi|\psi\rangle}, \qquad (4.9.33)$$

where

$$\hat{\rho}_{\text{Toep}} = \int_{\theta_0}^{\theta_0+2\pi} \langle\varphi|\hat{\rho}|\varphi\rangle |\varphi\rangle\langle\varphi| \, d\varphi, \qquad (4.9.34)$$

into (3.5.19). Essentially, for $\gamma = 2\pi$ and $\lambda = -(1 - |\zeta|^2)^{-1}$, $\zeta > 0$, the eigenvector is a coherent phase state $|\zeta\rangle$. Although this formulation is ours, the solution to this problem can be found in [Daeubler, Miller, Risken and Schoendorff (1993)].

4.9.2 Special states of the plane rotator

In the following, we shall treat intelligent states of the plane rotator and states defined by a variational problem and we shall demonstrate their relevance for the optical oscillator.

For the choice

$$(\mathrm{x}) \quad \hat{A} = \hat{N}, \quad \hat{B} = \widehat{\sin}\,\phi, \quad [\hat{N}, \widehat{\sin}\,\phi] = i\widehat{\cos}\,\phi, \tag{4.9.35}$$

in (3.5.24), we obtain the definition of the number–sine-of-phase intelligent state of the plane rotator. These states are at the same time the sine-of-rotation-angle-angular-momentum intelligent states. In conformity with the notation $\mathcal{H}_e$ for the enlarged Hilbert space of the harmonic oscillator, we shall use $|\psi\rangle_e$ for the the eigenstates. In [Carruthers and Nieto (1968)] this eigenvalue problem is solved in terms of the modified Bessel functions of the first kind of order $(N - \lambda)$,

$$|\psi\rangle_e = [I_0(2\gamma)]^{-\frac{1}{2}} \sum_{N=-\infty}^{\infty} I_{N-\lambda}(\cdot\gamma)|N\rangle, \quad {}_e\langle\psi|\psi\rangle_e = 1, \tag{4.9.36}$$

where $\lambda = \langle\hat{N}\rangle$ is an integer. The appropriate wave function is easily found in the angular representation and the angular distribution can be identified with the von Mises distribution [Rao (1973)]. The uncertainty relation (3.5.21) reads in this case

$$\langle(\Delta\hat{N})^2\rangle\langle(\Delta\widehat{\sin}\,\phi)^2\rangle \geq \frac{1}{4}\langle\widehat{\cos}\,\phi\rangle^2. \tag{4.9.37}$$

Following [Carruthers and Nieto (1968)], we have considered a symmetrical form of the uncertainty relation and—unlike [Carruthers and Nieto (1968)]—we have arrived at the relation [Lukš and Peřinová (1991)],

$$M_e \geq \frac{1}{4}, \tag{4.9.38}$$

where

$$M_e = \left[\langle(\Delta\hat{N})^2\rangle + \frac{1}{4}\right]\left[1 - |\langle\widehat{\exp}(i\phi)\rangle|^2\right]. \tag{4.9.39}$$

Due to this symmetry, the hope of attaining the equality in (4.9.38) is lost except in the number states $|N\rangle$, since it would require the fulfillment of both the conditions

$$\mathrm{cov}(\hat{N}, \widehat{\cos}\,\phi) = \mathrm{cov}(\hat{N}, \widehat{\sin}\,\phi) = 0, \tag{4.9.40}$$

formally rewritten

$$\mathrm{cov}(\hat{N}, \widehat{\exp}(i\phi)) = 0. \tag{4.9.41}$$

Compared with the ordinary uncertainty relation, the presence of $\frac{1}{4}$ in the first factor is striking, a correction for the discreteness of the variable $\hat{N}$. The second factor is the angular dispersion, free of variances as recommended in the statistical studies of directional data [Rao (1973)]. The uncertainty product (4.9.39) is useful [Lukš and Peřinová (1991, 1992), Lukš, Peřinová and Křepelka (1992a)] and open for further numerical analysis since no minimizing states of relation (4.9.38) are likely to exist except the number states $|N\rangle$. Part of [Bluhm, Kostelecký and Tudose (1995)] has been devoted to the construction of a wave packet, called a circular squeezed state, that minimizes the uncertainty product for the angular variables.

Replacing $|\psi\rangle \rightarrow \exp(i\gamma\widehat{\sin}\,\phi)|\psi\rangle_{\mathrm{e}}$ in (3.5.24), multiplying both the sides of this relation with $\exp(-i\gamma\widehat{\sin}\,\phi)$, and using the relation

$$\exp(-i\gamma\widehat{\sin}\,\phi)\hat{N}\exp(i\gamma\widehat{\sin}\,\phi) = \hat{N} - \gamma\widehat{\cos}\,\phi, \tag{4.9.42}$$

we arrive at the eigenvalue problem

$$\left[\hat{N} - \gamma\widehat{\exp}(-i\phi)\right]|\psi\rangle_{\mathrm{e}} = \lambda|\psi\rangle_{\mathrm{e}}. \tag{4.9.43}$$

It is not connected with any uncertainty relation, but is expected on the basis of symmetry. This problem has the solution

$$|\psi\rangle_{\mathrm{e}} = [I_0(2\gamma)]^{-\frac{1}{2}} \sum_{N=\lambda}^{\infty} \frac{\gamma^{N-\lambda}}{(N-\lambda)!}|N\rangle, \tag{4.9.44}$$

where λ is an integer. The wave function of this eigenstate is easily expressed in the angular representation. It determines that the angular probability distribution is that of von Mises [Rao (1973)].

The uncertainty product (4.9.39) for the state (4.9.36) and the state (4.9.44) reads

$$M_{\mathrm{e}}(\gamma) = \left[\frac{\gamma}{2}\frac{I_1(2\gamma)}{I_0(2\gamma)} + \frac{1}{4}\right]\left\{1 - \left[\frac{I_1(2\gamma)}{I_0(2\gamma)}\right]^2\right\} \tag{4.9.45}$$

and

$$M_{\mathrm{e}}(\gamma) = \left\{\gamma^2\left[1 - \left[\frac{I_1(2\gamma)}{I_0(2\gamma)}\right]^2\right] + \frac{1}{4}\right\}\left\{1 - \left[\frac{I_1(2\gamma)}{I_0(2\gamma)}\right]^2\right\}, \tag{4.9.46}$$

respectively. The absence of λ in these formulae is not only convenient for drawing graphs, but also a result of an interplay between ΔN in the formula (4.9.39) and the λ shift in the formulae (4.9.36) and (4.9.44). The product $M_{\mathrm{e}}(\gamma)$ is plotted versus the quantity V_{e},

$$V_{\mathrm{e}} \equiv V_{\mathrm{e}}(\gamma) = 1 - |\langle\widehat{\exp}(i\phi)\rangle|^2, \tag{4.9.47}$$

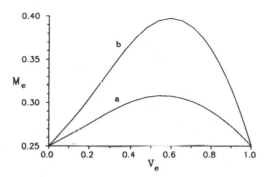

Figure 4.4: The uncertainty product $M_e \equiv M_e(\gamma)$ versus the phase dispersion $V_e(\gamma)$ for the $(\hat{N}, \widehat{i\sin}\,\phi)$ state (curve a) and the $(\hat{N}, -\widehat{\exp}(-i\phi))$ state (curve b).

which is for both the states

$$V_e(\gamma) = 1 - \left[\frac{I_1(2\gamma)}{I_0(2\gamma)}\right]^2, \tag{4.9.48}$$

in figure 4.4. In this picture we can see that the former state (4.9.36) is of better quality than the latter state (4.9.44), not only for the uncertainty relation (4.9.37), but also for (4.9.38).

The comparison of M_e values is rendered possible by either the parameter γ or the uncertainty $V_e(\gamma)$, which distinguish the states, and we compare, in fact, only the pairs of states with the same or nearly the same angle uncertainty V_e. The nonparametrized comparison between all the states from the two classes would be ambiguous.

A role of the analogy between the plane rotator and the optical oscillator in the solution of the phase problem will be expounded. This analogy can serve for the definition of optical oscillator states with the aid of the assimilation of plane rotator states.

The reduction of a pure state $|\psi\rangle_e$ onto a pure state $|\psi\rangle$ is defined as the normalized projection onto the original Hilbert space $\mathcal{H}$,

$$|\psi\rangle = \frac{\hat{I}|\psi\rangle_e}{\sqrt{{}_e\langle\psi|\hat{I}|\psi\rangle_e}}. \tag{4.9.49}$$

It is interesting that the state according to (4.9.44) is already a state of the original Hilbert space for $\lambda \geq 0$, it is not changed in the reduction. Also the intelligent state

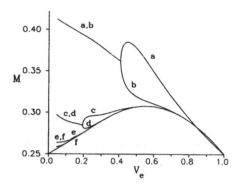

Figure 4.5: The uncertainty product $M \equiv M(\overline{N}, \gamma)$ versus the phase dispersion $V_e(\gamma)$ for $\overline{N} = 0, 1, 2, 3, 4, 5$ (curves a,b,c,d,e, and f, respectively).

according to (4.9.36) remains almost the same in the reduction for $\lambda \gg 0$. Therefore, a comparison has been performed of the reduced intelligent states with the Jackiw number–sine intelligent states [Lukš, Peřinová and Křepelka (1992a)]. Whereas the uncertainty product (4.9.45) in the intelligent state according to (4.9.36) does not depend on the mean number $_e\langle\psi|\hat{N}|\psi\rangle_e$, the reduction of these states makes the uncertainty product (4.9.22) for the reduced intelligent states dependent on the mean photon number $\langle\psi|\hat{N}|\psi\rangle = \langle\psi|\hat{n}|\psi\rangle \approx {}_e\langle\psi|\hat{N}|\psi\rangle_e$. This analysis has been performed in [Lukš, Peřinová and Křepelka (1992a)]. It is convenient to index the reduced intelligent states with $\overline{N} = {}_e\langle\psi|\hat{N}|\psi\rangle_e$. A similar indexing of the Jackiw states has been accomplished here with the relation (4.9.18). Substituting the state (4.9.20) into (4.9.22), we get the uncertainty product $M = M(\overline{N}, \gamma)$. An interplay between $\Delta\hat{n}$ in formula (4.9.22) and the λ shift in formula (4.9.20) results in a weak dependence of $M(\overline{N}, \gamma)$ on $\overline{N}$, but since λ takes on nonintegral values, the dependence on $\overline{N}$ is present. For $\overline{N}$ tending to infinity, $M(\overline{N}, \gamma)$ (the optical oscillator) converges to the uncertainty product $M_e(\gamma)$ according to (4.9.45) calculated for the state defined by the eigenvalue problem (4.9.35) (the plane rotator). The graphical illustration of $M \equiv M(\overline{N}, \gamma)$ versus $V_e(\gamma)$ in figure 4.5, is a bundle of curves with its centre at the point $(1, 0.25)$ corresponding to the number states $(\gamma = 0)$. The presence of complex eigenvalues for smaller values of $V_e(\gamma)$ makes itself evident by an identification of the pairs of branches appropriate to $(\overline{N}, \overline{N} + 1)$, for $\overline{N}$ even and small. The occurrence of real eigenvalues for larger values of $V_e(\gamma)$ manifests itself as a bifurcation and the branching curves approach a limit arc for $\overline{N}$ large, which is the curve a in figure 4.4.

In analogy to the quantum optical problem (3.5.19) with (4.9.29) to find a phase optimized state, we may search for the angle optimized states in the plane rotator physical system. Setting

$$(xi) \quad \hat{A} = \hat{N}, \quad \hat{B} = -\widehat{\cos}\,\phi, \quad [\hat{A}, \widehat{\cos}\,\phi] = -i\widehat{\sin}\,\phi, \tag{4.9.50}$$

in (3.5.19), we obtain an eigenvalue problem whose eigenvalues λ are all integers and the corresponding eigenvectors

$$|\psi\rangle_e = \sum_{N=-\infty}^{\infty} J_{N-\lambda}(\gamma)|N\rangle, \quad {}_e\langle\psi|\psi\rangle_e = 1. \tag{4.9.51}$$

From the expression for the angular representation ${}_e\langle\phi|\psi\rangle_e$ it follows that the phase distribution is uniform independent of λ and γ. Here we do not obtain an angle optimized state.

The eigenvalue problems for the Hermitian operators ensuing from the scheme (3.5.19) need not be interpreted as the stationarity (Fermat) conditions, but they may be related to quantum dynamics as the eigenvalue problems for Hamiltonians. For the choice

$$(xii) \quad \hat{A} = \hat{N}^2, \quad \hat{B} = -\widehat{\cos}\,\phi, \quad [\hat{N}^2, \widehat{\cos}\,\phi] = -i\widehat{\cos}\,\phi \tag{4.9.52}$$

and replacing $\gamma \rightarrow 2\gamma^2$, we arrive at the problem of eigenstates for the mathematical pendulum. In accordance with [Lukš, Peřinová and Křepelka (1992b)], we concentrate on the ground state $|p_{0,0,\gamma}\rangle_e$ of the Hamiltonian $\hat{H} = \hat{N}^2 - 2\gamma^2\widehat{\cos}\,\phi$,

$$|p_{0,0,\gamma}\rangle_e = \frac{1}{\pi\sqrt{2}} \int_{\theta_0}^{\theta_0+2\pi} ce_0\left(\frac{\phi}{2}, -4\gamma^2\right)|\phi\rangle_e \, d\phi, \tag{4.9.53}$$

where $ce_0(z, q)$ is the even, π-periodic Mathieu function appropriate to the minimum eigenvalue a_0 of the Mathieu operator

$$L\left[z, \frac{d}{dz}\right] = -\frac{d^2}{dz^2} + 2q\cos(2z), \tag{4.9.54}$$

where $q = -4\gamma^2$. The eigenvalue a_0 can be expressed in the form [Abramowitz and Stegun (1964)]

$$a_0 \equiv a_0(q) = -\frac{1}{2}q^2 + \frac{7}{128}q^4 - \frac{29}{2304}q^6 + \frac{68687}{18874368}q^8 + \ldots. \tag{4.9.55}$$

We remove the restriction on $\langle\hat{N}\rangle = 0$ and pref $\phi = 0$ by the use of the screw operator (cf. (4.6.79))

$$\hat{D}_{N\phi\theta_0}^{\text{screw}}(\overline{N}, \overline{\phi}) = \hat{G}_{N\phi\theta_0}(\overline{\phi}, -\overline{N}), \tag{4.9.56}$$

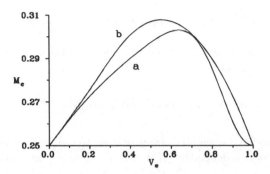

Figure 4.6: The uncertainty product M_e versus the phase dispersion $V_e(\gamma)$ of the $(\hat{N}, \widehat{i\sin}\phi)$ state for the state $|p_{\overline{N},\overline{\phi},\gamma}\rangle_e$ (curve a) and for the $(\hat{N}, \widehat{i\sin}\phi)$ state (curve b).

where the operator $\hat{G}_{N\phi_{\theta_0}}(\mu, \lambda)$ is defined in (4.6.50), and we introduce the states

$$|p_{\overline{N},\overline{\phi},\gamma}\rangle_e = \hat{D}^{\text{screw}}_{N\phi_{\theta_0}}(\overline{N}, \overline{\phi})|p_{0,0,\gamma}\rangle_e, \tag{4.9.57}$$

$$= \frac{1}{\pi\sqrt{2}} \int_{\theta_0}^{\theta_0+2\pi} \text{ce}_0\left(\frac{\phi - \overline{\phi}}{2}, -4\gamma^2\right)$$

$$\times \exp\left[-i\overline{N}\left(\phi - \frac{\overline{\phi}}{2}\right)\right]|\phi\rangle_e \, d\phi. \tag{4.9.58}$$

Here

$$\overline{N} = {}_e\langle p_{\overline{N},\overline{\phi},\gamma}|\hat{N}|p_{\overline{N},\overline{\phi},\gamma}\rangle_e,$$
$$\overline{\phi} = \text{Arg}\left\{{}_e\langle p_{\overline{N},\overline{\phi},\gamma}|\widehat{\exp}(i\phi)|p_{\overline{N},\overline{\phi},\gamma}\rangle_e\right\}. \tag{4.9.59}$$

In the system of the plane rotator the states $|p_{\overline{N},\overline{\phi},\gamma}\rangle_e$ provide smaller values of the uncertainty product (4.9.39) than the rotation-angle–angular-momentum states, which is demonstrated in figure 4.6. The comparison has been motivated by the approximations

$$\hat{N}^2 - 2\gamma^2\widehat{\cos}\phi \approx \hat{N}^2 + \gamma^2\hat{\phi}_{\theta_0}^2 - 2\gamma^2\hat{1}_e$$
$$\approx (\hat{N} - i\gamma\hat{\phi}_{\theta_0})(\hat{N} + i\gamma\hat{\phi}_{\theta_0}) - 2\gamma^2\hat{1}_e$$
$$\approx (\hat{N} - i\gamma\widehat{\sin}\phi)(\hat{N} + i\gamma\widehat{\sin}\phi) - 2\gamma^2\hat{1}_e, \tag{4.9.60}$$

where $\theta_0 \equiv \arg \gamma \pm \pi \pmod{2\pi}$. In analogy to the number–sine-of-phase intelligent states of the plane rotator, the reduction of the states $|p_{\overline{N},\bar{\vartheta},\gamma}\rangle_e$ has been considered in [Lukš, Peřinová and Křepelka (1992b)].

On assuming $\overline{N} \gg 0$, the special states of the plane rotator are relevant to the light-field mode. In the dependence on the parameters, the resulting states of the optical oscillator can be number squeezed (γ small), phase squeezed (γ large), as well as quasiclassical in the sense of being close to the Glauber coherent state. This statement has been demonstrated by the mean photon number, the photon-number distribution, the phase distribution, and by the quasidistribution $\Phi_A(\alpha)$. It can be expected that in the case of number-squeezed states (the crescent states), the states constructed in this way are appropriate to the physical system of the optical oscillator as their excellent conformity with the Jackiw crescent states proves.

4.9.3   Special states from the truncated Hilbert space of optical oscillator

It is instructive to start with the modifications of coherent states to the finite-dimensional Hilbert space $\mathcal{H}_s$. The binomial states [Stoler, Saleh and Teich (1985)] are appropriate here, which are identified with the SU(2) generalized coherent states from time to time, but as such they would be more appropriate to the case of two light modes. A mere truncation $|\alpha\rangle_\mathrm{I}$ of the Glauber coherent state has been introduced in [Kuang, Wang and Zhou (1993)] and studied in [Kuang, Wang and Zhou (1994), Opatrný, Miranowicz and Bajer (1996), Miranowicz, Opatrný and Bajer (1997)]. As the vacuum state belongs to any finite-dimensional space considered, the coherent state can be achieved also otherwise, by the action of the modified displacement operator on the vacuum [Bužek, Wilson-Gordon, Knight and Lai (1992)].

We will denote it by $|\alpha\rangle_\mathrm{II}$. In more detail

$$|\alpha\rangle_\mathrm{I} = \frac{1}{\sqrt{\mathrm{Prob}(n \leq s|\alpha)}} \hat{I}_s \exp(\alpha\hat{a}^\dagger - \alpha^*\hat{a})\hat{I}_s|0\rangle, \qquad (4.9.61)$$

where

$$\mathrm{Prob}(n \leq s|\alpha) = \exp(-|\alpha|^2) \sum_{n=0}^{s} \frac{|\alpha|^n}{n!}, \qquad (4.9.62)$$

$$|\alpha\rangle_\mathrm{II} = \exp\left[\hat{I}_s(\alpha\hat{a}^\dagger - \alpha^*\hat{a})\hat{I}_s\right]|0\rangle. \qquad (4.9.63)$$

Gangopadhyay (1994) has defined an s-phase coherent state, which can be characterized as $\hat{F}_s|\alpha\rangle_\mathrm{II}$ for $\theta_0 = 0$. He has defined a displaced s-phase coherent state, which is equal to $\hat{F}_s|\alpha, M\rangle_\mathrm{II}$ for $\theta_0 = 0$, where

$$|\alpha, M\rangle_\mathrm{II} = \exp\left[\hat{I}_s(\alpha\hat{a}^\dagger - \alpha^*\hat{a})\hat{I}_s\right]|M\rangle. \qquad (4.9.64)$$

Although the coherent state is a minimum-uncertainty state, this definition principle cannot be applied easily, because the eigenvalue problem $\hat{a}_s|\psi\rangle = \lambda|\psi\rangle$, where

$\hat{a}_{\mathbf{s}} = \hat{1}_{\mathbf{s}}\hat{a}\hat{1}_{\mathbf{s}}$, has no solutions. It is not excluded that a suitable modification of the operator $\hat{a}_{\mathbf{s}}$ could lead to an eigenvalue problem with interesting and important solutions. The difficulty lies in that we have a prior idea of the coherent states forming a two-dimensional continuum, whereas the number of intelligent states in the $(\mathbf{s}+1)$-dimensional Hilbert space can be only finite when the eigenvalues are not degenerate. Bužek, Wilson-Gordon, Knight and Lai (1992) have studied the statistical properties of the coherent states $|\alpha\rangle_{\mathrm{II}}$ and they have introduced the annihilation operator $\hat{p}_{\mathbf{s}}$ of the phase quanta (4.7.17), the s-phase quadratures $\mathrm{Re}\,\hat{p}_{\mathbf{s}}$, $\mathrm{Im}\,\hat{p}_{\mathbf{s}}$, and paid the attention to the case $\mathbf{s}=1$. The truncated coherent states have not only the second-order squeezing, but also higher-order squeezing with respect to the s-quadrature operators of the field under consideration [Kuang, Wang and Zhou (1994)]. In addition to the ordinary truncated coherent states $|\alpha\rangle_{\mathrm{I}}$ and $|\alpha\rangle_{\mathrm{II}}$, Kuang and Chen (1994a) have introduced s-phase coherent states using the creation operator $\hat{p}_{\mathbf{s}}^{\dagger}$ (4.7.17) of the phase quanta and have studied some properties of these states. Remarkably, these states are $\hat{F}_{\mathbf{s}}|\alpha\rangle_{\mathrm{I}}$ for $\theta_0 = 0$. The s-phase quadrature squeezing in the sense of [Wódkiewicz and Eberly (1985)] (see section 3.5), number–phase squeezing, and number–phase uncertainty relations for 1-phase coherent states have been investigated [Kuang and Chen (1994b)].

In the finite-dimensional Hilbert space $\mathcal{H}_{\mathbf{s}}$ a thorough analysis of the intelligent states has been performed with

$$\text{(xiii)} \quad \hat{A} = \hat{n}_{\mathbf{s}}, \quad \hat{B} = \hat{\phi}_{\theta_0,\mathbf{s}}, \tag{4.9.65}$$

and

$$\text{(xiv)} \quad \hat{A} = \hat{n}_{\mathbf{s}}, \quad \hat{B} = \left\{ \begin{array}{l} \cos(\hat{\phi}_{\theta_0,\mathbf{s}}), \\ \sin(\hat{\phi}_{\theta_0,\mathbf{s}}), \end{array} \right. \tag{4.9.66}$$

in (3.5.24) [Vaccaro and Pegg (1990a)]. The case of the cosine operator differs from that of the sine operator only unessentially. The properties of the $\hat{n}_{\mathbf{s}}$–$\cos(\hat{\phi}_{\theta_0,\mathbf{s}})$-intelligent states can be derived from those of the $\hat{n}_{\mathbf{s}}$–$\sin(\hat{\phi}_{\theta_0,\mathbf{s}})$-intelligent states. In the case of intelligent state according to (4.9.65), the matrix representations of the operator $(\hat{n}_{\mathbf{s}} + i\gamma\hat{\phi}_{\theta_0,\mathbf{s}})$ in the number state and s-phase state bases are very similar. This is connected with the fact that the Pegg–Barnett s-phase operator has been defined as a complementary one (in the original terminology, conjugate one) to the s-number operator. As a consequence, also the representations of the desired intelligent states in the s-number state and s-phase state bases are almost invariant with respect to the appropriate discrete Fourier transformation. Therefore, the coefficients of the intelligent states are discrete wrapped Gaussian in both the bases. For the $\hat{n}_{\mathbf{s}}$–$\sin(\hat{\phi}_{\theta_0,\mathbf{s}})$-intelligent states, a proximity to the number–sine intelligent states of the plane rotator can be expected and proved in a limiting procedure. It implies that for a negligible overlap with the vacuum the truncated Jackiw number–sine intelligent states will be approximately the $\hat{n}_{\mathbf{s}}$–$\sin(\hat{\phi}_{\theta_0,\mathbf{s}})$-intelligent states. Since $\sin(\hat{\phi}_{\theta_0,\mathbf{s}})$ is a circulant operator, no Hilbert-space limit of the $\hat{n}_{\mathbf{s}}$–$\sin(\hat{\phi}_{\theta_0,\mathbf{s}})$-intelligent states exists except it should be a number state [Vaccaro and Pegg (1990a)]. In characteristically quantum domain the $\hat{n}_{\mathbf{s}}$–$\sin(\hat{\phi}_{\theta_0,\mathbf{s}})$-intelligent states must deviate from the more natural truncated Jackiw number–sine intelligent states as indicated by Fujikawa (1995).

Passing to the variational problems, we substitute

$$(xv) \quad \hat{A} = \hat{n}_{\mathbf{s}}, \quad \hat{B} = -|0, \mathbf{s}\rangle\langle0, \mathbf{s}|, \tag{4.9.67}$$

where $|0, \mathbf{s}\rangle$ is the s-phase state $|\theta_m, \mathbf{s}\rangle$ on condition that $\theta_m = 0$, some m (equivalently, $\theta_0 \equiv 0 \pmod{\frac{2\pi}{\mathbf{s}+1}}$) in the scheme (3.5.19). This variational problem prescribes finding a vector for which the probability $\Phi_{\varphi\mathbf{s}}(0) \equiv \Phi_{\varphi\mathbf{s}}(\theta_m) = $ maximum [Shapiro, Shepard and Wong (1989)]. The eigenstate which is the solution of the problem reads

$$|\psi\rangle = A\sqrt{\mathbf{s}+1}(\hat{n}_{\mathbf{s}} - \lambda\hat{1}_{\mathbf{s}})^{-1}|0, \mathbf{s}\rangle, \tag{4.9.68}$$

where A is a normalization constant and $\lambda \neq 0, 1, 2, 3, \dots, \mathbf{s}$. More specifically, it can be assumed that $\lambda < 0$ is a negative integer as essentially done in [Shapiro, Shepard and Wong (1989)]. At least formally, the limiting procedure does not cause any difficulty. It appears, however, that in the course of the limiting procedure, either the condition of given averaged photon number cannot be fulfilled or the limiting procedure leads to a superposition of the vacuum with a hardly detectable phase component [Hradil and Shapiro (1992)]. Quite subconsciously, Nath and Kumar (1990) have studied the attenuation of s-phase state which changes the mean photon number from $\frac{\mathbf{s}}{2}$ to n, a constant. They have found the large-s limit of the photon-number distribution, which we relate to new pure or mixed states of radiation field. The number–s-phase intelligent states ($\mathbf{s} \to \infty$) can be approximated by wrapped Gaussian superpositions of coherent states lying on a circle [Adam, Janszky and Vinogradov (1991)].

Looking for physical states defined by the variational problem very similar to the previous one, but even more transparent, Summy and Pegg (1990) arrived at an eigenvalue problem, which we can obtain here by substituting

$$(xvi) \quad \hat{A} = \hat{n}_{\mathbf{s}}, \quad \hat{B} = \hat{\phi}_{\theta_0,\mathbf{s}}^2, \tag{4.9.69}$$

with $\theta_0 \approx -\pi$, into (3.5.19). Let us note that in [Summy and Pegg (1990)] we encounter an approximation

$$\hat{\phi}_{\theta_0,\mathbf{s}}^2 \approx \sin^2(\hat{\phi}_{\theta_0,\mathbf{s}}), \tag{4.9.70}$$

leading to the Airy function in the Fock-state representation. The phase behaviour of the Airy states has been explored using the measured phase operators [Vaccaro and Pegg (1994a)]. The approximation

$$\hat{\phi}_{\theta_0,\mathbf{s}}^2 \approx 4\sin^2\left(\frac{1}{2}\hat{\phi}_{\theta_0,\mathbf{s}}\right) \tag{4.9.71}$$

would lead us to the choice

$$\hat{A} = \frac{1}{2}\hat{n}_{\mathbf{s}}^2, \quad \hat{B} = -\cos(\hat{\phi}_{\theta_0,\mathbf{s}}), \tag{4.9.72}$$

where the introduction of $\hat{n}_s^2$ and the replacement $\gamma \rightarrow \gamma^2$ could be related to the considerations in [Lukš, Peřinová and Křepelka (1992b)].

Taking into account also the circularity in the number states, we can arrive at the choice

$$(\text{xvii}) \quad \hat{A} = -\left(\frac{s+1}{2\pi}\right)^2 \cos\left(\frac{2\pi}{s+1}\hat{n}_s\right), \quad \hat{B} = -\cos(\hat{\phi}_{\theta_0,s}) \tag{4.9.73}$$

in (3.5.19) with $\gamma \rightarrow \gamma^2$, which has been essentially done in [Opatrný (1995)].

It is not complicated to imagine a truncated version of the substitution (4.9.2) and to derive the generalized geometric states introduced in [Obada, Yassin and Barnett (1997)]. These states have been characterized by the quadrature Wigner function and the canonical phase distribution, and by the fluctuations in the phase and number.

4.9.4 SU(2) intelligent states

The two-mode states of light field with a definite photon-number sum can be of use in quantum optics. The SU(2) generalized coherent states are introduced as [Perelomov (1986), Brif and Mann (1996a)]

$$|j, \zeta\rangle = (1 + |\zeta|^2)^{-j} \sum_{n_1=0}^{2j} \binom{2j}{n_1}^{\frac{1}{2}} \zeta^{n_1} |n_1, 2j - n_1\rangle, \tag{4.9.74}$$

where j enjoys the property

$$(\hat{n}_s + \hat{1})|j, \zeta\rangle = (2j + 1)|j, \zeta\rangle, \tag{4.9.75}$$

with the photon-number sum operator $\hat{n}_s$ given in (4.8.47), and ζ is any complex number. Aragone, Guerri, Salamó and Tani (1974) have defined intelligent spin states using eigenvalue problems for the group SU(2). The eigenvalue problem (3.5.24) for the case

$$(\text{xvii}) \quad \hat{A} = \hat{J}_3, \quad \hat{B} = -\hat{J}_2, \quad [\hat{J}_3, -\hat{J}_2] = i\hat{J}_1, \tag{4.9.76}$$

has been treated by Brif and Mann (1996a). We assume the definite photon-number sum

$$(\hat{n}_s + \hat{1})|\psi\rangle = (2j + 1)|\psi\rangle. \tag{4.9.77}$$

The appropriate eigenvalues of the operator $(\hat{J}_3 - i\gamma\hat{J}_2)$ are

$$\lambda = m_0\sqrt{1 - \gamma^2}, \quad m_0 = -j, -j + 1, \ldots, j - 1, j. \tag{4.9.78}$$

From this it follows that $\langle \hat{J}_2 \rangle = 0$ provided that $|\gamma| \leq 1$. The eigenkets are given as

$$|\psi\rangle = \frac{1}{\sqrt{\mathcal{N}^{\text{SU}(2)}}} \sum_{n_1=0}^{2j} P_{j-n_1,-m_0}^j(x_{\text{SU}(2)})|n_1, 2j - n_1\rangle, \tag{4.9.79}$$

where

$$\mathcal{N}^{\mathrm{SU(2)}} = P^j_{m_0,m_0}\left(2x^2_{\mathrm{SU(2)}} - 1\right) = P^j_{m_0,m_0}\left(\frac{1+\gamma^2}{1-\gamma^2}\right), \qquad (4.9.80)$$

$$x_{\mathrm{SU(2)}} = \frac{1}{\sqrt{1-\gamma^2}}, \qquad (4.9.81)$$

and the special function $P^l_{mn}(x)$ [Vilenkin (1968)]

$$P^l_{mn}(x) = \sqrt{\frac{(l-m)!(l-n)!}{(l+m)!(l+n)!}}\; i^{-m-n}\left(\frac{1+x}{1-x}\right)^{\frac{m+n}{2}}$$

$$\times \sum_{s=\max(m,n)}^{l} \frac{(l+s)!(-1)^s}{(l-s)!(s-m)!(s-n)!}\left(\frac{1-x}{2}\right)^s. \qquad (4.9.82)$$

Replacing the angular-momentum operator $\hat{J}_2$ in the scheme (4.9.76) by the Luis–Sánchez-Soto sine operator $-\hat{S}_{\mathrm{d}}$, $\hat{S}_{\mathrm{d}} = \mathrm{Im}(\hat{U}_{\mathrm{d}})$, and substituting

$$(\mathrm{xviii}) \quad \hat{A} = \hat{J}_3, \; \hat{B} = \hat{S}_{\mathrm{d}} \qquad (4.9.83)$$

into (3.5.24), we nearly obtain (xiii) with $s = 2j$. This relationship is a parallel to the similarity between the generalized coherent states $|j, \zeta\rangle$ and the binomial states, we have mentioned in subsection 4.9.3.

4.9.5 SU(1,1) intelligent states

The two-mode states of light field with a definite photon-number difference can be of use in quantum optics, whenever the processes preserve the photon-number difference (4.8.54). The SU(1,1) generalized coherent states are introduced as [Perelomov (1986), Brif and Mann (1996a)]

$$|k, \zeta\rangle = (1 - |\zeta|^2)^k \sum_{n_2=0}^{\infty} \binom{n_2 + 2k - 1}{n_2}^{\frac{1}{2}} \zeta^{n_2} |n_2 + 2k - 1, n_2\rangle, \qquad (4.9.84)$$

where k enters the eigenvalue equation

$$\hat{n}_{\mathrm{d}}|k, \zeta\rangle = (2k - 1)|k, \zeta\rangle, \qquad (4.9.85)$$

and ζ is a complex number, $|\zeta| < 1$. The eigenvalue problems for the group SU(1,1) have been formulated by Vanden Bergh and DeMeyer (1978). The eigenvalue problem (3.5.24) for the case

$$(\mathrm{xix}) \quad \hat{A} = \hat{K}_3, \; \hat{B} = -\hat{K}_2, \; [\hat{K}_3, -\hat{K}_2] = i\hat{K}_1, \qquad (4.9.86)$$

has been treated by Brif and Mann (1996a, b). We assume the definite photon-number difference

$$\hat{n}_{\mathrm{d}}|\psi\rangle = (2k-1)|\psi\rangle. \tag{4.9.87}$$

The appropriate eigenvalues of the operator $(\hat{K}_3 - i\gamma\hat{K}_2)$ are

$$\lambda = m_0\sqrt{1+\gamma^2}, \quad m_0 = k, k+1, \ldots . \tag{4.9.88}$$

From this it follows that $\langle \hat{K}_2 \rangle = 0$. The eigenkets are given as

$$|\psi\rangle = \frac{1}{\sqrt{\mathcal{N}^{\mathrm{SU}(1,1)}}} \sum_{n_2=0}^{\infty} \sqrt{\frac{\Gamma(n_2+2k)\Gamma(m_0-k+1)}{\Gamma(m_0+k)n_2!}} \, i^{m_0-k-n_2}$$

$$\times \mathcal{P}^{k-1}_{n_2+k,m_0}(x_{\mathrm{SU}(1,1)})|n_2+2k-1, n_2\rangle, \tag{4.9.89}$$

where

$$\mathcal{N}^{\mathrm{SU}(1,1)} = \mathcal{P}^{k-1}_{m_0,m_0}\left(2x^2_{\mathrm{SU}(1,1)}-1\right) = \mathcal{P}^{k-1}_{m_0,m_0}\left(\frac{1-\gamma^2}{1+\gamma^2}\right), \tag{4.9.90}$$

$$x_{\mathrm{SU}(1,1)} = \frac{1}{\sqrt{1+\gamma^2}}, \tag{4.9.91}$$

with the special function $\mathcal{P}^l_{mn}(x)$ [Vilenkin (1968)]

$$\mathcal{P}^l_{mn}(x) = (-1)^{m-n}\Gamma(l+n+1)\Gamma(m-l)\left(\frac{x-1}{2}\right)^{\frac{m-n}{2}}\left(\frac{x+1}{2}\right)^{\frac{-m-n}{2}}$$

$$\times \sum_{s=\max(0,n-m)}^{\min(n-l-1,n+l)} \frac{(-1)^s}{s!\Gamma(l+n-s+1)\Gamma(n-l-s)\Gamma(m-n+s+1)}\left(\frac{x-1}{2}\right)^s. \tag{4.9.92}$$

The case with $\hat{A} = \hat{K}_1$, $\hat{B} = -\hat{K}_2$ cannot be obtained from the case (xix) by a mere permutation of subscripts and it has been treated by Brif and Ben-Aryeh (1994c) and Gerry and Grobe (1995b).

Replacing the angular-momentum operator $\hat{K}_2$ in the scheme (4.9.86) by the Susskind–Glogower sine operator $-\widehat{\sin}(\varphi_1+\varphi_2)$ and substituting

$$(\mathrm{xx}) \quad \hat{A} = \hat{K}_3, \; \hat{B} = \widehat{\sin}(\varphi_1+\varphi_2), \; [\hat{K}_3, \widehat{\sin}(\varphi_1+\varphi_2)] = i\widehat{\cos}(\varphi_1+\varphi_2), \tag{4.9.93}$$

into (3.5.24), we nearly obtain (iv). Early considerations of the SU(1,1) generalized coherent states for a single mode have been mentioned in subsection 4.9.1.

The states (4.9.79) and (4.9.89) have been treated as the $\hat{J}_2$–$\hat{J}_3$- and $\hat{K}_2$–$\hat{K}_3$-intelligent states, respectively, for nonclassical interferometry [Brif and Mann (1996a, b)]. Such states can be generated from the four-mode vacuum in two parametric down-converters whose idler beams are aligned, but only partially coupled, after the measurement of the photon numbers on two of the output modes. A suitable SU(2) or SU(1,1) transformation of the two other modes is assumed [Luis and Peřina (1996a)].

4.10 Quantum phase and phase-space quasidistributions

In accordance with [Tanaś, Miranowicz and Gantsog (1993), Leonhardt and Paul (1993c), Herzog, Paul and Richter (1993), Garraway and Knight (1993), Tanaś, Miranowicz and Gantsog (1996), Royer (1996)], we go through the quantum phase derived from phase-space quasidistributions. We have in mind the quasidistributions expounded in section 3.3. Especially, the quantum phase related to the usual diagonal P (Φ_N) function is conceivable and, indeed, it is the language of the laser theory, although the Φ_N function exists only in the sense of generalized functions for pure states and in many cases of mixed states. To the contrary, the positive P (Φ_N) function [Drummond and Gardiner (1980), Gilchrist, Gardiner and Drummond (1997)] is very useful for Monte Carlo computation and quantum trajectories, so that also the quantum phase can be studied in this approach as a numerical experiment rather than in terms of the operator-valued measures.

4.10.1 Single-mode optical fields

Let us recall the quasidistribution related to the s-ordering of field operators (cf. section 3.4)

$$\Phi_s(\alpha) = \text{Tr}\left\{\hat{\rho}\hat{\Phi}_s(\alpha)\right\}, \qquad (4.10.1)$$

where

$$\hat{\Phi}_s(\alpha) = \frac{1}{\pi^2} \int \exp(\alpha\beta^* - \alpha^*\beta)\hat{D}^{(s)}(\beta) \, d^2\beta, \qquad (4.10.2)$$

with the operator generating function (see (3.4.7), $\hat{D}^{(s)}(\beta)$ stands for $\hat{D}(\beta, s)$)

$$\hat{D}^{(s)}(\beta) = \exp\left(\frac{s}{2}|\beta|^2\right) \hat{D}(\beta). \qquad (4.10.3)$$

Let us repeat that for $s = -1$ we obtain the quasidistribution related to the antinormal ordering of field operators entering products to be averaged.

Using orderings, we can introduce s-ordered phase operators

$$\hat{\phi}_{-\pi}^{(s)} = \frac{1}{2i} \int (\ln \alpha - \ln \alpha^*)\hat{\Phi}_s(\alpha) \, d^2\alpha. \qquad (4.10.4)$$

For $s = -1$ we arrive at the concept (4.3.63).

An apparently more straightforward solution than the relation (4.3.61) is a definition of the kind

$$\hat{\phi}_{-\pi} \stackrel{?}{=} \frac{1}{2i}(\ln \hat{a} - \ln \hat{a}^\dagger). \qquad (4.10.5)$$

Almost the formula (4.10.5) can be found in [Lewis, Lawrence and Harris (1996)], whereas a modified reasoning has been presented in [Carrasco and Moya-Cessa (1997)].

The use of the formula (4.10.5) must be based on the existence of the operator $\ln \hat{a}$. We know that this should be accomplished on a substitution into (4.10.4) with the result independent of s. Nevertheless, we must substitute $\ln \alpha = \ln |\alpha| + i \mathrm{Arg}_{-\pi} \alpha$, which leads to the result dependent on s. This logical difficulty is apparently circumvented in [Lewis, Lawrence and Harris (1996)], where $\ln \hat{a}$ is supposed to be a suitable solution of the commutator equation $[\ln \hat{a}, \hat{a}^\dagger] = \frac{1}{\hat{a}}$. We know, however, that the existence of the operator $\frac{1}{\hat{a}}$ is as unclear as that of the operator $\frac{1}{\exp(i\varphi)}$. Even if we used the formula (4.3.33) and set $\widehat{\frac{1}{\exp(i\varphi)}} \overset{?}{=} \widehat{\exp}(-i\varphi)$, we would obtain

$$\frac{1}{\hat{a}} \overset{?}{=} \widehat{\exp}(-i\varphi)(\hat{n} + \hat{1})^{-\frac{1}{2}}, \tag{4.10.6}$$

but we let $\hat{a}^{(-)}$ denote the right-hand side. Then one observes that $\hat{a}\hat{a}^{(-)} = \hat{1}$, but $\hat{a}^{(-)}\hat{a} = \hat{1} - |0\rangle\langle 0|$ [Mehta, Roy and Saxena (1992), Roy and Mehta (1995)]. Neglecting blind alleys of the two papers, we comment that Lewis, Lawrence and Harris (1996) are close to characterize a result as $\hat{\phi}_{-\pi}^{(N)}$, while Carrasco and Moya-Cessa (1997) are convinced that they have obtained the operator $\hat{\phi}_{-\pi}^{(A)}$. Unfortunately, the derivation is ineffective.

To get closer to the important paper of Royer (1996), we perform an integration using the polar coordinates in (4.10.4),

$$\begin{aligned} \hat{\phi}_{-\pi}^{(s)} &= \int_{-\pi}^{\pi} \int_{0}^{\infty} \varphi \hat{\Phi}_s \left(\rho e^{i\varphi} \right) \rho \, d\rho \, d\varphi \\ &= \int_{-\pi}^{\pi} \varphi \hat{\Gamma}^{(s)}(\varphi) \, d\varphi, \end{aligned} \tag{4.10.7}$$

where the operator-valued density

$$\hat{\Gamma}^{(s)}(\varphi) = \int_{0}^{\infty} \rho \hat{\Phi}_s \left(\rho e^{i\varphi} \right) d\rho. \tag{4.10.8}$$

For $s = -1$ in (4.10.8) we rederive the operator (4.3.63). For $s = 0$ the resulting operator is essentially the operator introduced and studied in [Smith, Dubin and Hennings (1992), Dubin, Hennings and Smith (1994, 1995), Hennings, Smith and Dubin (1995a, b)]. Although the study of the Weyl ordering is very promising, the diagonal representation of $\hat{\phi}_{-\pi}^{(S)}$ has not been obtained yet, the variances of the eigenvalue of this operator in the Fock states differs from that in (4.3.106) by an oscillatory behaviour [Hennings, Smith and Dubin (1995a)]. A similar question arises for $s = -1$. Although the quantum phase can be obtained from an operational phase-space distribution, the operator $\hat{\phi}_{-\pi}^{(A)}$ in (4.3.63) cannot be the intrinsic phase operator contrary to [Englert and Wódkiewicz (1995), Englert, Wódkiewicz and Riegler (1995)]. The diagonal representation of the operator $\hat{\phi}_{-\pi}^{(A)}$ (the Turski operator) has not been obtained yet, although it could be interesting. Let us compare with the situation in the Toeplitz operators, where the diagonalization of the Garrison–Wong operator is known. An interesting limit theorem, a connecting relation between quasidistributions, and special commutator relations have been presented in [Englert, Wódkiewicz and Riegler (1995)].

We define a set of exponential phase operators as follows

$$
\widehat{\exp}^{(s)}(ik\varphi) = \int \frac{\alpha^k}{|\alpha|^k} \hat{\Phi}_s(\alpha) \, d^2\alpha
$$

$$
= \int_{\theta_0}^{\theta_0 + 2\pi} \exp(ik\varphi) \hat{\Gamma}^{(s)}(\varphi) \, d\varphi, \quad k = 0, 1, \ldots, \tag{4.10.9}
$$

$$
\widehat{\exp}^{(s)}(-ik\varphi) = [\widehat{\exp}^{(s)}(ik\varphi)]^\dagger, \quad k = 0, 1, \ldots. \tag{4.10.10}
$$

Quantum phase characteristics can be derived directly from a quantum phase distribution related to the s-ordering of field operators,

$$
P^{(s)}(\varphi) = \mathrm{Tr}\left\{\hat{\rho}\hat{\Gamma}^{(s)}(\varphi)\right\}. \tag{4.10.11}
$$

The importance of the distribution follows from the appeal to the quasidistributions of polar coordinates and action–angle variables,

$$
\Phi_s^{\rho\varphi}(\rho, \varphi) = \rho\Phi_s\left(\rho e^{i\varphi}\right), \tag{4.10.12}
$$

$$
\Phi_s^{W\varphi}(W, \varphi) = \frac{1}{2}\Phi_s\left(\sqrt{W}e^{i\varphi}\right). \tag{4.10.13}
$$

Substituting from (4.10.8) into (4.10.11) and exchanging the integrations, we obtain the marginal integrals

$$
P^{(s)}(\varphi) = \int_0^\infty \Phi_s^{\rho\varphi}(\rho, \varphi) \, d\rho \tag{4.10.14}
$$

$$
= \int_0^\infty \Phi_s^{W\varphi}(W, \varphi) \, dW. \tag{4.10.15}
$$

Let us remark that the radial integration in these formulae corresponds to more complicated transformations onto the canonical phase distribution

$$
P(\varphi) = \int_{\theta_0}^{\theta_0 + 2\pi} \int_0^\infty \langle\varphi|\hat{\Delta}(\rho e^{i\varphi'}, s)|\varphi\rangle \Phi_s^{\rho\varphi}(\rho, \varphi') \, d\rho \, d\varphi', \tag{4.10.16}
$$

$$
= \int_{\theta_0}^{\theta_0 + 2\pi} \int_0^\infty \langle\varphi|\hat{\Delta}(\sqrt{W}e^{i\varphi'}, s)|\varphi\rangle \Phi_s^{W\varphi}(W, \varphi') \, dW \, d\varphi'. \tag{4.10.17}
$$

Expressing the statistical operator in the number state basis, we write the quasidistribution (4.10.1) in the form

$$
\Phi_s\left(\rho e^{i\varphi}\right) = \sum_{n=0}^{\infty}{}' \sum_{k=-n}^{n} \rho_{n-k,n+k}\langle n+k|\hat{\Phi}_s\left(\rho e^{i\varphi}\right)|n-k\rangle, \tag{4.10.18}
$$

where the prime denotes that the summation step is $\frac{1}{2}$ and the matrix elements of the operator-valued density $\hat{\Phi}_s\left(\rho e^{i\varphi}\right)$ are

$$
\begin{aligned}
\langle n + k|\hat{\Phi}_s\left(\rho e^{i\varphi}\right)|n - k\rangle &= \frac{1}{\pi}\sqrt{\frac{(n - |k|)!}{(n + |k|)!}}\left(\frac{2}{1 - s}\right)^{2|k|+1} \\
&\quad \times \left(\frac{s + 1}{s - 1}\right)^{n-|k|} e^{i2k\varphi}\rho^{2|k|} \exp\left(-\frac{2\rho^2}{1 - s}\right) \\
&\quad \times L_{n-|k|}^{2|k|}\left(\frac{4\rho^2}{1 - s^2}\right).
\end{aligned}
\tag{4.10.19}
$$

Substituting from (4.10.18) into (4.10.12) and (4.10.14), we obtain that

$$
P^{(s)}(\varphi) = \sum_{n=0}^{\infty}{}' \sum_{k=-n}^{n} \rho_{n-k,n+k}\langle n + k|\hat{\Gamma}^{(s)}(\varphi)|n - k\rangle,
\tag{4.10.20}
$$

where

$$
\langle n + k|\hat{\Gamma}^{(s)}(\varphi)|n - k\rangle = \frac{1}{2\pi}e^{i2k\varphi}G^{(s)}(n - k, n + k),
\tag{4.10.21}
$$

with the particular matrix elements for $\varphi = 0$,

$$
\begin{aligned}
G^{(s)}(n - k, n + k) &= \left(\frac{2}{1 - s}\right)^n \sqrt{(n - k)!(n + k)!} \sum_{l=0}^{n-|k|}(-1)^l \\
&\quad \times \left(\frac{1 + s}{2}\right)^l \frac{\Gamma(n - l + 1)}{l!(n - k - l)!(n + k - l)!}.
\end{aligned}
\tag{4.10.22}
$$

The coefficients (4.10.22) can be rewritten in terms of the Jacobi polynomials [Tanaś, Miranowicz and Gantsog (1996), Leonhardt, Vaccaro, Böhmer and Paul (1995)]. The operator-valued density $\hat{\Gamma}^{(s)}(\varphi)$ must be reducible to its value $\hat{\Gamma}^{(s)}(0)$ and in the semiclassical regime its matrix representation with respect to the Fock basis must be Toeplitz. Hence, the s-parametrized and directly measured phase distributions are canonical in the semiclassical regime, they are weighted averages of pure canonical phase distributions in this regime [Leonhardt, Vaccaro, Böhmer and Paul (1995)].

Paul (1974) has proposed and analyzed theoretically the use of linear amplification to measure the quantum-optical phase of a light mode. The requirement of a simple behaviour with respect to amplification implies $s = -1$. Davidović, Lalović and Tančić (1994) have discussed a fact which bears on the minimum-noise phase-insensitive quantum amplification, namely, the unique property of the function $\Phi_A(\alpha)$ that $\lambda^2\Phi_A(\lambda\alpha)$, where $0 < \lambda < 1$, is also a Φ_A function. They have shown that the Wigner and P functions do not have this property. Garraway and Knight (1992) have shown that the London phase distribution and the marginal Wigner phase distribution agree for states with a small probability that the photon number is not large.

Hillery, Freyberger and Schleich (1995) have compared the properties of these quantum phase distributions, the marginal Q phase distribution, and the Vogel–Schleich proposal [Vogel and Schleich (1991)]. They have introduced the concept of quasiclassical phase states and have investigated large-amplitude classical states as falling into this class. They have gone to higher order of approximation of the phase representation of a coherent state. The operator-valued density can be associated also with the integrated intensity (the quantum action) by the formula [Abe (1996)]

$$\hat{P}_s(W) = \frac{1}{2} \int_{\theta_0}^{\theta_0+2\pi} \hat{\Phi}_s(\sqrt{W}\,e^{i\varphi})\,d\varphi. \tag{4.10.23}$$

The intensity–phase uncertainty relation based on the generalized Wehrl entropy has been derived from the s-parametrized quasidistributions in [Orłowski, Paul and Böhmer (1997)].

In the following, we present the quantum phase distribution related to the s-ordering for a coherent state and its generalizations, namely, a thermalized squeezed state and a displaced number state (the real states of optical field).

(i) The coherent state $|\xi\rangle$, $\xi = |\xi|\exp(i\tau)$, is described by the quasidistribution related to the s-ordering of field operators (section 3.4)

$$\Phi_s^{\text{coh}}(\alpha) = \frac{1}{\pi}\frac{2}{1-s}\exp\left(-\frac{2}{1-s}|\alpha-\xi|^2\right). \tag{4.10.24}$$

Substituting (4.10.24) into (4.10.12) and (4.10.14), we arrive at the s-parametrized phase distribution [Tanaś, Miranowicz and Gantsog (1993)],

$$P_{\text{coh}}^{(s)}(\varphi) = \frac{1}{2\pi}\exp(-\overline{S}^2)\left\{\exp(-\overline{C}^2) + \sqrt{\pi}\,\overline{C}[1+\text{erf}(\overline{C})]\right\}, \tag{4.10.25}$$

where

$$\overline{C} \equiv \overline{C}(\varphi,s) = \sqrt{\frac{2}{1-s}}|\xi|\cos(\varphi-\tau),$$

$$\overline{S} \equiv \overline{S}(\varphi,s) = \sqrt{\frac{2}{1-s}}|\xi|\sin(\varphi-\tau). \tag{4.10.26}$$

It is expected that in the strong-field limit the distribution (4.10.25) is Gaussian in the neighbourhood of the preferred phase τ even if $s \to 1$. In this limit, the first term in the curly bracket can be neglected. The behaviour of $\exp(\overline{S}^2)$ is similar to that of the probability density (4.4.94) for $\beta \equiv \tau \pmod{\pi}$. The factor $\overline{C}$ makes the peak at $\varphi \equiv \tau + \pi \pmod{2\pi}$ negative, but the factor $[1+\text{erf}(\overline{C})]$ erases this peak. In the case of the antinormal ordering ($s = -1$), the particular quantum phase distribution $P_{\text{coh}}^{(A)}(\varphi)$ motivated a resolution in [Lukš and Peřinová (1994)] based on the idea [Peřinová and Lukš (1990)] treated in section 5.4. It has been pointed out that the Mittag-Leffler function

$$E_{\frac{1}{2}}(z) = \sum_{n=0}^{\infty}{}' \frac{z^{2n}}{\Gamma(n+1)} \tag{4.10.27}$$

can be used in (4.10.25),

$$E_{\frac{1}{2}}(\overline{C}) = \exp(\overline{C}^2) + \exp(\overline{C}^2)\mathrm{erf}(\overline{C}), \tag{4.10.28}$$

where one term corresponds to the usual integral photon numbers and the other is connected to the half-odd photon numbers.

Quite simple formulae for the phase distribution related to the s-ordering have been derived for the ordinary squeezed states and the displaced number states and a detailed comparison with the appropriate probability densities $P(\varphi)$ (4.7.152) has been performed [Tanaś, Miranowicz and Gantsog (1993)].

(ii) Of high importance are the states of single-mode field described by the quasidistribution related to the s-ordering of the form

$$\Phi_s^{\mathrm{thsq}}(\alpha) = \frac{1}{\pi\sqrt{K(s)}} \exp\left\{ -\frac{B_s}{K(s)}|\alpha - \xi|^2 + \left[\frac{C^*}{2K(s)}(\alpha - \xi)^2 + \text{c. c.} \right] \right\}, \tag{4.10.29}$$

where

$$K(s) = B_s^2 - |C|^2, \quad K(0) \geq \frac{1}{4}, \tag{4.10.30}$$

and the coefficients B_s, C can be expressed as moments (section 3.8). For $K(0) = \frac{1}{4}$, this quantum state is pure and the appropriate quasidistribution with [Peřinová, Lukš and Kárská (1990)]

$$B_s = \frac{1-s}{2}|\mu|^2 + \frac{1+s}{2}|\nu|^2, \quad C = -\mu^*\nu, \tag{4.10.31}$$

corresponds to the ordinary squeezed state (two-photon coherent state $|\mu\xi + \nu\xi^*\rangle_g$). The s-parametrized phase distribution according to (4.10.14) reads

$$\begin{aligned} P_{\mathrm{thsq}}^{(s)}(\varphi) &= \frac{1}{2\pi\sqrt{K(s)L(\varphi,\varphi,s)}} \exp(-\overline{S}^2) \\ &\quad \times \left\{ \exp(-\overline{C}^2) + \sqrt{\pi}\,\overline{C}[1 + \mathrm{erf}(\overline{C})] \right\}, \end{aligned} \tag{4.10.32}$$

where

$$L(\varphi,\psi,s) = \frac{1}{K(s)} \left\{ B_s \cos(\varphi - \psi) - \left[\frac{C^*}{2} \exp[i(\varphi + \psi)] + \text{c. c.} \right] \right\}, \tag{4.10.33}$$

$$\begin{aligned} \overline{C} &\equiv \overline{C}(\varphi,\tau,s) = |\xi|\frac{L(\varphi,\tau,s)}{\sqrt{L(\varphi,\varphi,s)}}, \\ \overline{S} &\equiv \overline{S}(\varphi,\tau,s) = \epsilon|\xi|\sqrt{\frac{L(\varphi,\varphi,s)L(\tau,\tau,s) - L^2(\varphi,\tau,s)}{L(\varphi,\varphi,s)}}, \end{aligned} \tag{4.10.34}$$

with

$$\epsilon = \text{sgn}[\sin(\varphi - \tau)]. \tag{4.10.35}$$

Much work in quantum optics is devoted to the amplitude squeezing in the degenerate amplification process, which occurs on the condition that

$$C = -|C|\exp(i2\tau). \tag{4.10.36}$$

In this case the quantity (4.10.33) simplifies,

$$L(\varphi, \tau, s) = \frac{B_s + |C|}{K(s)}\cos(\varphi - \tau). \tag{4.10.37}$$

Particularly, for $s = 0$, the formula (4.10.37) comprises the major principal variance $2(B_S + |C|)$, which must be considered along with the interesting minor principal variance $2(B_S - |C|)$ in the study of quadrature squeezing. Focusing on the assumption (4.10.36), the derivation for an ordinary squeezed state can be found in [Tanaś, Miranowicz and Gantsog (1993)]. When $\xi = 0$, the condition (4.10.36) cannot be formulated, but in this thermalized squeezed vacuum case, the formula (4.10.32) simplifies, because $\overline{C}, \overline{S}$ vanish,

$$P_{\text{thsq}}^{(s)}(\varphi) = \frac{1}{2\pi}\frac{\sqrt{K(s)}}{(B_s + |C|)\sin^2(\varphi - \beta) + (B_s - |C|)\cos^2(\varphi - \beta)}, \tag{4.10.38}$$

where it is assumed that

$$C = |C|\exp(i2\beta). \tag{4.10.39}$$

Comparing with the formula (4.4.107), we recognize the two-peak phase distribution with the second-order preferred phase $\text{Pref}_{\theta_0}^{[2]}\Phi = \beta$ and the parameter

$$\exp(2r) = \sqrt{\frac{B_s + |C|}{B_s - |C|}}. \tag{4.10.40}$$

According to (4.4.109), the second-order dispersion of phase is of the form

$$D^{[2]}\Phi = \frac{1}{2}\frac{\sqrt{K(s)}}{B_s + \sqrt{K(s)}}. \tag{4.10.41}$$

(iii) The displaced number state $|\xi, M\rangle$ defined as [de Oliveira, Kim, Knight and Bužek (1990)]

$$|\xi, M\rangle = \hat{D}(\xi)|M\rangle, \tag{4.10.42}$$

where $|M\rangle$ is a number state and $\hat{D}(\xi)$ is the displacement operator, is described by the quasidistribution related to the s-ordering of field operators,

$$\begin{aligned}
\Phi_s^{\text{dn}}(\alpha) &= \frac{1}{\pi}\frac{2}{1-s}(-1)^M\left(\frac{1+s}{1-s}\right)^M\exp\left\{-\frac{2}{1-s}|\alpha - \xi|^2\right\} \\
&\times L_M\left(\frac{4|\alpha - \xi|^2}{1-s^2}\right).
\end{aligned} \tag{4.10.43}$$

Performing the integration (4.10.14) for the choice of the quasidistribution (4.10.43), we get the s-parametrized quantum phase distribution of an alternative form to [Tanaś, Miranowicz and Gantsog (1993)]

$$
\begin{aligned}
P_{\mathrm{dn}}^{(s)}(\varphi) &= \frac{1}{2\pi}(-1)^M \left(\frac{1+s}{1-s}\right)^M \exp\left(-\overline{S}^2\right) \\
&\times \sum_{m=0}^{M} L_{M-m}^{-\frac{1}{2}} \left(\frac{4\overline{S}^2}{1+s}\right) \Gamma\left(m+\frac{1}{2}\right) \\
&\times \sum_{j=0}^{m} \frac{\left(-\frac{4}{1+s}\right)^j}{j!(m-j)!\Gamma\left(j+\frac{1}{2}\right)} \left\{\Gamma(j+1,\overline{C}^2)\right. \\
&\left. +\overline{C}\Gamma\left(j+\frac{1}{2}\right) + |\overline{C}|\gamma\left(j+\frac{1}{2},\overline{C}^2\right)\right\},
\end{aligned}
\tag{4.10.44}
$$

where the incomplete gamma functions,

$$
\Gamma(z,x) = \int_x^\infty u^{z-1}\exp(-u)\,du, \quad \mathrm{Re}\,z > 0, \quad x > 0,
\tag{4.10.45}
$$

$$
\gamma(z,x) = \int_0^x u^{z-1}\exp(-u)\,du, \quad \mathrm{Re}\,z > 0, \quad x > 0,
\tag{4.10.46}
$$

whereas the Euler gamma function

$$
\Gamma(z) = \Gamma(z,0) = \gamma(z,\infty),
\tag{4.10.47}
$$

and $\overline{C}$, $\overline{S}$ are given in (4.10.26). Let us note that the s-parametrized phase distribution (4.10.25) of the coherent state can be obtained by setting $M = 0$ and using the values,

$$
\begin{aligned}
\Gamma(1,\overline{C}^2) &= \exp(-\overline{C}^2), \\
\Gamma\left(\frac{1}{2}\right) &= \frac{\sqrt{\pi}}{2}, \\
|\overline{C}|\gamma\left(\frac{1}{2},\overline{C}^2\right) &= \frac{\sqrt{\pi}}{2}|\overline{C}|\mathrm{erf}(|\overline{C}|) \\
&= \frac{\sqrt{\pi}}{2}\overline{C}\mathrm{erf}(\overline{C}).
\end{aligned}
\tag{4.10.48}
$$

4.10.2   Multimode optical fields

In the following we present the notation useful for multimode (J-mode) optical fields. Assuming for simplicity an equal ordering of field operators in the separate modes, we consider the quasidistribution related to the s-ordering of field operators,

$$
\Phi_s(\{\alpha_j\}) = \mathrm{Tr}\left\{\hat{\rho}\hat{\Phi}_s(\{\alpha_j\})\right\},
\tag{4.10.49}
$$

where

$$\hat{\Phi}_s(\{\alpha_j\}) = \frac{1}{\pi^{2J}} \int \cdots \int \prod_{j=1}^{J} \exp(\alpha_j \beta_j^* - \alpha_j^* \beta_j) \hat{D}^{(s)}(\{\beta_j\}) \, d^2\{\beta_j\}, \qquad (4.10.50)$$

with the operator generating function

$$\hat{n}^{(s)}(\{\beta_j\}) - \prod_{j=1}^{J} \hat{D}_j^{(s)}(\beta_j), \qquad (4.10.51)$$

$$d^2\{\beta_j\} = \prod_{j=1}^{J} d^2\beta_j. \qquad (4.10.52)$$

Of course,

$$\hat{D}_j^{(s)}(\beta_j) = \exp\left(\frac{s}{2}|\beta_j|^2\right) \exp\left(\beta_j \hat{a}_j^\dagger - \beta_j^* \hat{a}_j\right), \quad j = 1, \ldots, J. \qquad (4.10.53)$$

Also the operator-valued density (4.10.50) factorizes,

$$\hat{\Phi}_s(\{\alpha_j\}) = \prod_{j=1}^{J} \hat{\Phi}_{js}(\alpha_j), \qquad (4.10.54)$$

where

$$\hat{\Phi}_{js}(\alpha_j) = \frac{1}{\pi^2} \int \exp(\alpha_j \beta_j^* - \alpha_j^* \beta_j) \hat{D}_j^{(s)}(\beta_j) \, d^2\beta_j, \quad j = 1, \ldots, J. \qquad (4.10.55)$$

The definition of a phase operator considered in the relation (4.10.4) reads for the jth mode as

$$
\begin{aligned}
\hat{\phi}_{j,-\pi}^{(s)} \quad &- \quad \frac{1}{2i} \int \cdots \int (\ln \alpha_j - \ln \alpha_j^*) \hat{\Phi}_s(\{\alpha_j\}) \, d^2\{\alpha_j\}, \\
&= \quad \frac{1}{2i} \int (\ln \alpha_j - \ln \alpha_j^*) \hat{\Phi}_{js}(\alpha_j) \, d^2\alpha_j.
\end{aligned}
\qquad (4.10.56)
$$

As a generalization of the approach of Royer, we rewrite the formulae (4.10.56) in the form

$$
\begin{aligned}
\hat{\phi}_{j,-\pi}^{(s)} \quad &= \quad \int_{-\pi}^{\pi} \cdots \int_{-\pi}^{\pi} \varphi_j \hat{\Gamma}^{(s)}(\{\varphi_j\}) \, d\{\varphi_j\} \\
&= \quad \int_{-\pi}^{\pi} \varphi_j \hat{\Gamma}_j^{(s)}(\varphi_j) \, d\varphi_j,
\end{aligned}
\qquad (4.10.57)
$$

where the operator-valued densities

$$\hat{\Gamma}^{(s)}(\{\varphi_j\}) = \int_0^\infty \cdots \int_0^\infty \prod_{j=1}^{J} \rho_j \hat{\Phi}_s\left(\{\rho_j e^{i\varphi_j}\}\right) \, d\{\rho_j\}, \qquad (4.10.58)$$

$$\hat{\Gamma}_j^{(s)}(\varphi_j) = \int_0^\infty \rho_j \hat{\Phi}_{js}\left(\rho_j e^{i\varphi_j}\right) \, d\varphi_j, \quad j = 1, \ldots, J. \qquad (4.10.59)$$

Also the operator-valued density (4.10.58) has the factorization property,

$$\hat{\Gamma}^{(s)}(\{\varphi_j\}) = \prod_{j=1}^{J} \hat{\Gamma}_j^{(s)}(\varphi_j). \tag{4.10.60}$$

The modification of the definitions (4.10.9), (4.10.10) can lead to

$$\widehat{\exp}^{(s)}(ik\varphi_j) = \int \frac{\alpha_j^k}{|\alpha_j|^k} \hat{\Phi}_{js}(\alpha_j)\, d^2\alpha_j$$

$$= \int_{\theta_{0j}}^{\theta_{0j}+2\pi} \exp(ik\varphi_j)\hat{\Gamma}^{(s)}(\varphi_j)\, d\varphi_j, \quad k = 0, 1, \ldots, \tag{4.10.61}$$

$$\widehat{\exp}^{(s)}(-ik\varphi_j) = [\widehat{\exp}^{(s)}(ik\varphi_j)]^\dagger, \quad k = 0, 1, \ldots. \tag{4.10.62}$$

We assume that quantum phase characteristics can be derived directly from the joint quantum phase distribution related to the s-ordering of field operators

$$P^{(s)}(\{\varphi_j\}) = \mathrm{Tr}\left\{\hat{\rho}\hat{\Gamma}^{(s)}(\{\varphi_j\})\right\}. \tag{4.10.63}$$

To obtain a generalization of the formulae (4.10.14) and (4.10.15) for the multimode case, we define the quasidistributions for $2J$ polar coordinates and $2J$ action–angle variables

$$\Phi_s^{\{\rho_j,\varphi_j\}}(\{\rho_j,\varphi_j\}) = \Phi_s\left(\{\rho_j e^{i\varphi_j}\}\right) \prod_{j=1}^{J} \rho_j, \tag{4.10.64}$$

$$\Phi_s^{\{W_j,\varphi_j\}}(\{W_j,\varphi_j\}) = \frac{1}{2^J}\Phi_s\left(\{\sqrt{W_j}e^{i\varphi_j}\}\right). \tag{4.10.65}$$

Substituting from (4.10.58) into (4.10.63) and exchanging the J-fold integrations, we obtain the generalized marginal integrals

$$P^{(s)}(\{\varphi_j\}) = \int_0^\infty \ldots \int_0^\infty \Phi_s^{\{\rho_j,\varphi_j\}}(\{\rho_j,\varphi_j\})\, d\{\rho_j\} \tag{4.10.66}$$

$$= \int_0^\infty \ldots \int_0^\infty \Phi_s^{\{W_j,\varphi_j\}}(\{W_j,\varphi_j\})\, d\{W_j\}. \tag{4.10.67}$$

The statistical operator can be expressed in the number state basis yielding the quasidistribution (4.10.49) in the form

$$\Phi_s\left(\{\rho_j e^{i\varphi_j}\}\right) = \sideset{}{'}\sum_{n_1=0}^{\infty} \sum_{k_1=-n_1}^{n_1} \ldots \sideset{}{'}\sum_{n_J=0}^{\infty} \sum_{k_J=-n_J}^{n_J} \rho_{\{n_j-k_j\},\{n_j+k_j\}}$$

$$\times \langle\{n_j + k_j\}|\hat{\Phi}_s\left(\{\rho_j e^{i\varphi_j}\}\right)|\{n_j - k_j\}\rangle, \tag{4.10.68}$$

where the functions

$$\langle\{n_j + k_j\}|\hat{\Phi}_s\left(\{\rho_j e^{i\varphi_j}\}\right)|\{n_j - k_j\}\rangle = \prod_{j=1}^{J} \langle n_j + k_j|\hat{\Phi}_s\left(\rho_j e^{i\varphi_j}\right)|n_j - k_j\rangle. \tag{4.10.69}$$

It follows that the joint quantum phase distribution (4.10.63) can be decomposed,

$$P^{(s)}(\{\varphi_j\}) = \sum_{n_1=0}^{\infty}{}' \sum_{k_1=-n_1}^{n_1} \cdots \sum_{n_J=0}^{\infty}{}' \sum_{k_J=-n_J}^{n_J} \rho_{\{n_j-k_j\},\{n_j+k_j\}}$$

$$\times \langle\{n_j + k_j\}|\hat{\Gamma}^{(s)}(\{\varphi_j\})|\{n_j - k_j\}\rangle, \qquad (4.10.70)$$

where the functions

$$\langle\{n_j + k_j\}|\hat{\Gamma}^{(s)}(\{\varphi_j\})|\{n_j - k_j\}\rangle$$

$$= \frac{1}{(2\pi)^J} \prod_{j=1}^{J} G^{(s)}(n_j - k_j, n_j + k_j) \exp\left(i2\sum_{j=1}^{J} k_j\varphi_j\right). \qquad (4.10.71)$$

In relation to the single-mode and two-mode squeezed vacuum states, the canonical and phase-space quantum phase distributions behave differently. Whereas in the single-mode case the canonical quantum phase distribution has no closed form and the s-parametrized (phase-space quantum) phase distribution is of the form (4.10.38), in the two-mode case the canonical quantum phase distribution is given in section 4.8 and the s-parametrized (phase-space quantum) phase distribution can be obtained in a relatively simple form.

Substituting $J = 2$ into (4.10.70) and considering the matrix elements

$$\rho_{n_1-k_1,n_2-k_2,n_1+k_1,n_2+k_2} = \delta_{n_1 n_2}\delta_{k_1 k_2}\frac{(-\tanh r)^{2n_1}}{\cosh r}\exp(-i4k_1\eta), \qquad (4.10.72)$$

we expand the joint quantum phase distribution as

$$P^{(s)}(\varphi_1, \varphi_2) = \frac{1}{4\pi^2}\frac{1}{\cosh^2 r}\sum_{n_1=0}^{\infty}{}'(-\tanh r)^{2n_1}$$

$$\times \sum_{k_1=-n_1}^{n_1} \exp[i2k_1(\varphi_1 + \varphi_2 - 2\eta)]\left[G^{(s)}(n_1 - k_1, n_1 + k_1)\right]^2. \qquad (4.10.73)$$

Although this double sum cannot be expressed in a closed form, it is obvious that it depends on the phase sum $(\varphi_1 + \varphi_2)$ only and that this quantity plays a similar role as 2φ in the single-mode case. Substituting

$$\Phi_s(\alpha_1, \alpha_2) = \frac{1}{\pi^2\sqrt{L(s)}}\exp\left\{-\sum_{j=1}^{2} E_{js}|\alpha_j|^2 - (\alpha_1^*\alpha_2^* F_{12s} + \text{c. c.})\right\}, \qquad (4.10.74)$$

where

$$L(s) = B_{1s}B_{2s} - |C_{12}|^2,$$

$$E_{js} = \frac{B_{3-j,s}}{\sqrt{L(s)}}, \quad j = 1, 2,$$

$$F_{12s} = -\frac{C_{12}}{\sqrt{L(s)}}, \qquad (4.10.75)$$

with

$$
\begin{aligned}
B_{1s} &= B_{2s} = \frac{1-s}{2}\cosh^2 r + \frac{1+s}{2}\sinh^2 r, \\
C_{12} &= -e^{i2\eta}\cosh r\,\sinh r,
\end{aligned}
\tag{4.10.76}
$$

into (4.10.64) and (4.10.66), we arrive at an alternative form of the s-parametrized joint phase distribution (4.10.73),

$$
\begin{aligned}
P^{(s)}(\varphi_1,\varphi_2) &= \frac{1}{4\pi^2 E_{1s}^2\sqrt{L(s)}}\sum_{n_1=0}^{\infty}\left(\frac{-2|F_{12s}|}{E_{1s}}\right)^{n_1} \\
&\quad \times [\cos(\varphi_1+\varphi_2-2\eta)]^{n_1}\frac{\left[\Gamma\left(\frac{n_1}{2}+1\right)\right]^2}{n_1!}.
\end{aligned}
\tag{4.10.77}
$$

In the case of the joint phase distribution related to the antinormal ordering of field operators, one of the two sums in (4.10.73) can be simplified and we obtain the same result as from (4.10.77), namely,

$$
P^{(A)}(\varphi_1,\varphi_2)
$$

$$
= \frac{1}{4\pi^2\cosh^2 r}\sum_{n_1=0}^{\infty}{}'[-2\tanh r\,\cos(\varphi_1+\varphi_2-2\eta)]^{2n_1}\frac{[\Gamma(n_1+1)]^2}{(2n_1)!}.
\tag{4.10.78}
$$

4.10.3   Quantum phase difference/sum and SU(2)/SU(1,1) phase-space quasidistributions

The SU(2) Q function has proved useful for introducing phase distributions in angular-momentum system [Agarwal and Singh (1996)]. According to the Jordan–Schwinger representation, this approach can be transferred to the situation of the SU(2) two-mode states and to the definition of their phase-difference distributions. Strikingly enough, in this situation a discrete distribution of the phase difference is more familiar. But this approach to quantum phase difference or sum leads to a continuous distribution.

4.11   Phase properties of real states of optical field

Many kinds of the real states of optical field are encountered in the single-mode case and are treated in the connection with the s-parametrized phase distributions in this book (see section 4.10). Phase fluctuations in a one-mode squeezed state have been studied by Lynch (1987) using the measured phase operators. Sanders, Barnett and Knight (1986), Yao (1987), and Fan and Zaidi (1988) determined particular phase properties of the usual squeezed state using the Susskind–Glogower phase formalism. Using the Pegg–Barnett formalism, Vaccaro and Pegg (1989) re-examined this problem. They have compared the results for a squeezed vacuum state with those of Fan and Zaidi (1988), have considered also the superposition of s-phase states proportional to $\frac{1}{\sqrt{2}}(|\eta + \frac{\pi}{2}; s\rangle + |\eta - \frac{\pi}{2}; s\rangle)$, and have treated the phase probability density for the weakly squeezed vacuum. Grønbech-Jensen, Christiansen and Ramanujam (1989) have calculated expectation values and variances of the cosine and sine of phase and of the phase itself for general squeezed states using the Susskind–Glogower, Lerner–Lynch, Pegg–Barnett measured-phase-operator and Hermitian-phase-operator formalisms. The phase behaviour of "quasi-photon phase states" [Nath and Kumar (1991)] and squeezed number states [Tu and Gong (1993)] has been explored using the measured phase operators. The phase properties of binomial [Vidiella-Barranco and Roversi (1994)] and of binomial and negative binomial [Gantsog, Joshi and Tanaś (1994)] states have been investigated in the framework of the Pegg–Barnett formalism. The canonical phase statistics have been treated both approximately and numerically. A quality approximation of canonical phase distribution for a phase-squeezed state has been obtained by Collett (1993a, b) who determined also the constrained optimum of phase noise for fixed mean photon number. The early asymptotics of Caves (1981) is reflected by the geometrical uncertainty and rotational width of Freyberger and Schleich (1994). The phase properties of squeezed number states have been studied by Fan and Zaidi (1988), those of squeezed displaced number states by Mendaš and Popović (1995), and those of squeezed thermal states by Nath and Kumar (1996) using the Pegg–Barnett formalism. The number–phase uncertainty products, cf. the left-hand sides of the relations (4.9.25), (4.9.26), and (4.6.47) for displaced number states have been examined by Mendaš and Popović (1994). In [Gantsog and Tanaś (1996)] the problem of quantum phase fluctuations of the field generated in the lossless micromaser cavity for different initial atomic and field states has been discussed using the Pegg–Barnett phase formalism. Real states of two-mode optical fields are expounded in this section and we shift the emphasis on the canonical phase distributions, although the s-parametrized phase distributions are not neglected.

4.11.1   Single-mode and two-mode superposition states

In section 3.10 we have treated the distributions of the two canonically conjugate quadratures for the case of single-mode superpositions of two coherent states. Now we intend to investigate the distributions of other two canonically conjugate quantities (at least in a generalized sense), i. e., the photon number and quantum phase for superpositions of N coherent states. From the expansion (3.10.38) we see that the photon-number distribution can be expanded as

$$p(n) = \sum_{j=1}^{N} p_j p(n; \alpha_j) + \sum_{\substack{j,k=1 \\ j \neq k}}^{N} p_{jk} p(n; \alpha_j, \alpha_k), \qquad (4.11.1)$$

where the Poisson distribution

$$p(n; \alpha_j) = \frac{|\alpha_j|^{2n}}{n!} \exp(-|\alpha_j|^2) \tag{4.11.2}$$

and its generalization

$$p(n; \alpha_j, \alpha_k) = \frac{(\alpha_j \alpha_k^*)^n}{n!} \exp(-\alpha_j \alpha_k^*). \tag{4.11.3}$$

It can easily be seen that the expansion (4.11.1) is equivalent to the following one, which can be obtained in a more direct way,

$$p(n) = \frac{1}{n!} \left| \sum_{j=1}^{N} Z_j \alpha_j^n \exp\left(-\frac{1}{2}|\alpha_j|^2\right) \right|^2, \tag{4.11.4}$$

where Z_j are given in (3.10.34).

Similarly, the canonical phase distribution

$$P(\varphi) = \sum_{j=1}^{N} p_j P(\varphi; \alpha_j) + \sum_{\substack{j,k=1 \\ j \neq k}}^{N} p_{jk} P(\varphi; \alpha_j, \alpha_k), \tag{4.11.5}$$

where the canonical phase distribution of a coherent state

$$P(\varphi; \alpha_j) = \exp(-|\alpha_j|^2) \frac{1}{2\pi} \sum_{m=0}^{\infty} \sum_{n=0}^{\infty} \frac{\alpha_j^m \alpha_j^{*n}}{\sqrt{m!n!}} \exp[-i(m-n)\varphi] \tag{4.11.6}$$

and its generalization

$$P(\varphi; \alpha_j, \alpha_k) = \exp(-\alpha_j \alpha_k^*) \frac{1}{2\pi} \sum_{m=0}^{\infty} \sum_{n=0}^{\infty} \frac{\alpha_j^m \alpha_k^{*n}}{\sqrt{m!n!}} \exp[-i(m-n)\varphi]. \tag{4.11.7}$$

More directly, we obtain that

$$P(\varphi) = \frac{1}{2\pi} \left| \sum_{j=1}^{N} Z_j \exp\left(-\frac{1}{2}|\alpha_j|^2\right) \sum_{n=0}^{\infty} \frac{\alpha_j^n}{\sqrt{n!}} \exp(-in\varphi) \right|^2. \tag{4.11.8}$$

Whereas an analogue of (4.11.8) for the s-parametrized quantum phase description does not exist, the s-parametrized phase distribution can be written as

$$P^{(s)}(\varphi) = \sum_{j=1}^{N} p_j P^{(s)}(\varphi; \alpha_j) + \sum_{\substack{j,k=1 \\ j \neq k}}^{N} p_{jk} P^{(s)}(\varphi; \alpha_j, \alpha_k), \tag{4.11.9}$$

where the s-parametrized phase distribution of a coherent state is given in (4.10.25), (4.10.26) and its generalization consists of (4.10.25), where in general

$$\overline{C} = \frac{1}{2}\sqrt{\frac{2}{1-s}}\{|\alpha_k|\exp[i(\varphi - \tau_k)] + |\alpha_j|\exp[-i(\varphi - \tau_j)]\},$$

$$\overline{S} = \frac{1}{2i}\sqrt{\frac{2}{1-s}}\{|\alpha_k|\exp[i(\varphi - \tau_k)] - |\alpha_j|\exp[-i(\varphi - \tau_j)]\}. \quad (4.11.10)$$

Let us consider as a special case the superposition of two coherent states and remember the notation (3.10.49). As a rough approximation, some of superpositions of two coherent states can be compared with usual squeezed states. The conditions which the superpositions should fulfill are known, but they have not been formulated exactly. The interest in such an exact formulation appears to be limited by other very natural comparisons of the usual squeezed states with the phase-optimized states. To formulate the properties of number-squeezed superpositions of coherent states, we may rely on some properties recognized in the usual squeezed states. We may see an analogue of the squeezed vacuum state in the case $\alpha_1 + \alpha_2 = 0$, which has been mentioned in section 3.10. The appropriate canonical phase distribution has been compared with the s-parametrized phase distributions in [Garraway and Knight (1993), Bužek, Gantsog and Kim (1993), Hach III and Gerry (1993)]. We restrict ourselves to the case $\alpha_1 + \alpha_2 \neq 0$ and respect the requisite property for number squeezing. The requisite property is

$$\arg(\alpha_1 + \alpha_2) \equiv \tau \pmod{\pi}. \quad (4.11.11)$$

A suitable rotation of the phase space enables us to assume that $\frac{1}{2}(\alpha_1 + \alpha_2) > 0$ without loss of generality. As we have still two conditions (cf. (3.10.49)), we learn that

$$\alpha_{1,2} = \overline{\alpha}\exp(\mp i\overline{\varphi}), \quad (4.11.12)$$

where $\overline{\alpha} > 0$, $\overline{\varphi} \in [0, \frac{\pi}{2})$ are arbitrary. To arrive at a real representation in the number state basis, we choose $Z_1 = Z_2 > 0$. It is not simple to decide whether or under which conditions these "number-squeezed" states are sub-Poissonian, i. e., number squeezed in fact. The necessity of decision is suggested by the "number-squeezed" (amplitude-squeezed) ordinary squeezed states. There, an increase of the squeeze parameter can lead to the super-Poissonian behaviour [Peřinová, Křepelka and Peřina (1986)]. The "number-squeezed" superpositions of two coherent states can be encountered as n-cats in [Schaufler, Freyberger and Schleich (1994)]. The phase-narrowed superpositions of two coherent states are to be searched for among the superpositions obeying the condition

$$\arg[i(\alpha_1 + \alpha_2)] \equiv \tau \pmod{\pi}. \quad (4.11.13)$$

From the conditions (3.10.49) and (4.11.13) it follows that either one of α_j is 0 or $\arg \alpha_1 \equiv \arg \alpha_2 \pmod{\pi}$. Of course, we can characterize the phase-narrowed

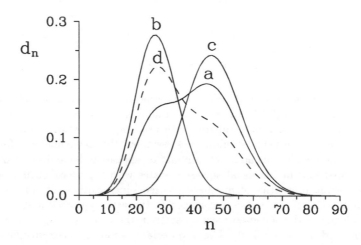

Figure 4.7: The coefficients d_n for the optimized superposition with $Z_1 < Z_2$ of the coherent states $|\alpha_1\rangle$, $|\alpha_2\rangle$, $\alpha_1 < \alpha_2$ (curve a), and for the coherent states $|\alpha_1\rangle$, $|\alpha_2\rangle$ (curve b, c, respectively). Curve d corresponds to the superposition with $Z_1 \geq Z_2$.

superpositions of two coherent states in a more interesting way, when we suppose that either one of α_j is 0 or arg $\alpha_1 \equiv$ arg α_2 (mod 2π). By a suitable rotation of the phase space and by the renumbering of complex amplitudes, if necessary, we can assume that $0 \leq \alpha_1 < \alpha_2$ without loss of generality. The real representation of these superpositions in the number state basis can be obtained assuming Z_1, $Z_2 > 0$. A related definition of a φ-cat is contained in [Schleich, Dowling, Horowicz and Varro (1990), Schaufler, Freyberger and Schleich (1994)]. The latter paper presents a recipe for the transmutation of a number cat into a phase cat. Also an optimized phase cat has been proposed there under the condition that $\alpha_1 + \alpha_2$ is constant, but $\alpha_2 - \alpha_1$ changes as well as Z_1 and Z_2. For $Z_1 < Z_2$ in the optimized phase cats, the graph of the representation of the states with the optimized α_1, α_2 by the coefficients d_n in the number state basis is more symmetrical for $Z_1 < Z_2$ than for $Z_1 \geq Z_2$ and it is super-Poissonian. It is illustrated in figure 4.7, where the coefficients d_n are plotted. The curve d for $Z_1 \geq Z_2$ shows more asymmetrical and less super-Poissonian situations. It is typical of the optimum that $Z_1 < Z_2$. This has been connected to the wedge shape of the phase-space distribution of other optimized phase states [Białynicki-Birula, Freyberger and Schleich (1993)]. A systematic method for finding optimized superpositions of more than two coherent states has been presented [Adam, Szabo and Janszky (1996)]. Vaccaro, Barnett and Pegg (1992) investigated the relationship between squeezing and reduced phase fluctuations for the strongly squeezed vacuum and the superposition of s-phase states

$$\frac{1}{2}\sqrt{\frac{s+1}{[\frac{s}{2}]+1}}\left(|\theta + \frac{\pi}{2}; s\rangle + |\theta - \frac{\pi}{2}; s\rangle\right).$$ (4.11.14)

They found that although squeezing the vacuum fluctuations of the electric field guarantees a more well-defined phase, reducing phase fluctuations is not sufficient for a squeezed electric field.

The number squeezing and the phase narrowing can be logically deduced from the physically less interesting properties, viz., the widening of the phase distribution and the stretching of the photon number distribution, respectively. Considered as processes the use of widening or stretching is limited by bifurcation in the widened or stretched distribution. It is obvious that of the two cases defined the second one leads to the stretching of the photon-number representation and therefore to the phase-representation narrowing as claimed. On the contrary, the first case seems to lead to the widening of the phase representation and then to the squeezing of the photon-number representation. This is a property of the Fourier transformation extended here to the Fourier synthesis of phase representation and to the evaluation of the Fourier coefficients. The behaviour of the representations determines that of the distributions. The bifurcation presents a limitation, because it corresponds to oscillations in the Fourier coefficient representation. For the reason that the oscillations may terminate the squeezing, but they are yet a nonclassical effect, they have been paid a great attention to [Schleich and Wheeler (1987a, b), Schleich, Horowicz and Varro (1989a, b), Schleich (1989), Milburn (1989), Schleich, Walls and Wheeler (1988), Milburn and Walls (1988)].

In the case of two modes described by the annihilation (creation) operators $\hat{a}_1$, $\hat{a}_2$ ($\hat{a}_1^\dagger$, $\hat{a}_2^\dagger$), we introduce new modes $\hat{a}_x$, $\hat{a}_y$,

$$\hat{a}_x = \frac{1}{\sqrt{2}}(\hat{a}_1 + \hat{a}_2),$$

$$\hat{a}_y = \frac{-i}{\sqrt{2}}(\hat{a}_1 - \hat{a}_2),$$ (4.11.15)

to define number-sum squeezed and phase-sum narrowed superpositions of four coherent states. We express the number-sum squeezed superposition as a tensor product of two superpositions,

$$|\psi\rangle = |\psi_x\rangle_x \otimes |\psi_y\rangle_y,$$ (4.11.16)

where the number-squeezed superposition

$$|\psi_x\rangle_x = Z_x^{(n)}\left(|\sqrt{2\alpha}\exp(-i\overline{\varphi})\rangle_x + |\sqrt{2\alpha}\exp(i\overline{\varphi})\rangle_x\right),$$ (4.11.17)

with

$$Z_x^{(n)} = \left\{2[1 + \cos[2\overline{\alpha}^2\sin(2\overline{\varphi})]\exp(-2\overline{\alpha}^2\sin^2\overline{\varphi})]\right\}^{-\frac{1}{2}},$$ (4.11.18)

and the even coherent state

$$|\psi_y\rangle_y = Z_y^{(n)} \left(| - i\sqrt{2}\overline{\alpha} \sin \overline{\varphi}\rangle_y + |i\sqrt{2}\overline{\alpha} \sin \overline{\varphi}\rangle_y \right), \qquad (4.11.19)$$

with

$$Z_y^{(n)} = \left\{ 2[1 + \exp(-4\overline{\alpha}^2 \sin^2 \overline{\varphi})] \right\}^{-\frac{1}{2}}. \qquad (4.11.20)$$

Returning to the original modes through the substitution

$$\hat{a}_1 = \frac{1}{\sqrt{2}}(\hat{a}_x + i\hat{a}_y),$$

$$\hat{a}_2 = \frac{1}{\sqrt{2}}(\hat{a}_x - i\hat{a}_y), \qquad (4.11.21)$$

we obtain the desired superposition in the form

$$|\psi\rangle = Z^{(n)} \sum_{s_x=\pm 1} \sum_{s_y=\pm 1} |\overline{\alpha} \exp(is_x\overline{\varphi}) - \overline{\alpha} \sin(s_y\overline{\varphi})\rangle_1$$
$$\otimes |\overline{\alpha} \exp(is_x\overline{\varphi}) + \overline{\alpha} \sin(s_y\overline{\varphi})\rangle_2, \qquad (4.11.22)$$

where

$$Z^{(n)} = Z_x^{(n)} Z_y^{(n)}. \qquad (4.11.23)$$

We search for phase-sum narrowed superpositions among tensor products (4.11.16), where the even coherent state

$$|\psi_x\rangle_x = Z_x^{(\varphi)} \left[\left| i\frac{\overline{\alpha}}{\sqrt{2}} \sin \gamma \right\rangle_x + \left| -i\frac{\overline{\alpha}}{\sqrt{2}} \sin \gamma \right\rangle_x \right], \qquad (4.11.24)$$

with

$$Z_x^{(\varphi)} = \left\{ 2[1 + \exp(-\overline{\alpha}^2 \sin^2 \gamma)] \right\}^{-\frac{1}{2}}, \qquad (4.11.25)$$

and the phase-narrowed superposition

$$|\psi_y\rangle_y = Z_y^{(\varphi)} \left[\left| -i\sqrt{2}\overline{\alpha} \left(1 - \frac{1}{2}\sin \gamma\right) \right\rangle_y + \left| -i\sqrt{2}\overline{\alpha} \left(1 + \frac{1}{2}\sin \gamma\right) \right\rangle_y \right], \qquad (4.11.26)$$

with $Z_y^{(\varphi)} = Z_x^{(\varphi)}$. The analogue of (4.11.22) for the phase-sum narrowed superposition reads

$$|\psi\rangle = Z^{(\varphi)} \sum_{s_x=\pm 1} \sum_{s_y=\pm 1} \left| \overline{\alpha} \left(1 + i\frac{s_x}{2} \sin \gamma - \frac{s_y}{2} \sin \gamma\right) \right\rangle_1$$
$$\otimes \left| \overline{\alpha} \left(-1 + i\frac{s_x}{2} \sin \gamma + \frac{s_y}{2} \sin \gamma\right) \right\rangle_2, \qquad (4.11.27)$$

where $Z^{(\varphi)} = [Z_x^{(\varphi)}]^2$.

The two kinds of two-mode superpositions of four coherent states have, in the first-mode factors, coherent states with complex amplitudes at the vertices of a square of side of $2\bar{\alpha}\sin\bar{\varphi}$ for the number-sum cat and of $\bar{\alpha}\sin\gamma$ for the phase-sum cat and of the centre at $\bar{\alpha}\cos\bar{\varphi}$ for the number-sum cat and at $\bar{\alpha}$ for the phase-sum cat. On the contrary, in the second-mode factors, the complex amplitudes are at the vertices of a square of the centre at $\bar{\alpha}\cos\bar{\varphi}$ for the number-sum cat and at $-\bar{\alpha}$ for the phase-sum cat.

The superposition states according to (4.11.16) are entangled in modes 1 and 2. In consequence, the reduced states in separate modes are noisy superposition states, but still ones according to the definition in section 3.10. The reduced number-sum cat in the first mode reads as

$$\hat{\rho}^{(1n)} = \text{Tr}_2\{|\psi\rangle\langle\psi|\} = \left(Z^{(n)}\right)^2 \sum_{s_x=\pm1}\sum_{s_y=\pm1}\sum_{s'_x=\pm1}\sum_{s'_y=\pm1} \rho^{(1n)}_{s_x s_y s'_x s'_y}$$

$$\times |\bar{\alpha}\exp(is_x\bar{\varphi}) - \bar{\alpha}\sin(s_y\bar{\varphi})\rangle_{11}\langle\bar{\alpha}\exp(is'_x\bar{\varphi}) - \bar{\alpha}\sin(s'_y\bar{\varphi})|, \quad (4.11.28)$$

where

$$\rho^{(1n)}_{s_x s_y s'_x s'_y} = \exp\left\{-\frac{\bar{\alpha}^2}{2}[(s'_y - s_y)^2 + (s'_x - s_x)^2]\sin^2\bar{\varphi} \right. \tag{4.11.29}$$

$$\left. + i\bar{\alpha}^2[(-s'_x s_y + s_x s'_y)\sin^2\bar{\varphi} + (-s'_x + s_x)\cos\bar{\varphi}\sin\varphi]\right\}.$$

Since $|\rho^{(1n)}_{s_x s_y s'_x s'_y}| < 1$ when at least one of $s'_x - s_x$, $s'_y - s_y$ does not vanish, the reduced state $\hat{\rho}^{(1n)}$ seems to be a mixed state.

The reduced number-sum cat in the second mode is of the form

$$\hat{\rho}^{(2n)} = \text{Tr}_1\{|\psi\rangle\langle\psi|\} = \left(Z^{(n)}\right)^2 \sum_{s_x=\pm1}\sum_{s_y=\pm1}\sum_{s'_x=\pm1}\sum_{s'_y=\pm1} \rho^{(2n)}_{s_x s_y s'_x s'_y}$$

$$\times |\bar{\alpha}\exp(is_x\bar{\varphi}) + \bar{\alpha}\sin(s_y\bar{\varphi})\rangle_{22}\langle\bar{\alpha}\exp(is'_x\bar{\varphi}) + \bar{\alpha}\sin(s'_y\bar{\varphi})|, \quad (4.11.30)$$

where

$$\rho^{(2n)}_{s_x s_y s'_x s'_y} = \rho^{(1n)}_{s_x, -s_y, s'_x, -s'_y}. \tag{4.11.31}$$

Neglecting the mode indices, changing appropriately the summation indices in (4.11.28) or (4.11.30), and taking into account (4.11.31), we arrive at the identity

$$\hat{\rho}^{(1n)} = \hat{\rho}^{(2n)}. \tag{4.11.32}$$

In complete analogy to (4.11.28) we obtain the reduced phase-sum cat in the separate modes in the form

$$\hat{\rho}^{(j\varphi)} = \left(Z^{(\varphi)}\right)^2 \sum_{s_x=\pm1}\sum_{s_y=\pm1}\sum_{s'_x=\pm1}\sum_{s'_y=\pm1} \rho^{(j\varphi)}_{s_x s_y s'_x s'_y}$$

$$\times \left| \overline{\alpha} \left[(-1)^{j-1} + i \frac{s_x}{2} \sin \gamma + (-1)^j \frac{s_y}{2} \sin \gamma \right] \right\rangle_j$$

$$\times_j \left\langle \overline{\alpha} \left[(-1)^{j-1} + i \frac{s'_x}{2} \sin \gamma + (-1)^j \frac{s'_y}{2} \sin \gamma \right] \right|, \qquad (4.11.33)$$

where

$$\rho^{(j\varphi)}_{s_x s_y s'_x s'_y} = \exp\left\{ i(-1)^j (s_x - s'_x) \frac{\overline{\alpha}^2}{2} \sin \gamma + \frac{\overline{\alpha}^2}{4} \sin^2 \gamma \left[-\frac{1}{2}(s'_x - s_x)^2 \right. \right.$$

$$\left. \left. -\frac{1}{2}(s'_y - s_y)^2 + i(-1)^{j-1}(s'_y s_x - s'_x s_y) \right] \right\} \qquad (4.11.34)$$

and $j = 1, 2$. Let us observe that

$$\rho^{(2\varphi)}_{s_x s_y s'_x s'_y} = \exp\left[i(s_x - s'_x)\overline{\alpha}^2 \sin \gamma \right] \rho^{(1\varphi)}_{s_x, -s_y, s_x, -s'_y}. \qquad (4.11.35)$$

Neglecting the mode index, changing appropriately the summation indices in (4.11.34) for $j = 1$ or $j = 2$, and taking into account the property (cf. (3.2.24)) of the displacement operator $\hat{D}(\alpha)$ from (3.2.2),

$$\hat{D}(\alpha)|\beta\rangle = \exp\left[\frac{1}{2}(\alpha\beta^* - \alpha^*\beta) \right] |\alpha + \beta\rangle, \qquad (4.11.36)$$

we arrive at the transformation

$$\hat{\rho}^{(2\varphi)} = \hat{D}(-2\overline{\alpha}) \hat{\rho}^{(1\varphi)} \hat{D}(2\overline{\alpha}). \qquad (4.11.37)$$

In the case of the two-mode superposition state, the relations (4.11.1), (4.11.2), (4.11.3) should be modified to those for the number-sum distribution. Similarly, the relations (4.11.5), (4.11.6), (4.11.7) should have analogues for the phase-sum distribution, but for the lack of place we do not present them.

The study of superposition states begins usually with their phase-space properties. This is easy in the single-mode superpositions, because the phase space is two-dimensional and the plots of quasidistributions are three-dimensional. For the two-mode superpositions, however, four-dimensional phase space is needed and the graphics "in the five-dimensional space" is not available.

4.11.2   Single-mode and two-mode ordinary squeezed states

The phase-space approach has been adopted in the study of the usual squeezed state in a single mode and it has proved to be especially fruitful with respect to the bifurcation of the canonical phase distribution [Schleich, Horowicz and Varro (1989b)]. For two-photon coherent states (cf. (3.5.34)) a notation can be introduced based on the property

$$|\tilde{\beta}\rangle_{(r,\eta)} = \hat{D}(\tilde{\beta})\hat{S}(r,\eta)|0\rangle, \qquad (4.11.38)$$

where $|0\rangle$ is the single-mode vacuum state and $\hat{S}(r,\eta)$ is the single-mode squeeze operator (cf. (3.5.42))

$$\hat{S}(r,\eta) = \exp\left[\frac{r}{2}\left(\hat{a}^2 e^{-i2\eta} - \hat{a}^{\dagger 2} e^{i2\eta}\right)\right],\tag{4.11.39}$$

with r being a nonnegative squeeze parameter and η being the phase of squeezing. Then the single-mode displacement operator $\hat{D}(\tilde{\beta})$ has been used, where $\tilde{\beta}$ is a complex displacement. It is also the coherent amplitude of the mode, because

$$_{(r,\eta)}\langle\tilde{\beta}|\hat{a}|\tilde{\beta}\rangle_{(r,\eta)} = \tilde{\beta}.\tag{4.11.40}$$

Introducing the phase τ of the coherent amplitude by the formula

$$\tilde{\beta} = |\tilde{\beta}|\exp(i\tau),\tag{4.11.41}$$

we can formulate two phase conditions. Firstly, the relation $\tau \equiv \eta \pmod{\pi}$ is the condition for the amplitude squeezing. Secondly, the relation $\tau \equiv \eta \pm \frac{\pi}{2} \pmod{\pi}$ is the condition for the phase narrowing. The two conditions are U(1)-invariant. They are not affected by the replacements $\tau \to \tau' = \tau - \omega t$, $\eta \to \eta' = \eta - \omega t$. From the relation

$$\hat{S}(r,\eta)\hat{D}(\beta)\hat{S}^{\dagger}(r,\eta) = \hat{D}(\tilde{\beta}),\tag{4.11.42}$$

where β is a given complex displacement and

$$\tilde{\beta} = \beta\cosh r - e^{i2\eta}\beta^*\sinh r,\tag{4.11.43}$$

it follows that also

$$|\tilde{\beta}\rangle_{(r,\eta)} = \hat{S}(r,\eta)\hat{D}(\beta)|0\rangle.\tag{4.11.44}$$

Writing (4.11.44) in the form

$$|\tilde{\beta}\rangle_{(r,\eta)} = \hat{S}(r,\eta)|\beta\rangle,\tag{4.11.45}$$

we see that the squeezing of a coherent state is a special case of the squeezing of an arbitrary state $|\psi(0)\rangle^{\mathrm{deg}}$, yielding

$$|\psi(r)\rangle^{\mathrm{deg}} = \hat{S}(r,\eta)|\psi(0)\rangle^{\mathrm{deg}}.\tag{4.11.46}$$

In general, the squeezing of an arbitrary state $\hat{\rho}^{\mathrm{deg}}(0)$ provides the description of the degenerate parametric amplification process in the interaction representation

$$\hat{\rho}^{\mathrm{deg}}(r) = \hat{S}(r,\eta)\hat{\rho}^{\mathrm{deg}}(0)\hat{S}^{\dagger}(r,\eta).\tag{4.11.47}$$

To the statistical operators $\hat{\rho}^{\mathrm{deg}}(0)$, $\hat{\rho}^{\mathrm{deg}}(r)$ we associate the Wigner functions

$$\Phi_S(\alpha,0) = \frac{1}{\pi}\mathrm{Tr}\left\{\hat{\rho}^{\mathrm{deg}}(0)\hat{\Delta}_S(\alpha)\right\},\tag{4.11.48}$$

where (cf. $\hat{\Delta}(\alpha, s)$ in (3.4.14))

$$\hat{\Delta}_S(\alpha) = 2\hat{D}(\alpha)(-1)^{\hat{a}^\dagger \hat{a}}\hat{D}^\dagger(\alpha) \tag{4.11.49}$$

is the appropriate version of the quantum ruler $|\alpha\rangle\langle\alpha|$, and

$$\Phi_S(\alpha, r) = \frac{1}{\pi}\text{Tr}\left\{\hat{\rho}^{\text{deg}}(r)\hat{\Delta}_S(\alpha)\right\}. \tag{4.11.50}$$

We consider an evolution operator $S(r, \eta)$ defined by the property

$$S(r, \eta)\Phi_S(\alpha, 0) \equiv \Phi_S(\alpha, r) = \Phi_S\left(\alpha\cosh r + e^{i2\eta}\alpha^*\sinh r, 0\right). \tag{4.11.51}$$

Although the phase representation and related concepts seem to be most funda-mental, the phase-space approach to the quantum phase problem seems to be most viable. Following [Schleich, Walls and Wheeler (1988), Schleich (1989)] many authors considered the marginal phase distribution of the Wigner function

$$P^{(S)}(\varphi, r) = \int_0^\infty \Phi_S\left(|\alpha|e^{i\varphi}, r\right)|\alpha|\,d|\alpha|. \tag{4.11.52}$$

A possible way how to interpret the introduction of the polar coordinates is the transformation of the phase plane, which changes the diverging rays through the origin into parallel lines and makes the concentric circles about the origin into parallel segments. This transformation changes the cigar-shaped three-dimensional graph of the amplitude-squeezed state into crescent-shaped plot. The horns of the graph are oriented in the direction of increasing $|\alpha|$. For shallow squeezing it can be seen that the phase distribution has a single peak. For a considerable squeezing the plot becomes horseshoe shaped. The plot may be designed so that it is confined in the phase interval $\left[\tau - \frac{\pi}{2}, \tau + \frac{\pi}{2}\right]$. It is obvious that such a case leads to two peaks of the phase distribution. A horseshoe-shaped Wigner function is exhibited by a massive particle in the linear potential and the polar coordinates are not needed in this case. It is very interesting that the three-dimensional plot of the Wigner function has oscillations inside the horseshoe and on its axis [Torres-Vega, Zúñiga-Segundo and Moralez-Guzmán (1996)]. These oscillations are in agreement with the behaviour of part of the photon-number distribution, which occurs inside the horseshoe due to the polar transformation, although this transformation itself cannot cause the oscillations. Of course, in the case of the above mentioned massive particle the oscillations resemble the interference occurring between two peaks of a superposition of two coherent states. The area-of-overlap and interference–in-phase-space concepts have been invented by Schleich and Wheeler (1987a, b). The phase-space-interference approach can be extended so that knowledge of the photon-number representation explains the appropriate quantum-phase-plane distribution [Gagen (1995)].

Substituting from (4.11.51) into (4.11.52), we obtain that

$$P^{(S)}(\varphi, r) = \int_0^\infty \Phi_S\left(|\alpha|\left[e^{i\varphi}\cosh r + e^{i2\eta}e^{-i\varphi}\sinh r\right], 0\right)|\alpha|\,d|\alpha|. \tag{4.11.53}$$

Performing the transformation of variable

$$|\tilde{\alpha}| = |\alpha| \left| e^{i\varphi}\cosh r + e^{i2\eta}e^{-i\varphi}\sinh r \right|, \qquad (4.11.54)$$

we obtain that

$$P^{(S)}(\varphi, r) = \frac{1}{\left| e^{i\varphi}\cosh r + e^{i2\eta}e^{-i\varphi}\sinh r \right|^2} P^{(S)}(f(\varphi), 0), \qquad (4.11.55)$$

where

$$f(\varphi) = \frac{1}{2i}\ln \frac{e^{i\varphi}\cosh r + e^{i2\eta}e^{-i\varphi}\sinh r}{e^{-i2\eta}e^{i\varphi}\sinh r + e^{-i\varphi}\cosh r}. \qquad (4.11.56)$$

Since

$$\left| \frac{d}{d\varphi}f(\varphi) \right| = \frac{|\alpha|^2}{|\tilde{\alpha}|^2}, \qquad (4.11.57)$$

we can rewrite (4.11.55) in the form

$$P^{(S)}(\varphi, r) = \left| \frac{d}{d\varphi}f(\varphi) \right| P^{(S)}(f(\varphi), 0). \qquad (4.11.58)$$

An initial coherent state $|\xi(0)\rangle$ can be described by a Wigner phase distribution $P_{\text{coh}}(\varphi, s - 0) - \Gamma_{\text{coh}}^{(s=0)}(\varphi)$ given in (4.10.25) [Tanaś, Miranowicz and Gantsog (1993)] and it evolves into the single-mode squeezed state described by a Wigner phase distribution $P_{\text{sq}}(\varphi, s = 0)$, which is a particular case of the distribution $P_{\text{thsq}}^{(s)}(\varphi)$ according to (4.10.32). The Wigner phase distribution $P_{\text{sq}}(\varphi, s = 0)$ under the restriction $\eta = 0$ and $\tau \equiv 0 \pmod{\pi}$ can be found in [Tanaś, Miranowicz and Gantsog (1993)].

The bifurcation condition reads as

$$\frac{\partial^2}{\partial\varphi^2}P^{(S)}(\varphi, r)\bigg|_{\varphi=\tau} = 0. \qquad (4.11.59)$$

The positive values of the second-order derivative correspond to a two-peak distribution, whereas the negative values of this derivative relate to a single-peak distribution. According to (4.11.58) and using that $f(\tau) = \tau$ in the case of amplitude squeezing, we arrive at the equivalent of the relation (4.11.59),

$$\frac{\frac{d^3}{d\varphi^3}f(\varphi)\big|_{\varphi=\tau}}{\left[\frac{d}{d\varphi}f(\varphi)\big|_{\varphi=\tau}\right]^3} = -\frac{\frac{\partial^2}{\partial\varphi^2}P^{(S)}(\varphi, 0)\big|_{\varphi=\tau}}{P^{(S)}(\tau, 0)} \qquad (4.11.60)$$

or

$$e^{4\tau} - 1 = \frac{\sqrt{2\pi}|\beta|\left(\frac{1}{2} + |\beta|^2\right) + 2|\beta|^2\exp(-2|\beta|^2)}{\sqrt{2\pi}|\beta| + \exp(-2|\beta|^2)}. \qquad (4.11.61)$$

A characteristic of good limiting properties with respect to $|\beta|$ is the simplified relation

$$e^{4r} = 1 + 2|\beta|^2, \tag{4.11.62}$$

from which it follows that the bimodal and unimodal cases correspond to the relations

$$e^{4r} > 1 + 2|\beta|^2, \tag{4.11.63}$$
$$e^{4r} < 1 + 2|\beta|^2, \tag{4.11.64}$$

respectively. Let us note that $\beta = \tilde{\beta} e^r$ in this case.

It is interesting to see the asymptotic form of the phase distribution for $r \to \infty$,

$$\begin{aligned}
P^{(S)}(\varphi, \infty) &= \overline{\delta}\left(\varphi - \tau - \frac{\pi}{2}\right) \int_{\tau}^{\tau+\pi} P^{(S)}(\varphi, 0)\, d\varphi \\
&+ \overline{\delta}\left(\varphi - \tau + \frac{\pi}{2}\right) \int_{\tau-\pi}^{\tau} P^{(S)}(\varphi, 0)\, d\varphi,
\end{aligned} \tag{4.11.65}$$

where $\overline{\delta}(\varphi)$ is the 2π-periodic Dirac delta function. A similar behaviour can be expected in the canonical phase distribution

$$P_{\mathrm{sq}}(\varphi) = \frac{1}{2\pi} \left| \sum_{n=0}^{\infty} c_n^{\mathrm{sq}} \exp(-in\varphi) \right|^2, \tag{4.11.66}$$

where the probability amplitudes

$$c_n^{\mathrm{sq}} = \langle n | \tilde{\beta} \rangle_{(r,\eta)} \tag{4.11.67}$$

or [Yuen (1976)]

$$c_n^{\mathrm{sq}} = \frac{1}{\sqrt{n!\cosh r}} \left(\frac{e^{i2\eta}}{2}\tanh r\right)^{\frac{n}{2}} \exp\left(-\frac{\tilde{\beta}^*\beta}{2\cosh r}\right) H_n\left(\frac{\beta}{\sqrt{e^{i2\eta}\sinh(2r)}}\right). \tag{4.11.68}$$

Here $H_n(x)$ are the Hermite polynomials. Except a conceivable factor $\exp(in\tau)$, the phase dependence of the probability amplitudes goes as $\tau - \eta$ (mod π). Besides this situation, when the parameter r varies and the initial value β is constant, we may consider a situation where the parameter r varies, but $\tilde{\beta}$ is constant. The asymptotic behaviour is similar in both the cases and the limit (4.11.65) is common. In the paper [Schleich, Horowicz and Varro (1989b)] r has been constant and β (or $\tilde{\beta}$ equivalently) has varied, and the canonical phase distribution has been studied.

In the study of single-mode and multimode fields, the immense importance of statistical operators which can be considered as Gaussian states has approved itself. In the intersection of Gaussian and pure states, Gaussian two-mode squeezed states have been introduced [Caves and Schumaker (1985), Schumaker and Caves (1985)]. A Gaussian two-mode squeezed state is defined by the relation

$$|\tilde{\beta}_1, \tilde{\beta}_2\rangle_{(r,\eta)} = \hat{D}(\tilde{\beta}_1, \tilde{\beta}_2)\hat{S}_{12}(r,\eta)|0,0\rangle, \tag{4.11.69}$$

where $|0,0\rangle$ is the two-mode vacuum state. It has been transformed by the two-mode squeeze operator [Caves and Schumaker (1985), Schumaker and Caves (1985)]

$$\hat{S}_{12}(r,\eta) = \exp\left[r\left(\hat{a}_1\hat{a}_2 e^{-i2\eta} - \hat{a}_1^\dagger\hat{a}_2^\dagger e^{i2\eta}\right)\right], \tag{4.11.70}$$

where r is a nonnegative squeeze parameter and η is the phase of the squeezing. Afterwards the two-mode displacement operator has been used,

$$D(\beta_1,\beta_2) = \hat{D}_1(\tilde{\beta}_1)\hat{D}_2(\tilde{\beta}_2), \tag{4.11.71}$$

where

$$\hat{D}_j(\tilde{\beta}_j) = \exp(\tilde{\beta}_j\hat{a}_j^\dagger - \tilde{\beta}_j^*\hat{a}_j), \quad j = 1,2, \tag{4.11.72}$$

with $\tilde{\beta}_j$ being complex displacements. They are also the coherent amplitudes of the modes, because

$$_{(r,\eta)}\langle\tilde{\beta}_1,\tilde{\beta}_2|\hat{a}_j|\tilde{\beta}_1,\tilde{\beta}_2\rangle_{(r,\eta)} = \tilde{\beta}_j, \quad j = 1,2. \tag{4.11.73}$$

We introduce the phases τ_j of the coherent amplitudes by the formulae

$$\tilde{\beta}_j = |\tilde{\beta}_j|\exp(i\tau_j), \quad j - 1,2. \tag{4.11.74}$$

Assuming that $|\tilde{\beta}_1| = |\tilde{\beta}_2| = |\tilde{\beta}|$, we can use two phase conditions to point out two situations. Firstly, the relation $\tau_1 + \tau_2 \equiv 2\eta \pmod{2\pi}$ is the condition for the number-sum squeezing. Secondly, the relation $\tau_1 + \tau_2 = 2\eta \pm \pi \pmod{2\pi}$ is the condition for the phase-sum narrowing. The two conditions are $U(1)\times U(1)$-invariant, i. e., they remain valid after the transformations, $\tau_1 \to \tau_1' = \tau_1 - \omega_1 t$, $\tau_2 \to \tau_2' = \tau_2 - \omega_2 t$, $\eta \to \eta' = \eta - \frac{1}{2}(\omega_1 + \omega_2)t$. Compare [Selvadoray, Kumar and Simon (1994)], where the concept of $U(1)\times U(1)$-invariant relative phase has been introduced. From the relation

$$\hat{S}_{12}(r,\eta)\hat{D}_j(\beta_j)\hat{S}_{12}^\dagger(r,\eta) = \hat{D}_j(\tilde{\beta}_j), \tag{4.11.75}$$

where β_j are given complex amplitudes and

$$\begin{aligned}\tilde{\beta}_1 &= \beta_1\cosh r - e^{i2\eta}\beta_2^*\sinh r,\\ \tilde{\beta}_2 &= -e^{i2\eta}\beta_1^*\sinh r + \beta_2\cosh r,\end{aligned} \tag{4.11.76}$$

it follows that also

$$|\tilde{\beta}_1,\tilde{\beta}_2\rangle_{(r,\eta)} = \hat{S}_{12}(r,\eta)\hat{D}(\beta_1,\beta_2)|0,0\rangle. \tag{4.11.77}$$

We can write (4.11.77) in the form

$$|\tilde{\beta}_1,\tilde{\beta}_2\rangle_{(r,\eta)} = \hat{S}_{12}(r,\eta)|\beta_1,\beta_2\rangle \tag{4.11.78}$$

to show that the two-mode squeezing of coherent states is a particular case of the two-mode squeezing of an arbitrary state $|\psi(0)\rangle$, which provides

$$|\psi(r)\rangle = \hat{S}_{12}(r,\eta)|\psi(0)\rangle. \tag{4.11.79}$$

Assuming a pure state described by the statistical operator $\hat{\rho}(0) = |\psi(0)\rangle\langle\psi(0)|$, we derive from (4.11.79) the description of the process of parametric down-conversion in the interaction representation

$$\hat{\rho}(r) = \hat{S}_{12}(r,\eta)\hat{\rho}(0)\hat{S}_{12}^{\dagger}(r,\eta). \tag{4.11.80}$$

To the statistical operators $\hat{\rho}(0)$, $\hat{\rho}(r)$ we attach the two-mode Wigner functions

$$\Phi_S(\alpha_1,\alpha_2,0) = \frac{1}{\pi^2}\mathrm{Tr}\left\{\hat{\rho}(0)\hat{\Delta}_S(\alpha_1,\alpha_2)\right\}, \tag{4.11.81}$$

where

$$\hat{\Delta}_S(\alpha_1,\alpha_2) = \hat{\Delta}_{1S}(\alpha_1)\hat{\Delta}_{2S}(\alpha_2), \tag{4.11.82}$$

with

$$\hat{\Delta}_{jS}(\alpha_j) = 2\hat{D}_j(\alpha_j)(-1)^{\hat{a}_j^{\dagger}\hat{a}_j}\hat{D}_j^{\dagger}(\alpha_j), \quad j = 1,2, \tag{4.11.83}$$

and

$$\Phi_S(\alpha_1,\alpha_2,r) = \frac{1}{\pi^2}\mathrm{Tr}\left\{\hat{\rho}(r)\hat{\Delta}_S(\alpha_1,\alpha_2)\right\}. \tag{4.11.84}$$

The evolution of the two-mode Wigner functions is described by the evolution operator $S_{12}(r,\eta)$ defined by the property

$$\begin{aligned}\Phi_S(\alpha_1,\alpha_2,r) &\equiv S_{12}(r,\eta)\Phi_S(\alpha_1,\alpha_2,0) \\ &= \Phi_S\left(\alpha_1\cosh r + e^{i2\eta}\alpha_2^*\sinh r, e^{i2\eta}\alpha_1^*\sinh r + \alpha_2\cosh r, 0\right).\end{aligned} \tag{4.11.85}$$

The joint photon-number distribution is related to the probability amplitudes

$$c_{n_1 n_2}^{\mathrm{sq}} = \langle n_1, n_2|\tilde{\beta}_1, \tilde{\beta}_2\rangle_{(r,\eta)}, \tag{4.11.86}$$

where the two-mode photon-number states $|n_1, n_2\rangle$ are tensor products of the number states $|n_1\rangle$, $|n_2\rangle$, or [Caves, Zhu, Milburn and Schleich (1991), Peřinová, Lukš and Křepelka (1996a)]

$$\begin{aligned}c_{n_1 n_2}^{\mathrm{sq}} &= \frac{(-e^{i2\eta}\sinh r)^p}{(\cosh r)^{q+1}}\sqrt{\frac{p!}{q!}}\beta_1^{n_1-p}\beta_2^{n_2-p}L_p^{q-p}\left(\frac{2\beta_1\beta_2}{e^{i2\eta}\sinh(2r)}\right) \\ &\quad \times \exp\left(-\frac{\tilde{\beta}_1^*\tilde{\beta}_1 + \tilde{\beta}_2^*\tilde{\beta}_2}{2\cosh r}\right),\end{aligned} \tag{4.11.87}$$

where

$$\begin{aligned}p &= \min(n_1, n_2), \\ q &= \max(n_1, n_2),\end{aligned} \tag{4.11.88}$$

and $L_p^{q-p}(x)$ are the Laguerre polynomials. Apart from the factor $\exp[i(n_1\tau_1 + n_2\tau_2)]$, the phase dependence of the probability amplitudes is confined to $\tau_1 + \tau_2 - 2\eta$ (mod 2π) (cf. [Selvadoray and Kumar (1997)]).

In the case of the number-sum squeezing, oscillations in the photon-number distribution occur. Caves, Zhu, Milburn and Schleich (1991) have extended the interference–in-phase-space concept from the single-mode case in order to explain these oscillations. These oscillations occur inside a parabola,

$$\frac{n_1 + n_2}{2} \geq \frac{(n_1 - n_2)^2}{16|\tilde{\beta}|^2} + |\tilde{\beta}|^2. \tag{4.11.89}$$

A slightly more general formula has been derived and presented by these authors.

Further investigation relates the oscillations in photon-number distribution with the shape of the joint phase distribution (peaks and ridges) [Lukš and Peřinová (1997d), Selvadoray and Kumar (1997), Lukš, Peřinová and Křepelka (1997)]. If $c_{n_1 n_2}$ are the probability amplitudes of any pure state $|\psi\rangle$, $c_{n_1 n_2} = \langle n_1, n_2 | \psi \rangle$, the phase representation of the state can be defined as [Agarwal, Chaturvedi, Tara and Srinivasan (1992)]

$$\psi(\varphi_1, \varphi_2) = \langle \varphi_1, \varphi_2 | \psi \rangle, \tag{4.11.90}$$

where

$$|\varphi_1, \varphi_2\rangle = |\varphi_1\rangle_1 \otimes |\varphi_2\rangle_2 \tag{4.11.91}$$

are the two-mode phase states. Substituting (4.11.91) and the sum

$$|\psi\rangle = \sum_{n_1=0}^{\infty} \sum_{n_2=0}^{\infty} c_{n_1 n_2} |n_1, n_2\rangle \tag{4.11.92}$$

into (4.11.90), we obtain the phase representation of the state as

$$\psi(\varphi_1, \varphi_2) = \frac{1}{2\pi} \sum_{n_1=0}^{\infty} \sum_{n_2=0}^{\infty} c_{n_1 n_2} \exp[-i(n_1\varphi_1 + n_2\varphi_2)],$$
$$\varphi_j \in [\theta_{0j}, \theta_{0j} + 2\pi), \quad j = 1, 2, \tag{4.11.93}$$

and derive the canonical joint phase distribution

$$P(\varphi_1, \varphi_2) = |\psi(\varphi_1, \varphi_2)|^2. \tag{4.11.94}$$

Now we introduce the phase representation of the Gaussian two-mode squeezed state,

$$\psi_{\text{sq}}(\varphi_1, \varphi_2) = \langle \varphi_1, \varphi_2 | \tilde{\beta}_1, \tilde{\beta}_2 \rangle_{(r,\eta)}, \tag{4.11.95}$$

and arrive at the appropriate canonical joint phase distribution,

$$P_{\text{sq}}(\varphi_1, \varphi_2) = |\psi_{\text{sq}}(\varphi_1, \varphi_2)|^2. \tag{4.11.96}$$

All phase properties are contained in the characteristic double sequence $\{\chi_{k_1 k_2}\}$, whose terms are

$$\chi_{k_1 k_2} = \int_{\theta_{01}}^{\theta_{01}+2\pi} \int_{\theta_{02}}^{\theta_{02}+2\pi} \exp[i(k_1\varphi_1 + k_2\varphi_2)] P_{\text{sq}}(\varphi_1, \varphi_2) \, d\varphi_1 \, d\varphi_2, \tag{4.11.97}$$

with k_1, k_2 integers. Using generalized Susskind–Glogower exponential phase operators

$$\widehat{\exp}[i(k_1\varphi_1 + k_2\varphi_2)] = \int_{\theta_{01}}^{\theta_{01}+2\pi} \int_{\theta_{02}}^{\theta_{02}+2\pi} \exp[i(k_1\varphi_1 + k_2\varphi_2)]$$
$$\times |\varphi_1, \varphi_2\rangle\langle\varphi_1, \varphi_2| \, d\varphi_1 \, d\varphi_2, \qquad (4.11.98)$$

we may write the terms of characteristic double sequence as the quantum average

$$\chi_{k_1 k_2} = {}_{(r,\eta)}\langle\tilde{\beta}_1, \tilde{\beta}_2|\widehat{\exp}[i(k_1\varphi_1 + k_2\varphi_2)]|\tilde{\beta}_1, \tilde{\beta}_2\rangle_{(r,\eta)}. \qquad (4.11.99)$$

Substituting from (4.11.96) and (4.11.95) into (4.11.97), we arrive at

$$\chi_{k_1 k_2} = \sum_{n_1=0}^{\infty} \sum_{n_2=0}^{\infty} c_{n_1 n_2}^{sq*} c_{n_1+k_1, n_2+k_2}^{sq}, \quad k_1 \geq 0, \quad k_2 \geq 0, \qquad (4.11.100)$$

$$\chi_{k_1 k_2} = \sum_{n_1=0}^{\infty} \sum_{n_2=0}^{\infty} c_{n_1, n_2-k_2}^{sq*} c_{n_1+k_1, n_2}^{sq}, \quad k_1 \geq 0, \quad k_2 \leq 0, \qquad (4.11.101)$$

$$\chi_{k_1 k_2} = \chi_{-k_1, -k_2}^{*}, \quad k_1 \leq 0. \qquad (4.11.102)$$

The marginal phase distributions in separate modes have characteristic sequences

$$\chi_k^{(1)} = \chi_{k0}, \quad \chi_k^{(2)} = \chi_{0k}, \qquad (4.11.103)$$

where k is an integer. In analogy, we take the phase sum and phase difference separately (mod 2π) and obtain their characteristic sequences in the form

$$\chi_k^{(+)} = \langle\widehat{\exp}[ik(\varphi_1 + \varphi_2)]\rangle = \chi_{kk}, \quad \chi_k^{(-)} = \langle\widehat{\exp}[ik(\varphi_1 - \varphi_2)]\rangle = \chi_{k,-k}. \qquad (4.11.104)$$

The phase-sum and phase-difference distributions can be obtained as

$$P_+(\varphi_+) = \frac{1}{2\pi} \sum_{k=-\infty}^{\infty} \chi_k^{(+)} \exp(-ik\varphi_+), \qquad (4.11.105)$$

$$P_-(\varphi_-) = \frac{1}{2\pi} \sum_{k=-\infty}^{\infty} \chi_k^{(-)} \exp(-ik\varphi_-). \qquad (4.11.106)$$

The two-mode squeezed vacuum $|0, 0\rangle_{(r,\eta)}$ is a particular photon-twin state. The photon-twin state is defined by the property

$$c_{n_1 n_2}^{twin} = \langle n_1, n_2|\psi\rangle_{twin} = \delta_{n_1 n_2} b_{n_1}, \qquad (4.11.107)$$

where b_{n_1} are any complex numbers fulfilling the normalization condition

$$\sum_{n_1=0}^{\infty} |b_{n_1}|^2 = 1. \qquad (4.11.108)$$

Such a state is written in the form

$$|\psi\rangle_{\text{twin}} = \sum_{n_1=0}^{\infty} b_{n_1} |n_1, n_2 = n_1\rangle, \qquad (4.11.109)$$

where $|n_1, n_2 = n_1\rangle \equiv |n_1\rangle_1 \otimes |n_1\rangle_2$. The photon-twin states have a simplified characteristic double sequence

$$\chi_{k_1 k_2}^{\text{twin}} = \delta_{k_1 k_2} \chi_{k_1 k_1}^{\text{twin}} = \delta_{k_1 k_2} \chi_{k_1}^{(\text{I})\text{twin}}. \qquad (4.11.110)$$

The phase-sum distribution has the characteristic sequence

$$\chi_k^{(+)\text{twin}} = \begin{cases} \sum_{n_1=0}^{\infty} b_{n_1}^* b_{n_1+k} & \text{for } k \geq 0, \\ \left(\chi_{-k}^{(+)\text{twin}}\right)^* & \text{for } k < 0. \end{cases} \qquad (4.11.111)$$

The characteristic sequence of the phase-difference distribution is found to be

$$\chi_k^{(-)\text{twin}} = \chi_{k,-k}^{\text{twin}} = \delta_{k,-k} \chi_k^{(+)\text{twin}} = \delta_{k0}. \qquad (4.11.112)$$

Substituting into the Fourier series (4.11.106), we derive that the photon-twin states have a uniform phase-difference distribution,

$$P_-^{\text{twin}}(\varphi_-) = \frac{1}{2\pi}. \qquad (4.11.113)$$

Substituting from (4.11.111) into (4.11.105), we obtain that

$$P_+^{\text{twin}}(\varphi_+) = \frac{1}{2\pi} \left| \sum_{n=0}^{\infty} b_n \exp(-in\varphi_+) \right|^2. \qquad (4.11.114)$$

In the framework of the quantum theory of measurement, the photon-twin state may result due to measurement of photon-number difference operator $\hat{a}_1^\dagger \hat{a}_1 - \hat{a}_2^\dagger \hat{a}_2$, when the outcome is $m = 0$. Let us recall that by the projection postulate the reduced state is

$$|\psi\rangle_{|m=0} = \left[\sum_{n_1=0}^{\infty} |c_{n_1 n_1}|^2 \right]^{-\frac{1}{2}} \sum_{n_1=0}^{\infty} c_{n_1 n_1} |n_1, n_1\rangle. \qquad (4.11.115)$$

Hence, the conditional distribution of phase sum reads as

$$P_+(\varphi_+|m=0) = \frac{1}{2\pi} \left[\sum_{n_1=0}^{\infty} |c_{n_1 n_1}|^2 \right]^{-1} \left| \sum_{n_1=0}^{\infty} c_{n_1 n_1} \exp(-in_1\varphi_+) \right|^2. \qquad (4.11.116)$$

Introducing the new modes $\hat{a}_x$, $\hat{a}_y$ according to (4.11.15), we observe that the substitution (4.11.21) leads to a disentanglement of the two-mode squeeze operator (4.11.70),

$$\cdot \ \hat{S}_{12}(r, \eta) = \hat{S}_x(r, \eta) \hat{S}_y(r, \eta), \qquad (4.11.117)$$

where $\hat{S}_x(r,\eta)$, $\hat{S}_y(r,\eta)$ are single-mode squeeze operators

$$\hat{S}_j(r,\eta) = \exp\left[\frac{r}{2}\left(\hat{a}_j^2 e^{-i2\eta} - \hat{a}_j^{\dagger 2} e^{i2\eta}\right)\right], \quad j = x, y. \tag{4.11.118}$$

A frequently used transformation

$$\begin{aligned}
\alpha_1 &= \frac{1}{\sqrt{2}}(\alpha_x + i\alpha_y), \\
\alpha_2 &= \frac{1}{\sqrt{2}}(\alpha_x - i\alpha_y),
\end{aligned} \tag{4.11.119}$$

or

$$\begin{aligned}
\alpha_x &= \frac{1}{\sqrt{2}}(\alpha_1 + \alpha_2), \\
\alpha_y &= \frac{-i}{\sqrt{2}}(\alpha_1 - \alpha_2),
\end{aligned} \tag{4.11.120}$$

applies to complex amplitudes of pure Gaussian states in the pair of modes 1, 2 and in modes x, y. Whereas in the former case the pure state $|\tilde{\beta}_1, \tilde{\beta}_2\rangle_{(r,\eta)}$ is entangled, in the latter case it is the tensor product $|\alpha_x\rangle_{(r,\eta)x} \otimes |\alpha_y\rangle_{(r,\eta)y}$.

In figure 4.8 the dependence of the conditional phase-sum distribution on the squeeze parameter r can be seen, which demonstrates a bifurcation for a certain value of the squeeze parameter. Contrary to the single-mode case [Schleich, Horowicz and Varro (1989b)], where the peaks in the phase distribution part into the antipodal positions for $r \to \infty$, in this case the bifurcation displayed in the phase window $[-\pi, \pi)$ is changed by another bifurcation, which means almost a fusion of two peaks. The preferred value of phase sum leaps in the squeezing process. Whereas for $r <$ 2.6295 $\mathrm{pref}(\phi|m = 0) \equiv 0$, for $r > 2.6295$ $\mathrm{pref}(\phi|m = 0) \equiv \pi \pmod{2\pi}$. The dispersion $D(\phi|m = 0)$ grows while $r < 2.6295$ and decreases when $r > 2.6295$.

In figure 4.9 the choice of window $[0, 2\pi)$ helps us understand that squeezing of phase-sum distribution proceeds without any limitation.

The joint phase distribution appropriate to the two-mode squeezed state $|3,3\rangle_{(2.5,0)}$ in figure 4.10 exhibits a decaying peak at the point $\varphi_1 = \varphi_2 = 0$. This characteristics of decrease has been derived on comparison with the joint phase distributions with $r < 2.5$. The ridges are developing approximately on the line $\varphi_1 + \varphi_2 \equiv \pi \pmod{2\pi}$ and especially two peaks are arising at $(\varphi_1, \varphi_2) = (\pm\pi, 0), (0, \pm\pi)$.

4.11.3   Miscellaneous two-mode states

There exist two-mode states which have joint canonical phase distribution in a closed form.

(i) Two-mode squeezed vacuum

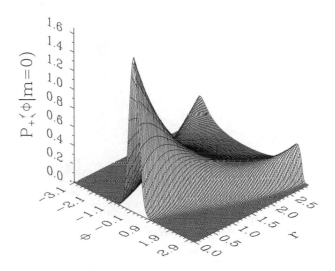

Figure 4.8: The conditional phase-sum distributions given that the photon number difference vanishes with $\phi \in [-\pi, \pi)$ and in the squeezed states $|3, 3\rangle_{(r,0)}$, $r \in [0, 3]$.

Barnett and Pegg (1990b) were first to derive the canonical phase distribution of the two-mode squeezed vacuum. Here we obtain the joint phase distribution (4.11.94) in the form [Lukš and Peřinová (1996)]

$$P(\varphi_1, \varphi_2) = \frac{1}{4\pi^2} \frac{1}{\cosh(2r) + \sinh(2r)\cos(\varphi_1 + \varphi_2 - 2\eta)}. \qquad (4.11.121)$$

The marginal phase distributions $P_j(\varphi_j)$, $j = 1, 2$, are uniform,

$$P_1(\varphi_1) = P_2(\varphi_2) = \frac{1}{2\pi}. \qquad (4.11.122)$$

It is a consequence of the fact that the reduced states in the separate modes are chaotic ((3.3.34)). Also the phase-difference distribution is uniform, $P_-(\varphi_-) = \frac{1}{2\pi}$ (see (4.11.113)). The uncertainty in phase difference is complementary to the certainty of number-difference vanishing ($n_1 = n_2$). The correlation between the phases is obvious from the fact that the conditional preferred phases are not constant (section 4.4)

$$\text{Pref}_{\theta_{01}}(\varphi_1|\varphi_2) \equiv \xi - \varphi_2 \pmod{2\pi}, \qquad (4.11.123)$$

where $\xi = 2\eta + \pi \pmod{2\pi}$,

$$\text{Pref}_{\theta_{02}}(\varphi_2|\varphi_1) \equiv \xi - \varphi_1 \pmod{2\pi}. \qquad (4.11.124)$$

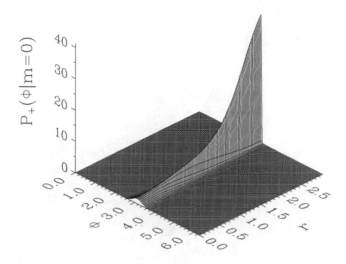

Figure 4.9: The conditional phase-sum distributions given that the photon-number difference vanishes with $\phi \in [0, 2\pi)$ and in the squeezed states $|3, -3\rangle_{(r,0)}$, $r \in [0, 3]$.

A use of conditional probability densities

$$P(\varphi_1|\varphi_2) = P(\varphi_2|\varphi_1) = 2\pi P(\varphi_1, \varphi_2) \tag{4.11.125}$$

has been assumed. The two conditional probability distributions are wrapped Cauchy distributions ((4.4.103)). The strength of the statistical dependence between the canonical phases is characterized by conditional dispersions,

$$D(\varphi_1|\varphi_2) = D(\varphi_2|\varphi_1) = \frac{1}{\cosh^2 r}. \tag{4.11.126}$$

The usual measure of the statistical dependence (mathematically, the covariance) has been used by Barnett and Pegg (1990b)]. They indicated the use of $\theta_{01} = \theta_{02} \equiv -\eta - \frac{\pi}{2}$ (mod π).

For the same purpose the codispersion can serve [Lukš and Peřinová (1996)],

$$\text{cod}(\varphi_1, -\varphi_2) = \langle \Delta \widehat{\exp}(i\varphi_1) \Delta \widehat{\exp}(i\varphi_2) \rangle, \tag{4.11.127}$$

where

$$\Delta \widehat{\exp}(i\varphi_j) = \widehat{\exp}(i\varphi_j) - \langle \widehat{\exp}(i\varphi_j) \rangle \hat{1}. \tag{4.11.128}$$

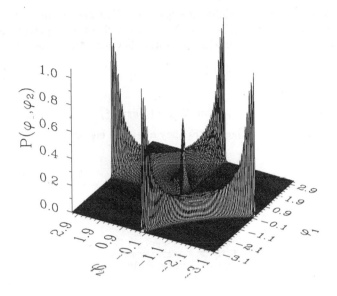

Figure 4.10: The joint phase distribution for $\varphi_1 \in [-\pi, \pi)$, $\varphi_2 \subset [-\pi, \pi)$ in the squeezed state $|3,3\rangle_{(2.5,0)}$. A concentration on the line $\varphi_1 + \varphi_2 = \pi \pmod{2\pi}$ is attained approximately.

In fact, for the two-mode squeezed vacuum

$$\text{cod}(\varphi_1, -\varphi_2) = \exp(i\xi)\tanh r. \qquad (4.11.129)$$

This expression enters the well-known series expansion,

$$|0,0\rangle_{(r,\eta)} = (\cosh r)^{-1} \sum_{n=0}^{\infty} [\exp(i\xi)\tanh r]^n |n,n\rangle. \qquad (4.11.130)$$

This striking coincidence is outweighed by the necessity of considering another codispersion

$$\text{cod}(\varphi_1, \varphi_2) = \langle \widehat{\Delta\exp}(i\varphi_1)\widehat{\Delta\exp}(-i\varphi_2)\rangle, \qquad (4.11.131)$$

where

$$\widehat{\Delta\exp}(-i\varphi_2) = [\widehat{\Delta\exp}(i\varphi_2)]^{\dagger}. \qquad (4.11.132)$$

Developing the observation that the phase-difference distribution is uniform, we obtain that

$$\text{cod}(\varphi_1, \varphi_2) = 0. \qquad (4.11.133)$$

(ii) Pair-coherent states
SU(1,1) intelligent states may be simultaneous $\hat{K}_1$–$\hat{K}_2$- or $\hat{K}_2$–$\hat{K}_3$-intelligent states

and eigenstates of the number-difference operator [Brif and Mann (1996a)]. The pair-coherent states have been defined by Agarwal (1986, 1988) as particular $\hat{K}_1-\hat{K}_2$-intelligent states $|\zeta, q\rangle$. As such intelligent states they have the property

$$\hat{a}_1\hat{a}_2|\zeta, q\rangle = \zeta|\zeta, q\rangle, \tag{4.11.134}$$

where

$$\zeta = -|\zeta|\exp(i2\eta) = |\zeta|\exp(i\xi), \tag{4.11.135}$$

and as the number-difference eigenstates they obey the eigenvalue equation

$$(\hat{a}_1^\dagger\hat{a}_1 - \hat{a}_2^\dagger\hat{a}_2)|\zeta, q\rangle = q|\zeta, q\rangle. \tag{4.11.136}$$

Since

$$|\zeta, q\rangle = \sum_{n=0}^{\infty} b_n \exp(in\xi)|n + q, n\rangle, \tag{4.11.137}$$

where

$$b_n = N_q\frac{|\zeta|^n}{\sqrt{n!(n + q)!}}, \tag{4.11.138}$$

with

$$N_q = \sqrt{\frac{|\zeta|^2}{I_q(2|\zeta|)}}, \tag{4.11.139}$$

the resulting joint probability distribution for the phases φ_1 and φ_2 of the two modes depends only on the sum of the phases,

$$P(\varphi_1, \varphi_2) = \frac{1}{(2\pi)^2}\left|\sum_{n=0}^{\infty} b_n \exp\left[in(\varphi_1 + \varphi_2 - \xi)\right]\right|^2. \tag{4.11.140}$$

For $q = 0$, the pair-coherent state is a photon-twin state, and (4.11.140) can be expressed in a simple form

$$P(\varphi_1, \varphi_2) = \frac{N_0^2}{(2\pi)^2} \exp[2|\zeta|\cos(\varphi_1 + \varphi_2 - \zeta)]. \tag{4.11.141}$$

As in the case of the two-mode squeezed vacuum, the conditional phase distributions are of interest and they are described by the probability densities according to (4.11.125). Here the two conditional probability densities are von Mises, but the conditional preferred phases are given again by the formulae (4.11.123) and (4.11.124). For any q the marginal phase distributions are given again by (4.11.122) and again the phase-difference distribution is uniform. Also the relation (4.11.133) holds and the analogue of (4.11.129) reads for $q = 0$

$$\text{cod}(\varphi_1, -\varphi_2) = \exp(i\xi)\frac{I_1(2|\zeta|)}{I_0(2|\zeta|)}. \tag{4.11.142}$$

In the case of arbitrary q, we obtain that the codispersion

$$\text{cod}(\varphi_1, -\varphi_2) = \exp(i\xi) \sum_{n=0}^{\infty} b_n b_{n+1}. \tag{4.11.143}$$

The advantage of the usual covariance (physically, the correlation coefficient) has been approved by Gantsog and Tanaś (1991a) (see also [Tanaś, Miranowicz and Gantsog (1990)]).
(iii) SU(2) and SU(1,1) beamsplitter states
In the Mach–Zehnder interferometry, the properties of two-mode states before the phase shifter and after a beamsplitter are of interest. The phase properties of the output states of an SU(2) beamsplitter have been characterized when two squeezed states, two number states, and a displaced two-mode squeezed state are on the input [Hillery, Zou and Bužek (1996)]. For the production of nonclassical states it is of importance that small fluctuation in the input photon-number difference is converted on small fluctuation in the output phase difference. Let us compare subsection 4.11.2, where a similar situation for the SU(1,1) beamsplitter with two coherent states on the input is considered.

4.12   Operational approach to quantum phase

The quantum statistical properties of optical fields can be generally deduced from photoelectric measurements giving the photocount distribution, its factorial moments or correlation functions. We assume that the signal mode represented by the annihilation operator $\hat{a}$ is superimposed on a local oscillator (reference) mode represented by the annihilation operator $\hat{b}$, using a lossless beamsplitter. If we denote its transmission and reflection amplitude coefficients for the input mode a by t and r, respectively, and those for the input mode b by t' and r', then the output annihilation and creation operators $\hat{c}$ and $\hat{d}$ are expressed as

$$\begin{aligned}
\hat{c} &= t\hat{a} + r'\hat{b}, \\
\hat{d} &= r\hat{a} + t'\hat{b},
\end{aligned} \tag{4.12.1}$$

and the unitarity of the transformation matrix (energy conservation law) requires that

$$|t|^2 + |r|^2 = 1, \quad |r'|^2 + |t'|^2 = 1, \quad tr'^* + rt'^* = 0. \tag{4.12.2}$$

A thorough analysis of a quantum-mechanical lossless beamsplitter has been presented by Campos, Saleh and Teich (1989). Here we assume that the beamsplitter is symmetrical, $t' = t$ and $r' = r$. For simplicity, we assume the input modes in the two-mode coherent state $|\alpha, \beta\rangle$. Two basic arrangements can be realized to measure effects in input signal mode (figure 4.11). In the ordinary homodyne detection $|r| \ll |t|$ and only one of the photodetectors (let us say P_1) is useful. In the balanced

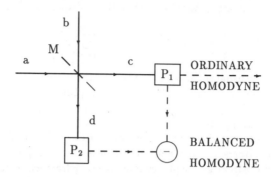

Figure 4.11: Scheme of homodyne detection; M is a beamsplitter, P_1 and P_2 are photodetectors.

homodyne detection, $|\mathbf{r}| = |\mathbf{t}| = \frac{1}{\sqrt{2}}$ and the output signal for processing is obtained as the difference of the two photodetector currents. The third condition in (4.12.2) is fulfilled by the phase condition $\arg \mathbf{r} - \arg \mathbf{t} = \frac{\pi}{2}$.

First let us consider ordinary homodyne detection providing for the mean number of detected photons

$$\langle \hat{c}^\dagger \hat{c} \rangle = |\mathbf{r}|^2 |\beta|^2 + |\mathbf{r}||\mathbf{t}||\beta|\langle \hat{P}(\bar{\tau}) \rangle + |\mathbf{t}|^2 \langle \hat{a}^\dagger \hat{a} \rangle, \tag{4.12.3}$$

where $\bar{\tau} = \arg \beta$ and the operator $\hat{P}(\bar{\tau})$ is defined as (cf. (3.10.51))

$$\hat{P}(\bar{\tau}) = i[\hat{a}^\dagger \exp(i\bar{\tau}) - \hat{a} \exp(-i\bar{\tau})]. \tag{4.12.4}$$

For the ordinary homodyne detection, it holds that $|\mathbf{r}||\beta| \gg |\mathbf{t}||\alpha|$. Under this assumption we obtain for the number of photons in the beam c

$$\langle \hat{n}_c \rangle = |\mathbf{r}|^2 |\beta|^2 + |\mathbf{r}||\mathbf{t}||\beta|\langle \hat{P}(\bar{\tau}) \rangle, \tag{4.12.5}$$

with the corresponding variance

$$\langle (\Delta \hat{n}_c)^2 \rangle = |\mathbf{r}|^2 |\beta|^2 [|\mathbf{r}|^2 + |\mathbf{t}|^2 \langle (\Delta \hat{P}(\bar{\tau}))^2 \rangle]. \tag{4.12.6}$$

In this way the mean rotated quadrature $\langle \hat{P}(\bar{\tau}) \rangle$ can be determined including its phase together with its fluctuations $\langle (\Delta \hat{P}(\bar{\tau}))^2 \rangle$. For a real photodetector, the quantum photoefficiency η has to be taken into account. We can subtract a constant level $|\mathbf{r}|^2 |\beta|^2$ in (4.12.5) thus obtaining for the mean number of counts relatively to this level

$$\langle \hat{m}_c \rangle = \eta |\mathbf{r}||\mathbf{t}||\beta|\langle \hat{P}(\bar{\tau}) \rangle = 2\eta |\mathbf{r}||\mathbf{t}||\alpha||\beta| \sin(\theta - \bar{\tau}), \tag{4.12.7}$$

where $\theta = \arg \alpha$. Hence, the phase of the signal is directly available. The dominant expression for the corresponding variance $\langle (\Delta \hat{m}_c)^2 \rangle$ is obtained on the basis of the Burgess variance theorem [Peřina (1991), section 3.7] in the form

$$\langle (\Delta \hat{m}_c)^2 \rangle = \eta |\mathbf{r}|^2 |\beta|^2 \{ 1 + \eta |\mathbf{t}|^2 [\langle (\Delta \hat{P}(\bar{\tau}))^2 \rangle - 1] \}. \tag{4.12.8}$$

In the coherent state (section 3.2), it holds that $\langle (\Delta \hat{P}(\bar{\tau}))^2 \rangle - 1 = 0$ and the first term in (4.12.8) represents the shot noise corresponding to the Poisson photon statistics (section 3.7). For squeezed quantum fluctuations, $\langle (\Delta \hat{P}(\bar{\tau}))^2 \rangle < 1$ and the photocount statistics are sub-Poissonian and, if $\langle (\Delta \hat{P}(\bar{\tau}))^2 \rangle \geq 1$, they are super-Poissonian. The deviation from the Poisson statistics is dependent on the local oscillator phase $\bar{\tau}$. This dependence may be cancelled when considering principal squeezing of vacuum fluctuations (section 3.10). The measurement is performed in such a way that the input is first blocked to determine the shot-noise level. The variance of the signal is then determined in relation to this shot-noise level.

Let us consider now the balanced homodyne detection proposed by Yuen and Chan (1983), firstly demonstrated by Abbas, Chan and Yee (1983) and further investigated by Schumaker (1984). Taking the difference of the mean photon numbers of the form (4.12.3) and

$$\langle \hat{d}^\dagger \hat{d} \rangle = |\mathbf{t}|^2 |\beta|^2 - |\mathbf{r}||\mathbf{t}||\beta| \langle \hat{P}(\bar{\tau}) \rangle + |\mathbf{r}|^2 \langle \hat{a}^\dagger \hat{a} \rangle, \tag{4.12.9}$$

we arrive at

$$\langle \hat{n}_{cd} \rangle = \langle \hat{c}^\dagger \hat{c} \rangle - \langle \hat{d}^\dagger \hat{d} \rangle = |\beta| \langle \hat{P}(\bar{\tau}) \rangle. \tag{4.12.10}$$

The great advantage is that in the balanced homodyne detection noise components of single photodetectors are cancelled and only the interference terms are conserved. The photon-number variance is expressed in the form

$$\langle (\Delta \hat{n}_{cd})^2 \rangle = |\beta|^2 \langle (\Delta \hat{P}(\bar{\tau}))^2 \rangle, \tag{4.12.11}$$

assuming that the local oscillator intensity is much greater than the signal intensity, $|\beta|^2 \gg |\alpha|^2$. Thus, the balancing also fully eliminates the local-oscillator noise. The corresponding mean number of counts and its variance can be obtained in the form

$$\langle \hat{m}_{cd} \rangle = \eta |\beta| \langle \hat{P}(\bar{\tau}) \rangle \tag{4.12.12}$$

and

$$\langle (\Delta \hat{m}_{cd})^2 \rangle = \eta |\beta|^2 \{ 1 + \eta [\langle (\Delta \hat{P}(\bar{\tau}))^2 \rangle - 1] \}. \tag{4.12.13}$$

The same discussion as for the ordinary homodyne detection in (4.12.8) can be performed for the photon statistics in this case.

Optical homodyne experiments have been performed by Walker and Carroll (1984, 1986) with eight-port detector. A theoretical study of such a measurement has been provided by Walker (1987). At that time, it was not possible to use experiments to test out alternative theories of quantum phase. Subsequently, the Pegg–Barnett formalism got ground [Pegg and Barnett (1988), Barnett and Pegg (1989)], it became possible to compare between the theories of canonical phase and the descriptions of feasible phase and, as a consequence, the results of Noh, Fougères and Mandel (1991, 1992a, b, 1993a, b, c, d) and Fougères, Noh, Grayson and Mandel (1994) have attracted attention of the quantum optical community. These authors have used an eight-port balanced homodyne detector in order to measure the relative phase between two classical or quantum light fields. Noh, Fougères and Mandel (1991) have considered separate measurement of the sine or the cosine of the phase difference between two light beams (scheme 1) and the complete setup (scheme 2). Scheme 1 does not markedly differ from that for the direct measurement of a quadrature. A complex optical scheme providing an indirect measurement of a quadrature of the field using a feedback mechanism has been proposed by D'Ariano and Sacchi (1997). A homodyne measurement of the cosine using scheme 1 will reduce the wave function of the transmitted beam which can be sampled and subjected to a measurement of the sine using another scheme 1 further reducing the wave function. Namely, the results of the sine measurements have the moduli determined by the cosine and their signs are indetermined [Franson (1994)]. There was a discussion on the normalization of the phase measurement of scheme 1 and the comparison of results with scheme 2 [Hradil and Bajer (1993), Noh, Fougères and Mandel (1993e)]. In analyzing scheme 2, Noh, Fougères and Mandel (1993e) have defined "operational" cosine and sine (trigonometric) operators. Although the experimental setup is somewhat different, the measurement scheme is given in figure 4.12.

We assume the use of four identical beamsplitters with complex amplitude reflectance r and transmittance t from one side and r', t' from the other side. We will denote the four beams by the subscripts $1, 2, 10, 20$. We assume the input state described by $\hat{\rho}$ and let $\hat{a}_k$ denote the photon annihilation operators for modes k, $k = 1, 2, 10, 20$. We distinguish three stages of the dynamics in the scheme, $l = 0, 1, 2$, $l = 0$ indicates the incident fields, $\hat{\rho}(0) = \hat{\rho}$, $\hat{a}_k(0) = \hat{a}_k$, $k = 1, 2, 10, 20$.

The first stage of the scheme is described by the relation

$$\hat{a}_k(1) = t\hat{a}_k(0) + r'\hat{a}_{k'}(0),$$
$$\hat{a}_{k'}(1) = r\hat{a}_k(0) + t'\hat{a}_{k'}(0), \qquad (4.12.14)$$

where $k = 1$, $k' = 10$ for one beamsplitter and $k = 20$, $k' = 2$ for the other beamsplitter. The coefficients r, t, r', t' have the properties

$$|r|^2 = |r'|^2 = |t|^2 = |t'|^2 = \frac{1}{2}, \quad tt'r^*r'^* = -\frac{1}{4}. \qquad (4.12.15)$$

It is assumed that the input mode 1 plays the role of the signal mode and that the input mode 2 is produced by local oscillator and its physical state $|\xi\rangle$, $\xi = |\xi| \exp(i\tau)$, is coherent, the other input modes being in the vacuum state. In the first stage, two highly correlated copies of the signal field are made, in fact, attenuated by the factor of two. By symmetry, the same proceeds for the local oscillator beam.

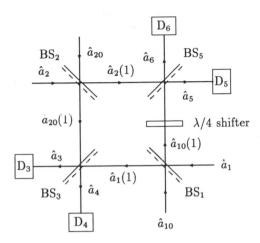

Figure 4.12: Outline of the eight-port interferometer in the field-operator evolution picture.

In the second stage, mode 10 passes through a $\frac{\lambda}{4}$ shifter to become canonically conjugate to mode 1. This stage is described by the relations

$$\hat{a}_k(2) = t\hat{a}_k(1) + r'\hat{a}_{k'}(1),$$
$$\hat{a}_{k'}(2) = r\hat{a}_k(1) + t'\hat{a}_{k'}(1), \qquad (4.12.16)$$

where $k = 20$, $k' = 1$, and

$$\hat{a}_k(2) = t\hat{a}_k(1) + ir'\hat{a}_{k'}(1),$$
$$\hat{a}_{k'}(2) = r\hat{a}_k(1) + it'\hat{a}_{k'}(1), \qquad (4.12.17)$$

where $k = 2$, $k' = 10$. The second stage serves measuring quadrature operators in modes 1, 10.

For simplicity, we introduce the output annihilation operators

$$\hat{a}_3 = \hat{a}_1(2), \; \hat{a}_4 = \hat{a}_{20}(2), \; \hat{a}_5 = \hat{a}_2(2), \; \hat{a}_6 = \hat{a}_{10}(2). \qquad (4.12.18)$$

Composing the indicated transformations and denoting [Bandilla (1993)]

$$v = 2tr'^*, \qquad (4.12.19)$$

we arrive at the relative photon-number operator,

$$\hat{n}_X(2) = \hat{n}_4 - \hat{n}_3 = \hat{a}_4^\dagger \hat{a}_4 - \hat{a}_3^\dagger \hat{a}_3$$
$$= \frac{1}{2}[(\hat{a}_1 + v^*\hat{a}_{10})(\hat{a}_2^\dagger + v^*\hat{a}_{20}^\dagger) + \text{H. c.}], \qquad (4.12.20)$$

which approximates the operator $\hat{J}_1$. The quantity

$$\hat{X}_{|\xi|}(0) = \frac{1}{|\xi|}(\hat{n}_4 - \hat{n}_3) = \frac{\hat{n}_X(2)}{|\xi|} \qquad (4.12.21)$$

is, in distribution, the quadrature $\text{Re}\left(\frac{\xi^*}{|\xi|}\hat{a}_1\right)$ of the signal mode in strong-local-oscillator limit. Considering the counts on the other detectors, we have the relative photon-number operator

$$\begin{aligned}
\hat{n}_Y(2) &= \hat{n}_5 - \hat{n}_6 = \hat{a}_5^\dagger\hat{a}_5 - \hat{a}_6^\dagger\hat{a}_6 \\
&= \frac{1}{2}[(-i\hat{a}_1 + iv^*\hat{a}_{10})(\hat{a}_2^\dagger - v^*\hat{a}_{20}^\dagger) + \text{H. c.}],
\end{aligned} \qquad (4.12.22)$$

which approximates the operator $-\hat{J}_2$. The quantity

$$\hat{Y}_{|\xi|}(0) = \frac{1}{|\xi|}(\hat{n}_5 - \hat{n}_6) = \frac{\hat{n}_Y(2)}{|\xi|} \qquad (4.12.23)$$

is, in distribution, the conjugate quadrature $\text{Im}\left(\frac{\xi^*}{|\xi|}\hat{a}_1\right)$ of the signal mode in strong-local-oscillator limit. On introducing the displaced mode

$$\Delta\hat{a}_2 = \hat{a}_2 - \xi\hat{1}, \qquad (4.12.24)$$

we arrive at the measured complex amplitude operator

$$\hat{\alpha}_{|\xi|}(0) = \hat{X}_{|\xi|}(0) + i\hat{Y}_{|\xi|}(0) = \frac{\xi^*}{|\xi|}\hat{a}_1 + v\frac{\xi}{|\xi|}\hat{a}_{10}^\dagger + \hat{N}_{|\xi|}(\tau, 0), \qquad (4.12.25)$$

where the discretization noise operator

$$\hat{N}_{|\xi|}(\tau, 0) = \frac{1}{|\xi|}\left[\hat{a}_1\Delta\hat{a}_2^\dagger + v\hat{a}_1^\dagger\hat{a}_{20} + v\hat{a}_{10}^\dagger\Delta\hat{a}_2 + v^{*2}\hat{a}_{10}\hat{a}_{20}^\dagger\right]. \qquad (4.12.26)$$

The measured cosine and sine can be related to the measured exponential phase operators

$$\widehat{\exp}_M[i(\varphi_1 - \varphi_2)] = \frac{\hat{n}_X(2) + i\hat{n}_Y(2)}{|\hat{n}_X(2) + i\hat{n}_Y(2)|} = \frac{\hat{a}_{|\xi|}(0)}{|\hat{a}_{|\xi|}(0)|}. \qquad (4.12.27)$$

This exponential phase operator is not defined for all elements of the Hilbert space $\mathcal{H} = \mathcal{H}_1 \otimes \mathcal{H}_{10} \otimes \mathcal{H}_2 \otimes \mathcal{H}_{20}$, where $\mathcal{H}_1, \mathcal{H}_{10}, \mathcal{H}_2, \mathcal{H}_{20}$ are Hilbert spaces of the respective input ports. The impossible division by zero is required on the subspace $\mathcal{H}_0 \subset \mathcal{H}$ spanned by the eigenstates with the zero complex amplitude,

$$\hat{n}_X(2)|\psi_0\rangle = \hat{n}_Y(2)|\psi_0\rangle = 0. \qquad (4.12.28)$$

The exponential phase operator $\widehat{\exp}_M[i(\varphi_1 - \varphi_2)]$ is defined on the orthogonal complement $\mathcal{H}_0^\perp$. The experimental procedure adopted in [Noh, Fougères and Mandel

(1992a)] solves this difficulty in a situation of repeated measurement. There the number of measurements corresponds only to those yielding the well defined phase. It is equivalent to the measurement of the phase in a state obtained as the projection onto the subspace $\mathcal{H}_0^\perp$.

Another solution could be not to discard the data leading to the division by zero, but instead of this, to generate a random value of phase. This approach is inherent in the proposal of Hradil (1993b) who, however, has enlarged the domain of the measured exponential phase operator onto the whole Hilbert space $\mathcal{H}$ to include also the subspace $\mathcal{H}_0$,

$$\widehat{\exp}_E[i(\varphi_1 - \varphi_2)] = \begin{cases} \widehat{\exp}_M[i(\varphi_1 - \varphi_2)] & \text{in } \mathcal{H}_0^\perp, \\ \hat{0} & \text{in } \mathcal{H}_0. \end{cases} \tag{4.12.29}$$

However, we may observe that a unitary exponential operator is not obtained. The measured phase distribution for the states from the subspace $\mathcal{H}_0^\perp$ is not influenced by this solution. In contrast, the phase distribution for the states from the subspace $\mathcal{H}_0$ is uniform, because all the elements of the characteristic sequence but one vanish. In general, a pure state which is a superposition of states from $\mathcal{H}_0^\perp$ and $\mathcal{H}_0$ yields a phase distribution which is a mixture of the phase distribution according to the proposal of Noh, Fougères and Mandel and the uniform one.

Let us note that the deliberate enlargement of the subspace $\mathcal{H}_0$ and the corresponding diminishment of the space $\mathcal{H}_0^\perp$ lead to a sharpening of the measured quantum phase distribution [Torgerson and Mandel (1997)].

In the limit of the strong field, the discretization noise operator can be neglected $\hat{N}_{|\xi|}(\tau, 0) \approx \hat{0}$ and the measured complex amplitude operator is a "feasible" complex amplitude operator

$$\hat{a}_{|\xi|}(0) \approx \hat{\alpha}(\tau, 0) = \hat{a}_1(\tau, 0) + v\hat{a}_{10}^\dagger(\tau, 0), \tag{4.12.30}$$

$$\hat{a}_k(\tau, 0) = \frac{\xi^*}{|\xi|}\hat{a}_k, \quad k = 1, 10. \tag{4.12.31}$$

In this limit, the measured exponential phase operator is a feasible phase operator,

$$\widehat{\exp}_M[i(\varphi_1 - \varphi_2)] \approx \widehat{\exp}[i(\phi_1 - \tau)], \tag{4.12.32}$$

where

$$\widehat{\exp}[i(\phi_1 - \tau)] = \frac{\hat{\alpha}(\tau, 0)}{|\hat{\alpha}(\tau, 0)|}. \tag{4.12.33}$$

Using the usual four-mode photodetection formula, we obtain the photon-number distribution

$$\begin{aligned} p(n_4, n_3, n_5, n_6) =\ & \int\int\int\int \exp\left(-|\alpha_4|^2 - |\alpha_3|^2 - |\alpha_5|^2 - |\alpha_6|^2\right) \\ & \times \frac{|\alpha_4|^{2n_4}|\alpha_3|^{2n_3}|\alpha_5|^{2n_5}|\alpha_6|^{2n_6}}{n_4! n_3! n_5! n_6!} \\ & \times \Phi_\mathcal{N}(\alpha_4, \alpha_3, \alpha_5, \alpha_6, 2)\, d^2\alpha_4\, d^2\alpha_3\, d^2\alpha_5\, d^2\alpha_6, \end{aligned} \tag{4.12.34}$$

where the quasidistribution Φ_N is related to the normal ordering of field operators describing the state $\hat{\rho}(2)$ on the output of the scheme. In the state evolution picture, the second stage is described by the relation

$$\hat{\rho}(2) = \hat{U}_3\hat{U}_5\hat{U}_{\frac{\lambda}{4}}\hat{\rho}(1)\hat{U}_{\frac{\lambda}{4}}^{\dagger}\hat{U}_5^{\dagger}\hat{U}_3^{\dagger}, \tag{4.12.35}$$

where

$$\hat{\rho}(1) = \hat{U}_1\hat{U}_2\hat{\rho}(0)\hat{U}_2^{\dagger}\hat{U}_1^{\dagger}, \tag{4.12.36}$$

$\hat{\rho}(0)$ being the input state of the scheme. Here the unitary operators

$$\hat{U}_{\frac{\lambda}{4}} = \exp\left(i\frac{\pi}{2}\hat{a}_{10}^{\dagger}\hat{a}_{10}\right),$$

$$\hat{U}_j = \exp\left(\frac{i}{\hbar}\hat{G}_j\right), \quad j = 1,2,3,5, \tag{4.12.37}$$

$$\hat{G}_j = \hbar\left[K\hat{a}_{k_j}^{\dagger}\hat{a}_{k_j} + K'\hat{a}_{k_j'}^{\dagger}\hat{a}_{k_j'} + (L^*\hat{a}_{k_j}^{\dagger}\hat{a}_{k_j'} + \text{H. c.})\right] \tag{4.12.38}$$

are introduced, where the compound indices k_j, k_j' are given by the table coding the scheme in [Noh, Fougères and Mandel (1993d)]

$$\begin{array}{ccc} j & k_j & k_j' \\ \hline 1 & 1 & 10 \\ 2 & 20 & 2 \\ 3 & 20 & 1 \\ 5 & 2 & 10. \end{array} \tag{4.12.39}$$

Here the coefficients K, K', L are of the form

$$\left.\begin{array}{c} K \\ K' \end{array}\right\} = \arg\left(\frac{t+t'}{2}\right) \mp i\frac{(t-t')}{2}\frac{\text{Cos}^{-1}\left|\frac{t+t'}{2}\right|}{\sqrt{1-\left|\frac{t+t'}{2}\right|^2}}\frac{(t+t')^*}{|t+t'|}, \tag{4.12.40}$$

$$L = -ir\frac{\text{Cos}^{-1}\left|\frac{t+t'}{2}\right|}{\sqrt{1-\left|\frac{t+t'}{2}\right|^2}}\frac{(t+t')^*}{|t+t'|} \tag{4.12.41}$$

provided by a simple inversion of formulae in [Peřinová, Lukš, Křepelka, Sibilia and Bertolotti (1991)].

Equivalently, in the field-operator evolution picture any first-stage operator reads

$$\hat{M}(1) = \hat{U}_2^{\dagger}\hat{U}_1^{\dagger}\hat{M}(0)\hat{U}_1\hat{U}_2, \tag{4.12.42}$$

which must be consistent with (4.12.14), but $\hat{M}(0)$ is any initial operator and the appropriate second-stage operator

$$\hat{M}(2) = \hat{U}_{\frac{\lambda}{4}}^{\dagger}\hat{U}_5^{\dagger}\hat{U}_3^{\dagger}\hat{M}(1)\hat{U}_3\hat{U}_5\hat{U}_{\frac{\lambda}{4}} \tag{4.12.43}$$

is consistent with (4.12.16) and (4.12.17). However, $\hat{\rho}(l) = \hat{\rho}$, $l = 0, 1, 2$. In this picture, the distribution of relative photon numbers

$$n_x = n_4 - n_3, \quad n_y = n_5 - n_6 \tag{4.12.44}$$

reads

$$
\begin{aligned}
p(n_x, n_y) = & \int \exp\left(-W_4 - W_3 - W_5 - W_6\right) \left(\frac{W_4}{W_3}\right)^{\frac{n_x}{2}} \\
& \times I_{n_x}\left(\sqrt{2W_4W_3}\right) \left(\frac{W_5}{W_6}\right)^{\frac{n_y}{2}} I_{n_y}\left(\sqrt{2W_5W_6}\right) \\
& \times \Phi_{\mathcal{N}}(\alpha_1, 0) \, d^2\alpha_1 \, d^2\alpha_2,
\end{aligned}
\tag{4.12.45}
$$

where

$$
\begin{aligned}
W_4 &= \frac{1}{4}|\alpha_1 + \alpha_2|^2, \quad W_3 = \frac{1}{4}|\alpha_1 - \alpha_2|^2, \\
W_5 &= \frac{1}{4}|\alpha_1 + i\alpha_2|^2, \quad W_6 = \frac{1}{4}|\alpha_1 - i\alpha_2|^2,
\end{aligned}
\tag{4.12.46}
$$

and $I_n(x)$ is the modified Bessel function. The input state on the four ports of the two input beamsplitters can be described by the quasidistribution related to the normal ordering of field operators

$$\Phi_{\mathcal{N}}(\alpha_1, \alpha_{10}, \alpha_2, \alpha_{20}, 0) = \Phi_{\mathcal{N}}(\alpha_1, \alpha_2, 0)\delta(\alpha_{10})\delta(\alpha_{20}), \tag{4.12.47}$$

where

$$\Phi_{\mathcal{N}}(\alpha_1, \alpha_2, 0) = \Phi_{\mathcal{N}}(\alpha_1, 0)\delta(\alpha_2 - \xi). \tag{4.12.48}$$

In the limit $|\xi| \to \infty$, the joint probability density of measured quadratures is the quasidistribution related to the antinormal ordering of field operators,

$$P\left(\frac{n_x}{|\xi|}, \frac{n_y}{|\xi|}\right) = |\xi|^2 p(n_x, n_y) \simeq \Phi_{\mathcal{A}}\left(\frac{n_x + in_y}{\xi}, 0\right). \tag{4.12.49}$$

An analysis of the eight-port homodyne detection scheme using phase-space functions (quasidistributions) has been performed by Freyberger and Schleich (1993), Vogel and Grabow (1993), Freyberger, Vogel and Schleich (1993a, b). To describe the process of simultaneous measurement of two noncommuting observables, Wódkiewicz (1984, 1986, 1988) has proposed a formalism based on an operational probability-density distribution which is equal to the convolution of the detected and displaced filtering Wigner functions. A particular choice of the state basis for the quantum ruler samples a specific type of accessible information concerning the system. The measurement of the operational phase-space distribution of Wódkiewicz is feasible in the experimental setup used by Noh, Fougères and Mandel (1993d). In this eight-port device, the signal state, which is going to be measured, is launched into one port of the first beamsplitter while the filter state is launched into the other port (it is the vacuum when the port is unused) of this beamsplitter. The $\Phi_\mathcal{A}$ function has been shown to be a phase-space propensity concerning the detected system and a filtering device described by a vacuum state [Burak and Wódkiewicz (1992)]. Banaszek and Wódkiewicz (1997) have derived exact expressions for the probability operator-valued measure and the corresponding operational operators of the balanced homodyne detection scheme with imperfect detectors. The comparison of quasidistributions with propensities as generalized quasidistributions obtained by quantum filtering is made easier by the property

$$\frac{1}{\pi} \int \hat{D}(\alpha)\hat{A}\hat{D}^\dagger(\alpha)\, d^2\alpha = \mathrm{Tr}\{\hat{A}\}\hat{1}, \qquad (4.12.50)$$

which holds for any trace-class operator $\hat{A}$. Hence, we obtain the resolution of the identity not only for $\hat{A} = |0\rangle\langle 0|$, the vacuum state, but also for $\hat{A} = \hat{\rho}_f$, the statistical operator of a quantum filter. Wünsche and Bužek (1997) have dealt with the reconstruction of statistical operators of quantum states from propensities. Bužek, Keitel and Knight (1995a) have introduced the concept of (the phase-space) sampling entropy and have shown that the Wehrl entropy represents a particular example of a sampling entropy when the quantum ruler is represented by coherent states. Braunstein, Caves and Milburn (1991) have rederived the Arthurs–Kelly measurement model [Arthurs and Kelly Jr. (1965)] for the $\Phi_\mathcal{A}$ function and generalized this result. They have considered a simple generalization of this model that yields the canonical form of the positive P ($\Phi_\mathcal{N}$) representation as its measurement statistics. In other words, more complicated quantum-mechanical detection schemes may lead to a use of the positive P representation [Drummond and Gardiner (1980)], which follows from the analysis of Braunstein, Caves and Milburn (1991). They have found that the measurement of position and momentum of systems with one degree of freedom can be described by a $\Phi_\mathcal{A}$ function when two output degrees of freedom are used and by the positive P representation in the case of four output degrees of freedom. These results have been generalized to s-ordered nondiagonal and diagonal measurements by de Oliveira (1992). The measured dispersion of phase has been compared with the dispersion computed from the canonical phase distribution and that computed from the Wigner-function–based phase distribution [Freyberger and Schleich (1993)]. The statistics of relative photon numbers in homodyne detection has been derived in [Vogel and Grabow (1993)], cf. earlier [Braunstein (1990)] for weak fields. There the strong local oscillator limit has been discussed with respect to the electric-field strength measurement taking into account the detector efficiency. The dispersion of phase difference between two microscopic (weak) fields theoretically obtained has been compared with that computed from experimental data. The joint count (relative photon-number) probability in an eight-port homodyne detector has been related to the $\Phi_\mathcal{A}$ function of an arbitrary input state in the strong local oscillator limit in

[Freyberger, Vogel and Schleich (1993a, b)]. Whereas the photocount distribution in [Freyberger, Vogel and Schleich (1993a)] is expressed in terms of the Glauber function $\langle \alpha | \hat{\rho} | \beta \rangle$, it is computed from the $\Phi_{\mathcal{N}}$ function in [Freyberger, Vogel and Schleich (1993b)] just as the formula (4.12.45). Using the P (normal) representation for the quantum states of light instead of a more noisy Q (antinormal) representation for the analysis of the Noh–Fougères–Mandel eight-port homodyne detection scheme, Riegler and Wódkiewicz (1994) have addressed the "noisy" representations of operational phase operators. To attribute the noise to these operators instead of to the states of light seems sometimes to be appropriate. Other notions which are relevant to the strong-field limit of the Noh–Fougères–Mandel scheme can be found in subsection 4.13.2. Obviously, the number-difference–phase uncertainty relation [Fan and Xiao (1997)] is an important complement to the theoretical analysis [Freyberger, Vogel and Schleich (1993b), Freyberger and Schleich (1993)]. Freyberger, Heni and Schleich (1995) have reviewed the operational quantum phase description of Noh, Fougères and Mandel and have shown that in the strong local oscillator limit it emerges as a two-mode theory of phase. The joint quadrature distribution is an appropriately scaled Q function. Thus, the two-mode theory of phase contains the quantum phase of Paul as a special case.

Another objection by Hradil (1993b) has been that the spectrum of the exponential phase-difference operator treated as a subset of the unit circle is not continuous. In fact, this spectrum is dense, but discrete. If the field contained at most s photons, the spectrum would not be dense effectively, it would consist of isolated points. To overcome this difficulty, Noh, Fougères and Mandel (1993a) have proposed to place a variable phase shifter τ in front of the input port. They have recommended to extract the phase difference from the shifted phase-difference measurements. This means an introduction of τ-dependent operators

$$\exp(-i\tau)\exp(-i\tau\hat{n}_1)\widehat{\exp}_M[i(\varphi_1 - \varphi_2)]\exp(i\tau\hat{n}_1) \approx \widehat{\exp}_M[i(\varphi_1 - \varphi_2)]. \qquad (4.12.51)$$

Actually, the eigenvalues of the exponential phase difference operator undergo a shift, but the phase-difference distribution remains phase unchanged as far as the spectrum allows it. A fine variation of the shift enables one to obtain values of the phase difference that cover almost the continuous range. Another solution is provided by the many-port homodyne detection of an optical phase [Raymer, Cooper and Beck (1993)]. The probability distribution of the phase difference between two anticorrelated quantum fields in dependence on the transmittance influencing this anticorrelation has been measured by Fougères, Monken and Mandel (1994). Torgerson and Mandel (1996) have demonstrated using histograms of the experimentally derived probability distributions that the quantum phase difference between two weak coherent fields has not the property $\varphi_1 - \varphi_2 = (\varphi_1 - \varphi_3) - (\varphi_2 - \varphi_3)$, $\varphi_j - \varphi_3$, $j = 1, 2$, being the phase differences measured against the field of a strong local oscillator, φ_3 being its phase. They have observed that the equality is attained in the Pegg–Barnett formalism, but they have not connected its violation with the disregard of the outcomes, which do not lead to a unique value of the phase. This censoring is needed less frequently when a strong local oscillator is used than in the direct approach. Hakioğlu, Shumovsky and Aytür (1994) have proposed and discussed an experimental setup for the measurement of all Stokes parameters mainly concentrating on the case of fully polarized coherent field.

The eight-port scheme of homodyne detection [Walker(1987)] can be analyzed on measurements of different physical quantities described by operators [Luis and Peřina (1996d)] and an interest in simultaneous measurement of some other operators can lead to the extension of the scheme [Luis and Peřina (1996e)]. The sine and cosine operators for the phase difference between two modes of a three-mode system, which resemble the operational cosine and sine, have been proposed [Shumovsky (1997)]. The analysis of the homodyne measurement can be accomplished for an N-mode light field. Differences of scaled photon counts (one reference channel) yield linear combinations of quadratures of the multimode field [Zucchetti, Vogel and Welsch (1996)]. The determination of the two quadratures of a single mode can be realized with a six-port scheme [Walker (1987)]. The operational observables have been discerned from the "intrinsic" ones in quantum mechanics [Englert and Wódkiewicz (1995)]. The phase operator of Turski (1972) has been identified as the intrinsic phase operator mistakenly. The nondemolition principle of the quantum measurement theory [Belavkin (1990)] can be applied not only to the ordinary quantum observables like the position of a quantum particle [Belavkin and Staszewski (1989)], but also to the phase of a quantum field [Belavkin and Bendjaballah (1994)]. This principle has allowed them to describe the time-continuous process (section 3.9) of quantum phase observations and also to study the stochastic dynamics of a quantum oscillator under such measurement.

4.13 Concepts of ideal and feasible phases

The problem of quantum phase measurement and that of simultaneous measurement of the conjugate position-coordinate and momentum operators in the harmonic oscillator have been treated by Helstrom (1976). Each problem can be understood and solved as the Naĭmark extension of the probability-operator measure. Whereas the explicit phase problem ends in the ideal phase concept, the position–momentum problem leads to the feasible quantum phase concept.

According to Davies (1976) (pages 129, 131), we can construct the Naĭmark extension both for the identity resolution

$$\int_{\theta_0}^{\theta_0+2\pi} \hat{\Gamma}(\varphi)\, d\varphi = \hat{1}, \tag{4.13.1}$$

where the phase probability-operator density

$$\hat{\Gamma}(\varphi) = |\varphi\rangle\langle\varphi|, \tag{4.13.2}$$

and for the identity resolution

$$\int_{-\infty}^{\infty} \int_{-\infty}^{\infty} \hat{\Phi}(x,y)\, dx\, dy = \hat{1}, \tag{4.13.3}$$

where the position–momentum probability-operator density

$$\hat{\Phi}(x,y) = \frac{A^2}{4\pi} \left| \frac{A}{2}(x+iy) \right\rangle \left\langle \frac{A}{2}(x+iy) \right|, \tag{4.13.4}$$

with a suitable constant A. Solving the problem of quantum phase measurement according to Davies (1976), we use the phase-state representations (4.6.31) and note that they are elements of the Hilbert space $L^2[\theta_0, \theta_0 + 2\pi)$. The enlargement $\mathcal{H}_e^\Gamma$ of the Hilbert space $\mathcal{H}$ can be identified with the whole Hilbert space $L^2[\theta_0, \theta_0 + 2\pi)$. We see that the probability-operator measure

$$\hat{\Pi}(I) = \int_I \hat{\Gamma}(\varphi)\, d\varphi, \quad I \subset [\theta_0, \theta_0 + 2\pi), \tag{4.13.5}$$

has the property

$$\hat{\Pi}(I)|\psi\rangle = \int_{\theta_0}^{\theta_0+2\pi} \left[\Pi_e^\Gamma(I)\psi \right](\varphi)|\varphi\rangle\, d\varphi, \tag{4.13.6}$$

where the projection-valued measure $\Pi_e^\Gamma(I)$ is defined as follows

$$\left[\Pi_e^\Gamma(I)\psi \right](\varphi) = j_I(\varphi)\psi(\varphi), \tag{4.13.7}$$

with $j_I(\varphi)$ the indicator of the set I.

The orthonormal functions

$$\langle \varphi|n \rangle = \frac{1}{\sqrt{2\pi}} \exp(-in\varphi), \quad n = 0, 1, 2, \ldots, \tag{4.13.8}$$

are not complete in $L^2[\theta_0, \theta_0+2\pi)$, but it is obvious how to extend this "basis". From the physical point of view, the identification of $\mathcal{H}_e^\Gamma$ with $L^2[\theta_0, \theta_0 + 2\pi)$ is not made. The Hilbert space $\mathcal{H}_e$ must be physically motivated as we have shown in section 4.6. The space $L^2[\theta_0, \theta_0+2\pi)$ appears as the space of phase-state representation which is based on the resolution of the identity (4.3.43) and is given in (4.6.32).

In the problem of simultaneous measurement of the position-coordinate and conjugate momentum operators in the harmonic oscillator, we consider the coherent state representations

$$\psi(x,y) = \frac{A}{2\sqrt{\pi}} \left\langle \frac{A}{2}(x+iy) \middle| \psi \right\rangle, \quad \text{any } \psi \in \mathcal{H}, \tag{4.13.9}$$

which are good candidates for wave functions as elements of the Hilbert space $L^2(\mathbb{R}^2)$. As above, the enlargement $\mathcal{H}_e^\Phi$ of the Hilbert space $\mathcal{H}$ can be identified with the whole Hilbert space $L^2(\mathbb{R}^2)$. We see that the probability-operator measure

$$\hat{\Pi}(I) = \int \int_I \hat{\Phi}(x,y)\, dx\, dy, \quad I \subset \mathbb{R}^2, \tag{4.13.10}$$

has the property

$$\hat{\Pi}(I)|\psi\rangle = \frac{A}{2\sqrt{\pi}} \int_{-\infty}^{\infty} \int_{\infty}^{\infty} \left[\Pi_{\mathrm{e}}^{\Phi}(I)\psi\right](x,y) \left|\frac{A}{2}(x+iy)\right\rangle dx\, dy, \qquad (4.13.11)$$

where the projection-valued measure $\Pi_{\mathrm{e}}^{\Phi}(I)$ is defined as follows

$$\left[\Pi_{\mathrm{e}}^{\Phi}(I)\psi\right](x,y) = j_I(x,y)\psi(x,y), \qquad (4.13.12)$$

with $j_I(x,y)$ the indicator of the set I. The orthonormal functions

$$\frac{A}{2\sqrt{\pi}}\left\langle \frac{A}{2}(x+iy)\Big|n\right\rangle = 2^n \frac{A}{\sqrt{\pi}}\frac{1}{\sqrt{n!}}\left[\frac{A}{2}(x-iy)\right]^n \exp\left[-\frac{A^2}{2}(x^2+y^2)\right] \qquad (4.13.13)$$

are not complete in $L^2(\mathbb{R}^2)$ and it is interesting to consider the complete orthogonal functions as follows [Lukš, Peřinová and Křepelka (1994a)]

$$\langle x,y|n_-,n_+\rangle = (-1)^{n_+} 2^{n_--n_+} \frac{A}{\sqrt{\pi}} \sqrt{\frac{n_+!}{n_-!}} \left[\frac{A}{2}(x-iy)\right]^{n_--n_+}$$

$$\times \exp\left(-\frac{A^2}{2}(x^2+y^2)\right) L_{n_+}^{n_--n_+}\left(A^2(x^2+y^2)\right) \quad \text{for } n_- \geq n_+, \qquad (4.13.14)$$

$$\langle x,y|n_-,n_+\rangle = \langle x,y|n_+,n_-\rangle^* \quad \text{for } n_+ \geq n_-, \qquad (4.13.15)$$

where the Laguerre polynomial is defined in (3.4.39). Particularly,

$$\frac{A}{2\sqrt{\pi}}\left\langle \frac{A}{2}(x+iy)\Big|n\right\rangle = \langle x,y|n_- = n, n_+ = 0\rangle. \qquad (4.13.16)$$

From the physical point of view, the identification of $\mathcal{H}_{\mathrm{e}}^{\Phi}$ with the Hilbert space of the wave functions $L^2(\mathbb{R}^2)$ is not necessary. The product Hilbert space $\mathcal{H}_{\mathrm{e}}^{\Phi} = \mathcal{H}_{-+}$ is used (cf. subsection 4.6.1).

A picture of the simultaneous measurement of the position coordinate and the conjugate momentum has been provided by Arthurs and Kelly Jr. (1965). The measuring procedure consists in instantaneously coupling the system under investigation to two properly prepared degrees of freedom, which are readout after the interaction. Under optimum conditions, the Φ_A function of the signal is measured. The inherent additional noise in quantum measurements has been formulated in the generalized Heisenberg uncertainty relation by Arthurs and Goodman (1988) who have derived that the Heisenberg uncertainty lower bound for noncommuting observables is multiplied by four to be used in the generalized Heisenberg uncertainty relation. A further development has been due to Braunstein, Caves and Milburn (1991), as reviewed by Stenholm (1992).

The quantum optical problem of simultaneous detection of quadratures is that of simultaneous measurement of a pair of conjugate observables. The quadratures obtained provide the action and phase-angle variables via a sort of polar transformation. Whereas this formulation is oriented to the processing of measured values, the same prescription leads to definitions of the measured action and phase-angle operators. A procedure of phase measurement has been put forward by Bandilla and Paul (1969, 1970) and reapproached by Schleich, Bandilla and Paul (1992). An attention to the amplification process with respect to the measurement of the phase and the number of quanta of an incoming electromagnetic wave has been paid as early as in 1961 [Louisell, Yariv and Siegman (1961)]. A report on phase measurement of a microscopic radiation field dates back to 1974 [Gerhardt, Büchler and Liftin (1974)]. It was the idea of Nieto (1977) that the analysis should be oriented to the phase difference. Essentially, he has used the symmetrical ordering of the Susskind–Glogower phase operators, which has caused a discrepancy for a near vacuum field. Typically, Lévy-Leblond (1977) has obtained a result related to the normal ordering of the exponential phase operators, which is for the near vacuum fields even slightly worse. A very similar experiment has been performed by Matthys and Jaynes (1980) and the description presented seems to relate to a simultaneous quadrature measurement. This phase-fluctuation measurement has been proposed by Gerhardt, Welling and Frölich (1973), but most subsequent analyses do not seem to have been based on : (i) the measure of phase fluctuation adopted by experimenters, (ii) the details of experiment. Gerry and Urbanski (1990) have completed the analysis which considers that the experimenters measured the phase difference of two modes in coherent states $|\alpha_1\rangle|\alpha_2\rangle$ with the same average photon number $\bar{n} = |\alpha_1|^2 = |\alpha_2|^2$. It still seems that they assess a suitable phase-difference operator for the description and use the Hermitian phase-difference operator of Pegg and Barnett. Based on the same philosophy, almost the same results have been obtained by Lynch (1990). A comprehensive paper has been devoted to the theory of phase measurement [Paul (1974)]. The importance of optical homodyne and heterodyne detection methods is reflected in the papers by Yuen and Shapiro (1978, 1980), Shapiro, Yuen and Machado Mata (1979). Feasible phase measurement schemes treated by Shapiro and Wagner (1984), Shapiro (1985), and Lai and Haus (1989) were based on the simultaneous detection of quadrature components using the heterodyne and homodyne detection techniques. Hradil (1990b) has attempted to derive the number–phase uncertainty relation for phase measurement via simultaneous detection of quadrature operators. The operator of complex amplitude playing the key role in the feasible schemes has been analyzed by Hradil (1992a). The relationship between the simultaneous measurement of quadratures and the $\Phi_{\mathcal{A}}$ quasidistribution and its generalization taking into account the efficiency of photodetectors has been clarified by Leonhardt (1993), Leonhardt and Paul (1993a, b, c, d, 1994a), and by Leonhardt, Vaccaro, Böhmer and Paul (1995). The uncertainty relations for realistic joint measurements of canonically conjugate quadratures have been derived by Leonhardt, Böhmer and Paul (1995). The connection of the simultaneous quadrature measurement with a two-dimensional rotationally symmetrical quantum harmonic oscillator has been illustrated and a comparison

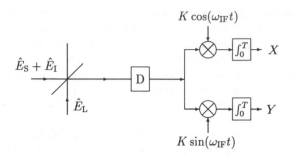

Figure 4.13: Outline of the heterodyne detection scheme.

with a canonical quantum phase system has been provided in [Lukš, Peřinová and Křepelka (1994a)]. Noisy simultaneous measurements of noncommuting observables in eight- and twelve-port homodyne detections with a suggestion for a reconstruction of the Wigner function have been discussed by Luis and Peřina (1996e).

As the experimental measurement of quadratures has been published regularly with the experimentally determined joint quadrature distributions, it became obvious that it would be feasible to acquire a sufficient number of measured distributions of rotated quadratures. Knowledge of the distributions of the rotated quadratures for all shifts within a π-interval is equivalent to the knowledge of the statistical operator [Vogel and Risken (1989)]. It resulted in the quantum tomography [Smithey, Beck, Raymer and Faridani (1993), Smithey, Beck, Cooper, Raymer and Faridani (1993)]. Part of the huge literature on the quantum state reconstruction which followed (e. g., [Leonhardt (1996)]) is still related to quantum phase theory.

4.13.1   Heterodyne detection of quadratures

In optical heterodyne detection [Teich (1968)], the signal field is combined on a beam-splitter with the field of a strong local oscillator laser whose frequency ω_L is offset by an intermediate frequency ω_{IF} from that of the signal field. From the analysis below it follows that the image band field at the frequency shift ω_{IF} from the local oscillator field and $2\omega_{IF}$ from the signal field is needed. Electrical filtering of the photocurrent serves selecting beat frequency components in the vicinity of ω_{IF} and yields an output containing a frequency translated replica of the signal field components that were coherent in space and time with the local oscillator field. A scheme for a quantum treatment of optical heterodyne detection of a single-mode signal field using ideal photon detector is shown in figure 4.13.

A detailed description of this scheme is provided by the electric field operators for quasi-monochromatic plane-wave pulsed signal, local oscillator, and image fields, respectively,

$$\hat{E}_j(x,y,t) = \hat{E}_j^{(+)}(x,y,t) + \hat{E}_j^{(-)}(x,y,t), \quad j = S, L, I, \tag{4.13.17}$$

where $(x,y) \in A_d$, the active region of the detector, whose area is A_d, $t \in [0, T]$, T being the pulse duration. The positive and negative frequency parts of the electric field operators read,

$$\hat{E}_j^{(+)}(x,y,t) = i\sqrt{\frac{\hbar\omega_j}{2A_d cT\epsilon_0}} \hat{a}_j \exp(-i\omega_j t), \tag{4.13.18}$$

$$\hat{E}_j^{(-)}(x,y,t) = \left[\hat{E}_j^{(+)}(x,y,t)\right]^\dagger, \tag{4.13.19}$$

where c is the velocity of light in the vacuum, ϵ_0 is the vacuum permittivity, and

$$\omega_S = \omega_L + \omega_{IF}, \quad \omega_I = \omega_L - \omega_{IF}, \tag{4.13.20}$$

with $\hat{a}_j$, $j = S, L, I$, the appropriate photon annihilation operators. For symmetry, we present the magnetic field operators

$$\hat{H}_j(x,y,t) = \hat{H}_j^{(+)}(x,y,t) + \hat{H}_j^{(-)}(x,y,t), \quad j = S, L, I, \tag{4.13.21}$$

where (μ_0 being the vacuum permeability)

$$\hat{H}_j^{(+)}(x,y,t) = \sqrt{\frac{\epsilon_0}{\mu_0}} \hat{E}_j^{(+)}(x,y,t), \tag{4.13.22}$$

$$\hat{H}_j^{(-)}(x,y,t) = \left[\hat{H}_j^{(+)}(x,y,t)\right]^\dagger. \tag{4.13.23}$$

Let us note that the power-density operator is given as

$$\hat{S}(x,y,t) = \hat{H}^{(-)}(x,y,t)\hat{E}^{(+)}(x,y,t) + \hat{E}^{(-)}(x,y,t)\hat{H}^{(+)}(x,y,t), \tag{4.13.24}$$

where

$$\hat{E}^{(+)}(x,y,t) = t\left[\hat{E}_S^{(+)}(x,y,t) + \hat{E}_I^{(+)}(x,y,t)\right] + r'\hat{E}_L^{(+)}(x,y,t), \tag{4.13.25}$$

$$\hat{E}^{(-)}(x,y,t) = \left[\hat{E}^{(+)}(x,y,t)\right]^\dagger, \tag{4.13.26}$$

$$\hat{H}^{(+)}(x,y,t) = \sqrt{\frac{\epsilon_0}{\mu_0}}\hat{E}^{(+)}(x,y,t), \tag{4.13.27}$$

$$\hat{H}^{(-)}(x,y,t) = \left[\hat{H}^{(+)}(x,y,t)\right]^\dagger. \tag{4.13.28}$$

Taking into account the observation that a photon detector does not respond to the power flux density, we introduce the reduced electric field operators

$$\hat{E}_j'(x,y,t) = \hat{E}_j'^{(+)}(x,y,t) + \hat{E}_j'^{(-)}(x,y,t), \quad j = S, L, I, \tag{4.13.29}$$

where (cf. [Cook (1982), Shapiro and Wagner (1984)])

$$\hat{E}_j''^{(+)}(x, y, t) = i\frac{1}{\sqrt{2A_d c T \epsilon_0}} \hat{a}_j \exp(-i\omega_j t), \tag{4.13.30}$$

$$\hat{E}_j''^{(-)}(x, y, t) = \left[\hat{E}_j''^{(+)}(x, y, t)\right]^\dagger. \tag{4.13.31}$$

The appropriate reduced magnetic field operators read

$$\hat{H}_j'(x, y, t) = \hat{H}_j''^{(+)}(x, y, t) + \hat{H}_j''^{(-)}(x, y, t), \quad j = S, L, I, \tag{4.13.32}$$

where

$$\hat{H}_j''^{(+)}(x, y, t) = \sqrt{\frac{\epsilon_0}{\mu_0}} \hat{E}_j''^{(+)}(x, y, t), \tag{4.13.33}$$

$$\hat{H}_j''^{(-)}(x, y, t) = \left[\hat{H}_j''^{(+)}(x, y, t)\right]^\dagger. \tag{4.13.34}$$

Now the photon-flux density operator is given as

$$\hat{S}'(x, y, t) = \hat{H}''^{(-)}(x, y, t)\hat{E}''^{(+)}(x, y, t) + \hat{E}''^{(-)}(x, y, t)\hat{H}''^{(+)}(x, y, t), \tag{4.13.35}$$

where

$$\hat{E}''^{(+)}(x, y, t) = t\left[\hat{E}_S''^{(+)}(x, y, t) + \hat{E}_I''^{(+)}(x, y, t)\right] + r'\hat{E}_L''^{(+)}(x, y, t), \tag{4.13.36}$$

$$\hat{E}''^{(-)}(x, y, t) = \left[\hat{E}''^{(+)}(x, y, t)\right]^\dagger, \tag{4.13.37}$$

$$\hat{H}''^{(+)}(x, y, t) = \sqrt{\frac{\epsilon_0}{\mu_0}} \hat{E}''^{(+)}(x, y, t), \tag{4.13.38}$$

$$\hat{H}''^{(-)}(x, y, t) = \left[\hat{H}''^{(+)}(x, y, t)\right]^\dagger. \tag{4.13.39}$$

We easily find that

$$A_d \int_0^T \exp(i\omega_{\mathrm{IF}} t)\hat{S}'(x, y, t)\, dt = \frac{|\xi|}{2}\left(\frac{\upsilon}{|\xi|}\hat{a}_L^\dagger \hat{a}_S + \frac{\upsilon^*}{|\xi|}\hat{a}_L \hat{a}_I^\dagger\right). \tag{4.13.40}$$

In the strong-field limit and on the assumption that $\upsilon\xi^* = |\xi|$ we obtain that

$$\frac{2}{|\xi|} A_d \int_0^T \exp(i\omega_{\mathrm{IF}} t)\hat{S}'(x, y, t)\, dt \approx \hat{a}_S + \hat{a}_I^\dagger. \tag{4.13.41}$$

Although the eight-port detection scheme is mainly intended for the homodyne detection (cf. section 4.12), we consider it also for the heterodyne detection without claim of the experimental feasibility. We modify the meaning of the field operators $\hat{a}_k(0), \hat{a}_k(1), \hat{a}_k(2)$, $k = 1, 2, 10, 20$, in the descriptions (4.12.14), (4.12.16), and (4.12.17) not to be the annihilation operators, but sums of three annihilation operators $\hat{a}_k(0) = \hat{a}_k(0, \omega_S) + \hat{a}_k(0, \omega_L) + \hat{a}_k(0, \omega_I)$, etc. Here $\hat{a}_k(0, \omega_j)$, $j = S, L, I$, are the photon annihilation operators describing the modes of frequencies ω_j. In more detail, we assume that $\hat{a}_k(0, \omega_j)$ fulfill appropriately modified beamsplitter equations (4.12.16) and (4.12.17). In other words, we assume for simplicity that $\mathbf{t}, \mathbf{r}', \mathbf{r}, \mathbf{t}'$ are independent of the considered frequencies. The input signal field is described by the annihilation operator $\hat{a}_1(0, \omega_S)$ and the local oscillator field described by $\hat{a}_2(0, \omega_L)$ is in the coherent state $|\xi(t)\rangle$, $\xi(t) = \xi(0) \exp[i\tau(t)]$, $\tau(t) = \tau(0) - \omega_L t$. In analogy to (4.12.30), we obtain the approximation by the strong-field limit ($|\xi(0)| \to \infty, \tau(0) = 0$)

$$\ddot{a}_{|\xi(0)|}(0) \approx \hat{\alpha}(\tau(t), 0) = \exp(i\omega_L t)\hat{a}_1(0) + v \exp(-i\omega_L t)\hat{a}_{10}^\dagger(0), \qquad (4.13.42)$$

where v is given in (4.12.19). We obtain the detected difference of intensities as

$$\hat{X}(\tau(t), 0) = \mathrm{Re}\left[\hat{\alpha}(\tau(t), 0)\right]. \qquad (4.13.43)$$

Introducing the slowly varying annihilation operators

$$\hat{A}_k(0, \omega_j) = \exp(i\omega_j t)\hat{a}_k(0, \omega_j), \qquad (4.13.44)$$

we expand the complex amplitude operator as

$$\begin{aligned}
\hat{\alpha}(\tau(t), 0) &= \exp(-i\omega_{\mathrm{IF}} t)\hat{A}_1(0, \omega_S) + \hat{A}_1(0, \omega_L) + \exp(i\omega_{\mathrm{IF}} t)\hat{A}_1(0, \omega_I) \\
&\quad + v\left[\exp(i\omega_{\mathrm{IF}} t)\hat{A}_{10}^\dagger(0, \omega_S) + \hat{A}_{10}^\dagger(0, \omega_L)\right. \\
&\quad \left. + \exp(-i\omega_{\mathrm{IF}} t)\hat{A}_{10}^\dagger(0, \omega_I)\right].
\end{aligned} \qquad (4.13.45)$$

Observing that

$$\begin{aligned}
\hat{X}(\tau(t), 0) &= \frac{1}{2}\left[\exp(-i\omega_{\mathrm{IF}} t)\hat{A}(0, \omega_{\mathrm{IF}}) + \hat{A}(0, 0)\right. \\
&\quad \left. + \exp(i\omega_{\mathrm{IF}} t)\hat{A}(0, -\omega_{\mathrm{IF}})\right],
\end{aligned} \qquad (4.13.46)$$

where

$$\begin{aligned}
\hat{A}(0, \omega_{\mathrm{IF}}) &= \hat{A}_1(0, \omega_S) + \hat{A}_1^\dagger(0, \omega_I) + v\hat{A}_{10}^\dagger(0, \omega_I) + v^*\hat{A}_{10}(0, \omega_S), \\
\hat{A}(0, 0) &= \hat{A}_1(0, \omega_L) + v\hat{A}_{10}^\dagger(0, \omega_L) + \mathrm{H.\,c.}, \\
\hat{A}(0, -\omega_{\mathrm{IF}}) &= \hat{A}^\dagger(0, \omega_{\mathrm{IF}}),
\end{aligned} \qquad (4.13.47)$$

we have a formal principle of the detection at the electronic stage. To the operators $\hat{X}(\tau(t), 0)$, $\hat{A}(0, \omega_{\mathrm{IF}})$ related are real- and complex-valued stochastic processes. The process $\hat{A}(0, \omega_{\mathrm{IF}})$ can be obtained in the electrical filtering of the intermediate frequency. In comparison with the original analysis by Shapiro and Wagner (1984), we obtain more complicated measured complex amplitude operator with additional noise due to the use of scheme 2 of Noh, Fougères, and Mandel instead of their simpler scheme 1. Leong and Shapiro (1986) have extended the results of Shapiro and Wagner (1984) to include signal and image band fields of a finite bandwidth. They have considered multimode (broadband) squeezed light.

4.13.2   Analogy between the feasible phase operator and the rotation-angle operator

The exposition of the formalism of the quantum optical phase has been provided commencing from the quantum mechanics of a massive particle [Lukš, Peřinová and Křepelka (1994a)]. The clockwise and anticlockwise components of the planar motion are shown to correspond to the signal and image modes in the heterodyne detection scheme. These components are also analogous to some modes in the homodyne detection scheme. Following the paper [Hradil (1993a)], but differently, the formalism of composite ideal phase operator [Rocca and Sirugue (1973), Ban (1991a, b, c, 1992, 1993), Lukš, Peřinová and Křepelka (1994a)] has been embodied in these detection schemes.

The study of the homodyne detection scheme is represented by the analysis of the simplest four-port scheme, i. e., the beamsplitter. Invoking for concreteness the eight-port homodyne scheme (section 4.12), we can identify the four-port scheme with beamsplitter 1. In connection with separate treatment of this beamsplitter, we replace the notation $\hat{a}_1(0), \hat{a}_{10}(0), \hat{a}_1(1), \hat{a}_{10}(1)$ by $\hat{a}_-, \hat{a}_+, \hat{a}_x, \hat{a}_y$, respectively. We conserve the assumption (4.12.15), i. e., the balanced beamsplitter. A generalization to the unbalanced beamsplitter can be found in [Leonhardt (1993)]. For simplicity, specific coefficients in relation (4.12.14) for $k = 1$, $k' = 10$ are chosen. For example, in [Hradil (1992a)]

$$\mathbf{t} = \mathbf{r} = \mathbf{r}' = -\mathbf{t}' = \frac{1}{\sqrt{2}}, \qquad (4.13.48)$$

whereas in [Leonhardt and Paul (1993b)]

$$\mathbf{t} = \mathbf{r} = -\mathbf{r}' = \mathbf{t}' = \frac{1}{\sqrt{2}}. \qquad (4.13.49)$$

We will use

$$\mathbf{t} = \mathbf{r}' = \frac{1}{\sqrt{2}}, \quad \mathbf{r} = -\mathbf{t}' = -\frac{i}{\sqrt{2}}. \qquad (4.13.50)$$

The transformation (4.12.14) with $k = 1$, $k' = 10$ and the choice (4.13.50) is encountered also when the negative and positive circular polarization components of the electric field vector are replaced by its x and y components. This transformation reads

$$\hat{a}_x = \frac{1}{\sqrt{2}}(\hat{a}_- + \hat{a}_+), \quad \hat{a}_y = -\frac{i}{\sqrt{2}}(\hat{a}_- - \hat{a}_+). \qquad (4.13.51)$$

The definition of the exponential rotation-angle operator

$$\widehat{\exp}(i\Phi) = \frac{\mathrm{Re}(\hat{a}_x) + i\mathrm{Re}(\hat{a}_y)}{\sqrt{[\mathrm{Re}(\hat{a}_x)]^2 + [\mathrm{Re}(\hat{a}_y)]^2}} \qquad (4.13.52)$$

presents no problem, because the quadratures $\mathrm{Re}(\hat{a}_x)$, $\mathrm{Re}(\hat{a}_y)$ commute, but the quantum phase related to the operator $\hat{a}_-$ is a more difficult concept, because the quadratures $\mathrm{Re}(\hat{a}_-)$, $\mathrm{Im}(\hat{a}_-)$ have the commutator $[\mathrm{Re}(\hat{a}_-), \mathrm{Im}(\hat{a}_-)] = \frac{i}{2}\hat{1}$. For the quadratures $\mathrm{Re}(\hat{a}_x)$, $\mathrm{Re}(\hat{a}_y)$, the uncertainty relations reads (cf. [Shapiro and Wagner (1984)])

$$\langle(\Delta\mathrm{Re}(\hat{a}_x))^2\rangle\langle(\Delta\mathrm{Re}(\hat{a}_y))^2\rangle \geq \frac{1}{16}, \tag{4.13.53}$$

similarly as [Peřina (1991), p. 118]

$$\langle(\Delta\mathrm{Re}(\hat{a}_-))^2\rangle\langle(\Delta\mathrm{Im}(\hat{a}_-))^2\rangle \geq \frac{1}{16}. \tag{4.13.54}$$

An excess noise origins from the mode $\hat{a}_+$. Of course, it is possible to assume that this mode is in the state of vacuum. This division into signal and noise can be seen from the relations

$$\mathrm{Re}(\hat{a}_x) = \frac{1}{\sqrt{2}}\mathrm{Re}\,\hat{\alpha}, \quad \mathrm{Re}(\hat{a}_y) = \frac{1}{\sqrt{2}}\mathrm{Im}\,\hat{\alpha}, \tag{4.13.55}$$

where the complex-amplitude operator

$$\hat{\alpha} = \hat{a}_- + \hat{a}_+^\dagger. \tag{4.13.56}$$

This operator can be used directly for reexpressing the exponential operator of the rotation angle (4.13.52) (cf. (4.6.18)),

$$\widehat{\exp}\,(i\Phi) = \frac{\hat{\alpha}}{|\hat{\alpha}|}, \tag{4.13.57}$$

where

$$|\hat{\alpha}| = \left(\hat{\alpha}\hat{\alpha}^\dagger\right)^{\frac{1}{2}} = \left(\hat{\alpha}^\dagger\hat{\alpha}\right)^{\frac{1}{2}}. \tag{4.13.58}$$

No problem appears, because the operator whose square root is taken is Hermitian positive definite in (4.13.58) unlike the expression

$$\widehat{\exp}\,(i\Phi) = \left(\frac{\hat{a}_- + \hat{a}_+^\dagger}{\hat{a}_-^\dagger + \hat{a}_+}\right)^{\frac{1}{2}}, \tag{4.13.59}$$

where the fraction

$$\begin{aligned}\left(\frac{\hat{a}_- + \hat{a}_+^\dagger}{\hat{a}_-^\dagger + \hat{a}_+}\right)^{\frac{1}{2}} &= \left(\hat{a}_-^\dagger + \hat{a}_+\right)^{-\frac{1}{2}}\left(\hat{a}_- + \hat{a}_+^\dagger\right)^{\frac{1}{2}} \\ &= \left(\hat{a}_- + \hat{a}_+^\dagger\right)^{\frac{1}{2}}\left(\hat{a}_-^\dagger + \hat{a}_+\right)^{-\frac{1}{2}}.\end{aligned} \tag{4.13.60}$$

In (4.13.60) the definition of the square root is more delicate and related to the number state basis of the input modes.

Using the number operators (4.6.28), we obtain the Susskind–Glogower exponential phase operators (4.6.22), (4.6.23) in the form

$$\widehat{\exp}\,(i\varphi_-) = \left(\hat{n}_- + \hat{1}\right)^{-\frac{1}{2}} \hat{a}_-, \quad \widehat{\exp}\,(i\varphi_+) = \left(\hat{n}_+ + \hat{1}\right)^{-\frac{1}{2}} \hat{a}_+. \tag{4.13.61}$$

The canonically conjugate operator to the feasible phase is the input photon-number difference ((4.6.27)), which can be expressed as

$$\hat{N} = \hat{n}_- - \hat{n}_+ = -i(\hat{a}_x \hat{a}_y^\dagger - \hat{a}_x^\dagger \hat{a}_y). \tag{4.13.62}$$

Let us assume that modes impinging on the input ports are in the state

$$\hat{\rho} = \int \Phi_N^{(-)}(\alpha_-) |\alpha_-\rangle_{--}\langle\alpha_-| \, d^2\alpha_- \otimes |0\rangle_{++}\langle 0|, \tag{4.13.63}$$

where $\Phi_N^{(-)}(\alpha_-)$ is interpreted as the quasidistribution related to the normal ordering of the reduced operators $\hat{a}_-$, $\hat{a}_-^\dagger$. Since the plus mode is in the vacuum state, it holds that $\hat{a}_+ \approx \hat{0}$, i. e., the operator $\hat{a}_+$ behaves approximately like the operator $\hat{0}$. Under the given assumption about $\hat{a}_+$ it holds that $\hat{N} \approx \hat{n}_-$. For the input modes, the resolution of identity reads

$$\hat{1} = \int_{\theta_0}^{\theta_0+2\pi} |\varphi\rangle_{--}\langle\varphi| \otimes \hat{1}_{a_+} \, d\varphi, \tag{4.13.64}$$

where the phase states $|\varphi\rangle_-$ are introduced in (4.6.25). Since $\hat{a}_+ \approx \hat{0}$, we use the second possibility in (4.13.60), the operators antinormally ordered in minus mode and the operators normally ordered in plus mode. This suggests the formal expansion

$$(\hat{a}_+^\dagger + \hat{a}_-)^{\frac{1}{2}}(\hat{a}_+ + \hat{a}_-^\dagger)^{-\frac{1}{2}} = \sum_{m=0}^{\infty}\sum_{m'=0}^{\infty} \binom{\frac{1}{2}}{m}\binom{-\frac{1}{2}}{m'} \hat{a}_+^{\dagger m} \hat{a}_-^{\frac{1}{2}-m} \hat{a}_+^{m'} \hat{a}_-^{\dagger(-\frac{1}{2}-m')}. \tag{4.13.65}$$

As $\hat{a}_+ \approx \hat{0}$, we retain the term with $m = m' = 0$ and observe that

$$(\hat{a}_+^\dagger + \hat{a}_-)^{\frac{1}{2}}(\hat{a}_+ + \hat{a}_-^\dagger)^{-\frac{1}{2}} \approx \hat{a}_-^{\frac{1}{2}} \hat{a}_-^{\dagger(-\frac{1}{2})}. \tag{4.13.66}$$

Confining ourselves to the minus mode, we obtain that

$$\langle n_- | \hat{a}_-^{\frac{1}{2}} \hat{a}_-^{\dagger(-\frac{1}{2})} | n'_-\rangle = \delta_{n_-+1,n'_-} \sqrt{(n_- + 1)_{\frac{1}{2}}(n'_- + 1)_{-\frac{1}{2}}}, \tag{4.13.67}$$

where

$$(x)_\nu = \frac{\Gamma(x + \nu)}{\Gamma(x)}. \tag{4.13.68}$$

This proves the identity with Paul's exponential phase operator [Paul (1974)]

$$\hat{E}_-^{(1)} \equiv \hat{a}_-^{\frac{1}{2}} \hat{a}_-^{\dagger(-\frac{1}{2})}. \tag{4.13.69}$$

More generally, the other exponential operators of Paul read

$$\hat{E}_-^{(k)} \equiv \hat{a}_-^{\frac{k}{2}} \hat{a}_-^{\dagger(-\frac{k}{2})}. \tag{4.13.70}$$

The role of the antinormal ordering is revealed by the mapping theorem for the reduced state $\hat{\rho}_- = {}_+\langle 0|\hat{\rho}|0\rangle_+$ and Paul's exponential phase operators,

$$\text{Tr}\left\{\hat{\rho}_- \hat{E}_-^{(k)}\right\} = \int \alpha_-^{\frac{k}{2}} \alpha_-^{*(-\frac{k}{2})} \Phi_{\mathcal{A}}^{(-)}(\alpha_-)\, d^2\alpha_-, \tag{4.13.71}$$

where the quasidistribution related to the antinormal ordering of the reduced operators $\hat{a}_-$, $\hat{a}_-^{\dagger}$,

$$\Phi_{\mathcal{A}}^{(-)}(\alpha_-) = \frac{1}{\pi} \hat{\rho}_-^{(N)}(\hat{a}_-, \hat{a}_-^{\dagger})|_{\hat{a}_- \to \alpha_-, \hat{a}_-^{\dagger} \to \alpha_-^*}. \tag{4.13.72}$$

To the assumption (4.13.63), the reduced operator is appropriate

$${}_+\langle 0|\widehat{\exp}(i\Phi)|0\rangle_+ = \hat{E}_-^{(1)}. \tag{4.13.73}$$

This proves that the Paul phase concept is a projection of the feasible (realistic) phase concept.

The components $\text{Re}\,\hat{\alpha}$, $\text{Im}\,\hat{\alpha}$ commute and their measurement provides the distribution

$$\Phi_{\text{meas}}(\alpha) = \Phi_{\mathcal{A}}(\alpha). \tag{4.13.74}$$

In general, we assume instead of the formula (4.13.63) that

$$\hat{\rho} = \hat{\rho}_- \otimes \hat{\rho}_+, \tag{4.13.75}$$

where

$$\hat{\rho}_j = \int \Phi_{s_j}^{(j)}(\alpha_j)\hat{\Delta}_{s_j}(\alpha_j)\, d^2\alpha_j, \quad j = -, +; \tag{4.13.76}$$

s_j characterizes the ordering of field operators. The measurement of components $\text{Re}\,\hat{\alpha}$, $\text{Im}\,\hat{\alpha}$ is described by the distribution

$$\Phi_{\text{meas}}(\alpha) = \int \Phi_{s_-}^{(-)}(\beta)\Phi_{s_+}^{(+)}(\alpha^* - \beta^*)\, d^2\beta \tag{4.13.77}$$

for $s_- + s_+ = 0$. Assuming that the statistical operator $\hat{\rho}$ is formed by a pure state and that $\Phi_{\mathcal{N}}^{(-)}(\alpha_-) = \Phi_{\mathcal{N}}(\alpha_-)$, $\Phi_{\mathcal{N}}^{(+)}(\alpha_+) = \Phi_{\mathcal{N}}(\alpha_+^*)$, with a suitable $\Phi_{\mathcal{N}}(\alpha_j)$, we obtain [Leonhardt and Paul (1993b)]

$$\Phi_{\text{meas}}(\alpha) = \frac{\pi}{4}\Phi_{\mathcal{S}}^2\left(\frac{\alpha}{2}\right). \tag{4.13.78}$$

Beside the simplest definition of the measured complex-amplitude operator (4.13.56), it is possible to consider also the operator

$$\text{Re}(\hat{a}_x) + i\text{Im}(\hat{a}_y) = \frac{1}{\sqrt{2}}\hat{a}_1. \tag{4.13.79}$$

In [Lukš, Peřinová and Křepelka (1994a)] $\hat{\alpha}_1$ stands for the present operator $\frac{1}{2}\hat{\alpha}$.

Two limiting procedures performed in [Lukš, Peřinová and Křepelka (1994a)] can be interpreted as an approximate equality of the measured phase to (i) the canonical phase of the signal on the assumption of strong signal mode and weak idler mode or symmetrically to (ii) the conjugate canonical phase of the idler mode for weak signal mode and strong idler mode of radiation. The exponential operator of the feasible phase can be approximated,

$$\widehat{\exp}(i\Phi) \approx (\hat{n}_- \geq \hat{n}_+)\widehat{\exp}(i\varphi_-) + (\hat{n}_- < \hat{n}_+)\widehat{\exp}(-i\varphi_+), \tag{4.13.80}$$

where the symbols in parantheses are operators diagonal in the minus–plus number state basis and enjoying the property

$$(\hat{n}_- \geq \hat{n}_+)|n_-, n_+\rangle = \begin{cases} |n_-, n_+\rangle & \text{if } n_- \geq n_+, \\ 0 & \text{otherwise}, \end{cases}$$

$$(\hat{n}_- < \hat{n}_+)|n_-, n_+\rangle = \begin{cases} |n_-, n_+\rangle & \text{if } n_- < n_+, \\ 0 & \text{otherwise} \end{cases} \tag{4.13.81}$$

and the Susskind–Glogower exponential phase operators are given in (4.13.61). The operator sum on the right-hand side of (4.13.80) enables us to consider any quantum superposition (Schrödinger cat) of the assumptions (i) and (ii). Motivated by the approximate equality (4.13.80), we define the ideal phase operator for the signal mode

$$\widehat{\exp}(i\Phi_{-+}) = (\hat{n}_- \geq \hat{n}_+)\widehat{\exp}(i\varphi_-) + (\hat{n}_- < \hat{n}_+)\widehat{\exp}(-i\varphi_+). \tag{4.13.82}$$

From this

$$\langle n_-, n_+|\widehat{\exp}(i\Phi_{-+})|n'_-, n'_+\rangle = \delta_{1,n'_- - n'_+ - (n_- - n_+)}\delta_{\min(n_-,n_+),\min(n'_-,n'_+)}. \tag{4.13.83}$$

Thus, we have obtained another definition of the unitary exponential phase operator $\widehat{\exp}(i\Phi_{-+})$ first considered by Rocca and Sirugue (1973), thoroughly studied by Ban and expressed by him in terms of the relative number states $\{|n, m\rangle\rangle, -\infty < n < \infty, m \geq 0\}$ [Ban (1991a, b, c, d, 1992)]. The relative number states are connected with the number states $|n_-, n_+\rangle$,

$$|n, m\rangle\rangle = \theta(n)|n_- = m + n, n_+ = m\rangle + \theta(-1 - n)|n_- = m, n_+ = m - n\rangle, \tag{4.13.84}$$

where

$$\theta(n) = \begin{cases} 1 & \text{for } n \geq 0, \\ 0 & \text{for } n < 0. \end{cases} \tag{4.13.85}$$

In this basis, the simultaneous ideal phase and minimum-number states have been introduced by Ban (1991a)

$$|\Phi, m\rangle\rangle = \frac{1}{\sqrt{2\pi}} \sum_{n=-\infty}^{\infty} \exp(in\Phi)|n, m\rangle\rangle. \tag{4.13.86}$$

Rewriting the relation (4.13.83) in Ban's basis, we easily obtain an instance of the more general relation

$$\langle\langle n, m|\widehat{\exp}(ik\Phi_{-+})|n', m'\rangle\rangle = \delta_{k,n'-n}\delta_{mm'}. \tag{4.13.87}$$

The operator $\widehat{\exp}(ik\Phi_{-+})$ has the property

$$[\hat{N}, \widehat{\exp}(ik\Phi_{-+})] = -k\widehat{\exp}(ik\Phi_{-+}), \quad \hat{N} = \hat{n}_- - \hat{n}_+, \tag{4.13.88}$$

which can be derived from the relations

$$[\hat{n}_\mp, \widehat{\exp}(\pm i\varphi_\mp)] = \mp\widehat{\exp}(\pm i\varphi_\mp). \tag{4.13.89}$$

Considering the resolution of identity in the form

$$\hat{1} = \int_{\theta_0}^{\theta_0+2\pi} \hat{1}_{N\Phi_{-+}} \otimes |\varphi\rangle\rangle\langle\langle\varphi| \, d\varphi, \tag{4.13.90}$$

where the phase-sum states

$$|\varphi\rangle\rangle = \frac{1}{\sqrt{2\pi}} \sum_{m=0}^{\infty} \exp(im\varphi)|m\rangle\rangle, \tag{4.13.91}$$

with $|n, m\rangle\rangle = |n\rangle_{N\Phi_{-+}} \otimes |m\rangle\rangle$, we may define any phase-sum operator

$$\hat{M}(\varphi) = \int_{\theta_0}^{\theta_0+2\pi} M(\varphi)\hat{1}_{N\Phi_{-+}} \otimes |\varphi\rangle\rangle\langle\langle\varphi| \, d\varphi. \tag{4.13.92}$$

The eigenstates of the operator $\widehat{\exp}(i\varphi)$ have been determined in [Agarwal (1993)].

The concept of the operator $(\hat{n}_- \geq \hat{n}_+)$ is intuitively clear and its eigenstates $|\psi_-\rangle$,

$$(\hat{n}_- \geq \hat{n}_+)|\psi_-\rangle = |\psi_-\rangle, \tag{4.13.93}$$

enable us to restrict the operators $\hat{M}(\Phi_{-+})$ to the operators $\hat{M}(\varphi_-)$ in the expectations,

$$\langle\psi_-|\hat{M}(\Phi_{-+})|\psi_-\rangle = \langle\psi_-|\hat{M}(\varphi_-)|\psi_-\rangle. \tag{4.13.94}$$

This is a generalization of the result stated by Ban (1993). Similarly, $(\hat{n}_- < \hat{n}_+)$ denotes the operator whose eigenstates $|\psi_+\rangle$,

$$(\hat{n}_- < \hat{n}_+)|\psi_+\rangle = |\psi_+\rangle, \tag{4.13.95}$$

allow us to restrict the operators $\hat{M}(\Phi_{-+})$ to the operators $\hat{M}(-\varphi_+)$ in the quantum averages

$$\langle\psi_+|\hat{M}(\Phi_{-+})|\psi_+\rangle = \langle\psi_+|\hat{M}(-\varphi_+)|\psi_+\rangle. \tag{4.13.96}$$

On the assumption (4.13.63), the reduced operator [Ban (1993)]

$$_+\langle 0|\widehat{\exp}(i\Phi_{-+})|0\rangle_+ = \widehat{\exp}(i\varphi_-). \tag{4.13.97}$$

We see that the Susskind–Glogower phase concept is the projection of the ideal canonical phase concept.

Hradil (1993a) has expressed the ideal phase operator in the form

$$\widehat{\exp}(i\Phi_{-+}) = \hat{u}\widehat{\exp}(i\Phi)\hat{u}^{\dagger}, \qquad (4.13.98)$$

where $\hat{u}$ is a suitable nonunitary operator, but $\hat{u}\hat{u}^{\dagger} = \hat{1}$. Ban (1994) has provided a similar expression

$$\widehat{\exp}(i\Phi_{-+}) = \hat{U}^{\dagger}\widehat{\exp}(i\Phi)\hat{U}, \qquad (4.13.99)$$

where $\hat{U}$ is a suitable unitary operator. A quantum theory of angle with an application to the measurement of the phase difference between the two oscillators has been presented by Shepard (1995).

4.13.3   Reconstruction of the statistical operator

A great attention has been paid to the reconstruction of quantum state of the field or of a particle (concretely, ion). The methods developed serve for an identification of a statistical operator using a quadrature distribution or the number state basis matrix elements. The most popular reconstruction is the optical homodyne tomography [Smithey, Beck, Raymer and Faridani (1993), Smithey, Beck, Cooper, Raymer and Faridani (1993)]. The input data of this procedure are the distributions of the rotated quadratures and the usual Wigner function for the field is obtained. Performing the appropriate transformations, we may arrive at the position-like quadrature representation and number-state representation. From this information the Pegg–Barnett phase distribution and the Wigner phase distribution and moments of the functions of the optical phase can be obtained [Beck, Smithey and Raymer (1993)]. All of the quantities involved in the uncertainty relation for the phase and photon number (cf. (4.7.72)) of a mode of the electromagnetic field have been determined experimentally when the field mode is in a weak coherent state [Smithey, Beck, Cooper and Raymer (1993)].

Using our notation, we may describe the reconstruction scheme as follows. The experiment provides the distributions $\Phi_S^Q(Q,\tau)$ or, which is the same, the distributions

$$\Phi_S^x(x,\tau) = 2\Phi_S^Q(2x,\tau). \qquad (4.13.100)$$

We have to determine the quadrature Wigner function $\Phi_S(\alpha)$ using the inverse Radon transformation [Natterer (1986)]. We may write the reconstruction as the single integral transformation, but the kernel will be a generalized function

$$\Phi_S(\alpha) = \frac{1}{4\pi^2}$$

$$\times \int_0^{\pi} \int_{-\infty}^{\infty} \Phi_S^x(x,\tau) \int_{-\infty}^{\infty} \exp\left\{i\xi\left[x - \frac{1}{2}\left(\alpha e^{-i\tau} + \alpha^* e^{i\tau}\right)\right]\right\} |\xi|\, d\xi\, dx\, d\tau. \qquad (4.13.101)$$

The integration domain $[0, \pi]$ is related to the property

$$\Phi_S^Q(Q, \tau + \pi) = \Phi_S^Q(-Q, \tau). \qquad (4.13.102)$$

However, we shall decompose the reconstruction into two steps. To see these steps more clearly, we assume that the statistical operator has been replaced with an operator which leads to the Wigner function

$$\Phi_S(\alpha) = \delta(\alpha - \beta). \qquad (4.13.103)$$

This choice is a good example of a violation of the Heisenberg uncertainty principle, i. e., we have not chosen a statistical operator. The substitution from (4.13.103) into (4.13.101) would lead to the multiplication of two generalized functions. So we decompose the complex exponential and change the order of integrations

$$\Phi_S(\alpha) = \frac{1}{4\pi^2}$$

$$\times \int_0^\pi \int_{-\pi}^\pi \left[\int_{-\infty}^\infty \Phi_S^x(x, \tau) \exp(i\xi x) \, dx \right] \exp\left[-\frac{i}{2}\xi \left(\alpha e^{-i\tau} + \alpha^* e^{i\tau} \right) \right] |\xi| \, d\xi \, d\tau. \quad (4.13.104)$$

Now the substitution from (4.13.103) into (4.13.104) is quite natural. Let us introduce the characteristic functions

$$C_S^x(\xi, \tau) = \int_{-\infty}^\infty \Phi_S^x(x, \tau) \exp(i\xi x) \, dx \qquad (4.13.105)$$

and denote

$$C(\zeta) = C_S^x(2|\zeta|, \text{Arg}_{-\pi} \zeta). \qquad (4.13.106)$$

Changing the variables $\zeta = 2|\zeta|$, $\tau = \text{Arg}_{-\pi} \zeta$ and using the notation (4.13.105) and (4.13.106), we can rewrite (4.13.104) in the form

$$\Phi_S(\alpha) = \frac{1}{\pi^2} \int C(\zeta) \exp(\alpha\zeta^* - \alpha^*\zeta) \, d^2\zeta. \qquad (4.13.107)$$

In our example,

$$\Phi_S(\alpha) = \frac{1}{\pi^2} \int \exp[(\alpha - \beta)\zeta^* - (\alpha^* - \beta^*)\zeta)] \, d^2\zeta = \delta(\alpha - \beta), \qquad (4.13.108)$$

which is the relation (4.13.103). We have arrived at the three-fold Fourier transformation (4.13.105), (4.13.107), which has been expounded in [Vogel and Risken (1989)].

Upon returning to (4.13.101), the image $\Phi_S(\alpha)$ can be written as a double integral

$$\Phi_S(\alpha) = \frac{1}{\pi^2} \int_0^\pi \int_{-\infty}^\infty \Phi_S^x(x, \tau) K\left(x - \frac{1}{2}\left(\alpha e^{-i\tau} + \alpha^* e^{i\tau} \right) \right) \, dx \, d\tau. \qquad (4.13.109)$$

The kernel $K(z)$ in (4.13.109) is given by

$$K(z) = -\frac{1}{2}\mathcal{P}\frac{1}{z^2}, \tag{4.13.110}$$

where $\mathcal{P}$ means that this generalized function is the limit of $\frac{1}{z^2}$, which has been replaced by $-\frac{1}{\varepsilon^2}$ in symmetrical ε-neighbourhoods of $z = 0$ as $\varepsilon \to +0$. Integrating equation (4.13.109) by parts, we obtain the filtering procedure

$$\Phi_S(\alpha) = \frac{1}{2\pi^2}\int_0^\pi \mathrm{V.\,p.}\int_{-\infty}^\infty \frac{\frac{\partial}{\partial x}\Phi_S^x(x,\tau)}{x - \frac{1}{2}\left(\alpha e^{-i\tau} + \alpha^* e^{i\tau}\right)}\,dx\,d\tau, \tag{4.13.111}$$

where V. p. stands for the Cauchy principal value. From the definition of the Wigner function it follows that the quadrature representation of the statistical operator with respect to an unrotated quadrature can be simply reconstructed from the rotated quadrature distributions [Kühn, Welsch and Vogel (1994)]. It is obvious that the reconstruction of the Wigner function is particularly interesting, because the Wigner random complex amplitude is unmeasurable. To the contrary, the eight-port homodyne detection scheme yields the Husimi random complex amplitude so that the reconstruction of the Husimi function is not so much necessary. Nevertheless, the usual reasoning that the Wigner random complex amplitude cannot exist, because some Wigner functions are negative somewhere, has been reverted. For the Gaussian states, which are known to have nonnegative Wigner functions, in fact for two-photon coherent states, a modified homodyne detection scheme yielding presumably the Wigner random complex amplitude has been studied [Bandilla and Ritze (1993)]. D'Ariano, Leonhardt and Paul (1995) have addressed the problem of experimentally 'sampling' a general matrix element $\langle\psi|\hat\rho|\varphi\rangle$. They have attempted to relate a bound on the quantum efficiency η of detectors to "resolutions" of basis vectors. They have also shown that experimental sampling is also possible for non-unit efficiency η at the detectors provided that η satisfies a lower bound. The final version [Leonhardt, Munroe, Kiss, Richter and Raymer (1996)] of the sampling method for the Fock basis was the result of some mathematical effort [D'Ariano, Leonhardt and Paul (1995), Leonhardt, Paul and D'Ariano (1995), Paul, Leonhardt and D'Ariano (1995), Richter (1996c)]. In the proposal of D'Ariano, Leonhardt and Paul (1995), a low efficiency of the detector is taken into account by the deconvolution woven in the reconstruction algorithm. Kiss, Herzog and Leonhardt (1995) has focused on the matrix representation of the statistical operator in the Fock basis and have derived the properties of a separate deconvolution procedure as those of the inverse generalized Bernoulli transformation. Numerical results for measured probabilities have been given and the mechanism underlying the generation of statistical errors in the measured matrix has been illustrated [D'Ariano (1995)]. Leonhardt and Paul (1994b) have pointed out some difficulties in reconstructing the Wigner function from experimentally determined smoothed distributions. These consist in a necessary limitation of the bandwidth of the characteristic function. They have confirmed that the reduced bandwidth affects our ability to distinguish between quantum superpositions and statistical mixtures [Dutra, Knight and Moya-Cessa (1993)]. The quadrature noise of homodyne tomography depends on the injected signal and can be low for a broadband squeezed vacuum with the phase of the squeezing locked to the local oscillator [D'Ariano, Mancini and Tombesi (1997)].

In the case of the truncated Hilbert space $\mathcal{H}_s$, it has been found that the number of rotated quadratures for accurate state reconstruction is $(s+1)$ when the equidistant phases are assumed [Leonhardt and Munroe (1996)].

A modification of the scheme for photon-number nondemolition measurement can yield the Wigner function of the field as a difference between two probabilities [Lutterbach and Davidovich (1997)]. Since the photon-number distribution is sufficient for the determination of the s-parametrized quasidistribution for $\alpha = 0$, the unbalanced homodyne detection scheme for the reconstruction of the distribution at each point of the phase space has been proposed [Wallentowitz and Vogel (1996)] and an analogous scheme has been adopted by Banaszek and Wódkiewicz (1996). It has been shown by Richter (1996b) that each normally ordered field moment can be obtained by averaging the rotated quadrature distributions with pattern functions, which are simply Hermite polynomials multiplied by an exponential phase factor. The reconstruction of the statistical operator from the rotated quadrature components via the normally ordered moments of the field has been investigated [Wünsche (1996a)]. It has been shown how arbitrary normally ordered moments of order n can be obtained from the rotated quadrature components for $n + 1$ discrete angles (phases).

The photon-number tomography has been proposed by Mancini, Tombesi and Man'ko (1997) as the reconstruction of the original statistical operator from the photon-number distributions of the displaced statistical operators. In these papers, unbalanced homodyne detection for measuring the quantum state of optical single-mode fields has been suggested. The photon-number distribution of the displaced signal can be used to determine the values of the s-parametrized quasidistributions. Opatrný and Welsch (1997) have presented a method for direct sampling of the Fock-basis matrix elements of the statistical operator and have shown that it is not necessary to displace by distinct moduli and it is sufficient to change only the phase.

The reconstruction of a pure state is specific. The infinite collection of the rotated quadrature distributions for the optical homodyne tomography can then be dramatically reduced, if a two-fold ambiguity of the solution is acceptable. Essentially, Pauli's problem [Pauli (1980)] of quantum mechanics is solved [Orłowski and Paul (1994)]. A similar problem for the number and phase distributions has been studied by Białynicka-Birula and Białynicki-Birula (1994) and Vaccaro and Barnett (1995). Beyond the homodyne detection, Bardroff, Mayr and Schleich (1995) have presented a scheme to measure the vector of a pure quantum state of the radiation field in a cavity. They have based the "quantum state endoscopy" on the interaction of a beam of two-level atoms with the resonant light field. The quantum state endoscopy is complementary to the quantum state tomography in the same sense as photon chopping of Paul, Törmä, Kiss and Jex (1996), who have proposed the use of a balanced $2N$-port as a new way to measure the pure quantum state of a single mode light field. Freyberger and Herkommer (1994) have proposed a simple setup based on the deflection of atoms from a single-mode radiation field. The proposed method uses a beam of two-level atoms and the atomic diffraction pattern resulting from a one-photon interaction provides information about the state of the field and also reveals the expectation value of the phase operator $\widehat{\exp}(i\phi)$. Advantages of using the two-photon interaction have been pointed out by Baseia, Vyas, Dantas and Bagnato (1994). Turning back to the quadratures, we can replace the two conjugate quadratures (the momentum-like and the position-like quadratures) by two quadratures differing by an infinitesimal rotation. In principle, this problem does not suffer from the ambiguity [Richter (1996a)].

The comparison of the optical homodyne tomography with the famous Pauli problem reveals that the latter involves the restriction of the space of physical states, viz., the states were not allowed to be mixed. It is possible to consider as an example a state space which includes pure states described by a finite number of parameters [Törmä (1996)]. Leonhardt and Jex (1994) have derived surprisingly simple relations between the Wigner function and any rotated quadrature distribution for quantum oscillator with random phase. The Jaynes principle of maximum entropy [Jaynes (1957a, b)] has been applied for the definition of classes of the quantum states of light on different observation levels in the framework of reconstructions of their Wigner functions [Bužek, Adam and Drobný (1996a, b)]. It has been demonstrated that the detection of quantum coherence depends on the choice of the observation level. The logical connection between three reconstruction schemes, namely, quantum Bayesian inference, reconstruction via the Jaynes principle of maximum entropy, and discrete quantum tomography, has been discussed in [Derka, Bužek, Adam and Knight (1996)].

Whereas in the first experiments the matrix elements in the number state basis with subsequent determination of quantum phase statistics have been obtained via the reconstructed Wigner function [Beck, Smithey and Raymer (1993), Beck, Smithey, Cooper and Raymer (1993), Smithey, Beck, Cooper and Raymer (1993)], recently methods of their direct sampling have been suggested and successfully applied [Vogel and Welsch (1994), D'Ariano, Machiavello and Paris (1994a, b)]. Similarly, Kühn, Welsch and Vogel (1994) have derived formulae for reconstructing the statistical operator in the position-like quadrature representation. A general method for calculating matrix elements of the statistical operator of a single light mode in optical homodyne tomography has been proposed [Leonhardt, Paul and D'Ariano (1995)]. The matrix elements are obtained by averaging pattern functions with respect to the homodyne data. The pattern functions for the coherent state and Fock representations have been presented.

Interesting experimental results in optical tomography of a highly squeezed vacuum state were presented by Schiller, Pereira, Breitenbach, Müller, White and Mlynek (1996). Quantum statistics and oscillations in the photon-number distributions were experimentally observed for squeezed vacuum generated by a continuous-wave optical parametric amplifier using homodyne detection and direct sampling of the single-mode matrix elements from the measured data [Schiller, Breitenbach, Pereira, Müller and Mlynek (1996)]. The subtle direct sampling of the statistical operator matrix elements in any fixed quadrature basis has been developed by Zucchetti, Vogel, Tasche and Welsch (1996). The attention has also been paid to the direct sampling of canonical phase distribution ((4.6.102))[Dakna, Knöll and Welsch (1997a)] and the Susskind–Glogower phase distributions [Dakna, Knöll and Welsch (1997b)]

$$P^{\cos}(\varphi) = \langle \cos \varphi | \hat{\rho} | \cos \varphi \rangle, \quad \varphi \in [0, \pi],$$
$$P^{\sin}(\varphi) = \langle \sin \varphi | \hat{\rho} | \sin \varphi \rangle, \quad \varphi \in \left[-\frac{\pi}{2}, \frac{\pi}{2} \right], \qquad (4.13.112)$$

where

$$| \cos \varphi \rangle = \sqrt{\sin \varphi} \, | C = \cos \varphi \rangle,$$
$$| \sin \varphi \rangle = \sqrt{\cos \varphi} \, | S = \sin \varphi \rangle, \qquad (4.13.113)$$

with $|C\rangle$, $|S\rangle$ defined in (4.5.32), (4.5.34), respectively.

Most direct method of sampling of a phase distribution is implicit in [Vogel and Schleich (1991)], where the collection $\{ \Phi_S^P(0, \varphi) \}$ has been proposed as π-periodical (unnormalized) phase distribution. The comparison of the Vogel–Schleich phase distribution with the Noh–Fougères–Mandel measurements using the first-order phase dispersion is not possible. Not being aware of the dispersion of the second order, Lynch (1993) has modified the usual dispersion to involve the absolute value of the function cos φ and has found an excellent agreement. Bužek and Adam (1995) have presented a study of SU(1,1) generalized coherent states in the framework of an operational approach to phase distributions of quantum states of a single mode of radiation. They discussed very appealing phase properties of the coherent phase states. They found a time independence of the atomic inversion in the Jaynes–Cummings model and a phase locking effect provided that initially the atomic phase is equal to the preferred phase of the generalized SU(1,1) coherent state under consideration. A generalization of the Vogel–Schleich formalism, which in principle overcomes the π-periodicity of the phase distribution, has been proposed using displaced squeezed states as operational phase states [Bužek and Hillery (1996)].

The optical homodyne tomography has been extended to the detection of multimode light. The problem of reconstruction of the quantum state of two correlated modes has been considered by M. G. Raymer, D. T. Smithey, M. Beck, M. Anderson, and D. F. McAlister. A detailed analysis of the N-mode case has been provided [Kühn, Welsch and Vogel (1995), Vogel and Welsch (1995), Zucchetti, Vogel and Welsch (1996)]. Generalizations of the direct sampling method of state reconstruction for separable and nonseparable two modes have been presented by Raymer,

McAlister and Leonhardt (1996). Reconstructing the density operator by using an overcomplete set of linearly transformed quadratures has been studied along with the two-mode tomography [D'Ariano, Mancini, Man'ko and Tombesi (1996)]. The N-mode fields present a convenient framework for optical homodyne experiments [Munroe, Boggavarapu, Anderson and Raymer (1995)] using local oscillator pulses that are short compared with the signal pulses. The train of these pulses is used to introduce a set of nonmonochromatic modes of the signal pulse at different times in it. Whereas this experiment provided only the photon-number statistics of the separate modes, the following theoretical study proposed determining their correlation properties [Opatrný, Welsch and Vogel (1997a, b)]. The above techniques of the quantum state reconstruction are linear and as such they hardly guarantee the positive definiteness of the statistical operator. The reconstruction of a pure state is not prone to this error. An algorithm for quantum state estimation based on the maximum-likelihood method has been suggested by Hradil (1997). The proposed method uses no other prior knowledge on the state, besides it is pure, and is appropriate to the cases where the data seem to be "underdetermining."

The Wigner function tomography, which allows one to measure also the Q function without performing repeated simultaneous quadrature measurements, and the simultaneous measurement of conjugate quadratures, which "ideally" leads to a Q function and thus it enables one to determine the Wigner function using an "ideal" deconvolution, have been treated in [D'Ariano (1997a, b), Tombesi (1997), Leonhardt (1997)]. Many papers devoted to the method of quantum state tomography have been collected along with other work on optical field state reconstruction in special issue of Journal of Modern Optics [Schleich and Raymer (1997)].

Chapter 5

Phase-shift measurements and phase dependence

In this chapter we treat the applicability of principles described in the above chapters to quantum optical measurements of a "classical" parameter. In particular, we discuss quantum interferometry, in which only a few photons or particles in an interferometer ensure its operation, and quantum nondemolition measurements, in which the optical field is detected without photon absorption. We expound some knowledge from the quantum estimation theory and treat quantum interferometry as a detection of phase shift. We summarize here knowledge how the short-time behaviour of nonlinear optical processes depends on the phases of initial coherent states when quantum description is adopted and complete it with results on the quantum phase dynamics in these processes. We also discuss the quantum interferometry using such particles as electrons, neutrons, atoms, and ions.

5.1   Quantum light interferometry

In chapter 2 we have described interference of strong classical optical beams and its relation to coherence. As we have seen the interference phenomenon is naturally explained in terms of classical wave picture, whereas it is not so obvious in quantum particle picture. Nevertheless, difficulties with quantum description of interference are balanced by deeper physical insight in this process based on principles of quantum theory. Assuming waves of probability amplitude, also particle-like theory of interference can be constructed, which is then capable to explain interference of low intensity beams, in which quantum phenomena play important role. Such an approach is a basis for explanation of photon interference as given by Dirac (1958) stating that "each photon interferes with itself. Interference between two different photons never occurs" (for a discussion, see [Mandel (1976)]). Such a point of view is necessarily related to standard two-beam interferometers detecting single photons when the interference is considered as providing the maximum visibility fringes. Taking into account partially coherent beams with the reduced visibility of interference pattern, broader space is open for explanation of single-photon interference. The observation of the interference pattern is related, from the viewpoint of quantum theory, to uncertainties in observables determined by commutation relations for the corresponding operators (time–energy, coordinate–momentum, photon number–phase, etc.). The wave–particle duality and complementarity principle play the principal role here. If a particle cannot be distinguished with respect to these uncertainties, the interference pattern may be observed, because the number of particles is uncertain in the measurement and a phase difference may be certain, which enables us to observe the interference. If particles in the interferometer can, in principle, be distinguished, a number state with determined number of particles is appropriate for the description, the corresponding phase difference has to be uncertain and the interference pattern is missing. Much broader varieties of interference and correlation phenomena are arising if higher-order correlation measurements of the Hanbury Brown–Twiss type are adopted and multiphoton detections are considered.

Consider a single-photon two-mode optical field described by the state

$$|\psi_1\rangle = \hat{b}^\dagger|0\rangle = \cos\theta\,|1,0\rangle + \sin\theta\,|0,1\rangle, \qquad (5.1.1)$$

where the photon creation operator is given by

$$\hat{b}^\dagger = \hat{a}_1^\dagger\cos\theta + \hat{a}_2^\dagger\sin\theta, \qquad (5.1.2)$$

θ being the angle between interfering beams described by the photon annihilation operators $\hat{a}_1$ and $\hat{a}_2$, respectively, and it holds that $[\hat{b},\hat{b}^\dagger] = \hat{1}$. The probability of detection of a photon at a space point $\mathbf{x}$ at time t in this field is proportional to the quantum expectation value of the operator of field intensity as follows

$$
\begin{aligned}
I(\mathbf{x},t) &= \cos^2\theta\,\langle 1,0|\hat{a}_1^\dagger\hat{a}_1|1,0\rangle + \sin^2\theta\,\langle 0,1|\hat{a}_2^\dagger\hat{a}_2|0,1\rangle \\
&\quad + \sin(2\theta)\cos\phi\,\langle 1,0|\hat{a}_1^\dagger\hat{a}_2|0,1\rangle \\
&= 1 + \sin(2\theta)\cos\phi, \qquad (5.1.3)
\end{aligned}
$$

where we have used the definition of annihilation and creation photon operators and the orthonormality of the number states; ϕ represents the phase difference of the mode fields. If $\theta = \pi/4$ we have the maximum fringe visibility and it is apparent that an interference pattern may be built up from a succession of single-photon interference events. This supports the interpretation that the interference arises from a single photon interfering with itself. In the case of two independent sources, the interference pattern is observable if one cannot distinguish, as a consequence of the Heisenberg uncertainty relations, from which source a given photon came. Localization (with the accuracy of the wavelength) of a photon at a space-time point $(\mathbf{x}, t)$ with the resolution better than one fringe width undetermines the photon momentum to such an extent that it is impossible to ascribe the photon to either of the two sources separately. This can nicely be demonstrated, for instance, by using an interference experiment with two atoms [Eichmann, Bergquist, Bollinger, Gilligan, Itano, Wineland and Raizen (1993)]. The authors used two ^{198}Hg$^+$ ions as sources of scattered light for the intereference of the Young type. They were able to observe interference fringes in the light scattered from two localized atoms driven by a weak laser field provided that both the atoms radiated from the same state, so that it was not possible to distinguish from which atom a given photon came to the observation screen. If the atoms radiated from different states, no interference pattern occurred and only a noise was observed. A theoretical description of this experiment has been developed by Wong, Tan, Collett and Walls (1997) treating the atoms as independent radiators synchronized by the phase of the incident laser field, using computer simulations and "which path" considerations.

Let us consider the n-photon state (the SU(2) generalized coherent state)

$$|\psi_n\rangle = \frac{\hat{b}^{\dagger n}}{\sqrt{n!}}|0\rangle = \sum_{j=0}^{n}(\cos\theta)^j(\sin\theta)^{n-j}\sqrt{\frac{n!}{j!(n-j)!}}\,|j,n-j\rangle. \qquad (5.1.4)$$

Assuming the two-photon state again under the phase condition $\theta = \pi/4$

$$|\psi_2\rangle = \frac{1}{2}[|0,2\rangle + \sqrt{2}|1,1\rangle + |2,0\rangle], \qquad (5.1.5)$$

then for the quantum expectation value of the intensity operator we have

$$\begin{aligned}
I(\mathbf{x},t) &= \frac{1}{4}[\langle 2,0|\hat{a}_1^\dagger\hat{a}_1|2,0\rangle + \langle 0,2|\hat{a}_2^\dagger\hat{a}_2|0,2\rangle \\
&\quad +2\langle 1,1|\hat{a}_1^\dagger\hat{a}_1|1,1\rangle + 2\langle 1,1|\hat{a}_2^\dagger\hat{a}_2|1,1\rangle \\
&\quad +4\sqrt{2}\cos\phi\,\langle 0,2|\hat{a}_2^\dagger\hat{a}_1|1,1\rangle] \\
&= 2[1+\cos\phi], \qquad (5.1.6)
\end{aligned}$$

providing again the maximum visibility pattern. If we consider the second-order interference and a single-photon detection, then in order to have a non-zero value of the second-order correlation function and observable interference, it is necessary that the states of single beams differ by one photon. Such a state containing n photons can be written in the form

$$|\psi_n\rangle = \frac{1}{\sqrt{2}}[|n,n-1\rangle + |n-1,n\rangle]. \qquad (5.1.7)$$

The two-beam interference leads to the interference law in the form (2.1.11), where the single-beam mean intensities are given as $I_j = |\alpha_j|^2(n-1/2)$, $j = 1,2$, α_j are the corresponding complex amplitudes of single beams and the degree of coherence is $\gamma = n\alpha_1^*\alpha_2/[(2n-1)|\alpha_1||\alpha_2|]$. Hence, such an n-photon optical field produces interference with the modulus of the degree of coherence (visibility) $|\gamma| = n/(2n-1)$,

i. e., such a field is partially coherent. In agreement with the above discussion, a single-photon field is coherent, because $|\gamma| = 1$ in this case, providing the maximum visibility (interference of a photon with itself) and the result is the same as with the coherent state involving cooperation of photons in an infinite superposition of weighted occupation number states. For $n \to \infty$, the modulus of the degree of coherence $|\gamma|$ tends to $1/2$ and the resulting field is partially coherent, reflecting the possibility of interference of different photons.

Single-photon interference was both experimentally and theoretically analyzed in a series of papers by Hariharan, Brown and Sanders (1993a, b), Hariharan, Roy, Robinson and O'Byrne (1993) and Hariharan, Fujima, Brown and Sanders (1993). They concluded that single-photon interference is appropriate to explain their experimental results and it is to be attributed to the formation of quantum superposition states. The role of vacuum state for this interference is crucial and was discussed by Agarwal and Hariharan (1993). Till now, all interference experiments including those with single photons of nonclassical light are in full agreement with quantum theory.

The simplest case to be followed is description of a beamsplitter specified by the intensity transmittance $\mathcal{T}$ and the intensity reflectance $\mathcal{R}$. The intensity of the transmitted beam is $I_t = \mathcal{T}I$ and the intensity of the reflected beam is $I_r = \mathcal{R}I = (1-\mathcal{T})I$, I being the incident light intensity. In terms of single photons, a probabilistic interpretation must be adopted: the probability that a photon is transmitted is equal to $\mathcal{T}$ and the probability that a photon is reflected is equal to $\mathcal{R} = 1 - \mathcal{T}$. With $\langle n \rangle$ denoting the mean number of photons incident on the beamsplitter during a detection time T, the mean number of transmitted photons is $\mathcal{T}\langle n \rangle$, whereas the number of reflected photons in that time is $\mathcal{R}\langle n \rangle = (1 - \mathcal{T})\langle n \rangle$. The photon-number statistics under the random partitioning by the beamsplitter is given as follows: if the incident beam consists of n photons, then the probability $p(m)$ that m photons are transmitted is equal to

$$p(m) = \frac{n!}{m!(n-m)!}\mathcal{T}^m(1 - \mathcal{T})^{n-m}, \quad m = 0, 1, ..., n. \quad (5.1.8)$$

The mean number of transmitted photons is $\langle m \rangle = \mathcal{T}n$ and for the photon variance we have

$$\langle (\Delta m)^2 \rangle = \mathcal{T}(1 - \mathcal{T})n = (1 - \mathcal{T})\langle m \rangle. \quad (5.1.9)$$

For large intensities, photons will be partitioned between the two transmitted and reflected beams according to classical optics of strong beams with the transmittance $\mathcal{T}$ and the reflectance $1 - \mathcal{T}$, respectively. If the photon-number distribution of the incident light is $p_0(n)$, we obtain the Bernoulli transformation ((3.7.3)) for the probability of finding m photons transmitted through the beamsplitter

$$p(m) = \sum_{n=m}^{\infty} \frac{n!}{m!(n-m)!}\mathcal{T}^m(1 - \mathcal{T})^{n-m}p_0(n), \quad m = 0, 1, 2, ... \quad (5.1.10)$$

It is easy to show that for typical non-pathological distributions important for description of real optical systems, such as the Poisson distribution, the Bose–Einstein distribution, the distribution describing the superposition of signal and noise, etc., the distributions $p(m)$ and $p_0(n)$ are of the same class, the former having the reduced mean number $\langle m \rangle = T \langle n \rangle$ by the transmittance T. On the other hand, for the number state $|N\rangle$, it holds that $p_0(n) = \delta_{nN}$ and the form of the distribution is substantially changed if $T < 1$. Equation (5.1.10) is also correct for the photodetection process, where the transmittance T is replaced by the photodetection efficiency η (section 3.7).

If a single photon is incident on the beamsplitter, anticorrelation of photon numbers occurs between outputs of the photodetectors placed in the beams. A general quantum theory of lossless beamsplitters was developed by Campos, Saleh and Teich (1989). Assume two separate photons are coincidently falling on the 50/50 beamsplitter along two different entrance paths. Let $p(n, m)$ be the probability that n and m photons are in the one and the other outgoing beams, respectively. Then for bosons (photons) exhibiting bunching effect, it holds that $p(2, 0) = p(0, 2) = 1/2, p(1, 1) = 0$, whereas for fermions exhibiting antibunching effect, $p(2, 0) = p(0, 2) = 0, p(1, 1) = 1$ as a consequence of the exclusion principle [Loudon (1989), Campos (1994)].

In quantum interferometry, low intensity beams are used so that quantum properties of light beams play substantial role. We can distinguish between single-photon quantum interferometry and two-photon quantum interferometry, involving pairs of photons from the nonlinear process of frequency down-conversion. If a strong pumping beam of a frequency ω_p is incident on a nonlinear crystal specified by the second order susceptibility $\chi^{(2)}$, we can describe this nonlinear interaction by the effective interaction Hamiltonian

$$\hat{H}_{\text{int}} = -\hbar g \hat{a}_1 \hat{a}_2 \exp(i\omega_p t - i\phi) - \text{H. c.,} \tag{5.1.11}$$

where g is a real coupling constant proportional to the susceptibility $\chi^{(2)}$ including also the real pumping amplitude, ϕ is the pumping phase, $\hat{a}_1$ and $\hat{a}_2$ are the annihilation operators of the sub-frequency modes possessing frequencies ω_1 and ω_2, respectively. Both the sub-frequency photons are strongly coupled by the energy and momentum conservation laws:

$$\omega_p = \omega_1 + \omega_2, \quad \mathbf{k}_p = \mathbf{k}_1 + \mathbf{k}_2. \tag{5.1.12}$$

This is a nondegenerate version of the optical parametric process of the second-subharmonic generation examined in section 3.8. The corresponding Heisenberg equations are

$$\frac{d\hat{a}_1}{dt} = -i\omega_1 \hat{a}_1 + ig\hat{a}_2^\dagger \exp(-i\omega_p t + i\phi),$$

$$\frac{d\hat{a}_2}{dt} = -i\omega_2 \hat{a}_2 + ig\hat{a}_1^\dagger \exp(-i\omega_p t + i\phi), \tag{5.1.13}$$

which are solved in the form

$$\hat{a}_1(t) = \exp(-i\omega_1 t)[\hat{a}_1(0)\cosh(gt) + i\hat{a}_2^\dagger(0)\exp(i\phi)\sinh(gt)],$$

$$\hat{a}_2(t) = \exp(-i\omega_2 t)[\hat{a}_2(0)\cosh(gt) + i\hat{a}_1^\dagger(0)\exp(i\phi)\sinh(gt)]. \tag{5.1.14}$$

From these solutions we see that the sub-frequency modes are not correlated in the sense of standard second-order interferometric measurements, because

$$\langle \hat{a}_1^\dagger(t)\hat{a}_2(t)\rangle = 0, \tag{5.1.15}$$

but they are correlated in the sense that

$$\langle \hat{a}_1(t)\hat{a}_2(t)\rangle = \frac{i}{2}\exp(-i\omega_p t + i\phi)\sinh(2gt) \neq 0, \quad t > 0. \tag{5.1.16}$$

This type of correlations is also reflected in higher-order correlation effects, such as coincidence of counts from two photodetectors, i. e., in the Hanbury Brown–Twiss correlations. However, if one of the down-converted beams is reflected from a phase-conjugating mirror, the standard correlations between the signal and idler beams can be observed again [Monken, Garuccio, Branning, Torgerson, Narducci and Mandel (1996)]. In real experiments spatial–temporal coherence properties of such quantum entangled light beams generated in parametric down-converters play important role [Joobeur, Saleh and Teich (1994), Joobeur, Saleh, Larchuk and Teich (1996), Ribeiro, Pádua, Machado da Silva and Barbosa (1994), Ribeiro and Barbosa (1996), Barbosa (1996), Řeháček and Peřina (1996a)].

If we start our discussion with single-photon quantum interferometry, we can mention single-photon experiments by Grangier, Roger and Aspect (1986), Aspect, Grangier and Roger (1989), and Aspect and Grangier (1991) demonstrating, by cumulative realizations of single-photon interference, the formation of classical interference picture, which is in agreement with the probabilistic interpretation of the wave function and with the principle of correspondence of quantum and classical descriptions of optical beams.

These authors used the Mach–Zehnder interferometer (figure 5.1) controlling that only one photon was inside the interferometer using a gate system. The path difference was discretized to 256 points of the distance $\lambda/50$. The integration time of photodetectors situated in both the beams was 1 s and 15 s in every channel. With increasing counting time the classical interference pattern was achieved from single detections with the visibility $C = 0.987 \pm 0.005$. As a reflection of the character of single photons, the anticorrelation of responses from both the photodetectors occurred, which was demonstrated by displaying that at the places of the maxima of counts at one photodetector there were minima at the other photodetector and vice versa. These experiments demonstrate the wave–particle duality for a single photon. They verify predictions of quantum theory in relation to other nonclassical effects, such as sub-Poissonian photon statistics, photon antibunching, squeezing of vacuum fluctuations, violation of classical and Bell's inequalities, oscillations of quantum origin in photocount distributions, collapses and revivals of atomic oscillations, etc.

New areas are open for investigations of optical properties of light beams by two-photon interferometry employing down-converted pairs of photons. Consider a lossless beamsplitter (figure 5.2) described by amplitude transmittance t and amplitude reflectance r and two coincidently incoming photons, each in one port. Then

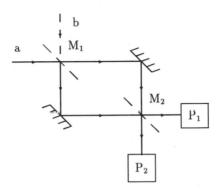

Figure 5.1: Scheme of Mach–Zehnder interferometer; a is a signal beam, b is a reference beam (if no reference beam is introduced, vacuum fluctuations are present in this entrance), M_1, M_2 are beamsplitters, and P_1, P_2 are photodetectors.

the outgoing field is described by the quantum state [Hong, Ou and Mandel (1987), Loudon (1989), Campos, Saleh and Teich (1989), Campos (1994)]

$$|\psi\rangle = (|t|^2 - |r|^2)|1, 1\rangle + i\sqrt{2}\, rt(|2, 0\rangle + |0, 2\rangle), \qquad (5.1.17)$$

which means that for the balanced beamsplitter possessing $r = t = 1/\sqrt{2}$, both the photons are present in one of the outgoing beams and the situation, where one photon is in each outcoming beam, is excluded.

This is illustrated in (figure 5.3) showing the use of Mach–Zehnder interferometer for two-photon interferometry. Detectors situated in beams outgoing from the interferometer will provide interference fringes if it is impossible to distinguish the paths in the interferometer along which the pairs of photons are propagating.

The effect of the beamsplitter on photon pairs generated by the parametric down-conversion can be discussed as follows [Hong, Ou and Mandel (1987), Ou and Mandel (1988a), Rarity and Tapster (1989), Sergienko, Shih and Rubin (1995), Sergienko, Shih, Pittman and Rubin (1996)]. When photon wave functions overlap, the coincidence count rate in the intensity correlations can be reduced when pairs of subfrequency photons are combined in the beamsplitter, which permits to perform measurements of the time interval between the two photons with sub-optical period accuracy in sub-picosecond range. The width of the coincidence reduction is related to the bandwidth of down-converted light and to the phase matching conditions in the nonlinear crystal producing down-converted beams. Since the photon position in space can be specified with the accuracy of a few wavelengths or a few light periodes in time, this method makes it possible to perform measurements in space and time with better accuracy, because the optical period does not represent an ultimate

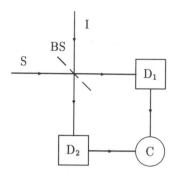

Figure 5.2: Interference of signal (S) and idler (I) beams on the beamsplitter BS and its coincidence detection; D_1, D_2 are photodetectors, and C is a coincidence counter.

limit to the accuracy of measurements. The signal and idler beams interfere on the beamsplitter and the resulting beams are detected behind the beamsplitter by two photodetectors and their responses are correlated in a coincidence counter.

In an experiment by Rarity and Tapster (1989), a beam from a crypton–ion laser operating at 413.4 nm wavelength was focused on a 15 mm-long crystal KD*P (deuterated potassium dihydrogen phosphate) using a 1 m-focal length lens. A pair of small apertures ensured that the phase matching condition was satisfied and that the matched pairs of photons were selected. The two created beams of the wavelength 826.8 nm were incident on a beamsplitter. The outputs of the beamsplitter were focused onto two photocounting avalanche photodiode detectors and the resulting photodetection pulse trains were fed to a single-bit correlator operating as a multichannel coincidence counter with 10 ns resolution. A computer acquired the correlator data and calculated the two-photon coincidence rate, which can be written in the form

$$R_{12} \propto 1 - C \exp\left[-\frac{(\delta x)^2}{2\sigma^2 \sin^2 \theta}\right], \qquad (5.1.18)$$

where C is the visibility of the coincidence pattern, $\delta x = c\delta\tau$ is the displacement introduced between the beams, σ is the halfwidth of the beam, and θ is the angle between the down-converted and pumping beams. The authors obtained the coincidence dip of the halfwidth 19 μm (63 fs) for a filter-free case with the visibility $C = 0.8$. They also performed a modified experiment in which the measurement was performed in the signal beam only adopting another beamsplitter in the beam and correlating outputs from the corresponding photodetectors. Clear enhancement of the coincidence rate was observed. Shih and Sergienko (1994) demonstrated this effect with the visibility $C = 0.97$. This effect is schematically demonstrated in figure 5.4.

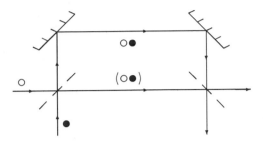

Figure 5.3: Interference of down-converted photon pairs in Mach–Zehnder interferometer.

Similar experiments can be performed with nondegenerate optical parametric process [Ou and Mandel (1988b), Rarity and Tapster (1990a)]. In this case, the signal and idler beams possess different frequencies, which leads to demonstration of additional beating oscillations in the coincidence rate and to demonstration of nonclassical and nonlocal effects [Larchuk, Campos, Rarity, Tapster, Jakeman, Saleh and Teich (1993), Larchuk, Teich and Saleh (1995)]. In this case, the coincidence rate is modified

$$R_{12} \propto 1 - C \exp\left[-\frac{(\delta x)^2}{2\sigma^2 \sin^2 \theta}\right] \cos\left(\frac{\omega_{\mathrm{p}} \delta x \Delta}{2c \tan \theta}\right), \qquad (5.1.19)$$

where ω_{p} is again the pump frequency, c is the light velocity, and Δ is the angular width of the slit situated in the down-converted beams. Here it is assumed that the resolving time of the coincidence counter is larger than the coherence time $1/\Delta\omega_{\mathrm{p}}$ of the pump radiation. Fitting the experimental data, the authors were able to obtain time resolution of about 40 fs with the visibility of the pattern $C = 0.84$. The sharp central minimum defines the centre of the photon wave packet. Its position can be determined with the accuracy of about 1 fs, regardless of the positional spread of the photon, so that the average position of the photon can be determined with the better time accuracy than the optical period. If one of the apertures in the down-converted beams is blocked, the effect is destroyed due to lack of overlap between the spectra of the photon pair. Thus, if the twin photons are incident onto a beamsplitter, the destructive interference occurs between quantum probability amplitudes corresponding to the final state consisting of one photon transmitted and the other one reflected and, consequently, neither two photons are transmitted or reflected. This means that both pair photons are propagating through one or the other arm of the interferometer to the first or the second detector.

Besides nonlocal interference and correlations, one can also observe nonlocal cancellation of dispersion effects [Steinberg, Kwiat and Chiao (1992a, b), Larchuk, Teich and Saleh (1995)] in propagation of single photons in an interferometer if pairs of

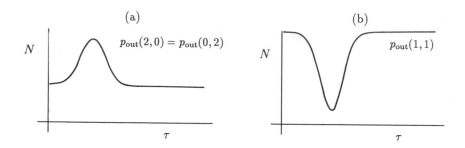

Figure 5.4: Probabilities $p_{out}(2,0), p_{out}(0,2)$ of detection of two down-converted photons in one output from the beamsplitter (a) and probability $p_{out}(1,1)$ that one photon is in one output and the other photon is in the other output (b); N is number of coincidences and τ time delay.

photons from a frequency down-converter are employed. Also nonlocal effects of the spectral filtering in one of the sub-frequency beams can be demonstrated [Chiao, Kwiat and Steinberg (1992)] in the other sub-frequency beam.

The experiment with down-converted photons in the Mach–Zehnder interferometer can demonstrate some complementarity of the second- and fourth-order interference effects, as given by Jaeger, Horne and Shimony (1993) and Jaeger, Shimony and Vaidman (1995) for one-particle interference visibilities C_j, $j = 1, 2$, and two-particle (coincidence) visibility C_{12} in the form of the following inequalities

$$C_j^2 + C_{12}^2 \leq 1, \quad C_j C_{12} \leq \frac{1}{2}, \quad j = 1, 2. \tag{5.1.20}$$

In relation to the correlation properties (5.1.15) and (5.1.16) Ou, Zou, Wang and Mandel (1990a) employed such an interferometer (figure 5.5). The two photons entering the input of the Mach–Zehnder interferometer were produced by down-conversion in a nonlinear crystal $LiIO_3$, which was pumped by an incident light beam at the wavelength of 351.1 nm from an argon–ion laser. Down-converted signal and idler photons of about 700 nm wavelength emerge at relative angles $\pm 7°$ from the crystal and provide the inputs for the beamsplitter of the interferometer, which can change its position. If it is in the symmetrical position, the two photons emerging from the output sides of the beamsplitter appear almost always together in one or the other arm of the interferometer and almost never in both arms simultaneously, because of the destructive interference as discussed above. The photons emerging from the input beamsplitter pass through the interferometer and interfere on the output beamsplitter, outputs of which are detected by two photodetectors. The path difference through the interferometer can be varied over a range of a few wavelengths by means of changing the position of output beamsplitter. After amplification and pulse shaping a coincidence counter yielded the coincidence rate. The experimental results provided no interference effect in dependence on the optical path difference in single counting

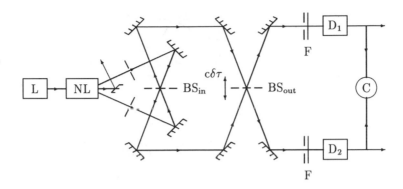

Figure 5.5: Scheme of Mach–Zehnder interferometer with down-converted photon beams; L is pump laser, NL nonlinear crystal, BS$_{in}$ input beamsplitter, BS$_{out}$ adjustable output beamsplitter, F are interference filters, D detectors, and C is coincidence counter. (Adapted from Ou, Zou, Wang and Mandel (1990a).)

rates of the detectors or if one of the inputs to the input beamsplitters was blocked. This reflects the fact that the interferometer was a bit unbalanced, so that the path difference exceeded the coherence length and, therefore, the second-order coherence could not appear. However, the coincidence rate R_{12} exhibited quantum interference with the visibility slightly less than unity. This is in contradiction with the classical result predicting no fourth-order interference in this case. For comparison the authors modified the experiment sending two coherent beams to the input beamsplitter by splitting a beam from He–Ne laser, so that both the beams were mutually coherent. They observed the second-order interference in dependence on the path difference in single photodetector rates, however there was not the fourth-order interference in the coincidence rate, because for such beams there are zero Hanbury Brown–Twiss correlations, in contrast with the above quantum case. This complementary behaviour of classical and quantum fields is a result of one-photon interference in the classical case and two-photon interference in quantum case of down-converted beams.

A new type of interesting intensity interferometer involving the correlated pairs of photons (or atomic cascades) was suggested by Franson (1989). Photon pairs from a down-converter enter two separate Mach–Zehnder interferometers (figure 5.6) and coincidences of output fields are registered. Frequency- and phase-matched photon pairs selected by signal and idler apertures pass through out-of-balance Mach–Zehnder interferometers to photodetectors, pulses of which are counted in a coincidence device. Variable phase shifts of the signal and idler beams ϕ_S and ϕ_I, respectively, are introduced into long arms of the interferometers. Such a two-photon interferometer makes it possible to observe nonlocal quantum interference effects, because energy and mo-

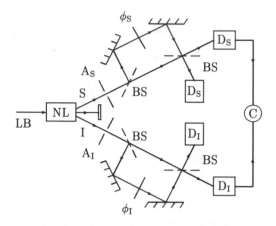

Figure 5.6: Franson two-photon interferometer; LB is laser pump beam, NL nonlinear crystal, S and I are signal and idler beams, A denote the corresponding apertures, BS are beamsplitters, D detectors, ϕ corresponding phases, and C is coincidence counter. (Adapted from Franson (1989).)

mentum conservation induce a phase coherence of the photon pair, which extends beyond the coherence length of the individual beams. This quantum entanglement is manifested in the Franson interferometer. The second-order interference effects are not observed provided that the individual interferometer path-length differences are greater than the coherence length of the individual pair-photon beams (removing the splitter in one of the interferometers, one can obtain the path information and this cancels the interference pattern), while the fourth-order coincidence interference can be observed. Such a device can exhibit interference with the visibility $C_{SI} = 1$, which represents a nonlocal quantum effect having no classical analogue.

Starting from equation (2.1.28), assuming $\langle I_1 \rangle = \langle I_2 \rangle$, and denoting $\langle I \rangle = \langle I_1 \rangle + \langle I_2 \rangle$, we obtain for coherent light that

$$\Gamma_{\mathcal{N}}^{(2,2)}(x_1, x_2, x_2, x_1) = \langle I \rangle^2$$
$$\times \left\{ 1 + \frac{1}{2} \cos[(\omega_1 - \omega_2)(t_1 - t_2) + (\mathbf{k}_1 - \mathbf{k}_2) \cdot (\mathbf{x}_2 - \mathbf{x}_1)] \right\} \quad (5.1.21)$$

and for chaotic light

$$\Gamma_{\mathcal{N}}^{(2,2)}(x_1, x_2, x_2, x_1) = \frac{3}{2} \langle I \rangle^2$$
$$\times \left\{ 1 + \frac{1}{3} \cos[(\omega_1 - \omega_2)(t_1 - t_2) + (\mathbf{k}_1 - \mathbf{k}_2) \cdot (\mathbf{x}_2 - \mathbf{x}_1)] \right\}. \quad (5.1.22)$$

This classical approach can be compared with the corresponding expressions for the intensity correlation functions of two different radiating atoms [Mandel (1983), Teich,

Saleh and Peřina (1984), Paul (1986)]. Considering completely quantum description, some fourth-order terms representing the probability of detection of two photons disappear, because the atom can radiate only one photon. Such terms contribute to the classical correlation function. Such different results from the quantum and classical approaches reflect the inability of the classical theory to describe correctly the particle properties. Compared to (5.1.21), we obtain in the quantum case

$$\Gamma_N^{(2,2)}(x_1, x_2, x_2, x_1) = \langle I \rangle^?$$
$$\times \{1 + \cos[(\omega_1 - \omega_2)(t_1 - t_2) + (\mathbf{k}_1 - \mathbf{k}_2) \cdot (\mathbf{x}_2 - \mathbf{x}_1)]\}, \qquad (5.1.23)$$

which means that the quantum description predicts twice the classical fourth-order visibility. This is crucial for experimental verification of quantum phenomena. We also see a natural result that the visibility for coherent fields is higher than for chaotic fields. Equation (5.1.23) demonstrates that the quantum character of light further increases the degree of order in light and the visibility of the fourth-order interference pattern. Moreover, this makes it possible to place both the detectors in such positions that both can simultaneously register the emission of a photon or to avoid any coincidences. These results can be generalized to two sources containing M and N atoms, respectively, for which the quantum correlation functions are

$$\Gamma_N^{(2,2)}(x_1, x_2, x_2, x_1) \propto N(N-1) + M(M-1)$$
$$+MN\{1 + \cos[(\omega_1 - \omega_2)(t_1 - t_2) + (\mathbf{k}_1 - \mathbf{k}_2) \cdot (\mathbf{x}_2 - \mathbf{x}_1)]\}, \qquad (5.1.24)$$

whereas for the classical correlations we obtain

$$\Gamma_N^{(2,2)}(x_1, x_2, x_2, x_1) \propto (M+N)^2 + M(M-1) + N(N-1)$$
$$+2MN \cos[(\omega_1 - \omega_2)(t_1 - t_2) + (\mathbf{k}_1 - \mathbf{k}_2) \cdot (\mathbf{x}_2 - \mathbf{x}_1)]. \qquad (5.1.25)$$

In the limiting case of $M, N \to \infty$ both the correlation functions tend to a classical value for chaotic light. The same result is reached when averaging with the Poisson distribution over both the sources. This means that in order to measure quantum properties, one has to control the number of emitting atoms.

Such large nonclassical modulations with the visibility exceeding the value 0.5 have been observed using the Franson interferometer by a number of authors [Ou, Zou, Wang and Mandel (1990a), Kwiat, Vareka, Hong, Nathel and Ciao (1990), Rarity, Tapster, Jakeman, Larchuk, Campos, Teich and Saleh (1990), Brendel, Mohler and Martienssen (1991, 1992), Shih, Sergienko and Rubin (1993a, b), Kwiat, Steinberg and Chiao (1993a, b)] and they clearly support the validity of quantum theory. A review of optical tests of quantum mechanics has been prepared by Chiao, Kwiat and Steinberg (1994).

Rarity and Tapster (1992) modified this two-photon Franson interferometer introducing into the signal arm, before the Mach–Zehnder interferometer, a multimode fibre-optical cable of the length 170 m. Such a modification may be interesting for

a more secure form of coding in terms of nonlocal correlations for optical communications, provided that the identical Mach–Zehnder interferometers are placed in
the transmitter and receiver. They have used a lithium iodate crystal pumped by a
helium–cadmium laser operating at 441.6 nm wavelength. The crystal axis was tilted
to produce 883.2 nm wavelength pair beams separated by an angle of about 30°. The
authors observed the coincidence rate modulations when changing the path-length
difference in the Mach–Zehnder interferometer with the visibility $C = 0.11$, which is
less than 0.5, but they concluded that the reduced visibility was caused by spatial
averaging in the interferometers and by mode scrambling in the optical fibre. This
experiment does not prove absolutely that the entangled states exist over the 170 m
detector separation, even if one can demonstrate that such a result is in agreement
with the quantum description and that any classical description is inadequate here
[Franson (1991a)]. Whereas the coherence length of single beams was a few microm-
eters, actual path-length differences to observe the interference pattern were changed
up to a few centimeters. Rarity and Tapster (1990b) demonstrated the visibility
slightly less than 1, but sufficient to show a violation of Bell's inequality. Rarity,
Burnett, Tapster and Paschotta (1993) realized high-visibility two-photon interference using a single-mode fibre in the interferometer, with the visibility $C = 0.9$, using
the entangled photon pairs and demonstrated nonlocal behaviour at the distance of
one kilometer. Franson (1991b) employed the two-photon interferometer, in which
the optical path length between the two Mach–Zehnder interferometers was 102 m.
The results of his measurements were in good agreement with quantum predictions
and are inconsistent with any classical theory. The visibility of modulation of the
coincidence rate was about $C = 0.31$ for close positions of the two Mach–Zehnder
interferometers, whereas the long-distance visibility was $C = 0.27$; no modulation
was observed in single-detector rates. This indicates that quantum correlations are
not reduced due to a collapse of the wave function over long distances, which is in
agreement with the information content of the wave function.

· Ou, Zou, Wang and Mandel (1990b) and Shih, Sergienko and Rubin (1993a) used
a modification of the Franson interferometer, in which the Mach–Zehnder interferometers were replaced by the Michelson interferometers (figure 5.7). Signal and idler
beams from the nonlinear crystal pass through independent Michelson interferometers and are detected by the coincidence counter. If the interferometers are set so that
the optical path differences are longer than the coherence length of the single beams,
the second-order coherence cannot be observed. However, if the optical paths of both
the interferometers are approximately equal, the fourth-order interference can occur
arising from the frequency and wave-vector correlations of photon pairs generated under conditions (5.1.12). When the time window of the coincidence device is shorter
than the optical path difference, the registration time of one photon traversing the
long path L and of the other one following the short path S in the interferometers
is outside the coincidence window and they will not be registered by the coincidence
counter. The resulting state of the light in the interferometer is

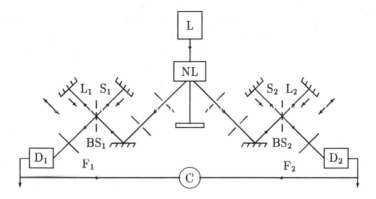

Figure 5.7: Scheme of two-photon interferometer involving the Michelson interferometers in down-converted beams; L is laser, NL nonlinear crystal, BS are beamsplitters, F spectral filters, D detectors, C is coincidence counter, L and S denote the corresponding long and short paths. (Adapted from Ou, Zou, Wang and Mandel (1990b).)

$$|\psi\rangle = |\psi(L_1, L_2)\rangle + |\psi(S_1, S_2)\rangle + |\psi(L_1, S_2)\rangle + |\psi(L_2, S_1)\rangle, \qquad (5.1.26)$$

which gives the maximum visibility 0.5, and one cannot distinguish the measurement from the classical one. However, the coincidence time window shorter than the optical path difference of the interferometers cancels the last two terms in (5.1.26), which leads to the coincidence interference pattern of the maximum visibility $C = 1$. Shih, Sergienko and Rubin (1993a) used a single-mode cw argon laser of the wavelength 351.1 nm to pump a KDP nonlinear crystal to produce the optical parametric down-conversion. Both the degenerate and nondegenerate photon pairs were employed with wavelengths 702.2 nm, and 632.8 nm and 788.7 nm, respectively. Single rates and their correlations were measured. When both the path differences ΔL_1 and ΔL_2 were shorter than the coherence length of the beams, the visibilities 0.97 and 0.82 for the interference of the fourth and second orders were observed, respectively. When ΔL_1 and ΔL_2 were greater than the coherence length, the second-order interference was not observed and the fourth-order coincidence measurements provided the interference patterns with the visibility 0.59.

Further demonstrations of quantum correlations of down-converted photon pairs in quantum interferometers were reported by Rarity, Tapster, Jakeman, Larchuk, Campos, Teich and Saleh (1990), Shih, Sergienko and Rubin (1993a), and Kwiat, Steinberg and Chiao (1993a). Rarity, Tapster, Jakeman, Larchuk, Campos, Teich and Saleh (1990) adopted the arrangement of the Mach–Zehnder interferometer as given in figure 5.8, the corresponding theory was presented by Campos, Saleh and Teich (1990). The coincidence probabilities were determined at the output ports of a beamsplitter of a Mach–Zehnder interferometer, where two photons were incident on each port and the spectral composition of wave packets was arbitrary. For uncorrelated and distinguishable input photons, the fourth-order interference can be understood in terms of particle or wave behaviour of each photon acting independently. For highly correlated photon pairs arising in the spontaneous process of down-conversion, the fourth-order interference at the output of the interferometer is determined by the joint quantum nature of pairs and it cannot be described by the particle or wave properties of single photons.

Rarity, Tapster, Jakeman, Larchuk, Campos, Teich and Saleh (1990) used a krypton–ion laser producing light of 413.4 nm wavelength as the pumping for a KD*P nonlinear crystal, which produced sub-frequency photons of the wavelength 826.8 nm incident on the input ports of the Mach–Zehnder interferometer (figure 5.8). The photons were directed to a beamsplitter cube BS_1 and then redirected onto a lower portion of the beamsplitter cube BS_2 by two right-angle prisms, thus forming a folded Mach–Zehnder interferometer. The beamsplitter cube and prisms were mounted together as a unit on a translation stage. The distance between one of the prisms and the beamsplitter cube can be adjusted within the unit by a combined micrometer and piezoelectric positioner to introduce an adjustable path-length difference time delay between the arms of the interferometer. The photons at the output ports of the interferometer were then focused by means of two lenses onto two Si avalanche photodiodes. After discrimination, amplification, and shaping, the outputs of the photodiodes were fed to digital correlator that provided the coincidence rate and the individual rates. A computer controlled the experiment and collected the data. The authors observed nonclassical coincidence interference with a spatial period equal to the wavelength of pump photons with the visibility 0.62 for path-length differences both smaller and greater than the single-photon coherence length. This demonstrates the quantum correlation nature of two-photon states produced by the parametric down-conversion process. The interference is expected to disappear when the path-length difference in the interferometer exceeds the coherence length of the pump beam. For independent photons, the interference pattern was lost for path-length differences exceeding the coherence length of either photon. Larchuk, Campos, Rarity, Tapster, Jakeman, Saleh and Teich (1993) also realized interference effects with quantum correlated photons of different colors with centre-wavelength differences about 40 nm (800 nm and 840 nm wavelengths of down-converted beams) making use of the same arrangement. The difference frequency of the photons appears directly in the spectrum of the coincidence rate. Such a phase-sensitive behaviour can reveal centre-frequency differences from zero to hundreds of THz. They have observed a continuation of pump frequency oscillations for path-length differences exceeding substantially the second-order coherence lengths, which again represents a fully quantum nonlocal effect observed in strongly nondegenerate optical parametric

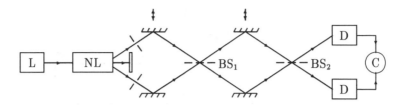

Figure 5.8: Two-photon coincidence experiment; L is laser, NL nonlinear crystal, BS are beam-splitters, D detectors, C is coincidence counter. (Adapted from Rarity, Tapster, Jakeman, Larchuk, Campos, Teich and Saleh (1990).)

process. By adjusting the angle of the input mirrors, the photon paths within the interferometer can be made to have properties of full spatial overlapping or no spatial overlapping, equivalent to the case that each photon is sent into a separate interferometer. The output pulses from the detectors were counted for 1 s to provide a measure of the rate of photon detections at each of the ports and the sequence of standardized pulses from the two detectors was passed through a 10 ns gate and counted for 1 s to provide the coincidence rate. High visibility coincidence interference near unity was observed.

Some new interesting information about quantum interference, its interpretation, and relation between quantum interference and path distinguishability of photons can be gained using two identical down-converters [Ou, Wang, Zou and Mandel (1990), Zou, Wang and Mandel (1991), Wang, Zou and Mandel (1991a)] or one can adopt one down-converter in an arrangement of counterpropagating beams, in which the pump, signal, and idler beams are reflected back to the nonlinear crystal by means of mirrors [Herzog, Rarity, Weinfurter and Zeilinger (1994), Herzog, Kwiat, Weinfurter and Zeilinger (1995)]. In one of these experiments (figure 5.9) two coherent pump waves fall on two identical nonlinear crystals, down-converted signal and idler beams from the two crystals are mixed by two beamsplitters and the single and coincidence counting rates are measured. The counting rate depends on the phase difference between the two coherent pump waves and the interference of the down-converted photons with the physical vacuum can be examined, which can be used for the demonstration of violation of Bell's inequalities and for the observation of nonlocal quantum effects.

If a continuum of sub-frequency modes is taken into acount, the quantum state describing the down-conversion can be written in the form [Ou, Wang and Mandel (1989)], on the assumption that the directions of the signal and idler beams are defined by apertures,

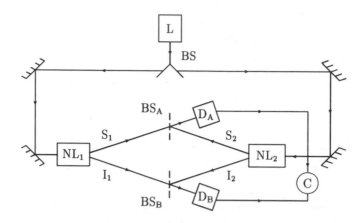

Figure 5.9: Scheme for observing interference of down-converted photons from two nonlinear crystals; L is laser, NL are nonlinear crystals, S signal beams, I idler beams, D detectors, BS beamsplitters, and C coincidence counter. (Adapted from Ou, Wang, Zou and Mandel (1990).)

$$
\begin{aligned}
|\psi(t)\rangle &= \exp\left[-\frac{i}{\hbar}\int_0^t \hat{H}_{\text{int}}(t')\,dt'\right]|0\rangle \approx M|0\rangle_{\text{S}}|0\rangle_{\text{I}} \\
&+\eta V\delta\omega t \sum_{\omega',\omega''} \Phi(\omega',\omega'')\text{sinc}\left(\frac{\omega'+\omega''-\omega}{2}t\right)\exp\left(i\frac{\omega'+\omega''-\omega}{2}t\right) \\
&\times|\omega'\rangle_{\text{S}}|\omega''\rangle_{\text{I}},
\end{aligned}
\tag{5.1.27}
$$

where the interaction Hamiltonian is a multimode generalization of the Hamiltonian given in (5.1.11), M is a coefficient related to the physical vacuum, $\delta\omega$ is the mode spacing, η represents an efficiency coefficient, V is the pump field amplitude, $\Phi(\omega',\omega'')$ is a spectral weighting function, symmetrical in ω' and ω'' and peaked at $\omega'=\omega''=\omega/2$, which is normalized so that $2\pi\sum_{\omega'}|\Phi(\omega',\omega-\omega')|^2\delta\omega = 1$. From the normalization of the state $|\psi(t)\rangle$ it follows that $|M|^2+\eta^2|V|^2t = 1$. For a weak down-conversion, $|\eta Vt|^2 \ll 1$, so that $|M| \approx 1$. In a simplified case, neglecting sum over modes, we can write for quantum states $|\psi_1\rangle_1$ and $|\psi_2\rangle_2$ of the down-converted beams in the interaction picture, produced by the nonlinear crystals NL1 and NL2

$$
\begin{aligned}
|\psi_1\rangle_1 &= M_1|0\rangle_{\text{S}_1}|0\rangle_{\text{I}_1} + \eta_1 V_1|\omega_1\rangle_{\text{S}_1}|\omega_2\rangle_{\text{I}_1}, \\
|\psi_2\rangle_2 &= M_2|0\rangle_{\text{S}_2}|0\rangle_{\text{I}_2} + \eta_2 V_2|\omega_1\rangle_{\text{S}_2}|\omega_2\rangle_{\text{I}_2}
\end{aligned}
\tag{5.1.28}
$$

and for the annihilation operators we have in the interaction picture

$$
\begin{aligned}
\hat{A}_{\text{A}}^{(+)} &\propto \hat{a}_{\text{S}_1} + i\hat{a}_{\text{S}_2}, \\
\hat{A}_{\text{B}}^{(+)} &\propto \hat{a}_{\text{I}_1} + i\hat{a}_{\text{I}_2}.
\end{aligned}
\tag{5.1.29}
$$

We obtain the rates of single detectors

$$R_j = {}_1\langle\psi_1|_2\langle\psi_2|\hat{A}_j^{(-)}\hat{A}_j^{(+)}|\psi_2\rangle_2|\psi_1\rangle_1$$
$$= |\eta_1 V_1|^2 + |\eta_2 V_2|^2, \quad j = \mathrm{A,B}, \tag{5.1.30}$$

without any interference pattern. The two-photon coincidence counting rate is obtained as

$$R_{\mathrm{AB}} = {}_1\langle\psi_1|_2\langle\psi_2|\hat{A}_{\mathrm{A}}^{(-)}\hat{A}_{\mathrm{B}}^{(-)}\hat{A}_{\mathrm{B}}^{(+)}\hat{A}_{\mathrm{A}}^{(+)}|\psi_2\rangle_2|\psi_1\rangle_1$$
$$= |\eta_1 V_1 M_2 - \eta_2 V_2 M_1|^2$$
$$\approx |\eta_1 V_1|^2 + |\eta_2 V_2|^2 - 2|\eta_1 V_1||\eta_2 V_2|\cos[\arg(V_1) - \arg(V_2)]. \tag{5.1.31}$$

Thus, the two-photon coincidence counting rate exhibits modulation in dependence on the phase difference $\arg(V_1) - \arg(V_2)$. It is easy to show that a more realistic treatment involving the other parameters of the nonlinear crystals and detectors leads to the same conclusion [Ou, Wang and Mandel (1989)]. The experiment by Ou, Wang, Zou and Mandel (1990) fully confirmed these conclusions. They have observed the cosinusoidal modulations, although not exactly with the expected visibility $C = 1$, which was caused by various imperfections in the measurement. These conclusions can be explained in terms of the relation between interference and indistinguishability of photon paths. If it is impossible to determine in which down-converter the photons originate, then the corresponding probability amplitudes for the two paths have to be summed in order to obtain the resulting detection probability and, consequently, the interference pattern arises. On the other hand, if there is, in principle, a possibility to determine the source of photons, the interference is missing. In the above arrangement, there is no way to determine the source of each photon pair without introducing disturbances in the coincidence measurement. However, if we consider the interference of only signal or only idler photons, we can remove one of the beam-splitters without affecting the corresponding interference and then we can identify the source of photons, the indistinguishability is lost and no interference on single detectors can occur. It is not necessary to perform such auxiliary measurement, only its possibility excludes the interference pattern. Worth noting is the substantial role of the vacuum state reflected by the factors M_1, M_2 involved in (5.1.28) and (5.1.31). This quantum effect is really a result of the superposition of the vacuum state of the field and the two-photon state from down-converters.

Zou, Wang and Mandel (1991) and Wang, Zou and Mandel (1991a) realized experimentally the arrangement of two down-converters pumped by mutually coherent beams and related by the idler beam (figure 5.10). An idler photon emitted spontaneously from the first nonlinear crystal serves as stimulating idler input to the second nonlinear crystal. On detecting the signal beams from the first and the second crystals, one can show that the first idler photon can induce a stimulated down-conversion process acting as an optical amplifier. We are interested in the conditions under which the interference of signal beams from both the crystals can be observed. As shown by Ou, Wang and Mandel (1990), the counting rate of the detector registering the signal beam from the second crystal is given by the expression

$$R_2 = |\eta_2 V_2|^2 \left[1 + \frac{1}{T_R} \int_0^{T_R} |\gamma_{12}(t_2 + \tau_1 - \tau_2)|^2 \, dt_2 \right], \qquad (5.1.32)$$

where η_2 is the conversion efficiency of the crystal, V_2 is again the complex amplitude of the pumping field incident on the crystal, T_R is the resolving time of the detector, γ_{12} is the second-order degree of coherence related to the spectral properties of both the signal beams, and τ_1 and τ_2 are the times of propagation between nonlinear crystals and detectors. The first term in (5.1.32) represents spontaneous emission from the second crystal, whereas the second term is due to stimulated emission from the second crystal induced by the idler beam from the first crystal. The authors used an argon–ion laser of the wavelength 351.1 nm pumping two similar 25 mm long crystals of LiIO$_3$ producing the idler beams I$_1$ and I$_2$ at 632.8 nm wavelength and the signal beams S$_1$ and S$_2$ at the 788.7 nm wavelength, which interfere at the beamsplitter BS$_2$ and are registered by the detector D$_S$, whereas the idler beam I$_2$ is detected by the detector D$_I$. The interference filters IF$_S$ and IF$_I$ were centred at 788.7 nm and 632.8 nm, respectively, and the coherence length of down-converted beams was about 0.3 mm. The neutral density filter placed between the nonlinear crystals NL$_1$ and NL$_2$ to the idler beam I$_1$ can be used to change the strength of the connection of both the nonlinear crystals. The authors observed the interference effect as a function of displacements of the beamsplitter BS$_2$ in the counting rate of the signal detector D$_S$, when the transmittance $|t| = 0.91$ of the filter NF, with the visibility of about 0.3, which was linearly decreasing with decreasing transmittance; the interference fringes were missing for $|t| = 0$, when the idler beam I$_1$ was blocked. A simple description of this interference effect provides the visibility of fringes

$$C = \frac{2|f_1||f_2|\sqrt{\langle I_1\rangle\langle I_2\rangle}|\gamma_{12}|}{|f_1|^2\langle I_1\rangle + |f_2|^2\langle I_2\rangle}|t|, \qquad (5.1.33)$$

where $\langle I_1\rangle$ and $\langle I_2\rangle$ are the mean intensities of the two pumping beams, $|\gamma_{12}|$ is their degree of partial coherence at the two crystals, and $|f_1|^2$ and $|f_2|^2$ are fractions of the incident pumping photons down converted at the crystals. Hence, this is a method of controlling the degree of mutual coherence between two light beams, changing the transmittance $|t|$ leaving the light intensities unchanged, because we have here induced coherence that is not accompanied by induced emission. The coincidence rate R_{SI} exhibited modulation of the visibility of about 0.53 for $|t| = 0.91$. The interpretation of these results is only quantal. If the beam I$_1$ is blocked, then, in principle, the detector D$_I$ is able to determine, because the signal and idler photons are emitted together, from which signal beam a given photon comes to the signal detector D$_S$. Whenever a detection by D$_S$ is accompanied by a simultaneous detection by D$_I$, the signal photon comes from NL$_2$ and, whenever a detection by D$_S$ is not accompanied

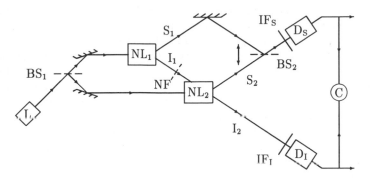

Figure 5.10: Scheme for interference of signal beams from two down-converters; L is laser, NL nonlinear crystals, BS beamsplitters, S and I signal and idler beams, IF interference filters, D detectors, C coincidence counter, and NF neutral density filter. (Adapted from Zou, Wang and Mandel (1991).)

by a detection of D_I, the signal photon comes from NL_1. The consequence is that the possibility of this distinguishing between the signals S_1 and S_2 rules out the interference pattern in the counting rate R_S, regardless of the auxiliary measurement by means of D_I is actually performed or not. We can conclude that blocking the idler beam, the interference of the signal beams is not missing because of creation of uncontrollable disturbance of the system or as a consequence of the uncertainty principle. It is a consequence of obtaining which-path information that destroys coherence and interference. This experiment can be described in a completely quantum statistical way solving the corresponding Heisenberg equations and calculating the quantum characteristic function for the multimode field [Řeháček and Peřina (1996b)], which enables us to obtain the photon-number probability distribution for combined signal modes and phase-difference distribution function. The conclusion is that increasing strength of the connection of both the down-converters via the idler mode leads to broadening the photon-number distribution and to narrowing the phase distribution (decrease of the phase uncertainty), which supports the formation of the interference fringes, in qualitative agreement with the measurement [Noh, Fougères and Mandel (1993c), Fougères, Monken and Mandel (1994)].

The same arrangement can be adopted to demonstrate that the visibility of interference fringes formed by signal photons in the down-conversion process can be controlled by inserting differential time delays between the idler photons, which is again a consequence of the quantum entanglement between signal and idler photons [Zou, Grayson, Barbosa and Mandel (1993)]. Although two nonlinear crystals acting as parametric down-converters are excited by two mutually coherent pumping light beams, there is no mutual coherence between the generated signal beams or idler beams. However, the interference effects can be observed between two signal beams when two nonlinear crystals are related by the idler beams, as discussed above. The authors inserted a time delay plate instead of the absorber to the idler beam. When the detection time T exceeds the coherence time of the down-converted light, the interference at the signal detector D_S cannot occur, because now one can distinguish the source of each photon using an auxiliary detector D_I with high efficiency in the idler mode. This detector does not disturb in any way the interference of the signal beams. If the detector D_S registers a photon in coincidence with a photon at D_I, the photons were emitted at NL_2. If a detection by D_S is followed by a detection at D_I after a delay T, then both the photons come from NL_1. This eliminates any observable interference. Moreover, it does not matter if the time delay is introduced to I_1 rather than to S_1, which is actually interfering, because both the signal and idler photons are emitted simultaneously in the entangled state. In the case that the detection time T is less than the coherence time, the visibility of interference fringes will be reduced, but they will occur. The effect of the change of the state in the idler beam I_1 is changing the states in the signal beams S_1 and S_2 as a kind of nonlocal connection between beams.

If the detection time T is much larger than the coherence time, the interference is missing in the space–time domain, but it is observable in the frequency domain, as discussed in section 2.1. Taking this into account, Zou, Grayson and Mandel (1992) modified the above arrangement with two down-converters connected by the idler beam inserting a glass plate in the signal beam S_1 introducing an optical path difference of about 1.5 mm between the arms, which exceeds the coherence length of 0.3 mm. Thus, the interference effect in light falling on the signal detector D_S was lost. However, insertion of the scanning Fabry–Perot interferometer with resolution of about 40 GHz in front of the detector D_S reveals the spectral modulation according to the spectral second-order interference law (2.1.5). If the idler beam I_1 is blocked, the interference disappears as a result of distinguishability of signal photons in the beams S_1 and S_2. Further, the authors removed the glass plate from the beam S_1 and placed it in between the nonlinear crystals into the stimulating idler beam I_1, which caused the absence of the interference fringes in the intensity and the same spectral modulation as before, even though neither of the two interfering signals S_1 and S_2 is delayed or directly disturbed by the insertion of the plate. Hence, we have again nonclassical and nonlocal quantum effect connected with the quantum correlated signal and idler photons. The absence of the interference in the total intensity measurement by means of the signal detector D_S is again explained by the distinguishability of paths from NL_1 and NL_2 to D_S if an efficient detector D_I of the idler beam is applied. The spectral interference effect, however, cannot be ruled out, because it is measured with high-resolution interferometer, the passband of which is much narrower than $1/T$, which makes the time delay T unresolvable. However, this interference disappears if the idler beam I_1 is blocked.

Grayson, Zou, Branning, Torgerson and Mandel (1993) realized the similar ex-

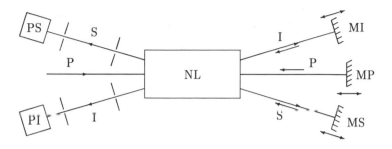

Figure 5.11: Interference of two down-conversion realizations; NL is nonlinear crystal, P pumping beams, S and I are signal and idler beams, MP, MS, and MI are corresponding mirrors, and PS and PI photodetectors. (Adapted from Herzog, Rarity, Weinfurter and Zeilinger (1994).)

periment with two down-converters related by the idler beam situated in an optical cavity, providing a possible time delay for the idler beams. Whether the idler photon is to be delayed or not determines whether the corresponding signal photon behaves like a wave exhibiting interference or it behaves like a particle. This experiment can be considered as a kind of the delayed-choice experiment examining the interference pattern provided that conditions for the interference were created after a photon has chosen one of the interference paths. The results are again a manifestation of the entanglement of the signal and idler photons and they demonstrate that the diagonal statistical operator reflects not only what is known about the physical system, but what is knowable in principle. Using polarization elements or additional interferometers inside the idler beams I_1 and I_2, one can measure the Berry–Pancharatnam topological phase [Grayson, Torgerson and Barbosa (1994)] and a quantum eraser can be created permitting to restore interference erasing distinguishability, because the decision to measure wave-like or particle-like behaviour may be delayed until the irreversible process in the detector occurs [Kwiat, Steinberg and Chiao (1994)].

Experiments of a similar kind were performed by Herzog, Rarity, Weinfurter and Zeilinger (1994) and Herzog, Kwiat, Weinfurter and Zeilinger (1995) using only one nonlinear crystal producing down-converted photons, but pumping and down-converted beams were reflected back by mirrors creating pairs of indistinguishable beams being able to interfere (figure 5.11). The change of the position of any mirror in the signal, idler or pumping beam changes the interference patterns in all beams simultaneously, which demonstrates nonlocal interference described in the single beams by the simple interference law $1 + \cos(\phi_P - \phi_S - \phi_I)$, ϕ_P, ϕ_S, and ϕ_I being the pump phase, signal phase, and idler phase, respectively. This behaviour of light beams represents the interference of two possible ways of photon emissions. This arrangement can be employed for nice demonstration of the relation of coherence, interference, and indistinguishability of optical paths [Herzog, Kwiat, Weinfurter and

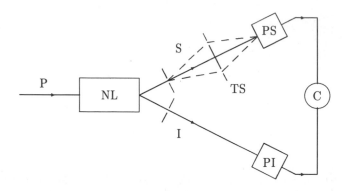

Figure 5.12: Down-conversion combined with two-slit experiment; P is pump beam, NL is nonlinear crystal, TS two-slit screen, S and I are signal and idler beams, PS and PI corresponding photodetectors, C is coincidence counter. (Adapted from Ribeiro, Pádua, Machado da Silva and Barbosa (1994).)

Zeilinger (1995)]. An interpretation of mirrors effects in induced coherence in this arrangement was presented by Ribeiro and Barbosa (1997). If distinguishability is introduced and the interference is missing, a suitable measurement can erase this distinguishability and the interference can again be recovered.

The nonlocal quantum properties can also be demonstrated using down-converted beams when a double slit is placed into the signal beam (figure 5.12) [Ribeiro, Pádua, Machado da Silva and Barbosa (1994), Řeháček and Peřina (1996a)]. In this experiment the degree of coherence can be controlled in the signal beam by the width of slit placed in the idler beam and the fourth-order interference can occur regardless of the fact that linear dimensions of the second-order coherence area are smaller than the distance between slits as a consequence of the quantum entanglement of photon pairs. For this case the fourth-order van Cittert–Zernike theorem can be applied.

A class of interesting experiments demonstrating nonlocal quantum behaviour of down-converted beams can employ the stimulated process of down-conversion. Wang, Zou and Mandel (1991b) realized an experiment for interference of idler beams from two parametric down-converters provided that the signal beams are stimulated by mutually coherent injected beams, so that a stimulated process of down-conversion is performed (figure 5.13). The down-converters consisted of two similar 2.5 cm long crystals of $LiIO_3$, which were pumped by light of the wavelength 351.1 nm from an argon–ion laser L_1. The beamsplitter BS_1 directed the pump beam to the nonlinear crystals NL_1 and NL_2. Apertures A_1, A'_1 and A_2, A'_2 selected directions of the down-converted light photons corresponding to signal beams of the wavelength 632.8 nm. The idler beams had the wavelength 788.7 nm. An interference filter IF limited the

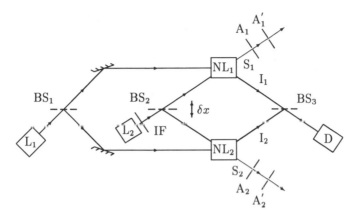

Figure 5.13: Scheme of experiment with two stimulated down-conversion processes; L are lasers, BS beamsplitters, NL nonlinear crystals, A and A' apertures, S and I are signal and idler beams, IF is filter, D detector, and δx path delay. (Adapted from Wang, Zou and Mandel (1991b).)

spectral bandwidth $\Delta\omega$ of the signal beam to about 3×10^{10} Hz of light from a He–Ne laser L_2 oscillating at the wavelength 632.8 nm, which was split by the beamsplitter BS_2 giving inducing fields to the signal beams S_1 and S_2. The generated idler beams I_1 and I_2 were combined at the beamsplitter BS_3 and the resulting field was detected by a detector D in a 5 s time interval. Changing displacement of the beamsplitter BS_2, the interference pattern can be observed. Typical photon counting rates were of the order of 10^4 s^{-1}. It is obvious from (5.1.14) that for the case of stimulated emission, the correlation (5.1.15) is no longer zero, as a result of the stimulated emission specified by the initial complex amplitudes ξ_{S_1}, $\xi_{S_2} \neq 0$, $\xi_{I_1} = \xi_{I_2} = 0$. The authors observed the interference pattern of the visibility $C = 0.7$. The degree of coherence of the idler beams increases with increasing occupation number of photons per mode and, if stimulated emission in the signal beams substantially exceeds spontaneous emission, generated idler fields are fully coherent. This is again a nice demonstration of nonlocal quantum behaviour of optical beams generated in the nonlinear process of frequency down-conversion.

A similar experiment was performed by Sergienko, Shih and Rubin (1993), who used a KDP crystal for nondegenerate down-conversion pumped by argon–ion laser, using a beam from a He–Ne laser to stimulate the signal beam. The idler beam (not stimulated) was introduced to a Michelson interferometer (figure 5.14). Denoting the mean spontaneous intensity by $\langle I_{\mathrm{spont}} \rangle$ and the mean intensity induced by the input signal by $\langle I_{\mathrm{stim}} \rangle$, they were able to observe the interference pattern in the idler beam of the visibility $C = 0.8$ provided that $\langle I_{\mathrm{spont}} \rangle / \langle I_{\mathrm{stim}} \rangle \ll 1$ even if the optical path difference in the interferometer was much larger than the coherence length of

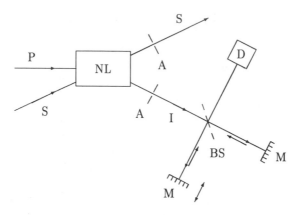

Figure 5.14: Stimulated down-conversion combined with the Michelson interferometer; P, S, and I are pump, signal, and idler beams, respectively, NL is nonlinear crystal, A are apertures, M mirrors, BS is beamsplitter, and D photodetector. (Adapted from Sergienko, Shih and Rubin (1993).)

the spontaneous idler radiation. The visibility was determined by the input coherent signal. No interference was observed with only spontaneous radiation from the down-conversion process when $\langle I_{\mathrm{spont}} \rangle / \langle I_{\mathrm{stim}} \rangle \gg 1$.

Similar experiment was realized by Ribeiro, Pádua, Machado da Silva and Barbosa (1995), in which a double slit was situated in one of the down-converted beams (signal beam) instead of the Michelson interferometer, while the other down-converted beam (idler beam) was stimulated by He–Ne laser (figure 5.15). Then the degree of coherence and visibility of interference fringes produced by the beam transmitted through the double slit can be controlled by aligning an auxiliary laser with the other beam of the same wavelength and varying its power. In this case the degree of coherence of the source is varied directly by the inducing laser intensity. The authors employed a $LiIO_3$ crystal to produce parametric down-converted light, which was pumped by an argon–ion laser beam (351.1 nm), of the wavelengths 632.8 nm (stimulated) and 788.7 nm (diffracted by the double slit). The interference pattern was registered by a photomultiplier D at the entrance of which scans were performed with the use of a 300 μm slit. An interference filter IF placed at the photomultiplier having bandwidth 10 nm and centred at 788.7 nm provided almost monochromatic light. Pulses from the photomultiplier were sent to photon counters and to a computer, where data were processed. The inducing laser intensity was controlled by neutral filters placed in front of the crystal. The authors demonstrated the increase of visibility of interference fringes with increasing stimulating intensity, which provides a way of controlling the degree of spatial coherence without disturbing the beam. Thus, the Young double-slit experiment with light produced in the stimulated down-conversion

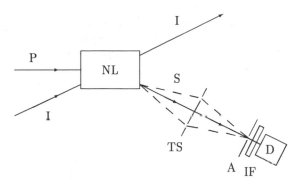

Figure 5.15: Stimulated down-conversion combined with double-slit experiment; P, S, and I are pump, signal, and idler beams, respectively, NL is nonlinear crystal, TS two-slit screen, A aperture, IF interference filter, and D photomultiplier. (Adapted from Ribeiro, Pádua, Machado da Silva and Barbosa (1995).)

provides coherence and interference pattern as functions of mean-photon occupation number per mode (degeneracy parameter) and makes it possible to control them by means of the conjugated photon beam.

Some modifications of the cascade of two down-converters connected with the idler beam were discussed by Ryff (1995) giving further arguments for the relation between induced coherence, interference, and indistinguishability. Improvement of measurement accuracy of the field phase using an interferometer involving two down-converters were discussed by Brif and Ben-Aryeh (1996). The possibility to describe the two aligned optical parametric down-conversion processes in terms of SU(2) and SU(1,1) coherent states was shown by Luis and Peřina (1996a). The quantum Zeno effect in the spontaneous parametric down-conversion was demonstrated by Luis and Peřina (1996c). Some interesting features of the process of optical parametric down-conversion can be seen in the Wigner representation [Casado, Marshall and Santos (1997), Casado, Fernández-Rueda, Marshall, Risco-Delgado and Santos (1997)]. Indirect interference of two modes with different frequencies on the single-photon level by means of the parametric up-conversion was discussed by Dušek (1997), with the possibility to distinguish pure and mixed quantum states. If the parametric down-conversion process is realized with two non-collinear pump beams, four-photon entangled states for higher-order tests of quantum theory may be generated [Tewari and Hariharan (1997)].

The use of the two-photon interferometry can give interesting information about principles of quantum theory, providing various tests of violation of Bell's and the classical inequalities [Peřina, Hradil and Jurčo (1994), Walls and Milburn (1994), Mandel and Wolf (1995), Tapster, Rarity and Owens (1994)], which in all cases supports the validity of quantum theory. An experimental test of violation of local realism in quantum mechanics without Bell's inequalities was given by Boschi, De Martini and Di Giuseppe (1997). Similarly, three-particle interferometers can be considered [Greenberger, Horne, Shimony and Zeilinger (1990), Shih and Rubin (1993), Pittman (1995), Pittman, Shih, Strekalov, Sergienko and Rubin (1996), Krenn and Zeilinger (1996), Zeilinger, Horne, Weinfurter and Zukowski (1997), Banaszek and Knight (1997), Zukowski, Zeilinger and Horne (1997), Zheng and Guo (1997b)] including a possibility to use atom–photon pairs [Kurtsiefer, Dross, Voigt, Ekstrom, Pfau and Mlynek (1997)], which have not been realized till now, making it possible to test quantum principles without statistical inequalities. Also nonlocality of a single photon can be examined [Hardy (1994), Home and Agarwal (1995), Freyberger (1995), Garuccio (1995)]. Essential violation of local realism in microscopic world without using Bell's inequalities was experimentally shown by Di Giuseppe, De Martini and Boschi (1997). Interferometric Bell-state analysis was experimentally demonstrated by Michler, Mattle, Weinfurter and Zeilinger (1996), using two-photon interference effects. A theory of two-photon entanglement in polarization parametric down-conversion has been developed by Rubin, Klyshko, Shih and Sergienko (1994). Two-photon quantum entanglement can be used for optical imaging [Pittman, Shih, Strekalov and Sergienko (1995)], with a possibility to develop a kind of two-photon geometric optics [Pittman, Strekalov, Klyshko, Rubin, Sergienko and Shih (1996)], and for experimental tests of Bell's inequalities in space and time [Pittman, Shih, Sergienko and Rubin (1995)]. Polarization entanglement of photon pairs provides a high-intensity source of photon pairs giving high visibility fringes and permits to observe a violation of Bell's inequalities by over 100 standard deviations in less than 5 minutes [Kwiat, Mattle, Weinfurter and Zeilinger (1995)], which can further be improved by means of a postselection procedure [Strekalov, Pittman, Sergienko, Shih and Kwiat (1996)]. Two-photon interference is a practical tool for experimental analysis of femtosecond fluorescence pulses [Li, Baba and Matsuoka (1997)]. Two-photon interference can differ from interference of two photons, as demonstrated by Pittman, Strekalov, Migdall, Rubin, Sergienko and Shih (1996). Theory of two-photon entanglement for spontaneous parametric down-conversion driven by a narrow pump pulse was developed by Keller and Rubin (1997) and spectral information and distinguishability in down-conversion with a broadband pump were discussed by Grice and Walmsley (1997). Optimal quantum measurements for phase-shift estimation in optical interferometry can be realized using an active interferometer containing two parametric amplifiers [Sanders, Milburn and Zhang (1997)].

Besides nonlocal interference, nonlocal diffraction [Christanell, Weinfurter and Zeilinger (1993), Strekalov, Sergienko, Klyshko and Shih (1995), Shih, Sergienko, Pittman, Strekalov and Klyshko (1996)], and nonlocal dispersion can be demonstrated [Steinberg, Kwiat and Chiao (1992a), Larchuk, Teich and Saleh (1995)] and applied.

Quantum interference law can have important applications in quantum cryptography and teleportation [Ekert, Rarity, Tapster and Palma (1992), Goldenberg and

Vaidman (1995), Sun, Mazurenko and Fainman (1995), Braunstein and Mann (1995), Huttner, Imoto, Gisin and Mor (1995)]. A review of quantum phenomena in optical interferometry was provided by Hariharan and Sanders (1996).

5.2 Nondemolition measurements

Quantum measurements performed on a physical system disturb the system with respect to quantum principles and introduce uncontrollable quantum fluctuations into it. However, in principle, there exist dynamical variables of a quantum system which can remain undisturbed by the measurement, because the measurement may introduce the disturbance and the corresponding quantum noise into the conjugate variables of the system. Such a dynamical variable can be called a quantum nondemolition variable and the corresponding measurement is a quantum nondemolition measurement [Braginsky and Vorontsov (1974), Unruh (1978, 1979), Hollenhorst (1979), Braginsky, Vorontsov and Thorne (1980), Caves, Thorne, Drever, Sandberg and Zimmermann (1980), Milburn and Walls (1983a, b), Imoto, Haus and Yamamoto (1985), Teich and Saleh (1988), Braginsky and Khalili (1992), Reynaud, Heidmann, Giacobino and Fabre 1992, Meystre (1992)]. In the Heisenberg picture, the operator $\hat{A}$ represents the quantum nondemolition variable if

$$[\hat{A}(t), \hat{A}(t')] = \hat{0} \tag{5.2.1}$$

for all times t, t'. In such a case the repeated measurement of $\hat{A}$ will yield the same outcome. This will occur if $\hat{A}$ commutes with the total hamiltonian $\hat{H}$ of the system,

$$[\hat{A}, \hat{H}] = \hat{0}. \tag{5.2.2}$$

This means that, in the Schrödinger picture, if a system satisfying (5.2.2) starts from an eigenstate of $\hat{A}$, it remains in this eigenstate and $\hat{A}$ is really quantum nondemolition variable. Sometimes one speaks of a back-action evading variable represented by $\hat{A}$ if the interaction Hamiltonian $\hat{H}_{\text{int}}$ for the measurement depends only on the operator $\hat{A}$, so that it is unaffected by the interaction with the measurement device. This method was originally developed for a mechanical harmonic oscillator assumed to be in interaction with gravitational waves, in order to improve the signal-to-noise ratio of detecting devices.

As an example, the Kerr effect can serve, which is caused by the intensity dependence of the refractive index of the nonlinear medium. Considering a two-mode optical field interacting with a nonlinear medium possessing the Kerr nonlinearity, the Hamiltonian of the system can be written in the form

$$\hat{H} = \hbar\omega_1 \hat{a}_1^\dagger \hat{a}_1 + \hbar\omega_2 \hat{a}_2^\dagger \hat{a}_2 + \hbar g_1 \hat{a}_1^{\dagger 2} \hat{a}_1^2 + \hbar g_2 \hat{a}_2^{\dagger 2} \hat{a}_2^2 + \hbar\kappa \hat{a}_1^\dagger \hat{a}_1 \hat{a}_2^\dagger \hat{a}_2, \tag{5.2.3}$$

where ω_1 and ω_2 are the corresponding free frequencies, g_1 and g_2 are single-mode Kerr coupling constants, and κ is an intermodal Kerr coupling constant; these coupling constants are proportional to the third-order susceptibility in the medium. Taking into account that the normally ordered fourth-order moments can be expressed in terms of the number operators as follows $\hat{a}_j^{\dagger 2}\hat{a}_j^2 = \hat{n}_j(\hat{n}_j - 1)$, $j = 1, 2$, we see that

$$[\hat{n}_j, \hat{H}] = \hat{0}, \quad j = 1, 2, \tag{5.2.4}$$

and the photon numbers n_j are constants of motion and therefore quantum nondemolition variables. They can be measured as well as their moments and the photon statistics without affecting subsequent measurements, because these quantities remain unchanged in time. If also losses in the radiation modes are included, this interaction is described by the Heisenberg–Langevin equations

$$
\begin{aligned}
\frac{d\hat{a}_1}{dt} &= -(i\omega_1 + \frac{\gamma_1}{2})\hat{a}_1 - i2g_1\hat{n}_1\hat{a}_1 - i\kappa\hat{a}_1\hat{n}_2 + \hat{L}_1, \\
\frac{d\hat{a}_2}{dt} &= -(i\omega_2 + \frac{\gamma_2}{2})\hat{a}_2 - i2g_2\hat{n}_2\hat{a}_2 - i\kappa\hat{n}_1\hat{a}_2 + \hat{L}_2,
\end{aligned}
\tag{5.2.5}
$$

where γ_j are the damping constants of the radiation modes and $\hat{L}_j$ are the corresponding Langevin forces (see, e. g., [Peřina (1991), chapter 7], section 3.9). Quantum statistical properties of the system are completely described by the generalized Fokker–Planck equation for the quasidistribution $\Phi_{\mathcal{A}}(\alpha_1, \alpha_2, t)$ related to the antinormal ordering of field operators as follows

$$
\begin{aligned}
\frac{\partial \Phi_{\mathcal{A}}}{\partial t} &= \sum_{j=1}^{2} \left[i2g_j|\alpha_j|^2\alpha_j\frac{\partial \Phi_{\mathcal{A}}}{\partial \alpha_j} + ig_j\alpha_j^2\frac{\partial^2 \Phi_{\mathcal{A}}}{\partial \alpha_j^2} + \text{c. c.} \right] \\
&\quad + i\kappa\left[|\alpha_2|^2\alpha_1\frac{\partial \Phi_{\mathcal{A}}}{\partial \alpha_1} + |\alpha_1|^2\alpha_2\frac{\partial \Phi_{\mathcal{A}}}{\partial \alpha_2} + \alpha_1\alpha_2\frac{\partial^2 \Phi_{\mathcal{A}}}{\partial \alpha_1\partial \alpha_2} - \text{c. c.} \right] \\
&\quad + \sum_{j=1}^{2} \left[(i\omega_j + \frac{\gamma_j}{2})\frac{\partial}{\partial \alpha_j}(\alpha_j\Phi_{\mathcal{A}}) + \text{c. c.} + \gamma_j(\langle n_{\mathrm{R}}^{(j)}\rangle + 1)\frac{\partial^2 \Phi_{\mathcal{A}}}{\partial \alpha_j\partial \alpha_j^*} \right], \tag{5.2.6}
\end{aligned}
$$

where $\langle n_{\mathrm{R}}^{(j)}\rangle$ are the mean photon numbers of reservoir oscillators. If only nonlinear dynamics is considered and losses are neglected, the operator solutions are easily expressed as

$$
\begin{aligned}
\hat{a}_1(t) &= \exp[-i\omega_1 t - i2g_1 t\hat{n}_1(0) - i\kappa t\hat{n}_2(0)]\hat{a}_1(0), \\
\hat{a}_2(t) &= \exp[-i\omega_2 t - i2g_2 t\hat{n}_2(0) - i\kappa t\hat{n}_1(0)]\hat{a}_2(0), \tag{5.2.7}
\end{aligned}
$$

which clearly demonstrates that only the phase factors evolve in this nonlinear process, whereas the photon-number statistics are phase and time independent. Also the generalized Fokker–Planck equation (5.2.6) can be solved in a closed non-Gaussian form [Peřina, Horák, Hradil, Sibilia and Bertolotti (1989)]

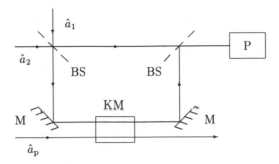

Figure 5.16: Scheme of nondemolition measurement using the Mach–Zehnder interferometer with Kerr medium, $\hat{a}_1$ and $\hat{a}_2$ describe signal beams, $\hat{a}_p$ describes the probe beam, KM is Kerr medium, P photodetector, BS are beamsplitters, and M mirrors.

$$
\Phi_A(\alpha_1, \alpha_2, t) = \frac{1}{\pi^2} \exp(-|\alpha_1|^2 - |\alpha_2|^2 - |\xi_1|^2 - |\xi_2|^2)
$$
$$
\times \sum_{k,l,m,n=0}^{\infty} \frac{(\alpha_1 \xi_1^* e^{i\omega_1 t})^k (\alpha_1^* \xi_1 e^{-i\omega_1 t})^l (\alpha_2 \xi_2^* e^{i\omega_2 t})^m (\alpha_2^* \xi_2 e^{-i\omega_2 t})^n}{k!\,l!\,m!\,n!}
$$
$$
\times \exp\{ig_1 t[k(k-1) - l(l-1)] + ig_2 t[m(m-1) - n(n-1)]
$$
$$
+ i\kappa t(km - ln)\}, \tag{5.2.8}
$$

provided that the radiation field is in the initial coherent state $|\xi_1\rangle|\xi_2\rangle$. For further discussion of this subject we refer the reader to a review by Peřinová and Lukš (1994).

We can now interpret mode 1 as a signal mode (s) and mode 2 as a probe mode (p) and we see that then both the operators $\hat{n}_s$ and $\hat{n}_p$ are quantum nondemolition variables of back-action evading type, which are constants of motion. The interaction between modes affects the phases of the optical field, but not photon numbers. We see that the probe field affects the phase of the signal field and also the signal field causes an intensity-dependent phase itself, and similar is the effect of the signal field on the probe field and the probe field on itself. Hence, the Kerr effect makes it possible to measure intensity or photon number of the signal field without disturbing the photon number. This can be used in the Mach–Zehnder interferometer, where one can put the Kerr medium in one of the arms of the interferometer (figure 5.16) [Yamamoto, Imoto and Machida (1986), Sanders and Milburn (1989)]. A probe wave propagating through the Kerr medium can be used for nondemolition photon-number measurements, which will introduce a phase shift into the probe wave with respect to the presence of a photon in the signal wave or its absence. However, such a nondemolition measurement does not mean that the path of a photon can

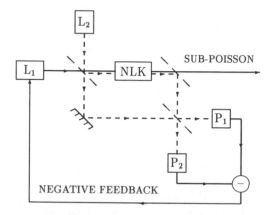

Figure 5.17: Generation of sub-Poissonian light using quantum nondemolition measurement of the laser output photon number to control the laser excitation rate; L_1 is a signal laser, L_2 probe laser, NLK nonlinear Kerr medium, and P are photodetectors.

be determined without affecting the interference pattern. As analyzed by Mandel and Wolf (1995) in section 22.6, the fringe visibility can be unity only when the signal-to-noise ratio is zero, in which case one cannot know any information about the presence of a photon in the arm. On the other hand, if this ratio is large, the fringe visibility vanishes, which reflects a complementary behaviour described by an uncertainty relation. Thus, the idea that interference is a manifestation of indistinguishability of the photon paths is pointed out.

The possibility to use the Kerr effect for measurement of the photon number in the signal mode without photon annihilation following the phase of a probe wave passed through the Kerr medium also provides information about the change of nonlinear index of refraction in dependence on the signal photon number in the medium. Precision in the signal photon-number measurement causes an increased uncertainty in the corresponding phase variable, as indicated in the photon-number–phase uncertainty relation ((4.9.23))

$$\left[\langle(\Delta\hat{n})^2\rangle + \frac{1}{4}\right] D\varphi \geq \frac{1}{4}, \qquad (5.2.9)$$

where the phase dispersion $D\varphi = 1 - |\langle\widehat{\exp}(i\varphi)\rangle|^2$ has been chosen as the measure of phase uncertainty. Yamamoto, Imoto and Machida (1986) suggested to use such a quantum nondemolition measurement at the output of a semiconductor laser for a negative feedback to control the rate of excitation of the laser (figure 5.17). This permits to produce sub-Poisson light beams.

Besides this application of quantum nondemolition measurements for generation of sub-Poissonian light, various which-path experiments can be suggested [Kärtner and Haus (1993)] for the investigation of collapse of the wave function during a measurement, and single photons can be detected in an open resonator by atomic beam deflection, including spontaneous emission and losses [Matsko, Vyatchanin, Mabuchi and Kimble (1994)]; also solitons can be adopted for quantum nondemolition measurements [Drummond, Breslin and Shelby (1994)]. Back-action evading measurements for quantum nondemolition detection and quantum optical taping using optical parametric amplifier were performed by Pereira, Ou and Kimble (1994). Repeated quantum nondemolition measurements can also be realized [Alter and Yamamoto (1995), Bencheikh, Levenson, Grangier and Lopez (1995)]. Quantum nondemolition measurement of intensity-difference fluctuations using a nondegenerate parametric oscillator was discussed by Harrison and Walls (1996). Interesting experimental results for the meaning of path information in quantum interferometry were presented by Zeilinger, Herzog, Horne, Kwiat, Mattle and Weinfurter (1996). Experimental realization of a quantum eraser using down-conversion process was performed by Monken, Branning and Mandel (1996). Using a single-photon interference experiment together with a quantum nondemolition measurement scheme for which-path information, a precision phase measurement was discussed by Ou (1996). Quantum nondemolition measurements using cold trapped atoms were performed by Roch, Vigneron, Grelu, Sinatra, Poizat and Grangier (1997). Continuous quantum nondemolition measurements using monolithic degenerate optical parametric amplifier were realized by Bruckmeier, Schneider, Schiller and Mlynek (1997). Interesting quantum interference effects can be obtained in the process of down-conversion with optical pulses [Ou (1997a), Di Giuseppe, Haiberger, De Martini and Sergienko (1997)].

5.3   Phase-dependent measurements

The quantum phase problem can be illuminated in the context of quantum estimation theory, especially in the context of the quantum estimation of a phase shift. As the phase shift is a "classical" quantity, the quantum estimation theory appears to be interwoven with the quantum measurement theory (section 5.1). From this it could also be deduced that at a certain level of analysis a concrete quantum measurement could be separated from a residual classical estimation theory.

Let us consider the statistical operator $\hat{\rho}(\boldsymbol{\theta})$, which depends on the parameters $\theta_1, \theta_2, \ldots, \theta_m$ to be estimated. We assume that $\boldsymbol{\theta} = (\theta_1, \theta_2, \ldots, \theta_m)$ belongs to an m-dimensional space Θ. Quantum estimation consists in inferring some c-number estimates $\tilde{\theta}_1, \tilde{\theta}_2, \ldots, \tilde{\theta}_m$ of the parameters and is described by a probability-operator measure $\hat{\Pi}(\Delta)$, where Δ is a particular region of the parameter space Θ. In the Bayesian formulation of the estimation theory, we must provide the prior probability density function $z(\boldsymbol{\theta})$ of the parameters and a function $C(\tilde{\boldsymbol{\theta}}, \boldsymbol{\theta})$ to assess the cost of errors in the estimates. These estimates will be random variables and the probability that they lie in the region Δ will be given by

$$\text{Prob}(\tilde{\boldsymbol{\theta}} \in \Delta | \boldsymbol{\theta}) = \text{Tr}\left\{\hat{\rho}(\boldsymbol{\theta})\hat{\Pi}(\Delta)\right\}, \tag{5.3.1}$$

when $\boldsymbol{\theta}$ represents the true values of the parameters. The appropriate probability density reads

$$P(\tilde{\boldsymbol{\theta}}|\boldsymbol{\theta}) = \text{Tr}\left\{\hat{\rho}(\boldsymbol{\theta})\frac{\hat{\Pi}(d^m\tilde{\boldsymbol{\theta}})}{d^m\tilde{\boldsymbol{\theta}}}\right\}. \tag{5.3.2}$$

The average cost incurred is

$$\begin{aligned}
\overline{C} &= \int_\Theta \int_\Theta z(\boldsymbol{\theta})C(\tilde{\boldsymbol{\theta}},\boldsymbol{\theta})P(\tilde{\boldsymbol{\theta}}|\boldsymbol{\theta})\, d^m\boldsymbol{\theta}\, d^m\tilde{\boldsymbol{\theta}} \\
&= \text{Tr}\left\{\int_\Theta \hat{W}(\tilde{\boldsymbol{\theta}})\hat{\Pi}(d^m\tilde{\boldsymbol{\theta}})\right\}, \tag{5.3.3}
\end{aligned}$$

upon introducing the Hermitian risk operator

$$\hat{W}(\tilde{\boldsymbol{\theta}}) = \int_\Theta z(\boldsymbol{\theta})C(\tilde{\boldsymbol{\theta}},\boldsymbol{\theta})\hat{\rho}(\boldsymbol{\theta})\, d^m\boldsymbol{\theta}. \tag{5.3.4}$$

It is seen that the best estimator, i. e., the best probability-operator measure, is that for which the average cost is least. The equations for the optimum estimator [Helstrom (1976)] have the form

$$\left[\hat{W}(\tilde{\boldsymbol{\theta}}) - \hat{\Upsilon}\right]\frac{\hat{\Pi}(d^m\tilde{\boldsymbol{\theta}})}{d^m\tilde{\boldsymbol{\theta}}} = \hat{0}, \tag{5.3.5}$$

$$\hat{W}(\tilde{\boldsymbol{\theta}}) - \hat{\Upsilon} \geq \hat{0}, \tag{5.3.6}$$

where the Lagrange operator $\hat{\Upsilon}$ is given by

$$\begin{aligned}
\hat{\Upsilon} &= \int_\Theta \hat{W}(\tilde{\boldsymbol{\theta}})\hat{\Pi}(d^m\tilde{\boldsymbol{\theta}}) \\
&= \int_\Theta \frac{\hat{\Pi}(d^m\tilde{\boldsymbol{\theta}})}{d^m\tilde{\boldsymbol{\theta}}}\hat{W}(\tilde{\boldsymbol{\theta}})\, d^m\tilde{\boldsymbol{\theta}}. \tag{5.3.7}
\end{aligned}$$

The two forms differing by the ordering indicate that the Lagrange operator is Hermitian rather than that $\hat{W}(\tilde{\boldsymbol{\theta}})$ and $\hat{\Pi}(d^m\tilde{\boldsymbol{\theta}})/d^m\tilde{\boldsymbol{\theta}}$ commute.

The maximum-likelihood estimation can be analyzed according to the above scheme for the delta-function cost

$$C(\tilde{\boldsymbol{\theta}},\boldsymbol{\theta}) = -\prod_{k=1}^m \delta(\tilde{\theta}_k - \theta_k). \tag{5.3.8}$$

Denoting

$$\hat{W}_-(\tilde{\boldsymbol{\theta}}) = -\hat{W}(\tilde{\boldsymbol{\theta}}) = z(\tilde{\boldsymbol{\theta}})\hat{\rho}(\tilde{\boldsymbol{\theta}}), \tag{5.3.9}$$

we write the equations (5.3.5), (5.3.6) in the form

$$\left[\hat{\Upsilon}_- - \hat{W}_-(\tilde{\boldsymbol{\theta}})\right]\frac{\hat{\Pi}(d^m\tilde{\boldsymbol{\theta}})}{d^m\tilde{\boldsymbol{\theta}}} = \hat{0}, \tag{5.3.10}$$

$$\hat{\Upsilon}_- - \hat{W}_-(\tilde{\boldsymbol{\theta}}) \geq \hat{0}, \tag{5.3.11}$$

with the Lagrange operator

$$\hat{\Upsilon}_- = \int_\Theta \hat{W}_-(\tilde{\boldsymbol{\theta}})\hat{\Pi}(d^m\tilde{\boldsymbol{\theta}}). \tag{5.3.12}$$

Let us consider the physical system of quantum rotator (section 4.6) and the statistical operators

$$\hat{\rho}(\theta) = \exp(i\theta\hat{N})\hat{\rho}(0)\exp(-i\theta\hat{N}), \quad \theta \in [-\pi, \pi), \tag{5.3.13}$$

where $\hat{\rho}(0)$ corresponds to an initial state. Now we shall determine the maximum-likelihood estimator of the angle θ in (5.3.13) when the quantum rotator is in a pure state initially,

$$\hat{\rho}(\theta) = |\psi(\theta)\rangle\langle\psi(\theta)|, \quad |\psi(\theta)\rangle = \exp(i\theta\hat{N})|\psi(0)\rangle. \tag{5.3.14}$$

We use the delta-function cost for $m = 1$. Supposing that nothing is known in advance about the true value of the displacement parameter θ, we assign to it a uniform prior probability density function

$$z(\theta) = \frac{1}{2\pi}, \quad \theta \in [-\pi, \pi). \tag{5.3.15}$$

Seeking a solution as an unnormalized pure state

$$\frac{\hat{\Pi}(d\tilde{\theta})}{d\tilde{\theta}} = |\xi(\tilde{\theta})\rangle\langle\xi(\tilde{\theta})|, \tag{5.3.16}$$

we write the optimization equations (5.3.10), (5.3.11), with (5.3.12) as

$$\left[\hat{\Upsilon}_- - \frac{1}{2\pi}\hat{\rho}(\tilde{\theta})\right]|\xi(\tilde{\theta})\rangle = 0, \tag{5.3.17}$$

$$\hat{\Upsilon}_- - \frac{1}{2\pi}\hat{\rho}(\tilde{\theta}) \geq \hat{0}, \tag{5.3.18}$$

with

$$\hat{\Upsilon}_- = \frac{1}{2\pi}\int_{-\pi}^{\pi}\hat{\rho}(\tilde{\theta})|\xi(\tilde{\theta})\rangle\langle\xi(\tilde{\theta})|\,d\tilde{\theta}. \tag{5.3.19}$$

We proceed by what is called a covariant measurement and assume that [Holevo (1982)]

$$|\xi(\tilde{\theta})\rangle = \exp(i\tilde{\theta}\hat{N})|\xi(0)\rangle, \tag{5.3.20}$$

where $|\xi(0)\rangle$ remains to be determined. The Lagrange operator is diagonal with respect to the number state basis,

$$
\begin{aligned}
\hat{\Upsilon}_- &= \langle\psi(0)|\xi(0)\rangle \sum_{N=-\infty}^{\infty} \langle N|\psi(0)\rangle\langle\xi(0)|N\rangle |N\rangle\langle N| \\
&= \langle\xi(0)|\psi(0)\rangle \sum_{N=-\infty}^{\infty} \langle\psi(0)|N\rangle\langle N|\xi(0)\rangle |N\rangle\langle N|.
\end{aligned}
\tag{5.3.21}
$$

The appropriate formula for the maximum likelihood can be easily applied,

$$
\begin{aligned}
-\overline{C} &= \mathrm{Tr}\left\{\hat{\Upsilon}_-\right\} \\
&= |\langle\psi(0)|\xi(0)\rangle|^2.
\end{aligned}
\tag{5.3.22}
$$

Since $\hat{\Pi}([-\pi,\pi)) = \hat{1}_e$, we obtain that

$$
|\langle N|\xi(0)\rangle|^2 = \frac{1}{2\pi}.
\tag{5.3.23}
$$

From this,

$$
|\langle N|\xi(\tilde{\theta})\rangle|^2 = \frac{1}{2\pi},
\tag{5.3.24}
$$

or the covariant measurement yields the uniform distribution of the desired estimator in the number states which occur. Of course, the phase of the coefficients $\langle N|\xi(0)\rangle$ is to be chosen with the Hermiticity of $\hat{\Upsilon}_-$ in mind

$$
\arg[\langle N|\xi(0)\rangle] = \arg[\langle N|\psi(0)\rangle] + \text{constant}, \quad \text{if } \langle N|\psi(0)\rangle \neq 0.
\tag{5.3.25}
$$

By (5.3.22), we obtain that $-\overline{C} \geq \frac{1}{2\pi}$ and the equality is attained only for the number states. In the condition (5.3.18), we can set $\tilde{\theta} = 0$. Having rewritten the relation

$$
\left\langle\chi\left|\left[\hat{\Upsilon}_- - \frac{1}{2\pi}\hat{\rho}(0)\right]\right|\chi\right\rangle \geq 0
\tag{5.3.26}
$$

in the form of the Schwartz inequality for the kets $|\chi\rangle'$, $|\psi(0)\rangle'$,

$$
|\chi\rangle' = \sum_{N=-\infty}^{\infty} \sqrt{|\langle N|\psi(0)\rangle|}\,\langle N|\chi\rangle|N\rangle,
\tag{5.3.27}
$$

$$
|\psi(0)\rangle' = \sum_{N=-\infty}^{\infty}{}^{(0)} \frac{\langle N|\psi(0)\rangle}{\sqrt{|\langle N|\psi(0)\rangle|}}|N\rangle,
\tag{5.3.28}
$$

we prove that the formulae

$$
\langle N|\xi(0)\rangle = \begin{cases} \dfrac{1}{\sqrt{2\pi}}\dfrac{\langle N|\psi(0)\rangle}{|\langle N|\psi(0)\rangle|}, & \text{if } \langle N|\psi(0)\rangle \neq 0, \\[2mm] \text{defined in part,} & \text{if } \langle N|\psi(0)\rangle = 0, \end{cases}
\tag{5.3.29}
$$

determine essentially the maximum-likelihood estimator. The summation in (5.3.28) extends only to the terms where division is defined. Introducing a diagonal operator

$$\hat{A}(0) = \sum_{N=-\infty}^{\infty}{}^{(0)} \mathrm{Arg}_{-\pi}[\langle N|\psi(0)\rangle]|N\rangle\langle N|, \qquad (5.3.30)$$

we may write a maximum-likelihood estimator in the compact form

$$|\xi(\tilde{\theta})\rangle = \exp[i\,\hat{A}(0)]|\phi = \tilde{\theta}\rangle_c, \qquad (5.3.31)$$

where the rotation-angle states are given in (4.3.41). In the plane rotator system it is possible to define the unitary operator of the estimation

$$\widehat{\exp}(i\tilde{\theta}) = \int_{-\pi}^{\pi} \exp(i\tilde{\theta})|\xi(\tilde{\theta})\rangle\langle\xi(\tilde{\theta})|\,d\tilde{\theta} \qquad (5.3.32)$$

$$= \exp[i\hat{A}(0)]\widehat{\exp}(i\phi)\exp[-i\hat{A}(0)], \qquad (5.3.33)$$

with the unitary rotation-angle operator $\widehat{\exp}(i\phi)$ given in section 4.3 (the formula (4.3.40)).

The formula (5.3.31) is the solution of the estimation problem for the rotation angle with the initial pure state (5.3.14) and a more general cost function. The only condition is [Holevo (1982), p. 176] that the cost function is of the form $C(\tilde{\theta}, \theta) = C_d(\tilde{\theta} - \theta)$, where $C_d(\tilde{\theta} - \theta)$ is an even 2π-periodic function satisfying the relation

$$\int_0^{2\pi} C_d(\varphi)\cos(k\varphi)\,d\varphi \le 0, \quad k = 1, 2, \ldots . \qquad (5.3.34)$$

Let us remark that for the estimation of a displacement parameter the quadratic cost function is of importance. With respect to the peculiarity of the rotation angle, the quadratic cost function should be modified to the cost function

$$C_{\sin}(\tilde{\theta}, \theta) = \mu_P^2(\tilde{\theta}, \theta) \qquad (5.3.35)$$

$$= 4\sin^2\left(\frac{\tilde{\theta} - \theta}{2}\right), \qquad (5.3.36)$$

with the chord metric $\mu_P(\tilde{\theta}, \theta)$ given in (4.4.31). The explicit 2π-periodicity of the cost function may be useful also for the maximum-likelihood estimation,

$$C_{\bar{\delta}}(\tilde{\theta}, \theta) = -\bar{\delta}(\tilde{\theta} - \theta). \qquad (5.3.37)$$

It is interesting to note that the theory of the optimum estimation of the phase shift in the system of harmonic oscillator is of the same form as the above theory for the amount of rotation in the system of the plane rotator. The only difference is that the prescription (5.3.31) has no analogue to generate orthogonal states. We observe that the partial phase states [Pegg and Barnett (1989)] can be written in the form

$$|\psi(0)\rangle = \exp[i\overline{\varphi}(0)\hat{n}]\sum_{n=0}^{\infty}|\langle n|\psi(0)\rangle||n\rangle, \qquad (5.3.38)$$

where $\overline{\varphi}(0) \in [-\pi, \pi)$. In this case, $\hat{A}(0) = \overline{\varphi}(0)\hat{n}$, so that the unitary estimation operator

$$
\begin{aligned}
\widehat{\exp}(i\tilde{\theta}) &= \widehat{\exp}\{i[\varphi - \overline{\varphi}(0)]\} \\
&= \exp[i\overline{\varphi}(0)\hat{n}]\widehat{\exp}(i\varphi)\exp[-i\overline{\varphi}(0)\hat{n}].
\end{aligned} \tag{5.3.39}
$$

To the contrary, phase-admixed states [Vourdas and Bishop (1989)] on the input would lead to an appropriately involved unitary estimation operator.

A restriction to the space $\Psi_s(\mathcal{H}_s)$ (see section 4.7) does not cause any substantial change of the theory. In this connection, let us note that the Pegg–Barnett orthogonal states arise in the multiple hypothesis testing [Helstrom (1976), p. 100]. From the viewpoint of decision theory, the multiple hypothesis testing has much in common with the theory of signal detection. An application of the homodyne detection for the decision between four coherent states phases $\frac{\pi}{2}$ apart has been performed by Helstrom, Charbit and Bendjaballah (1987). Ban (1997) has approached the homodyne-direct receiver or Kennedy's receiver [Kennedy (1973), Helstrom (1976), Shapiro (1980)]. He has obtained the quantum detection operators and an average probability of error for any binary quantum-state signal. He has taken into account the quantum efficiency η of a photodetector used in the receiver. He has investigated, in detail, the binary phase-shift keyed signal including thermal noise.

The theory complicates when two light modes are supposed, but still a phase shift is estimated. Let us begin with the phase shift in the first mode. For definiteness, we rewrite the second relation in (5.3.14) as

$$
|\psi(\theta)\rangle = \exp(i\theta\hat{n}_1)|\psi(0)\rangle. \tag{5.3.40}
$$

Similarly, the relation (5.3.20) becomes

$$
|\xi(\tilde{\theta})\rangle = \exp(i\tilde{\theta}\hat{n}_1)|\xi(0)\rangle, \tag{5.3.41}
$$

where $|\xi(0)\rangle$ is an unknown unnormalized ket. As a generalization of (5.3.23), we obtain that

$$
\langle\xi(0)|n_1\rangle_{11}\langle n_1|\xi(0)\rangle = \frac{1}{2\pi}, \quad \text{if } {}_1\langle n_1|\psi(0)\rangle \neq 0. \tag{5.3.42}
$$

On the left-hand side in (5.3.42) the scalar product in the second mode is taken. Of course, the relation (5.3.42) is only a property of the solution,

$$
{}_1\langle n_1|\xi(0)\rangle = \frac{1}{\sqrt{2\pi}}\frac{{}_1\langle n_1|\psi(0)\rangle}{\sqrt{\langle\psi(0)|n_1\rangle_{11}\langle n_1|\psi(0)\rangle}}, \quad \text{if } {}_1\langle n_1|\psi(0)\rangle \neq 0. \tag{5.3.43}
$$

The reduction to the Susskind–Glogower phase system in the first mode is possible for the input states with the property

$$
|\psi(0)\rangle = |\psi_1(0)\rangle_1 \otimes |\psi_2(0)\rangle_2. \tag{5.3.44}
$$

Let us proceed with the case, where the phase shifts in the two modes have the same magnitude, but the opposite sense. We rewrite the second relation in (5.3.14) with the aid of the photon-number difference operator $\hat{n}_d$ given in (4.8.54)

$$|\psi(\theta)\rangle = \exp\left(i\frac{\theta}{2}\hat{n}_d\right)|\psi(0)\rangle. \tag{5.3.45}$$

Accordingly, the relation (5.3.20) becomes

$$|\xi(\tilde{\theta})\rangle = \exp\left(i\frac{\tilde{\theta}}{2}\hat{n}_d\right)|\xi(0)\rangle, \tag{5.3.46}$$

where $|\xi(0)\rangle$ will be given below. As an analogue of the relation (5.3.42), we note that

$$\langle\xi(0)|\hat{P}_d(n_d)|\xi(0)\rangle = \frac{1}{2\pi}, \quad \text{if } \langle\psi(0)|\hat{P}_d(n_d)|\psi(0)\rangle \neq 0, \tag{5.3.47}$$

where

$$\hat{P}_d(n_d) = \sum_{n_1=n_d}^{\infty} \hat{P}(n_1, n_1 - n_d), \quad n_d \geq 0,$$
$$\hat{P}_d(n_d) = \sum_{n_2=-n_d}^{\infty} \hat{P}(n_d + n_2, n_2), \quad n_d < 0, \tag{5.3.48}$$

with

$$\hat{P}(n_1, n_2) = |n_1, n_2\rangle\langle n_1, n_2|. \tag{5.3.49}$$

The solution is characterized by its properties

$$\hat{P}_d(n_d)|\xi(0)\rangle = \frac{1}{\sqrt{2\pi}} \frac{\hat{P}_d(n_d)|\psi(0)\rangle}{\sqrt{\langle\psi(0)|\hat{P}_d(n_d)|\psi(0)\rangle}}, \quad \text{if } \langle\psi(0)|\hat{P}_d(n_d)|\psi(0)\rangle \neq 0. \tag{5.3.50}$$

The reduction to a result of quantum phase theory is possible when the input state is of the form

$$|\psi(0)\rangle = \sum_{N=-\infty}^{\infty} \sum_{m=0}^{\infty} c_N^{(d)} c_m^{(\min)}|N, m\rangle, \tag{5.3.51}$$

where the relative number states have been defined in (4.13.84). When the quantum estimation theory is applied, the case of single mode and that of quantum rotator reappear in the two-mode case with minor changes. We have seen that in the first example the Susskind–Glogower phase states reappear with the replacement of the ket $\frac{1}{\sqrt{2\pi}}|n\rangle$ by the ket $|n_1\rangle_1 \otimes {}_1\langle n_1|\xi(0)\rangle$. Similarly, in the second example, the rotation-angle states should be found after the replacement of the ket $\frac{1}{\sqrt{2\pi}}|N\rangle$ by the ket $\hat{P}_d(n_d = N)|\xi(0)\rangle$. Although we have recognized the Rocca–Sirugue–Ban quantum phase formalism in a particular case, in general, no such a two-mode phase formalism can be derived in the framework of the quantum estimation theory.

The quantum interferometry enjoys a copious literature on two-mode schemes. Apart from the details of the description, the fact that in one arm of the interferometer the phase shift is generated by the corresponding photon-number operator (cf. (5.3.40)) entails that the Susskind–Glogower phase states should be realized in this device as a means and the result of optimization of the phase-shift estimation. In this connection another kind of optimization is formulated. It is interesting also when the optimization of the first kind cannot be accomplished as it ought to be. It is the optimization of the input state $|\psi(0)\rangle$. In quantum interferometry, the quantum estimation theory cannot be used, as the probability-operator measure is known, e. g., the operator $\hat{J}_3$ is measured ($\hat{J}_{3\,out}$ in the Heisenberg picture). Especially, the moment method is used. This consists in equating the photon-number difference to its expectation value where the parameter θ has been replaced by the estimator $\tilde{\theta}$,

$$n_{\mathrm{d}} = \langle n_{\mathrm{d\,out}}\rangle|_{\theta\to\tilde{\theta}}. \tag{5.3.52}$$

When the random variable $\tilde{\theta}$ is expressed explicitly from the relation (5.3.52), the definition of this estimator is complete. This classical technique has a quantal reformulation,

$$\hat{\tilde{J}}_{3\,out} = \langle \hat{J}_{3\,out}\rangle\Big|_{\theta\to\hat{\tilde{\theta}}}. \tag{5.3.53}$$

In this manner, the estimator of the phase shift has not been derived, because the information acquired suffices only for the estimation of unsigned deviation from a "reference" phase. The quality of the estimator can be evaluated more easily according to a linearized estimator whose variance reads

$$\langle (\Delta\hat{\tilde{\theta}})^2 \rangle = \frac{\langle (\Delta \hat{J}_{3\,out})^2 \rangle}{\left(\frac{\partial \langle \hat{J}_{3\,out}\rangle}{\partial\theta} \right)^2}. \tag{5.3.54}$$

However, we can appreciate the simplification we obtain on setting $\theta = 0$. The variance $\langle (\Delta\hat{\tilde{\theta}})^2 \rangle|_{\theta=0}$ then determines the minimum detectable phase shift (squared) and, therefore, the (squared) sensitivity of an interferometer. In case the linearization of the estimator is not realiable, perhaps, the estimation should be based on another expectation value, e. g., we could equate

$$n_{\mathrm{d}}^2 = \langle n_{\mathrm{d\,out}}^2\rangle\Big|_{\theta\to\tilde{\theta}}, \quad \text{etc.} \tag{5.3.55}$$

The determination of optimal states is still a nonclassical problem. Squeezed input states have presented a suitable generalization of the input coherent state and they proved useful for the optimized interferometry [Caves (1981)]. The description of an interferometer has not been restricted only to unitary transformations and losses have also been taken into account in [Gea-Banacloche and Leuchs (1987)].

Rather obviously, the optimization of the average cost given in (5.3.3) with respect to the input state $\hat{\rho}(0)$ when the estimator is known leads to the eigenvalue problems

$$z(0)\hat{C}(0)\hat{\rho}(0) = \left[\lambda\hat{1} + \sum_{k=1}^{K} \lambda_k \hat{B}_k(0) \right] \hat{\rho}(0), \tag{5.3.56}$$

where the cost operator has been introduced,

$$\hat{C}(0) = \int_\Theta C(\tilde{\boldsymbol{\theta}}, 0)\hat{\Pi}(d^m\tilde{\boldsymbol{\theta}}), \tag{5.3.57}$$

$\hat{B}_k$, $[\hat{B}_k, \hat{N}] = \hat{0}$ ($[\hat{B}_k, \hat{n}_1] = \hat{0}$, $[\hat{B}_k, \hat{n}_d] = \hat{0}$), enter the subsidiary conditions

$$\mathrm{Tr}\left\{\hat{\rho}(\boldsymbol{\theta})\hat{B}_k\right\} = \overline{B}_k, \tag{5.3.58}$$

with $\overline{B}_k$ known, and λ, λ_k, $k = 1, \ldots, K$, are the appropriate Lagrange multipliers.

On restriction to the states with fixed photon-number sum and respecting a co-variant measurement, we formulate the eigenvalue problem

$$\frac{1}{2\pi}\hat{C}_n(0)|\psi_n(0)\rangle = \lambda_n(0)|\psi_n(0)\rangle, \tag{5.3.59}$$

where the cost operator

$$\hat{C}_n(0) = \int_{-\pi}^{\pi} C(\tilde{\theta}, 0)\hat{\Pi}_n(d\tilde{\theta}). \tag{5.3.60}$$

Here the probability-operator measures $\hat{\Pi}_n(d\tilde{\theta})$ have the property

$$\hat{\Pi}_n(d\tilde{\theta}) = |\xi_n(\tilde{\theta})\rangle\langle\xi_n(\tilde{\theta})|d\tilde{\theta} \tag{5.3.61}$$

and the kets arise according to the recipe

$$|\xi_n(\tilde{\theta})\rangle - \exp[i\hat{A}_n(0)]|\varphi_- = \tilde{\theta}\rangle_n, \tag{5.3.62}$$

where the auxiliary operator

$$\hat{A}_n(0) = \sum_{n_1=0}^{n} \chi_{n_1, n-n_1}|n_1, n-n_1\rangle\langle n_1, n-n_1| \tag{5.3.63}$$

and the phase-difference ket

$$|\varphi_-\rangle_n = \frac{1}{\sqrt{2\pi}}\sum_{n_1=0}^{n}\exp(in_1\varphi_-)|n_1, n-n_1\rangle. \tag{5.3.64}$$

For two important instances of a cost function presented in (5.3.37), (5.3.36), we obtain from (5.3.60) the cost operators

$$\hat{C}_n^{(\delta)}(0) = -\hat{\Pi}_n(0), \tag{5.3.65}$$

$$\hat{C}_n^{(\sin)}(0) = 2\exp[i\hat{A}_n(0)][\hat{1}_n^{SU(2)} - \widehat{\cos}_n(\varphi_-)]\exp[-i\hat{A}_n(0)], \tag{5.3.66}$$

respectively, where the operator $\widehat{\cos}_n(\varphi_-)$ is the projection of the Susskind–Glogower phase-difference cosine operator $\widehat{\cos}(\varphi_-) \equiv \widehat{\cos}(\varphi_1 - \varphi_2)$,

$$\widehat{\cos}_n(\varphi_-) = \hat{1}_n^{SU(2)}\widehat{\cos}_n(\varphi_-)\hat{1}_n^{SU(2)}, \tag{5.3.67}$$

with the operator $\hat{I}_n^{SU(2)}$ defined in (4.8.60). The minimizing eigenkets (eigenvectors) of the cost operators (5.3.65), (5.3.66) read [Myška (1996)]

$$\left|\psi_n^{(\bar{\delta})}(0)\right\rangle = \frac{1}{\sqrt{n+1}} \sum_{n_1=0}^{n} \exp(i\chi_{n_1,n-n_1})|n_1, n-n_1\rangle, \tag{5.3.68}$$

$$\left|\psi_n^{(\sin)}(0)\right\rangle = \sqrt{\frac{2}{n+2}} \sum_{n_1=0}^{n} \exp(i\chi_{n_1,n-n_1}) \sin\left[(n_1+1)\theta_{n,\frac{1}{2}}\right] |n_1, n-n_1\rangle, \tag{5.3.69}$$

respectively, corresponding to the eigenvalues

$$\lambda_n^{(\bar{\delta})}(0) = -\frac{n+1}{4\pi^2}, \tag{5.3.70}$$

$$\lambda_n^{(\sin)}(0) = \frac{1}{\pi}\left(1 - C_{n,\frac{1}{2}}\right), \tag{5.3.71}$$

where the quantities $\theta_{n,\frac{1}{2}}$, $C_{n,\frac{1}{2}}$ are introduced in (4.8.71), (4.8.70), respectively. Bondurant and Shapiro (1984) have classified the phase-sensing interferometers as difference-direct detection and homodyne detection interferometers. They have derived that both detection schemes reach the standard quantum limit on position-measurement sensitivity at roughly the same average photon number. Multiplying devices can surpass the standard quantum limit to position sensing. The quantum phase measurement process has been investigated in the situation leading to the Susskind–Glogower probability-operator measure by Shapiro, Shepard and Wong (1989) and Shapiro and Shepard (1991).

Assuming for simplicity that $\langle n|\psi(0)\rangle \geq 0$, Shapiro, Shepard and Wong (1989) used the reciprocal likehood (cf. V_M (4.4.45))

$$\delta\tilde{\theta} = \frac{1}{P(\tilde{\theta}|\tilde{\theta})}, \tag{5.3.72}$$

where

$$\tilde{\theta} = \arg\max_{-\pi \leq \theta \leq \pi} P(\tilde{\theta}|\theta) \text{ for } -\pi \leq \tilde{\theta} < \pi, \tag{5.3.73}$$

as a performance measure in the maximum-likelihood estimation of the phase shift. They addressed the problem of estimation for a single mode of the radiation field under an average-photon-number constraint. Since this constraint was not sufficient for obtaining well-behaved solution, they restricted the Hilbert space of the input (cf. the formulations (4.9.28) and (4.9.67)). The solution is of the form (4.9.68), with $\lambda = -r$, $r \geq 1$. They performed a comparison of their optimized phase measurement with the optimized squeezed state interferometry [Caves (1981)], where the value $\Delta\tilde{\theta} \equiv \langle(\Delta\hat{\tilde{\theta}})^2\rangle^{\frac{1}{2}} \approx \overline{N}^{-1}$, $\overline{N}$ being the number of detected photons, beats the coherent state standard rms error $\Delta\tilde{\theta} \approx \overline{N}^{-\frac{1}{2}}$, and stated that their optimized phase measurement might be capable of an accuracy of their measure $\delta\tilde{\theta} \sim \overline{N}^{-2}$. A careful comparison of two different input states would require the use of the same measure. Paradoxically enough, the proposed measure $\delta\tilde{\theta}$ would be more useful when applied to the optimized squeezed state than to the Shapiro–Shepard–Wong state. The explanation is that the optimum state density exhibits a very narrow spike, which makes an impression of a high performance, whereas the baseline (heavy tail) of the distribution spoils it. For completeness, we mention that work has also been done of obtaining states of minimum s-phase variance [Summy and Pegg (1990)] and minimum phase dispersion [Bandilla, Paul and Ritze (1991)].

The problem of the state with minimum phase dispersion can be traced back in the formulation (5.3.56), (5.3.58) for the cost function (5.3.36) and the subsidiary condition

$$\text{Tr}\left\{\hat{\rho}(0)\hat{n}\right\} = \overline{n}. \tag{5.3.74}$$

Similarly, the problem of Shapiro–Shepard–Wong corresponds to the cost function (5.3.37) and the subsidiary condition (5.3.74). The interest of Shapiro, Shepard and Wong (1989) in the Susskind–Glogower probability-operator measure (in their terminology, SG (exponential phase) operator) has been reflected in the quantum description of optical phase [Hall (1991)] and in the proof that canonical phase detection is superior to all other shift-invariant phase detection methods (e. g., heterodyne phase detection) in its ability to resolve signals under any energy constraint [Hall and Fuss (1991)].

The phase-accuracy estimates for the Shapiro–Shepard–Wong extremal state have been reexamined from the viewpoint of measures of phase uncertainty [Schleich, Dowling and Horowicz (1991), Hradil (1992b), Hradil and Shapiro (1992), Braunstein (1992b, 1994)]. Retaining the assumption that $r = 1$, Schleich, Dowling and Horowicz (1991) treated in detail the central peak of the phase distribution. Replacing the spike with the best fitting Gaussian and showing that the area underneath the peak rapidly vanishes, they reinforced an observation of Shapiro, Shepard and Wong (1989). They used a periodic phase uncertainty measure, which is only π-periodic instead of being 2π-periodic (compare our comment concerning the optimized states of Summy and Pegg in section 4.9).

The dispersion and the peak likelihood are related to the cost functions (5.3.36) and (5.3.37), respectively, and it is argued [Hradil (1992b)] that the Shapiro–Shepard–Wong state explicitly demonstrates that the dispersion is more relevant to the phase measurement than the reciprocal peak likelihood. Having applied the uncertainty relation (4.6.47) in the least advantageous case of $\hat{\phi}_{0,s}$ (the maximum peak at the origin), Hradil (1992b) has derived the relation between the photon-number variance and the peak likelihood. An analysis of the case $0 < r < 1$, which had been neglected in [Shapiro, Shepard and Wong (1989)], has been performed by Hradil and Shapiro (1992). It has been shown that the corresponding optimum states again exhibit the infinite peak likelihood, nevertheless they convey the zero phase information in the infinite-s limit, because in this limit the probability distribution becomes uniform.

Work of Shapiro (1992) indicates that estimation of phase shifts can be improved, in principle, via multimode phase correlations. Braunstein (1992b, 1994) has attacked the question of the quantum limits to precision phase measurement by considering the schemes that split the total energy available into small packets so that they require a subsequent data analysis of the multiple measurements. He has derived the sensitivity of maximum-likelihood estimation for symmetrical schemes near the optimal energy split. For states having highly non-Gaussian noise [Braunstein (1992b)] and the near-vacuum states [Braunstein (1994)] these schemes do not beat the single measurement sensitivity of squeezed state interferometry. Particularly, Monte Carlo simulations of the Shapiro–Shepard–Wong scheme have been run [Braunstein (1992b), Braunstein, Lane and Caves (1992), Lane, Braunstein and Caves (1993)] and their results (strengthened by general mathematical analysis based on the Cramér–Rao lower bound and the Fisher information) suggest a limiting phase accuracy slightly worse than $\overline{N}^{-1}$. It has been concluded that the reciprocal peak likelihood is not a good measure of the phase sensitivity. Any distribution $P(\tilde{\theta}|\theta)$, which does not change when equal numbers are subtracted from both $\tilde{\theta}$ and θ, has been called translationally invariant. The Fisher information [Cramér (1946)]

$$F(\theta) = -\int_{-\pi}^{\pi} P(\tilde{\theta}|\theta)\frac{\partial^2 \ln P(\tilde{\theta}|\theta)}{\partial\,\theta^2}\,d\tilde{\theta} \qquad (5.3.75)$$

for the translationally invariant 2π-periodic distributions can be calculated using the fiducial distribution $P(\tilde{\theta}|0)$,

$$F(0) = -\int_{-\pi}^{\pi} P(\tilde{\theta}|0)\frac{d^2 \ln P(\tilde{\theta}|0)}{d\,\theta^2}\,d\tilde{\theta}, \qquad (5.3.76)$$

and its maximization subject to a constraint on the mean photon number provides the optimized confidence intervals [Braunstein (1992c)] characterizing the sensitivity of multiple phase measurements. It was demonstrated by D'Ariano and Paris (1997) that in M-path interferometric measurements the phase sensitivity rescales as $\Delta\tilde{\theta} \approx M^{-1}$.

Various measures of phase resolution for angular distributions of single-mode fields including variance, entropy, confidence half-width, and reciprocal peak likelihood have been reviewed and their properties summarized by Hall (1993). The asymptotic energy bounds for the phase resolution have been derived valid not only for the canonical phase detection, but for all covariant methods of phase-shift estimation, which represent fundamental limits to phase resolution. With respect to the Shapiro–Shepard–Wong scheme, it has been found that the reciprocal peak likelihood cannot be interpreted as a direct measure of angular uncertainty, as it may vanish or be arbitrarily small even if the phase distribution is close to a uniform distribution. Another approach to this problem using the Shannon information as a robust performance criterion for phase detection has been applied by Jones (1993). In order to assess the performance of a detection function $P(\tilde{\theta}|\theta)$, a prior probability density $z(\theta)$ standing for the initial information about the phase (parameter) θ

has been stipulated. The expected information gain for the outcome $\tilde{\theta}$ is $\langle \Delta I \rangle$, the mutual information,

$$\langle \Delta I \rangle = \int_{-\pi}^{\pi} \int_{-\pi}^{\pi} P(\tilde{\theta}, \theta) \ln \left[\frac{P(\tilde{\theta}, \theta)}{P(\tilde{\theta}) z(\theta)} \right] d\tilde{\theta} \, d\theta, \qquad (5.3.77)$$

where the posterior probability density $P(\tilde{\theta}, \theta)$ according to Bayes' rule,

$$P(\theta|\tilde{\theta}) P(\tilde{\theta}) = P(\tilde{\theta}, \theta) = P(\tilde{\theta}|\theta) z(\theta). \qquad (5.3.78)$$

A new optimal phase measurement problem has arisen, namely, subject to the prior information $z(\theta)$ to choose the input state and the probability-operator measure so as to maximize the expected information gain $\langle \Delta I \rangle$.

Quantum bounds to mutual information have been developed and applied by Hall (1997) based on three techniques: Mappings between joint-measurement and communication correlation contexts, a duality relation for quantum ensembles and quantum measurements, and an information exclusion principle [Hall (1995)]. Special attention has been devoted to the measurements with nonorthogonal states, such as canonical phase detection and ideal heterodyne detection [see also Hall (1994)].

A new scheme for measuring phase shifts at the Heisenberg limit of $\frac{1}{n}$ rad has been proposed by Holland and Burnett (1993). For two Fock states with the same photon number n as inputs to a 50 . 50 beamsplitter (an SU(2) interferometer), nonlocal quantum correlations are generated between the fields in the two arms of the interferometer, which allows greater resolution than for any interferometer driven by a coherent field. The Holland–Burnett proposal for the SU(2) interferometers has been reapproached by Sanders, Milburn and Zhang (1997) together with a simple initial state for the SU(1,1) interferometry using the optimal positive operator-valued measure.

Two-mode phase communication circumventing the Paley–Wiener constraint of single-mode phase measurement has been proposed by Shapiro (1993). The motivation has come from the commuting observables representation of the phase measurement [Shapiro and Shepard (1991)], in which the non-Hermitian operator on the joint state space $\mathcal{H}_S \times \mathcal{H}_C = \mathcal{H}_0 \cup \mathcal{H}_0^{\perp}$, where $\mathcal{H}_0 \equiv \text{span}\{|n_S\rangle_S |n_C\rangle_C, \min(n_S, n_C) = 0\}$ and $\mathcal{H}_0^{\perp}$ is the orthogonal complement of $\mathcal{H}_0$ in $\mathcal{H}_S \times \mathcal{H}_C$,

$$\widehat{\exp}(i\tilde{\theta}) - \widehat{\exp}(i\varphi)_S \otimes |0\rangle_{CC}\langle 0| + |0\rangle_{SS}\langle 0| \otimes [\widehat{\exp}(i\varphi)_C]^{\dagger}, \qquad (5.3.79)$$

where $|0\rangle_S$ and $|0\rangle_C$ are the vacuum states, and $\widehat{\exp}(i\varphi)_S$ and $\widehat{\exp}(i\varphi)_C$ are the Susskind–Glogower exponential phase operators of the signal and apparatus modes, respectively, which can be measured. The $\widehat{\exp}(i\tilde{\theta})$ measurement gives the maximum-likelihood estimate of the c-number-conjugate shift θ when a number-product vacuum state is used as the input. The $\widehat{\exp}(i\tilde{\theta})$ measurement performed on a number-product vacuum state, $|\psi\rangle_{\text{in}} \in \mathcal{H}_0$, which has undergone phase-conjugate modulation θ yields the maximum-likelihood estimate of θ, on the assumption that $|\psi\rangle_{in}$ has an $\mathcal{H}_0$ number representation that is real valued and non-negative. The relationship between the operator $\widehat{\exp}(i\tilde{\theta})$ and the two-mode work of Ban (1992) has been examined (compare also section 4.6) and a generalization to a complete sequence of maximum-likelihood phase-conjugate estimation problems has been performed. In this case the joint state space $\mathcal{H}_S \times \mathcal{H}_C$ decomposes into an infinite sequence of entangled-state Hilbert spaces

$$\mathcal{H}_S \times \mathcal{H}_C = \sum_{m=0}^{\infty} \mathcal{H}_m, \tag{5.3.80}$$

where the space $\mathcal{H}_m$ has complete orthogonal number and phase kets,

$$|n\rangle_m = |\max(n+m,m)\rangle_S| - \min(n-m,-m)\rangle_C, \quad n = 0, \pm 1, \pm 2, \ldots, \tag{5.3.81}$$

and

$$\left|e^{i\tilde{\theta}}\right\rangle_m = \sum_{n=-\infty}^{\infty} e^{in\tilde{\theta}}|n\rangle_m, \quad \tilde{\theta} \in [-\pi, \pi), \tag{5.3.82}$$

respectively. The phase kets $\left|e^{i\tilde{\theta}}\right\rangle_m$, $m = 0, 1, \ldots, \infty$, are the eigenkets of the operator

$$\begin{aligned}
\widehat{\exp}(i\tilde{\theta})_m &= [\widehat{\exp}(i\varphi)_S]^{\dagger m}[\widehat{\exp}(i\varphi)_S]^{m+1} \otimes |m\rangle_{CC}\langle m| \\
&+ |m\rangle_{SS}\langle m| \otimes [\widehat{\exp}(i\varphi)_C]^{\dagger(m+1)}[\widehat{\exp}(i\varphi)_C]^m
\end{aligned} \tag{5.3.83}$$

in the space $\mathcal{H}_m$. Then, on $\mathcal{H}_S \times \mathcal{H}_C$, the Ban operator

$$\widehat{\exp}(i\tilde{\theta})_B = \sum_{m=0}^{\infty} \widehat{\exp}(i\tilde{\theta})_m \tag{5.3.84}$$

generates the probability-operator measure

$$\frac{\hat{\Pi}_B(d\tilde{\theta})}{d\tilde{\theta}} = \frac{1}{2\pi} \sum_{m=0}^{\infty} \left|e^{i\tilde{\theta}}\right\rangle_{m\,m}\left\langle e^{i\tilde{\theta}}\right|, \quad \tilde{\theta} \in [-\pi, \pi). \tag{5.3.85}$$

It has been found that this probability-operator measure does not obey the condition (5.3.17) for any state $|\psi\rangle_{in}$ with nonzero components in two or more of the $\{\mathcal{H}_m\}$.

The optimal phase measurement scheme for two optical modes [Holland and Burnett (1993), Shapiro (1993)] in arbitrary input states has been determined by Sanders and Milburn (1995) using quantum parameter estimation theory [Helstrom (1976), Holevo (1982), Milburn, Chen and Jones (1994)]. Optimal phase measurements with the same mean photon number $\overline{N}$ of the input states, reveal an asymptotic $\overline{N}^{-1}$ decrease in the phase standard deviation $\Delta\varphi = \langle\widehat{\Delta\varphi^2}\rangle^{\frac{1}{2}}$ [D'Ariano and Paris (1994)]. Milburn, Chen and Jones (1994) have addressed optimal estimation of a squeeze parameter in the sense of quantum estimation theory. Motivated by the circular-phase shift estimation, they have referred to the squeeze parameter as the hyperbolic-phase shift. The heterodyne detection of the two-mode phase has been analyzed in [D'Ariano and Sacchi (1995)]. If the eigenstate of the heterodyne photocurrent is impinged into the heterodyne detector, the scheme achieves the ideal sensitivity. A quantum heterodyne detection system has been described that affords K-ary phase-based digital communication at zero error probability [Shapiro (1995)]. An entangled two-mode quantum state for a heterodyne receiver has been proposed using the two-dimensional position representation.

The actual problem of high accuracy interferometry is the improvement of the phase sensitivity, i. e., the optimization of the minimum detectable phase shift $\Delta\tilde{\theta}$ for a given mean total number $\overline{N}$ of photons passing through phase shifters. It has been shown [Yurke, McCall and Klauder (1986)] that the $SU(2)$ interferometers can achieve a phase sensitivity $\Delta\tilde{\theta} \sim \overline{N}^{-1}$ provided that the light entering the input ports is prepared in a two-mode squeezed state. Yurke (1986) has shown that the $SU(2)$ interferometers are present in the fermion interferometry also and that the phase sensitivity can approach $\overline{N}^{-1}$ provided that the numbers of the fermions in the two input beams are almost equal. The $SU(1,1)$ interferometers can achieve this sensitivity even when the vacuum fluctuations enter the input ports. Holland and Burnett (1993) have considered the reduction of the uncertainty in $\Delta\tilde{\theta}$ in an $SU(2)$ interferometer with two Fock states with equal numbers of photons. Hillery and Mlodinow (1993) have proposed to use the $SU(2)$ intelligent states of the two-mode light field for increasing the precision of the $SU(2)$ interferometric measurements. Bandyopadhyay and Rai (1995) have shown that squeezing in mode 1 of a two-mode coherent state leads to the squeezing of the normalized quadrature $\langle(\Delta\hat{J}_y)^2\rangle/|\langle\hat{J}_z\rangle|$. The accuracy of the $SU(1,1)$ interferometers can be improved by using the two-mode $SU(1,1)$ coherent states, which are simultaneously the $SU(1,1)$ intelligent states, when the photon-number difference between the modes is large [Brif and Ben-Aryeh (1996)] and by using the $SU(1,1)$ minimum-uncertainty states [Brif and Mann (1996b)]. The $SU(2)$ and $SU(1,1)$ interferometers with intelligent states of light at the input have been studied in detail [Brif and Mann (1996a)] and exact analytic expressions for the phase sensitivity have been derived. The so-called Heisenberg limit, $\overline{N}^{-1}$, as the fundamental quantum limit to precision phase measurement, has been studied [Ou (1997b)].

In realistic quantum phase measurements [Noh, Fougères and Mandel (1991, 1992a, b, 1993a), Holland and Burnett (1993), Hradil (1993b)], the statistics of continuous phase shift should be inferred from the knowledge of measured (discrete) phase-like output. A shift-invariant measurement (named the inference) has been suggested by [Noh, Fougères and Mandel (1993c)]. Using the relation (5.3.37), we find the conditional probability distribution of the phase difference as

$$P(\varphi_- | - \tau) = \sum_k p_k(-\tau)\overline{\delta}(\varphi - \varphi_k + \tau), \qquad (5.3.86)$$

where $p_k(-\tau)$ is the conditional probability of measuring the value φ_k when the phase shift τ is applied. The averaging over τ leads to

$$
\begin{aligned}
P(\varphi_-) &= \frac{1}{2\pi} \int_0^{2\pi} P(\varphi_- | -\tau)\, d\tau \\
&= \frac{1}{2\pi} \sum_k p_k(\varphi_- - \varphi_k).
\end{aligned}
\tag{5.3.87}
$$

In the framework of the phase-shift estimation, $P(\tilde{\theta}) \equiv P_{\mathrm{NFM}}(\tilde{\theta}|0)$. In general (compare with Hradil (1995), who has different signs),

$$
\begin{aligned}
P_{\mathrm{NFM}}(\tilde{\theta}|\theta) &= P_{\mathrm{NFM}}(\tilde{\theta} - \theta|0), \\
&= \frac{1}{2\pi} \sum_k p_k(\tilde{\theta} - \theta - \varphi_k).
\end{aligned}
\tag{5.3.88}
$$

Two conditional probability distributions of inferred phase shift as an alternative to the shift-invariant measurement [Noh, Fougères and Mandel (1993c)] have been proposed by Hradil (1995). Firstly, the conditional probability of inferring the phase shift $\tilde{\theta}$ when the true value is θ reads

$$
P_1(\tilde{\theta}|\theta) = \sum_k p(\tilde{\theta}|\varphi_k) p_k(\theta),
\tag{5.3.89}
$$

where $p(\tilde{\theta}|\varphi_k)$ is the conditional probability of phase shift after a single measurement without any prior knowledge and $p_k(\theta)$ is the conditional probability of measuring the value φ_k when θ is true. Secondly, when the measurement of φ_{k_j} is performed repeatedly n times, $j = 1, 2, \ldots, n$, under identical conditions, the measured output φ_{k_j} is interpreted as the estimate of the phase shift with the probability distribution $p(\tilde{\theta}|\varphi_{k_j})$. The probability distribution after the nth measurement is of the form

$$
P_2(\tilde{\theta}|\theta) \propto \prod_{j=1}^{n} p(\tilde{\theta}|\varphi_{k_j})
\tag{5.3.90}
$$

and the phase shift can be determined using the maximum-likelihood estimation [Braunstein (1992b), Braunstein, Lane and Caves (1992), Lane, Braunstein and Caves (1993)]. The phase-shift estimation based on (5.3.89) and (5.3.90) has been modified even for a very low number of detected particles and used in neutron interferometry [Hradil, Myška, Peřina, Zawisky, Hasegawa and Rauch (1996)]. Besides the phase shift, the properties of the source and interferometer (the mean number of particles and the visibility) can be inferred from the measured data. The problem of optimizing measurements on finite ensembles has been addressed by Massar and Popescu (1995) and described in the case of spin-$\frac{1}{2}$ particles. It has been shown that optimal measurements cannot be realized by separate measurements on each particle. D'Ariano, Macchiavello and Paris (1995) have analyzed the detection of the phase shift of a single mode of the field. They considered both the cases of joint measurements (achieved by a double homodyne detection in the closed scheme) and independent measurements (corresponding to the double homodyne scheme in its open version). They found that the open scheme achieved the best sensitivity for states whose Wigner function is factorized into the product of two distributions of the quadratures. The θ-parametrized probability distribution of the estimated phase shift $\tilde{\theta}$ can be approximated by a formula which is based on information-theoretical considerations and depends on the registered data only via $\tilde{\theta}$ [Hradil (1995)]. Applications of this formula can be found in [Hradil, Myška, Opatrný and Bajer (1996)]. D'Ariano, Macchiavello, Sterpi and Yuen (1996) have proposed a scheme for amplifying small phase shifts which reduces the bit-error rate and increases the information retrieved from the measurement. The best performance is achieved by phase-coherent states (and the ideal probability-operator measure), but good results are also obtained in the practical situation of coherent states with heterodyne phase detection. A special approach is needed to confirm the confidence of experimentalists that feedback assisted homodyne detection provides a convenient way to measure phase shifts [D'Ariano, Paris and Seno (1996)]. The scheme of conjugate quadratures has been compared with the usual single-homodyne detection according to the actual performance of the feedback.

Paris (1997) has addressed interferometry as a binary decision problem and has derived lower bounds to the minimum detectable phase shift for some phase-enhanced states of the radiation field (coherent states, phase-coherent states, squeezed coherent states) in terms of photon-number fluctuations of the signal mode carrying the phase information.

The total photon number is conserved when the light passes through the lossless interferometer. Therefore, it is tempting to perform independently the optimization of the parameter estimation in each of the linear subspaces of the two-mode Hilbert space pertinent to a photon-number sum. In this case it is also possible to maintain the basic situation of the estimation (cf. the above remark on the multiple hypothesis testing) with the phase shift $\theta \in [0, 2\pi)$, but to restrict the analysis to the probability operator measures $\hat{\Pi}_n(\Delta)$, support of which consists of only $(n+1)$ discrete values. For definiteness, we assume that these values are [Luis and Peřina (1996b)]

$$\tilde{\theta}_{n,k} = \frac{2\pi k}{n+1}, \quad k = 0, 1, \ldots, n. \tag{5.3.91}$$

The resolution of the identity $\hat{1}_n^{SU(2)}$ is generated by the operators

$$\hat{\Pi}_n(\tilde{\theta}_{n,k}) = |\xi_n(\tilde{\theta}_{n,k})\rangle\langle\xi_n(\tilde{\theta}_{n,k})| \tag{5.3.92}$$

and the property (5.3.20) is modified

$$|\xi_n(\tilde{\theta}_{n,k})\rangle = \exp(i\tilde{\theta}_{n,k}\hat{n}_1)|\xi_n(0)\rangle. \tag{5.3.93}$$

Here

$$|\xi_n(0)\rangle = \sum_{n_1=0}^{n} \langle n_1, n - n_1|\xi_n(0)\rangle|n_1, n - n_1\rangle, \tag{5.3.94}$$

with [Luis and Peřina (1996b)]

$$\langle n_1, n - n_1 | \xi_n(0) \rangle = \frac{1}{\sqrt{n+1}} \exp(i\chi_{n_1, n-n_1}). \qquad (5.3.95)$$

In more detail,

$$\langle n_1, n - n_1 | \xi_n(\tilde{\theta}_{n,k}) \rangle = \frac{1}{\sqrt{n+1}} \exp(i\chi_{n_1, n-n_1}) \exp(in_1\tilde{\theta}_{n,k}). \qquad (5.3.96)$$

From the equal spacing (5.3.91) it follows that the operators (5.3.92) are orthogonal projections. The average cost, which represents the performance measure, is given by a modified relation (5.3.3)

$$\overline{C} = \text{Tr} \left\{ \sum_{n=0}^{\infty} \sum_{k=0}^{n} \hat{W}(\tilde{\theta}_{n,k}) \hat{\Pi}_n(\tilde{\theta}_{n,k}) \right\}, \qquad (5.3.97)$$

with $\hat{W}(\tilde{\theta}_{n,k})$ defined in (5.3.4). It is plausible that for the family of input pure states (5.3.14), the formulae (5.3.92), (5.3.93), and (5.3.94) determine an almost optimum solution to the problem $\overline{C} = \min$. Nevertheless, for these probability-operator measures and the cost functions (5.3.36) and (5.3.37), the search for the optimum input states is comprised in [Luis and Peřina (1996b)].

5.4   Phase dependence and phase dynamics in nonlinear optical processes

The study of nonlinear interactions in nonlinear and quantum optics is still of interest, although it used to assume a coherent input field and to derive a short-time quantum dynamics. It not only reinterpreted the knowledge how classical intensities depend on the initial phase as how mean photon numbers depend on the phases of the initial coherent state, but it also considered the effect of the phases of initial modes on other photon-number statistics. The determination of the phase behaviour for all quantum descriptions faces the ambiguity of quantum phase distributions. If the canonical quantum phase distribution is the choice, numerical calculations ensue as the only response to the absence of some exact closed formulae. Several results illustrate this ambiguity and convince of a similarity of quantum phase distributions. The comparison with results based on classical theory is not excluded in the dynamics of the preferred phase for an initial coherent state. A particular approach is needed as concerns the phase dynamics of ℓ-photon coherent states.

5.4.1   Phase dependence

In nonlinear and quantum optics of the sixties, the study of nonlinear interactions assumed a coherent input field and involved considerations of the effect of the phases of initial modes on the radiation properties in the process, for instance, on the photon-number statistics and, rudimentarily, on the phase dynamics. In the case of the second-subharmonic generation (degenerate parametric amplification) described by the Hamiltonian (cf. (3.8.10))

$$\hat{H}_{\text{rad}} = \hbar\omega \left(\hat{a}^\dagger \hat{a} + \frac{1}{2}\hat{1} \right) - \frac{1}{2}\hbar g \left[\hat{a}^2 \exp(i2\omega t - i\psi) + \text{H. c.} \right], \qquad (5.4.1)$$

where the frequency of the signal field is ω, the frequency of pumping is 2ω, and the phase of the pump complex amplitude is ψ, whereas φ is the phase of the complex amplitude of the initial coherent state $|\xi\rangle$, $\xi \neq 0$, the maximum antibunching of photons occurs for $2\varphi - \psi \equiv -\frac{\pi}{2}$ (mod 2π) [Stoler (1974), Mišta and Peřina (1977a, b)]. The onset of the oscillatory behaviour of the photon-number distribution has been recorded by Mišta, Peřinová, Peřina and Braunerová (1977). This effect is connected to the quadrature squeezing as demonstrated by Peřina and Bajer (1990), compare also section 4.11 for the results of W. P. Schleich and coworkers. Whereas the oscillations lead ultimately to the super-Poissonian photon statistics, there is no limitation of this kind on the quadrature squeezing.

In the nondegenerate version of the optical parametric process with the classical pumping described by the Hamiltonian

$$\hat{H}_{\text{rad}} = \sum_{j=1}^{2} \hbar\omega_j \left(\hat{a}_j^\dagger \hat{a}_j + \frac{1}{2}\hat{1} \right) - \hbar q \left[\hat{a}_1 \hat{a}_2 \exp(i\omega t - i\psi) + \text{H. c.} \right], \qquad (5.4.2)$$

where the frequencies of the signal and idler modes are ω_1, ω_2, respectively, the pumping frequency is $\omega = \omega_1 + \omega_2$, and the phase of the pump complex amplitude is ψ. Denoting the initial coherent state by $|\xi_1, \xi_2\rangle$ and considering the more general case of the down-conversion than the parametric generation when $\xi_1 = \xi_2 = 0$ or the amplification process ($\xi_1 \neq 0$, $\xi_2 = 0$), we introduce the phases $\varphi_j = \arg \xi_j$, $j = 1, 2$. The sub-Poissonian effect in the distribution of the total photon number is maximum when $\varphi_1 + \varphi_2 - \psi \equiv -\frac{\pi}{2}$ (mod 2π) [Mišta and Peřina (1977a, b), Peřinová and Peřina (1981), Paul and Brunner (1981)]. The comparison with the degenerate case is made easier by the assumption $|\xi_1| = |\xi_2|$. The effect of oscillations is complicated as demonstrated by variety of figures [Mišta and Peřina (1978)]. The quadrature squeezing can occur only in a suitably defined compound mode [Peřina (1991)] (cf. section 4.11, the formula (4.11.15)).

To be able to describe the experiment of Ou, Wang, Zou and Mandel (1990), we modify the Hamiltonian (5.4.2) to the form

$$\hat{H}_{\text{rad}} = \sum_{j=\text{s,i}} \hbar\omega_j \sum_{k=1}^{2} \left(\hat{a}_{jk}^\dagger \hat{a}_{jk} + \frac{1}{2}\hat{1} \right) - \hbar g \sum_{k=1}^{2} \left[\hat{a}_{sk} \hat{a}_{ik} \exp(i\omega t - i\psi_k) + \text{H. c.} \right], \quad (5.4.3)$$

where the subscript $j=$s,i means the signal and idler modes and the subscript k labels distinct nonlinear crystals. Using two beamsplitters, they have arrived at a measurement scheme sensitive to the phase difference $\psi_1 - \psi_2$ between the two pump beams.

When the quantum pumping is considered and the phase-matching condition is removed, the previous model becomes the three-mode parametric optical process,

$$\hat{H}_{\text{rad}} = \sum_{j=1}^{3} \hbar\omega_j \left(\hat{a}_j^\dagger\hat{a}_j + \frac{1}{2}\hat{1}\right) - \hbar\left(g\hat{a}_1\hat{a}_2\hat{a}_3^\dagger + \text{H. c.}\right), \qquad (5.4.4)$$

where the frequencies of single modes ω_j, $j = 1, 2, 3$, satisfy the condition $\omega_3 = \omega_1+\omega_2$. The phase of the coupling constant is denoted by ψ. Assuming the initial coherent state $|\xi_1, \xi_2, \xi_3\rangle$ and the more general process than the parametric amplification, $\xi_2 = 0$, and parametric generation, $\xi_1 = \xi_2 = 0$, we introduce the phases $\varphi_j = \arg \xi_j$, $j = 1, 2, 3$. The anticorrelation between the signal and idler modes 1 and 2 is maximum if $\varphi_1+\varphi_2+\psi-\varphi_3 \equiv -\frac{\pi}{2} \pmod{2\pi}$ up to the first order in t [Peřina (1991) and references therein, Peřinová and Peřina (1978a, b)]. As a result, the squeezing in the sum of photon numbers of anticorrelated modes occurs if $\varphi_1 + \varphi_2 - \psi \equiv -\frac{\pi}{2} \pmod{2\pi}$. The sum-frequency mode exhibits the sub-Poissonian behaviour in $(gt)^6$ in the process of sum-frequency generation. For the sum of photon numbers in modes 1 and 2, the sub-Poissonian behaviour has been found under the same phase condition assuming that $|\xi_1| = |\xi_2|$ [Peřina, Bajer, Křepelka and Hradil (1987)]. The maximum squeezing in one of the quadratures of the compound mode $(1,2)$ is attained up to the second order in t for $\varphi_1 + \varphi_2 \equiv 0 \pmod{2\pi}$, but in the other for $\varphi_1 + \varphi_2 \equiv \pm\pi \pmod{2\pi}$ in the process of sum-frequency generation $\xi_3 = 0$ [Peřina, Peřinová and Koďousek (1984), Koďousek and Peřina (1984)]. In general, the assumption $|\xi_j| > |\xi_3|$, $j = 1, 2$, can be made and the squeezing in the Q quadrature in the compound mode $(j, 3)$ can occur for $\varphi_j + \varphi_3 \equiv 0 \pmod{2\pi}$. The maximum squeezing in the sum-frequency mode $(\xi_3 = 0)$ can occur up to t^4 in the Q quadrature for $\varphi_1 + \varphi_2 \equiv 0 \pmod{\pi}$ and in the P quadrature for $\varphi_1 + \varphi_2 \equiv \pm\pi \pmod{\pi}$. For a general stimulated process, the maximum squeezing in mode 3 can be attained up to t^3 in the Q quadrature if $\varphi_1 + \varphi_2 + \varphi_3 \equiv -\frac{\pi}{2} \pmod{2\pi}$ and in the P quadrature if $\varphi_1 + \varphi_2 + \varphi_3 \equiv \frac{\pi}{2} \pmod{2\pi}$. For the spontaneous sub-frequency generation $(\xi_1 = \xi_2 = 0)$, this effect is maximum if $\varphi_3 \equiv \pm\pi \pmod{2\pi}$ in one of the quadratures and if $\varphi_3 \equiv 0 \pmod{2\pi}$ in the other up to the fourth order in time. A sub-Poissonian behaviour after one "quasiperiod" in three-wave interaction has been discovered for the initial coherent states and has been made possible within quasiperiod via the replacement of the initial coherent state in the signal mode by a Kerr state respecting a phase condition [Bandilla, Drobný and Jex (1995)].

If we cannot distinguish between the signal and idler modes, the degenerate model is appropriate

$$\hat{H}_{\text{rad}} = \sum_{j=1}^{2} \hbar\omega_j \left(\hat{a}_j^\dagger\hat{a}_j + \frac{1}{2}\hat{1}\right) - \hbar\left(g\hat{a}_1^2\hat{a}_2^\dagger + \text{H. c.}\right), \qquad (5.4.5)$$

where $\omega_2 = 2\omega_1$. In the stimulated subharmonic generation process with the initial coherent state $|\xi_1, \xi_2\rangle$, $\xi_j = |\xi_j| \exp(i\varphi_j)$, $j = 1, 2$, the subharmonic mode (mode 1) exhibits the sub-Poissonian behaviour if $2\varphi_1 - \varphi_2 + \psi \equiv -\frac{\pi}{2}$ (mod 2π), where ψ is the phase of the coupling constant [Peřinová and Peřina (1978c)]. In the case of the second-harmonic generation, which may be stimulated, $\xi_2 \neq 0$, and must be significant, $|\xi_1| \gg |\xi_2|$, no phase condition applies and we have number squeezing in the fundamental mode independent of the initial phases. Also quadrature squeezing properties of the degenerate process are analogous to those for the nondegenerate process [Mandel (1982a, b), Peřina, Peřinová and Koďousek (1984)]. The fundamental mode (mode 1) can exhibit squeezing in the quadrature Q or P with respect to the initial phase φ_1 up to t^2 in the process of the second-harmonic generation ($\xi_2 = 0$) and the effect is maximum if $\varphi_1 \equiv 0$ (mod 2π) or $\varphi_1 \equiv \pm\pi$ (mod 2π), respectively. It corresponds to the sub-Poissonian statistics of this mode. Squeezing can also occur in the second subharmonic generation ($\xi_1 = 0$) and its maximum amount is attained in the quadrature Q or P for $\varphi_2 \equiv \frac{\pi}{2}$ (mod 2π) or $\varphi_2 \equiv -\frac{\pi}{2}$ (mod 2π), respectively. Maximum squeezing appears in the compound mode field $(1,2)$ in the second-harmonic generation ($\xi_2 = 0$) in the quadrature Q or P up to t^2 if $\varphi_1 \equiv 0$ (mod π) or $\varphi_1 \equiv \pm\pi$ (mod π), respectively. In the second-subharmonic generation ($\xi_1 = 0$), this effect is maximum in the Q or P quadrature up to the first order in t if $\varphi_2 \equiv \frac{\pi}{2}$ (mod 2π) or $\varphi_2 \equiv -\frac{\pi}{2}$ (mod 2π), respectively. The sum-frequency mode (mode 2) can exhibit squeezing in higher powers of t. Provided that $\xi_2 = 0$, the maximum squeezing occurs in the Q or P quadrature up to t^4 if $\varphi_1 \equiv \pm\frac{\pi}{4}$ (mod $\frac{\pi}{2}$) or $\varphi_1 = 0$ (mod $\frac{\pi}{2}$), respectively. For the stimulated process ($\xi_1 \neq 0$, $\xi_2 \neq 0$), the appropriate phase condition for the maximum Q- or P-quadrature squeezing in t^3 reads $2\varphi_1 + \varphi_2 \equiv -\frac{\pi}{2}$ (mod 2π) or $2\varphi_1 + \varphi_2 \equiv \frac{\pi}{2}$ (mod 2π), respectively, and for the subharmonic generation ($\xi_1 = 0$), the respective phase condition $\varphi_2 \equiv \frac{\pi}{2}$ (mod π) or $\varphi_2 \equiv 0$ (mod 2π) is to be fulfilled for the maximum squeezing to occur in the Q or P quadrature up to t^4 in the sum-frequency mode (mode 2).

The Raman scattering of intense classical laser light with complex classical amplitude $e_L(t)$ with a frequency ω_L comprises an infinite Markovian system of phonons and is described by the Hamiltonian

$$\hat{H} = \sum_{j=L,S,A} \hbar\omega_j \left(\hat{a}_j^\dagger \hat{a}_j + \frac{1}{2}\hat{1} \right) + \sum_l \hbar\omega_{Vl} \hat{a}_{Vl}^\dagger \hat{a}_{Vl}$$
$$- \hbar \sum_l \left[g_l e_L(t) \hat{a}_S^\dagger \hat{a}_{Vl}^\dagger + \kappa_l^* e_L(t) \hat{a}_{Vl} \hat{a}_A^\dagger + \text{H. c.} \right], \qquad (5.4.6)$$

where $\hat{a}_S$, $\hat{a}_A$, $\hat{a}_{Vl}$ are the annihilation operators corresponding to the Stokes, anti-Stokes, and the vibration phonon modes with the frequencies $\omega_S, \omega_A, \omega_{Vl}$, respectively, and g_l and κ_l are the Stokes and anti-Stokes coupling constants. After the elimination of the phonon reservoir system, the resonance condition $2\omega_L \approx \omega_S + \omega_A$ and the initial two-mode coherent state $|\xi_S, \xi_A\rangle$ are assumed. Pursuing the nonclassical behaviour of the total photon number in the Stokes and anti-Stokes modes, we encounter the maximum anticorrelation effect up to the first order of t for $2\varphi_L - \varphi_S - \varphi_A + \psi_S - \psi_A \equiv 0$ (mod 2π), where φ_L, φ_S, φ_A, ψ_S, ψ_A are the phases of $e_L(0)$, ξ_S, ξ_A, $g_l = g(\omega_V)$, $\kappa_l = \kappa(\omega_V)$, with $\omega_V = \omega_L - \omega_S = \omega_A - \omega_L$, respectively [Peřina (1981a, b)].

Restricting ourselves to a single vibration mode and assuming again the strong laser field and absorbing $e_L(t)$ by the coupling constants $ge_L(t) \to g\exp(-i\omega_L t + i\varphi_L)$, $\kappa^* e_L(t) \to \kappa^* \exp(-i\omega_L t + i\varphi_L)$, we have the effective Hamiltonian of the Brillouin scattering

$$\hat{H}_{\text{int}} = -\hbar \left[g\hat{a}_S^\dagger \hat{a}_V^\dagger \exp(-i\omega_L t + i\varphi_L) + \kappa^* \hat{a}_V \hat{a}_A^\dagger \exp(-i\omega_L t + i\varphi_L) + \text{H. c.} \right]. \quad (5.4.7)$$

The initial three-mode coherent state $|\xi_S, \xi_A, \xi_V\rangle$ is assumed and $g > 0$, $\kappa > 0$ for simplicity. The sub-Poissonian behaviour occurs in separate modes with initially equal mean photon numbers under the joint conditions $\varphi_S + \varphi_V - \varphi_L \equiv -\frac{\pi}{2}$ (mod 2π) and $2\varphi_L - \varphi_S - \varphi_A \equiv \pi$ (mod 2π), where φ_S, φ_A, φ_V are the phases of ξ_S, ξ_A, ξ_V, respectively [Pieczonková and Peřina (1981), Pieczonková (1982a, b)]. Periodic squeezing effect of the photon-number sum of the Stokes and anti-Stokes modes occurs for the joint conditions $\varphi_S + \varphi_V - \varphi_L \equiv -\frac{\pi}{2}$ (mod 2π) and $2\varphi_L - \varphi_S - \varphi_A \equiv 0$ (mod 2π). The same effect for the total photon number of three modes under consideration appears if the conditions $\varphi_S + \varphi_V - \varphi_L \equiv \pm\frac{\pi}{2}$ (mod 2π) and $2\varphi_L - \varphi_S - \varphi_A \equiv -\pi$ (mod 2π) are valid.

If also the laser field is described as a quantum mode, the Hamiltonian of the Raman scattering process takes the form [Walls (1970)]

$$\hat{H}_{\text{int}} = -\hbar \left(g\hat{a}_L \hat{a}_S^\dagger \hat{a}_V^\dagger + \kappa^* \hat{a}_L \hat{a}_V \hat{a}_A^\dagger + \text{H. c.} \right). \quad (5.4.8)$$

For the initial coherent state $|\xi_L, \xi_S, \xi_A, \xi_V\rangle$ and $g = \kappa$, the maximum sub-Poissonian effect arises in the laser mode under the condition $2\varphi_L - \varphi_S - \varphi_A \equiv 0$ (mod 2π) and for $|\xi_S| > |\xi_A|$ up to the order of t^2 [Tänzler and Schütte (1981)]. The photon-number sum of the laser and phonon modes and that of the Stokes and phonon modes exhibit the maximum anticorrelation up to the first order of t for $\varphi_L + \varphi_V - \varphi_A - \psi_A \equiv -\frac{\pi}{2}$ (mod 2π) and $\varphi_S + \varphi_V - \varphi_L - \psi_S \equiv -\frac{\pi}{2}$ (mod 2π), respectively [Szlachetka, Kielich, Peřina and Peřinová (1979, 1980)]. The maximum anticorrelation between the laser and Stokes modes occurs in stimulated scattering up to the second order of t if $\Delta\varphi = 2\varphi_L - \varphi_S - \varphi_A + \psi_S - \psi_A \equiv \pi$ (mod 2π). The similar maximum effect between the Stokes and anti-Stokes modes appears for $\Delta\varphi \equiv 0$ (mod 2π). A generalization of the analysis by Tänzler and Schütte (1981), but focused on the quadrature squeezing, has been provided by Peřina, Peřinová and Koďousek (1984) and it has been found that it is advantageous to choose $\varphi_S + \varphi_A - \psi_S + \psi_A \equiv 0$ (mod 2π) and $|g||\xi_S| > |\kappa||\xi_A|$ in order to observe squeezing in the laser mode up to t^2, here $\varphi_L = 0$. Whereas there is no quadrature squeezing expected in the compound modes (L, S), (L, V), and (A, V), the number squeezing arises in t^2 provided that $2\varphi_L - \varphi_S - \varphi_A + \psi_S - \psi_A \equiv \pi$ (mod 2π) and $|g||\xi_S| > |\kappa||\xi_A|$. The maximum quadrature-squeezing effect occurs in the compound mode (L, A) in t^2 if $\varphi_S + \varphi_A - \psi_S + \psi_A \equiv 0$ (mod 2π), $\varphi_L + \varphi_A \equiv 0$ (mod 2π), and $|g||\xi_S| > |\kappa||\xi_A|$. In this situation also the condition for the number squeezing, which reads $2\varphi_L - \varphi_S - \varphi_A + \psi_S - \psi_A \equiv 0 \pmod{2\pi}$, is met. In the compound mode (S, A), the maximum quadrature squeezing arises provided that $2\varphi_L + \varphi_S - \varphi_A \equiv 0 \pmod{2\pi}$ and $|g| < |\kappa|$. The number squeezing is maximum on the phase condition $\varphi_S + \varphi_A - \psi_S + \psi_A \equiv 0 \pmod{2\pi}$, but $|g||\xi_S| < |\kappa||\xi_A|$ is assumed. The compound mode (S, V) exhibits the maximum squeezing in the Q quadrature in t when $\varphi_L + \psi_S \equiv \frac{\pi}{2}$ $\pmod{2\pi}$.

The nondegenerate hyper-Raman scattering can occur when two modes of laser light incident onto the scattering medium are sufficiently intense. This process is described by the effective Hamiltonian

$$\hat{H} = \sum_{j=1,2,S,A,V} \hbar\omega_j \left(\hat{a}_j^\dagger \hat{a}_j + \frac{1}{2}\hat{1}\right) - \hbar\left(g\hat{a}_1\hat{a}_2\hat{a}_S^\dagger\hat{a}_V^\dagger + \kappa^*\hat{a}_1\hat{a}_2\hat{a}_A^\dagger\hat{a}_V + \text{H. c.}\right), \quad (5.4.9)$$

where the resonance conditions $\omega_S = \omega_1 + \omega_2 - \omega_V$ and $\omega_A = \omega_1 + \omega_2 + \omega_V$ are fulfilled. For the light modes initially coherent and the phonon mode initially chaotic, the maximum anticorrelation between the laser and scattered modes (cases (1S), (2S), (1A), (2A)) occurs up to t^2 if $\Delta\varphi = 2\varphi_1 + 2\varphi_2 + \psi_S - \psi_A - \varphi_S - \varphi_A \equiv \pi \pmod{2\pi}$, where φ_j are the phases of the initial complex field amplitudes ξ_j, $j = 1, 2$ [Peřinová, Peřina, Szlachetka and Kielich (1979a), Peřina (1991), p. 314]. The maximum strength of the same effect for the field (SA) is reached if $\Delta\varphi \equiv 0 \pmod{2\pi}$. In the spontaneous scattering ($\xi_S = \xi_A = 0$), we obtain the maximum squeezing of the Q quadrature for the compound mode $(1, 2)$ on the condition that $\varphi_1 + \varphi_2 = 0 \pmod{2\pi}$.

The degenerate hyper-Raman scattering is described by the effective Hamiltonian

$$\hat{H} = \sum_{j=L,S,A,V} \hbar\omega_j \left(\hat{a}_j^\dagger \hat{a}_j + \frac{1}{2}\hat{1}\right) - \hbar\left(g\hat{a}_L^2\hat{a}_S^\dagger\hat{a}_V^\dagger + \kappa^*\hat{a}_L^2\hat{a}_A^\dagger\hat{a}_V + \text{H. c.}\right) \quad (5.4.10)$$

and by the simplified resonance conditions $\omega_S = 2\omega_L - \omega_V$, $\omega_A = 2\omega_L + \omega_V$. Initial chaotic phonons and coherent photons are assumed again. The degeneration is reflected in the form of the conditions, which are formulated with $\Delta\varphi = 4\varphi_L + \psi_S - \psi_A - \varphi_S - \varphi_A$ [Peřinová, Peřina, Szlachetka and Kielich (1979b)]. The maximum anticorrelation between the laser and scattered modes (cases (LS), (LA)) occurs up to t^2 if $\Delta\varphi \equiv \pi \pmod{2\pi}$, whereas again the maximum anticorrelation between the Stokes and anti-Stokes photons (the field (SA)) is attained if $\Delta\varphi \equiv 0 \pmod{2\pi}$. Let us return to the initial conditions from [Peřina, Peřinová and Koďousek (1984)], where also the simplifying assumption $\xi_S = \xi_A = 0$ has been made. The maximum squeezing effect in the Q quadrature of the laser mode can be achieved for $\varphi_L \equiv 0$ $\pmod{\pi}$.

The four-wave mixing process is described by the Hamiltonian

$$\hat{H}_{\text{rad}} = \sum_{j=1}^{4} \hbar\omega_j \left(\hat{a}_j^\dagger \hat{a}_j + \frac{1}{2}\hat{1}\right) + \hbar\left(g\hat{a}_1^\dagger\hat{a}_2^\dagger\hat{a}_3\hat{a}_4 + \text{H. c.}\right), \quad (5.4.11)$$

where ω_1, ω_2 (ω_3, ω_4) are the frequencies of the signal (pump) modes. The frequency-resonance condition for the process to be effective reads $\omega_1 + \omega_2 = \omega_3 + \omega_4$. The incident beams are in a four-mode coherent state $|\xi_1, \xi_2, \xi_3, \xi_4\rangle$, $\xi_j = |\xi_j| \exp(i\varphi_j)$, $j = 1, 2, 3, 4$. The maximum number-sum squeezing in the signal modes appears if $\Delta\varphi = \varphi_3 + \varphi_4 - \varphi_1 - \varphi_2 + \psi \equiv -\frac{\pi}{2} \pmod{2\pi}$, where ψ is the phase of the coupling constant g. The maximum number-sum squeezing in the pump modes occurs if $\Delta\varphi \equiv \frac{\pi}{2} \pmod{2\pi}$. This behaviour is related to the short-length solution [Peřina, Peřinová, Sibilia and Bertolotti (1984), Bajer and Peřina (1985)]. The maximum squeezing in the Q quadrature of the compound signal mode $(1, 2)$ is achieved up to the first order of z (length of propagation) for $\varphi_3 + \varphi_4 + \psi \equiv -\frac{\pi}{2} \pmod{2\pi}$ and this effect for the compound pump mode occurs for $\varphi_1 + \varphi_2 - \psi \equiv -\frac{\pi}{2} \pmod{2\pi}$.

The same order of nonlinearity as the four-wave mixing is characteristic of the nonlinear phase shift which is experienced by the light propagating in a Kerr medium. The simplest nonlinear oscillator model comprises the Hamiltonian

$$\hat{H} = \hbar\omega\left(\hat{a}^\dagger\hat{a} + \frac{1}{2}\hat{1}\right) + \hbar\kappa\hat{a}^{\dagger 2}\hat{a}^2, \tag{5.4.12}$$

where ω is the frequency of light and κ is a real constant for the intensity dependence. We assume that the mode is in the coherent state $|\xi\rangle$ initially. The maximum short-time (short-length) squeezing in the Q quadrature is achieved for $\varphi \equiv -\frac{\pi}{4} \pmod{2\pi}$, φ being the phase of the complex field amplitude ξ.

Many of these results can be generalized to nonlinear couplers composed of linear and nonlinear waveguides or two nonlinear waveguides connected by evanescent waves. The second-order nonlinear media [Peřina and Peřina Jr. (1995, 1996)] and the third-order nonlinear media [Korolkova and Peřina (1996, 1997)] have been considered.

The time development of a physical system in the Schrödinger picture can be described by the evolution equation

$$\frac{d}{dt}\hat{\rho}(t) = \hat{\hat{L}}\hat{\rho}(t), \tag{5.4.13}$$

where $\hat{\hat{L}}$ is a superoperator,

$$\hat{\hat{L}}\hat{\rho}(t) = -\frac{i}{\hbar}[\hat{H}, \hat{\rho}(t)]. \tag{5.4.14}$$

In the Heisenberg picture, an operator $\hat{M}(t)$ to be averaged evolves according to the equation

$$\frac{d}{dt}\hat{M}(t) = \hat{\hat{L}}^\dagger\hat{M}(t), \tag{5.4.15}$$

where

$$\hat{\hat{L}}^\dagger\hat{M}(t) = -\frac{i}{\hbar}[\hat{M}(t), \hat{H}]. \tag{5.4.16}$$

Let us remember the advantageous notation ((3.5.9)),

$$\mathrm{cov}(\hat{A}, \hat{B}) = \frac{1}{2}[\langle \Delta \hat{A} \Delta \hat{B} \rangle + \langle \Delta \hat{B} \Delta \hat{A} \rangle]. \tag{5.4.17}$$

Using the formal solution for the equation (5.4.15), we obtain

$$\hat{A}(t) = \sum_{n=0}^{\infty} \frac{t^n}{n!} \hat{\hat{L}}^{\,tn} \hat{A}(0), \quad \hat{B}(t) = \sum_{n=0}^{\infty} \frac{t^n}{n!} \hat{\hat{L}}^{\,tn} \hat{B}(0). \tag{5.4.18}$$

Although the same formal expansion works for $\hat{M}(t) = \hat{A}(t)\hat{B}(t)$, we will multiply the expansions (5.4.18) and obtain

$$\hat{A}(t)\hat{B}(t) = \sum_{n=0}^{\infty} \frac{t^n}{n!} \sum_{j=0}^{n} \binom{n}{j} \hat{\hat{L}}^{\,tj} \hat{A}(0) \hat{\hat{L}}^{\,t(n-j)} \hat{B}(0). \tag{5.4.19}$$

Averaging both sides of equations (5.4.18) and multiplying, we obtain a similar result

$$\langle \hat{A}(t) \rangle \langle \hat{B}(t) \rangle = \sum_{n=0}^{\infty} \frac{t^n}{n!} \sum_{j=0}^{n} \binom{n}{j} \langle \hat{\hat{L}}^{\,tj} \hat{A}(0) \rangle \langle \hat{\hat{L}}^{\,t(n-j)} \hat{B}(0) \rangle. \tag{5.4.20}$$

Performing in addition the exchange $\hat{A} \leftrightarrow \hat{B}$ in (5.4.19) and substituting into (5.4.17), we obtain that the covariance

$$\mathrm{cov}(\hat{A}(t), \hat{B}(t)) = \sum_{n=0}^{\infty} \frac{t^n}{n!} \sum_{j=0}^{n} \binom{n}{j} \mathrm{cov}\left(\hat{\hat{L}}^{\,tj} \hat{A}(0), \hat{\hat{L}}^{\,t(n-j)} \hat{B}(0) \right). \tag{5.4.21}$$

Particularly, the variance can be obtained from the expansion

$$\langle (\Delta \hat{M}(t))^2 \rangle = \sum_{n=0}^{\infty} \frac{t^n}{n!} \sum_{j=0}^{n} \binom{n}{j} \mathrm{cov}\left(\hat{\hat{L}}^{\,tj} \hat{M}(0), \hat{\hat{L}}^{\,t(n-j)} \hat{M}(0) \right). \tag{5.4.22}$$

The formulae (5.4.21) and (5.4.22) can be used also in the situation when the super-operators $\hat{\hat{L}}$ and $\hat{\hat{L}}^{\dagger}$ are generalized in order to take into account the dissipation. The short-time solutions, which served for the derivation of phase conditions, have been obtained basically from the foregoing expansions truncated after several first terms.

5.4.2 Phase dynamics

The determination of the phase behaviour for all the quantum descriptions of the non-linear optical dynamics is a tremendous task. Using computers, one can encounter a monotony of approach, which at last leads to a decline of the initial great interest. As a consequence, the work done is not finished yet. Nevertheless, there are some nice results, some of which will be reproduced in what follows. The exposition of real states of light field (section 4.11) has much in common with and even just contains the dynamics of nonlinear parametric optical processes studied mainly from the viewpoint of the quantum phase. Braunstein and McLachlan (1987) studied "generalized squeezed" states appropriate for the description of the third- and fourth-subharmonic generation. Braunstein and Caves (1990), among others, associated the statistics of heterodyne detection for these states with the Φ_A function and the appropriate antinormal phase distribution. The refined analysis has been based on the consideration of the quantum pump [Gantsog and Tanaś (1991b), Gantsog, Tanaś and Zawodny (1991a), Tanaś, Gantsog and Zawodny (1991a, b), Drobný and Jex (1992)]. For the process described by the Hamiltonian (5.4.5), typical initial conditions for the second-harmonic generation have been adopted, a coherent state in the fundamental mode and the vacuum in the second harmonic mode. The direct solution of the dynamics on the computer does not restrict the treatment to the coherent state, but allows to treat any partial phase state [Pegg and Barnett (1989)] and even any pure state of the fundamental mode [Tanaś, Gantsog and Zawodny (1991b)]. The joint phase probability distribution $P(\varphi_1, \varphi_2)$ derived by the standard procedure of the Pegg–Barnett phase formalism has been determined at a finite number of times and its behaviour has confirmed the anticipations based on the classical theory. The graphical illustration for $|g|t = 0$ shows that the joint phase distribution is a product of the phase distribution for the coherent state of mode 1 and the uniform phase distribution for the vacuum state of mode 2. For $|g|t = 0.2$, this factorization is still approximately valid, but the distribution of the fundamental mode is broadened, while a peak of the phase distribution in the second harmonic mode starts to grow. The phase of the peak fulfills the phase relation $2\varphi_1 - \varphi_2 + \psi \equiv -\frac{\pi}{2}$ (mod 2π), which is known (see above) to provide the maximum sub-Poissonian behaviour when the second-harmonic generation is stimulated by a weak coherent beam in the second harmonic mode. For later times, the joint phase distribution cannot be factorized and it splits into two peaks. For still later times, more and more peaks appear in the phase distribution $P(\varphi_1, \varphi_2)$. Since the marginal distribution of the phase in the fundamental mode may have two peaks only, it appears that the marginal distribution of the second harmonic mode becomes more uniform. Nevertheless, the phase variances for both the modes tend asymptotically to the value $\frac{\pi^2}{3}$ of the uniformly distributed phase. Partial revivals of the phase structure have been observed during the evolution [Drobný and Jex (1992)]. The intracavity second-harmonic generation has been interesting for a quantum counterpart of the Hopf bifurcation. Above some critical value of the pump field, the dynamics of the photon numbers of modes treated semiclassically becomes unstable and enters a self-pulsing regime [McNeil, Drummond and Walls (1978)]. The dynamics of the phase has been investigated numerically using the positive P (Φ_N) representation [Gevorkyan and Maloyan (1997)].

The parametric down-conversion process with quantum pump is a second-subharmonic generation process. The initial condition is the vacuum state in the subharmonic mode and the coherent state in the pump mode. The distinction between the second-harmonic generation process and that of the second-subharmonic generation consists in that the creation of one photon of the energy $\hbar 2\omega$ requires the annihilation of two photons of the fundamental mode, whereas the annihilation of one photon of the pump mode means the creation of two photons of the second subharmonic mode. The two-photon generation leads to two-photon states. When the phase is discussed [Gantsog, Tanaś and Zawodny (1991b, 1993), Jex, Drobný and Matsuoka (1992), Tanaś and Gantsog (1992a, b)], the illustration for $|g|t = 0$ shows that the joint phase distribution is a product of the uniform phase distribution of the vacuum state in mode 1 and the phase distribution for the coherent state in mode 2. For $|g|t = 0.2$, this factorization is still possible approximately. The distribution of the pump mode is broadened, whereas two peaks of the phase distribution of the second subharmonic mode emerge. The phases of the new peaks obey the phase relation $2\varphi_1 - \varphi_2 + \psi \equiv \frac{\pi}{2}$ (mod 2π). For some time, the factorization is possible, but a second peak develops also in the phase distribution of the pump mode. It is possible, at least for short times, to treat and present the phase statistics using the phase windows independent of t. Relative to the chosen windows, the mean phase and phase variance have been calculated using the Pegg–Barnett formalism. The long time effect of the quantum fluctuations of the pump mode is the randomization of the phase distribution for both signal and pump modes. This randomization process does not lead invariably to the uniform phase distribution, but it has turned out that at least partial revivals of the phase structure occur during the evolution [Gantsog (1992), Gantsog, Tanaś and Zawodny (1993)].

The impeded second-harmonic generation has been described by the generator of spatial progression [Peřinová and Lukš (1995), Peřinová, Lukš, Křepelka, Sibilia and Bertolotti (1996)]

$$\hat{G}(z) = \hat{G}_0(z) + \hat{G}_{int}(z), \tag{5.4.23}$$

where

$$\hat{G}_0(z) = \sum_{j=1}^{2} \hbar k_j \hat{a}_j^\dagger(z)\hat{a}_j(z), \tag{5.4.24}$$

with $k_j > 0$ being wave vectors, $k_j \approx \frac{\omega_j}{c}$, c is the light velocity, and

$$\hat{G}_{int}(z) = \hbar[g^* \hat{a}_1^{\dagger 2}(z)\hat{a}_2(z) + \text{H. c.}], \tag{5.4.25}$$

with g being the coupling constant. The phase properties of the process depend on the phase mismatch $\Delta k = k_2 - 2k_1$. Considering the progression up to the second order in z, we may evaluate the contribution of different terms, among them a Kerr-like one [Peřinová and Lukš (1995)]. Studying the argument of $\langle \hat{A}_1(z) \rangle$, where $\hat{A}_1(z) = \hat{a}_1(z)\exp(-ik_1 z)$, we may encounter a phase shift characteristic of cubic or Kerr-like behaviour of the fundamental mode. In addition to the usual codirectional propagation, the contradirectional propagation has also been considered. A unitary input–output transformation up to the second order in z has been derived for this new situation revealing the cubic behaviour of the annihilation operator in the fundamental mode [Peřinová, Lukš, Křepelka, Sibilia and Bertolotti (1996)].

The harmonic oscillator has the obvious property that the photon-number distribution does not vary during the evolution. The same property is shared by the model of a nonlinear Kerr medium, which is called the anharmonic oscillator. One can say that such an oscillator has purely phase properties. Of course, it does not exclude the study of quadrature squeezing, because also the quadrature operators do not commute with the number operator and their distributions change during the Kerr evolution.

Supposing simply the initially coherent light interacting with a lossless Kerr medium modelled as the third-order nonlinear oscillator described by the Hamiltonian (5.4.12), we observe that the phase distribution is broadened at the beginning of the evolution (the phase diffusion). Although the shape of the distribution does not change significantly, soon peaks emerge, which can be related to the superpositions of the microscopically distinguishable states (see section 3.10). The initial state of the lossless nonlinear oscillator revives after a certain time interval (the period) [Milburn and Holmes (1986)].

An explanation of the origin of the periodicity of the initial state has been based on an analysis of the quasidistribution $\Phi_{\mathcal{A}}(\alpha, t)$ oriented to its relation to the matrix elements of the state in the number state basis [Peřinová and Lukš (1990)]. The matrix elements $\rho_{nm}(t)$, which determine the quasidistribution $\Phi_{\mathcal{A}}(\alpha, t)$, assume the simple form

$$\rho_{nm}(t) = \exp[i\kappa(m-n)(m+n-1)t]\rho_{nm}(0). \qquad (5.4.26)$$

It is clear that the matrix elements $\rho_{nm}(t)$ exhibit the shortest common period $\bar{t} = \frac{\pi}{\kappa}$. This period is exemplified by the "fundamental" matrix element

$$\rho_{02}(t) = \exp(2i\kappa t)\rho_{02}(0), \qquad (5.4.27)$$

and the matrix elements (5.4.26) are the harmonics. All relevant quantum statistics share this period.

Following Milburn (1986), Milburn and Holmes (1986), and Daniel and Milburn (1989), we consider the classical dynamics of the lossless third-order nonlinear oscillator as described by the classical equation for the complex amplitude

$$\alpha^*(t) = \exp(i2\kappa|\alpha(0)|^2 t)\alpha^*(0). \qquad (5.4.28)$$

A probability distribution of the complex variable $\alpha(t)$ is considered. It is evident that the rotation shear, which occurs in the classical description, may be modified on the assumption of discrete values for the intensity $|\alpha(t)|^2$ and classical revivals result for $|\alpha(t)|^2 = 0, \frac{1}{2}, 1, \frac{3}{2}, 2, \ldots$.

Let us consider the particular case of the formula (4.10.14) for $s = -1$,

$$\Phi_A(|\alpha|\exp(i\varphi), t) = \frac{1}{\pi}\sum_{n=0}^{\infty}{}' \sum_{k=-n}^{n} \frac{\rho_{n-k,n+k}(t)}{\sqrt{(n-k)!(n+k)!}}$$
$$\times \exp(i2k\varphi)|\alpha|^{2n}\exp(-|\alpha|^2). \tag{5.4.29}$$

This quasidistribution can be rewritten in the form [Peřinová and Lukš (1990)]

$$\Phi_A(|\alpha|\exp(i\varphi), t) = \sum_{n=0}^{\infty}{}' \frac{2}{\Gamma(n+1)}|\alpha|^{2n}\exp(-|\alpha|^2)\Phi_A(n,\varphi,t), \tag{5.4.30}$$

where

$$\Phi_A(n,\varphi,t) = \frac{\Gamma(n+1)}{2\pi}\sum_{k=-n}^{n} \frac{\rho_{n-k,n+k}(t)}{\sqrt{(n-k)!(n+k)!}}\exp(i2k\varphi),$$
$$n = 0, \frac{1}{2}, 1, \frac{3}{2}, 2, \ldots. \tag{5.4.31}$$

$\sum'$ means that the summation step is $\frac{1}{2}$. The variable n assumes not only the eigenvalues of the number operator, but also half-odd values. Another flaw of the function $\Phi_A(n,\varphi,t)$ may be the property

$$\Phi_A(n,\varphi+\pi, t) = (-1)^{2n}\Phi_A(n,\varphi,t), \tag{5.4.32}$$

which contradicts the expected smoothness of the quasidistribution. As it holds that

$$\int_0^{2\pi} \Phi_A(n,\varphi,t)\, d\varphi = \begin{cases} p(n) & \text{for } n \text{ integer,} \\ 0 & \text{for } n \text{ half-odd,} \end{cases} \tag{5.4.33}$$

$$\sum_{n=0}^{\infty} p(n) = 1, \tag{5.4.34}$$

the functions $\Phi_A(n,\varphi,t)$, $n = 0, \frac{1}{2}, 1, \frac{3}{2}, 2, \ldots$, form a quasidistribution of number and the $\hat{a}$–$\hat{a}^\dagger$-antinormal phase. Substituting from (5.4.30) into (4.10.14) (section 4.10), we obtain the $\hat{a}$–$\hat{a}^\dagger$-antinormal phase distribution

$$P^{(A)}(\varphi, t) = \sum_{n=0}^{\infty}{}' \Phi_A(n,\varphi,t). \tag{5.4.35}$$

By analogy with the classical Liouville equation, the functions $\Phi_A(n,\varphi,t)$ satisfy the partial differential equations

$$\frac{\partial}{\partial t}\Phi_A(n,\varphi,t) = \kappa(2n-1)\frac{\partial}{\partial\varphi}\Phi_A(n,\varphi,t). \tag{5.4.36}$$

The equations have solutions

$$\Phi_A(n,\varphi,t) = \Phi_A(n,\varphi+\kappa(2n-1)t, 0), \tag{5.4.37}$$

which evoke the picture of the quasidistributions $\Phi_A(n,\varphi,t)$ rotating in a clockwise direction with the angular velocities $\kappa(2n-1)$. The quasidistribution $\Phi_A(n,\varphi,t)$ for $n=\frac{1}{2}$ does not move and that for $n=1$ rotates with angular velocity κ, but according to (5.4.32) it consists of two identical parts. Therefore, the circular frequency is 2κ and the period is $\frac{\pi}{\kappa}$. The sum

$$\sum_{n\,\text{integer}} \Phi_A(n,\varphi,t) \tag{5.4.38}$$

has the period $\frac{\pi}{\kappa}$. The quasidistribution $\Phi_A(n,\varphi,t)$ for $n=\frac{3}{2}$ rotates with the angular velocity 2κ and also the sum

$$\sum_{n\,\text{half-odd}} \Phi_A(n,\varphi,t) \tag{5.4.39}$$

has the period $\frac{\pi}{\kappa}$ again. From this, the $\hat{a}$–$\hat{a}^\dagger$-antinormal phase distribution $P^{(A)}(\varphi,t)$ has the period $\frac{\pi}{\kappa}$. Let us assume that the phase distribution $P^{(A)}(\varphi,t)$ has a peak at $\varphi=\overline{\varphi}$ and that this can be connected to the following property of the quasidistribution $\Phi_A(n,\varphi,t)$. All the functions $\Phi_A(n,\varphi,t)$, $n>0$, have a positive peak at $\overline{\varphi}$ and $\Phi_A(n,\varphi,t)$, $n=1,2,3,\ldots$, have another positive peak at $\overline{\varphi}+\pi$, but the functions $\Phi_A(n,\varphi,t)$, $n=\frac{1}{2},\frac{3}{2},\ldots$, have a negative dip at $\overline{\varphi}+\pi$. We determine the shape of the graph of the distribution $P^{(A)}(\varphi,t)$ at $t=M\frac{\pi}{2\kappa}$, where M is an odd number. For n integral, $\kappa(2n-1)t\equiv\pm\frac{\pi}{2}$ (mod π). From this, the sum (5.4.38) has positive peaks at $\overline{\varphi}\pm\frac{\pi}{2}$ (mod 2π). For n half-odd, $\kappa(2n-1)t\equiv 0$ (mod π) if $n\equiv\frac{1}{2}$ (mod 2) and $\kappa(2n-1)t\equiv\pm\pi$ (mod 2π) if $n\equiv\frac{3}{2}$ (mod 2). From this, the sum (5.4.39) is approximately zero at $\overline{\varphi}$, $\overline{\varphi}\pm\pi$ and, in general, this sum almost vanishes. Therefore, the distribution $P^{(A)}(\varphi,\frac{M\pi}{2\kappa})$ exhibits positive peaks at $\overline{\varphi}\pm\frac{\pi}{2}$. The quantum coherence is sensitive to dissipation, with the result that the quasidistribution $\Phi_A(n,\varphi,t)$ may not rotate with velocities which preserve the resonance of motion.

The phase properties of coherent state evolving in the conservative third-order nonlinear oscillator have been studied in the framework of the Pegg–Barnett approach [Gerry (1990), Gantsog and Tanaś (1991c, d), Tanaś, Gantsog, Miranowicz and Kielich (1991)], the Susskind–Glogower approach [Gerry (1987), Gantsog and Tanaś (1991c), Agarwal, Chaturvedi, Tara and Srinivasan (1992)], and of the formalism based on the measured phase operators [Lynch (1988), Gantsog and Tanaś (1991b)]. The results of the above approaches to the phase problem have been compared [Gantsog and Tanaś (1991c)]. The phase properties of coherent light interacting with a lossy Kerr medium have been analyzed in the framework of the Pegg–Barnett and Susskind–Glogower phase formalisms [Gantsog and Tanaś (1991d)]. It has been shown that the damping accelerates the randomization of phase at the beginning of the evolution and removes the quantum periodicity of the process. The damping is useful for illustration of the differences between the two formalisms in the long-time limit. Since the damping diminishes the mean photon number in the mode and brings the field to a state close to the vacuum, the difference between the results of the two phase approaches is significant. Number–phase properties of optimum displaced Kerr state have been studied in the framework of the formalism of antinormally ordered exponential phase operators [Peřinová, Vrana and Lukš (1995)]. The range of the random phase variable has been chosen according to the evolving preferred phase in order to keep the peak near the centre of the interval $[\theta_0, \theta_0+2\pi)$. The preferred phase has been involved in the rotated quadrature operators, which enables one to compare their distributions with those of the photon number and quantum phase. Since in the optimum displaced Kerr states there is a similarity between the number–phase pair and the principal quadratures, it follows that the two phase-centred quadrature operators approach the principal quadrature operators during the evolution.

From the Schrödinger equation for a conservative third-order nonlinear oscillator,

$$i\hbar\frac{\partial}{\partial t}|\psi_N(t)\rangle = \hat{H}_N|\psi_N(t)\rangle, \quad \hat{H}_N = \hbar\kappa\hat{a}^{\dagger 2}\hat{a}^2, \tag{5.4.40}$$

describing the evolution of the system in the interaction picture, it is possible to obtain, supposing an initial coherent state, a generalized coherent state [Titulaer and Glauber (1966), Stoler(1971b), Vourdas and Bishop (1989)]

$$|\psi_N(t)\rangle = |\xi(0), \{\varphi_{Nn}\}\rangle = \exp\left(-\frac{1}{2}|\xi(0)|^2\right)\sum_{n=0}^{\infty}\frac{\xi^n(0)}{\sqrt{n!}}\exp(i\varphi_{Nn})|n\rangle, \tag{5.4.41}$$

where

$$\varphi_{Nn} = -\kappa t n(n-1). \tag{5.4.42}$$

When the coefficients $\exp(i\varphi_{Nn})$ have the property of k-periodicity, $k > 0$,

$$\exp(i\varphi_{N(n+k)}) = \exp(i\varphi_{Nn}) \tag{5.4.43}$$

for $n = 0, 1, \ldots, \infty$, the generalized coherent state (5.4.41) can be expressed as a discrete superposition of coherent states [Białynicka-Birula (1968), Tombesi and Mecozzi (1987)]

$$|\psi_N(t)\rangle = \sum_{j=1}^{k} a_{Nj}|\exp(i\theta_{Nj})\xi(0)\rangle. \tag{5.4.44}$$

The formulae for the phases θ_{Nj} and the coefficients a_{Nj}, $j = 1, \ldots, k$, are to be found in papers devoted to the propagation of the coherent light in a Kerr medium [Yurke and Stoler (1986), Miranowicz, Tanaś and Kielich (1990), Tanaś (1991)]. For a suitable propagation length in the medium, superpositions of both even and odd numbers of coherent states may arise [Blow, Loudon and Phoenix (1993)]. General rules for the creation of superpositions of a certain number of coherent states have been derived and the maximum number of well-separable states for a given initial mean photon number has been estimated. When the evolution time is a fraction $\frac{L}{\overline{N}}$ of the period, where L and $\overline{N}$ are mutually prime integers, then the state is a superposition of $\overline{N}$ phase shifted versions of the initial state and the phase shifts are

$-\frac{\pi}{\overline{N}}k$, $k = 1, 3, \ldots, 2\overline{N} - 1$, for $\overline{N}$ even and $-\frac{\pi}{\overline{N}}k$, $k = 0, 2, \ldots, 2\overline{N} - 2$, for $\overline{N}$ odd [Peřinová and Křepelka (1993)].

For the generalized coherent state (5.4.41), we determine the quasidistribution $\Phi_{AN}(\alpha, t)$,

$$
\begin{aligned}
\Phi_{AN}(\alpha, t) &= |\langle \alpha | \psi_N(t) \rangle|^2 \\
&= \exp[-|\alpha|^2 - |\xi(0)|^2] \left| \sum_{n=0}^{\infty} \frac{[\alpha^* \xi(0)]^n}{n!} \exp(i\varphi_{Nn}) \right|^2. \quad (5.4.45)
\end{aligned}
$$

This quasidistribution illustrates well the generation of superposition states [Miranowicz, Tanaś and Kielich (1990), Tanaś (1991)]. It exhibits regular structures when the component states are separate. The canonical phase distribution $P(\varphi, t)$ according to (4.6.102) and the $\hat{a}$–$\hat{a}^\dagger$-antinormal phase distribution $P^{(A)}(\varphi, t)$ indicate distinctly the superpositions of coherent states, because both of them exhibit the k-tuple rotational symmetry in the case of superposition of k states [Tanaś, Gantsog, Miranowicz and Kielich (1991), Gantsog and Tanaś (1991c), Peřinová and Lukš (1994)]. The k-tuple rotational symmetry is almost perfect, but not exact, because the superpositions are not just k-photon states [Peřinová, Lukš and Křepelka (1997a)]. Bifurcation into a number of peaks has been examined using two different measures of phase uncertainty, namely, the generalized Paul–Bandilla phase uncertainty and the entropy [Vaccaro and Orłowski (1995)]. The damping of the superpositions is fast and may be treated adequately through the first-order perturbation theory [Braunstein (1992a), Peřinová and Lukš (1994), Tanaś, Miranowicz and Gantsog (1996)].

Phase properties of the squeezed and displaced number states [Král (1990a)] propagating in a lossy Kerr medium have been studied in terms of the $\hat{a}$–$\hat{a}^\dagger$-antinormal phase distribution $P^{(A)}(\varphi, t)$. The canonical phase distribution has been used for the illustration of distinct superpositions of phase-shifted squeezed and displaced number states [Peřinová and Křepelka (1993)]. The phase properties of the displaced number states evolved in a lossless Kerr medium have been described by the phase distribution and variances of not only the phase angle, but also the cosine and sine Pegg–Barnett operators [Leng, Du and Gong (1995)]. Sanders (1992) has taken the Pegg–Barnett s-phase state for the initial state and has illustrated that also this state evolves in a Kerr medium into a finite superposition of some s-phase states at the rational multiples of the period.

The phase properties may always be related to the generation of macroscopically distinguishable quantum states in the $(2k - 1)$th-order nonlinear oscillator, $k \geq 2$. Such superpositions have been considered already by Yurke and Stoler (1986). The generation of superposition states for an initial coherent state and for the symmetrical version of the Hamiltonian, $\hat{H} = \hbar \kappa (\hat{a}^\dagger \hat{a})^k$, has been studied by Tombesi and Mecozzi (1987). The Kerr dynamics of the phase in the case $k = 2$ expressed by plots of the Wigner function can be found in [Atakishiyev, Chumakov, Rivera and Wolf (1996)]. The formation of discrete superpositions of coherent states in the course of evolution in the k-photon $((2k - 1)$th-order) nonlinear oscillator with the explicitly normally ordered Hamiltonian has also been studied [Paprzycka and Tanaś (1992)]. Exact analytical formulae for the superposition coefficients have been obtained. In contrast to the third-order nonlinear oscillator, the superposition components enter the superposition with different amplitudes, which makes the superposition less symmetrical. The phase distributions for the resulting states illustrate the amount of symmetry and show the number of components. A study of driven damped anharmonic oscillator initially in a coherent state has been devoted to the quantum phase properties, such as mean phase and phase variance [Enzer and Gabrielse (1997)]. In this case the anharmonic oscillator does not model vibrations of an optical mode, but those of an electron.

The system of two coupled oscillators can be realized as the elliptically polarized light propagating in a Kerr medium described by the Hamiltonian [Tanaś and Kielich (1983, 1984)]

$$\hat{H} = \overline{\alpha}(\omega)(\hat{a}_1^\dagger \hat{a}_1 + \hat{a}_2^\dagger \hat{a}_2) + \frac{1}{2}\left\{4\overline{\gamma}_1(\omega)\hat{a}_1^\dagger \hat{a}_1 \hat{a}_2^\dagger \hat{a}_2 \right.$$
$$\left. +[\overline{\gamma}_2(\omega) + \overline{\gamma}_3(\omega)]\left(\hat{a}_1^{\dagger 2}\hat{a}_1^2 + \hat{a}_2^{\dagger 2}\hat{a}_2^2 + 2\hat{a}_1^\dagger \hat{a}_1 \hat{a}_2^\dagger \hat{a}_2\right)\right\}, \qquad (5.4.46)$$

where $\hat{a}_1$ and $\hat{a}_2$ are the annihilation operators for the circularly right- and left-polarized modes of the field, respectively. The quantities $\overline{\alpha}(\omega)$, $\overline{\gamma}_j(\omega)$, $j = 1, 2, 3$, are related to the expectation values of the polarization and hyperpolarization tensors over all possible orientations of the molecule.

The above quantum model for propagation of the elliptically polarized light in a Kerr medium has been applied to partially polarized light for initial coherent and number states [Agarwal and Puri (1989, 1990)]. States have been proved to be generated which are macroscopic superpositions of coherent states. It has been found that if the input field is completely polarized, the quantum effects result in partial polarization of the output field, which contradicts the prediction of the classical theory. The energies in separate modes, the correlation between two modes, higher-order correlations, and the mean photon number have been studied in relation to the input states. The input photon statistics are found to have a considerable effect on the dynamics.

The phase properties of the elliptically polarized light propagating in a Kerr medium have been investigated in the framework of the Pegg–Barnett formalism [Gantsog and Tanaś (1991b, e), Tanaś and Gantsog (1991)]. The Hermitian phase operator concept has been generalized to two-mode fields and the joint phase distribution of the two orthogonal modes describing the elliptically polarized field has been computed. In addition, marginal phase distributions as well as the expectation values and variances of the phases have been determined and their dynamics studied. It has been found that in the course of the propagation, a correlation between the phases of the two modes comes into being. The degree of correlation depends substantially on the asymmetry parameter d of the medium, $d = f(\overline{\gamma}_1(\omega), \overline{\gamma}_2(\omega), \overline{\gamma}_3(\omega))$. The highest correlation occurs for $d = \frac{1}{2}$. This correlation reduces the variance of the difference of the phase operators for the separate modes. If $d = \frac{1}{2}$, neither the variance of the phase-difference operator nor the degree of the light polarization is affected during the propagation. However, for $d \neq \frac{1}{2}$, this variance rises rapidly to $\frac{1}{2}$, which is the value for the randomly distributed phase difference. To the randomization of the phase difference there corresponds a degradation of the degree of light polarization. The phase properties of the elliptically polarized light in a Kerr medium confirm the results predicting this degradation [Agarwal and Puri (1989), Tanaś and Kielich (1990)].

The formation of discrete superpositions of coherent states has been considered for elliptically polarized light propagating in a nonlinear Kerr medium [Gantsog and Tanaś (1991b)]. It has been shown that superpositions with any number of components can be obtained if the evolution time is taken as a fraction $\frac{L}{N}$ of the period, where L and $\overline{N}$ are mutually prime integers. Exact analytical formulae for finding the superposition coefficients have been given. It has been shown that the coupling between the two circularly polarized components of the elliptically polarized light caused by the asymmetry of the nonlinear properties of the medium can suppress the number of components if the asymmetry parameter takes on appropriate values. The phase distribution for the two-mode field exhibits a well-resolved, multi-peak structure, clearly indicating the generation of the discrete superpositions of coherent states.

Quantum fluctuations in the Stokes parameters, which are the expectation values of the Hermitian Stokes operators, have been studied for strong light propagating in a lossless and damped Kerr medium [Tanaś and Kielich (1990), Tanaś and Gantsog (1992c, d)]. In the case without losses [Tanaś and Kielich (1990), Tanaś and Gantsog (1992c)], the periodic behaviour of quantum evolution of the light polarization is revealed explicitly. Although the variances of the Stokes operators are reminiscent of the field operator variances determining squeezing, they have been shown not to fall below the level for a coherent state. The signal-to-noise ratio for the measurement of the Stokes parameters is reduced by quantum field fluctuations. All polarization parameters depend crucially on the asymmetry parameter of the nonlinear medium. Quantitative assessment of the destructive role of linear damping in the quantum evolution of the field has been performed [Tanaś and Gantsog (1992d)].

The phase properties of light propagating in a two-mode Kerr medium have been reconsidered [Luis, Sánchez-Soto and Tanaś (1995)] from the viewpoint of the Hermitian phase-difference operator introduced by Luis and Sánchez-Soto (1993b, 1995), which is based on the polar decomposition of the Stokes operators. The predictions of the Pegg–Barnett formalism have been compared with those of the Luis–Sánchez-Soto phase-difference formalism. The difference found, which is manifest for weak fields, is related to the difference between the distributions. Whereas in the Pegg–Barnett formalism the limiting procedure provides a continuous phase distribution, the Luis–Sánchez-Soto phase-difference operator is related to the discrete distribution of this quantity. A quantum theory of cross-phase modulation for short propagation distances has been developed by Blow, Loudon and Phoenix (1994). The system composed of two single-mode electromagnetic fields copropagating through a lossless Kerr medium has been studied with respect to phase properties [De Moura and Lyra (1996)]. The phase correlation between the two modes measured as the codispersion (4.4.135) has been shown to exhibit a rich pattern of collapses and revivals. The canonical phase distributions of separate modes manifest not only the self-phase modulation, but also the cross-phase modulation. The number of distinguishable components in the marginal distributions depends on the parameters of the evolution. Using the Pegg–Barnett formalism, Dung and Shumovsky (1992) have discussed the effects of cavity damping on the properties of the field phase in the Jaynes–Cummings model. They have compared the canonical phase distribution with those obtained by integrating the Φ_A function and the Wigner function over the real amplitude. The phase dynamics of a field mode in the interaction with one and two two-level atoms has been investigated using the Pegg–Barnett formalism [Eiselt and Risken (1991), Jex, Matsuoka and Koashi (1993)]. Analytical and approximate expressions for the phase probability distribution have been derived and the mean value of the phase and its variance have been calculated. Considering continuous measurement scheme using a three-level atom system, it has been found that the phase distribution of one component of two-mode coherent state can be narrowed [Agarwal, Scully and Walther (1993)]. Both the process of quantum nondemolition measurement (cf. section 3.9) and the phase diffusion associated with it have been analyzed in a systematic manner [Herzog and Paul (1993)]. Paul (1991) has analyzed the measurement proposed by Brune, Haroche, Lefevre, Raimond and Zagury (1990). His results have indicated that the phase destruction (diffusion) in quantum nondemolition measurement of the photon number can also be formulated as the decay of the expectation values $|\langle \widehat{\exp}[i(m-n)\varphi] \rangle|$. Herzog and Richter (1994) derived a simple analytical expression for the decay of the off-diagonal matrix elements of the field statistical operator in the description of a quantum nondemolition measurement of the photon number according to Holland, Walls and Zoller (1991). With the help of this expression, they investigated the phase destruction (diffusion) of the field due to the measurement process. Vaccaro and Pegg (1994b) have studied phase properties of optical linear amplifiers choosing to use the phase formalism of Pegg and Barnett. Assuming that the mean photon numbers are at least of the order 10, they have derived a diffusion equation which is satisfied by the phase probability density. The phase-diffusion coefficient is made to depend on the mean photon number. So, they have used the strong field assumption to reduce suitably the statistical operator. The phase distributions of an initially coherent single-mode field interacting with one, two, three, and four identical atoms have been investigated using the Pegg–Barnett phase formalism [Hu (1995)].

5.4.3   Phase dynamics of ℓ-photon coherent states in a Kerr medium

Multiphoton special states have been introduced in two closely related ways. Multiphoton squeezed states have been constructed [D'Ariano, Rasetti and Vadacchino (1985)] using the Brandt–Greenberg multiphoton annihilation and creation operators $\hat{A}_\ell, \hat{A}_\ell^\dagger$ [Brandt and Greenberg (1969)],

$$
\hat{A}_\ell = \hat{a}^\ell \left(\left[\frac{\hat{n}}{\ell} \right] \frac{(\hat{n} - \ell\hat{1})!}{\hat{n}!} \right)^{\frac{1}{2}},
$$

$$
\hat{A}_\ell^\dagger = \left(\left[\frac{\hat{n}}{\ell} \right] \frac{(\hat{n} - \ell\hat{1})!}{\hat{n}!} \right)^{\frac{1}{2}} \hat{a}^{\dagger\ell}, \tag{5.4.47}
$$

where [ ] denotes the integer part. The appropriate multiphoton squeezed states have been defined using a unitary transformation of the vacuum state so they can be classified as generalized group theoretical coherent states [Perelomov (1972)]. They can be expressed in the form

$$
|\alpha^\ell, \ell\rangle^{\mathrm{WH}} = \exp\left(-\frac{1}{2} |\alpha|^{2\ell} \right) \sum_{n'=0}^\infty \frac{\alpha^{n'\ell}}{\sqrt{n'!}} |n'\ell\rangle, \tag{5.4.48}
$$

where α^ℓ is a squeeze parameter and WH stands for the Weyl–Heisenberg group. According to the original construction, these states should be named the multiphoton squeezed vacuum states for $\ell > 1$ and the coherent state for $\ell = 1$. Instead, these states are called multiphoton coherent states uniformly. This terminology is justified also by the fact that the multiphoton coherent state can be physically realized as the output state of a noiseless photon amplifier with an input coherent state [Yuen (1986), D'Ariano (1992)]. Two-photon squeezed vacuum according to (5.4.48) differs from the usual squeezed vacuum for $\ell = 2$. The phase and statistical properties of the multiphoton states (5.4.48) have been studied by Luis and Sánchez-Soto (1993c). Sukumar (1989a) proceeded only in a slightly different manner to introduce other multiphoton squeezed vacuum states for $\ell > 1$ and the coherent phase state for $\ell = 1$.

Another possibility is to define multiphoton coherent states as the eigenstates of the generalized (multiphoton) annihilation operator [Barut and Girardello (1971), Bužek and Jex (1989), Bužek, Jex and Quang (1990)]. The simplest choice is the operator $\hat{a}^\ell$ as the generalized ℓ-photon annihilation operator and then the ℓ-photon coherent states are eigenstates $|\alpha^\ell, \ell\rangle$ of this operator,

$$
\hat{a}^\ell |\alpha^\ell, \ell\rangle = \alpha^\ell |\alpha^\ell, \ell\rangle. \tag{5.4.49}
$$

The eigenvalue problem (5.4.49) has ℓ kinds of solutions [Xin, Zhao, Hirayama and Matumoto (1994)] and we will call the ℓ-photon coherent states such solutions, which can be expanded in the Fock states $|n'\ell\rangle$.

They can be expressed in the form [Bužek and Jex (1989), Bužek, Jex and Quang (1990)]

$$|\alpha^\ell, \ell\rangle = \left[\sum_{j=0}^\infty \frac{|\alpha|^{2j\ell}}{(j\ell)!}\right]^{-\frac{1}{2}} \sum_{n'=0}^\infty \frac{\alpha^{n'\ell}}{\sqrt{(n'\ell)!}}|n'\ell\rangle. \tag{5.4.50}$$

The ℓ-dimensional eigenspace belonging to the degenerate eigenvalue α^ℓ is spanned by the eigenvectors

$$|\alpha^\ell, m \bmod \ell\rangle = \left[\sum_{j=0}^\infty \frac{|\alpha|^{2(j\ell+m)}}{(j\ell+m)!}\right]^{-\frac{1}{2}} \sum_{n'=0}^\infty \frac{\alpha^{n'\ell+m}}{\sqrt{(n'\ell+m)!}}|n'\ell+m\rangle, \tag{5.4.51}$$

where $m = 0, 1, \ldots, \ell - 1$. The resolution of the identity in terms of the eigenstates (5.4.51) appropriate to all complex numbers α^ℓ can be formulated, but its formulation simplifies when unnormalized states are used [Xin, Zhao, Hirayama and Matumoto (1994)]. The Glauber coherent state $|\alpha\rangle$ belongs to this eigenspace, the eigenstate (5.4.50) recovers as $|\alpha^\ell, \ell\rangle = |\alpha^\ell, 0 \bmod \ell\rangle$. Especially, for $\ell = 2$, we obtain the Schrödinger cat states $|\alpha_+\rangle = |\alpha^2, 0 \bmod 2\rangle$, $|\alpha_-\rangle = |\alpha^2, 1 \bmod 2\rangle$, which have long been known as the even and odd coherent states (3.2.32) and (3.2.33), respectively. Shanta, Chaturvedi, Srinivasan, Agarwal and Mehta (1994) have obtained a large class of multiphoton annihilation operator ($\hat{A}$) eigenstates by constructing an operator $\hat{B}$ such that $[\hat{A}, \hat{B}^\dagger] = [\hat{B}, \hat{A}^\dagger] = \hat{1}_{sect}$ of the Hilbert space, where $\hat{1}_{sect}$ is a restricted identity operator or a projection operator onto a sector. It is interesting to see what will evolve from the initial ℓ-photon coherent state [Bužek and Jex (1989), Bužek, Jex and Quang (1990)] in a Kerr medium. Of course, when the dissipation is not taken into account, the ℓ-photon state remains ℓ-photon for all times. This entails that the use of the first-order preferred phase and also the value of the first-order dispersion is vain. Higher-order preferred values and phase dispersions have been applied for these input states [Peřinová, Lukš and Křepelka (1997a)], but these phase characteristics can be used also when the input is not ℓ-photon.

An ℓ-photon coherent state on the input

$$|\psi(0)\rangle = \sum_{n'=0}^\infty c'_{n'}(0)|n'\ell\rangle, \tag{5.4.52}$$

where

$$c'_{n'}(0) = \left[\sum_{j=0}^\infty \frac{|\xi(0)|^{2j\ell}}{(j\ell)!}\right]^{-\frac{1}{2}} \frac{[\xi(0)]^{n'\ell}}{\sqrt{(n'\ell)!}}, \tag{5.4.53}$$

with $\xi(0)$ being the complex field amplitude, evolves into an ℓ-photon Kerr state

$$|\psi(t)\rangle = \sum_{n'=0}^\infty c'_{n'}(t)|n'\ell\rangle, \tag{5.4.54}$$

where

$$c'_{n'}(t) = \exp[-i\kappa t n'\ell(n'\ell - 1)]c'_{n'}(0). \tag{5.4.55}$$

The Kerr-evolved preferred phase and phase dispersion of the kth order (introduced in (4.4.75) and (4.4.78), respectively) are of the forms

$$\left(\text{pref}^{[k]}\varphi\right)(t) = \frac{1}{k}\arg\left[\langle\widehat{\exp}(ik\varphi)\rangle(t)\right], \tag{5.4.56}$$

$$\left(D^{[k]}\varphi\right)(t) = \frac{1}{k^2}\left[1 - |\langle\widehat{\exp}(ik\varphi)\rangle(t)|^2\right], \tag{5.4.57}$$

where

$$\langle\widehat{\exp}(ik\varphi)\rangle(t) = \begin{cases} \sum\limits_{j=0}^{\infty} c'^{*}_{j}(t)c'_{j+\frac{k}{\ell}}(t) & \text{for } \ell|k, \\ 0 & \text{for } \ell \nmid k. \end{cases} \tag{5.4.58}$$

Although the input state is a superposition of ℓ (one-photon) coherent states,

$$|\psi(0)\rangle = \frac{1}{\ell}Z\sum_{j=0}^{\ell-1}\left|\exp\left(-i\frac{2\pi j}{\ell}\right)\xi(0)\right\rangle, \tag{5.4.59}$$

where

$$Z = \exp\left(\frac{1}{2}|\xi(0)|^2\right)\left[\sum_{j=0}^{\infty}\frac{|\xi(0)|^{2j\ell}}{(j\ell)!}\right]^{-\frac{1}{2}}, \tag{5.4.60}$$

the general result still applies that in the unitary Kerr evolution, superposition states emerge. Rather obviously, just from the superposition principle it follows that these superpositions consist of one-photon coherent states. At the fraction $\frac{L}{N}$ of the period $\frac{\pi}{\kappa}$, where L and $\overline{N}$ are relatively prime integers, the superposition is of the form

$$\left|\psi\left(\frac{L\pi}{\overline{N}\kappa}\right)\right\rangle = Z\sum_{j=0}^{2K-1}\tilde{v}_j\left|\exp\left(-\frac{i\pi j}{K}\right)\xi(0)\right\rangle, \tag{5.4.61}$$

where

$$\tilde{v}_j = \frac{1}{2K}\sum_{n'=0}^{2\frac{K}{\ell}-1}\exp\left[-i\kappa t n'\ell(n'\ell - 1)\right]\exp\left(i\frac{\pi j n'\ell}{K}\right), \tag{5.4.62}$$

and K is the least common multiple of L and $\overline{N}$, $K = \text{lcm}(L, \overline{N})$.

The usefulness of the kth-order preferred phase (5.4.56) and the phase dispersion of the kth order (5.4.57) depends on whether or not ℓ divides k and will be illustrated in the following figures.

The dispersion of phase in figure 5.18 is almost independent of time for $\xi(0)$ small ($\lesssim 1$). In fact, the precise time independence occurs for the vacuum state, $\xi(0) = 0$. For these nearly vacuum states, the dispersion of phase decreases from its maximum value 1 down to almost one half. At least from this plot it can be deduced that the

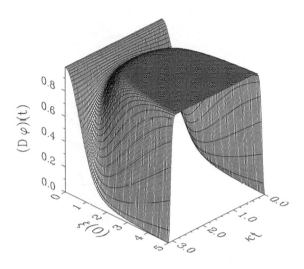

Figure 5.18: The Kerr evolution of the phase dispersion ($k=1$) versus the initial complex field amplitude $\xi(0) \in [0,5]$ ($\ell = 1$), $\kappa t \in [0, \pi]$.

immediately following values of $\xi(0)$ exhibit strong dependence on time. For these values of $\xi(0)$, the dispersion of phase is already small enough, but the dispersion at half a period does not yet return to its maximum value. The evolved state for $t = \frac{\pi}{2\kappa}$ is a Yurke–Stoler superposition of two coherent states. The phase dispersion of such superposition states can be high due to the presence of two antipodal coherent terms. It can be seen, however, that the presence of a significant interference term reduces the dispersion. With increasing $\xi(0)$, the relative weight of the interference term fades out, $|\xi(0)| \gtrsim 2$. For greater values of $\xi(0)$, the remaining interesting property is that the quasiclassical part, where the phase distribution spreads, is getting shorter, $t \simeq 0$, $t \geq 0$. By symmetry, the same is valid for the typically quantum behaviour, for $t \simeq \frac{\pi}{\kappa}$ and $t \leq \frac{\pi}{\kappa}$, when the phase distribution shrinks.

The preferred value of the phase enables us to perform a comparison with the classical model, which does not grow into the statistical character of the radiation. Figure 5.19, when restricted to $t \in [0, \frac{\pi}{\kappa}]$, provides a correction to our picture of quasiclassical behaviour derived from figure 5.18. The quasiclassical part ends where the staircase in the picture begins. Again, in comparison with figure 5.18, the perfect symmetry is not present, but the time period when the phase distribution shrinks can be established and it follows after the staircase. Although this stage bears a similarity with the quasiclassical stage, the appropriate part of the plot has also a

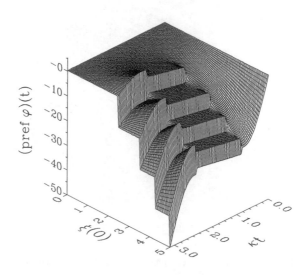

Figure 5.19: The phase shift of the light field characterized by the preferred value of the phase ($k = 1$) versus the initial complex field amplitude $\xi(0) \in [0,5]$ ($\ell = 1$), $\kappa t \in [0, \pi]$.

form of staircase. This is connected with the fact that

$$(\text{pref } \varphi)\left(\frac{\pi}{\kappa}\right) \equiv (\text{pref } \varphi)(0) \pmod{2\pi}. \tag{5.4.63}$$

After the phase unwrapping, it holds that

$$(\text{pref } \varphi)\left(\frac{\pi}{\kappa}\right) = \arg[\xi(0)] - 2\pi k[\xi(0)], \tag{5.4.64}$$

where $k(\alpha)$ is an integral-valued function, $k(\alpha) \geq 0$. The staircase is situated at about $t = \frac{\pi}{2\kappa}$, which can be accounted for by the high dispersion of phase in this stage. We can say that at the rest of the time period, the quasiclassical behaviour does not allow the light mode to "drive" the preferred value of the phase to its "prescribed" value $0 \pmod{2\pi}$. Let us remark that the phase unwrapping causes that the preferred phase is not periodic as usual, but periodic of the second kind. Let us summarize the symmetry properties,

$$(D\varphi)\left(\frac{\pi}{\kappa} - t\right) = (D\varphi)(t), \tag{5.4.65}$$

$$(\text{pref } \varphi)\left(\frac{\pi}{\kappa} - t\right) = 2\arg[\xi(0)] - (\text{pref } \varphi)(t) - 2\pi k[\xi(0)]. \tag{5.4.66}$$

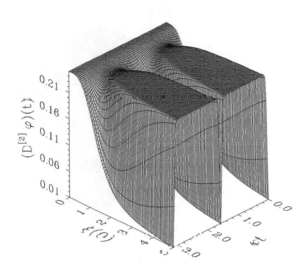

Figure 5.20: The Kerr evolution of the second-order phase dispersion versus the initial complex field amplitude $\xi(0) \in [0, 5]$ ($\ell - 1$) and $\kappa t \in [0, \pi]$.

The higher-order dispersions can serve as detectors of the Yurke–Stoler superposition states, where the number of coherent components is appropriate. As a consequence of a great simplicity of Kerr's evolution modelled by an anharmonic oscillator, three-dimensional plots of phase dispersions of higher (kth) order resemble figure 5.18 split into k equal parts. For example, for $k = 4$ from the above theory it follows that three dips resulting after splitting correspond to superpositions of four, two, four coherent states at $\kappa t = \frac{\pi}{4}, \frac{\pi}{2}, \frac{3\pi}{4}$, respectively. Similarly as in figure 5.18, the division of values of $\xi(0)$ into three parts remains valuable also for $k > 1$ as is obvious from figure 5.20. For small $\xi(0)$, the near vacuum states, the phase dispersion of any order is almost independent of time. Therefore, for these $\xi(0)$ the splitting into k equal parts related to the time evolution has a negligible effect. For $\xi(0) = 5$ already for $k = 1$, a switch between a minimum and a maximum is rather fast and this enhances with increasing k, when the maximum value of the dispersion of phase has been proposed as $\frac{1}{k^2}$. Since the end value $\xi(0) = 5$ is quite large, only for $k \geq 3$ can it be seen that the superposition states do not lead to complete revivals of the low dispersion of the phase. We did not consider $\xi(0)$ greater than 5, because the phase distribution in coherent states with so large a complex amplitude is the Dirac delta function. The appropriate dispersion of the phase is zero. We remember also that the interference terms of the Yurke–Stoler superpositions cease contributing to

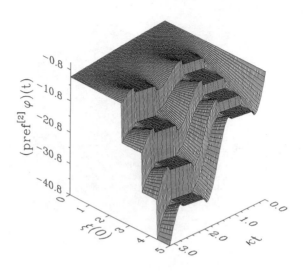

Figure 5.21: The phase shift imparted to the one-photon coherent state with the complex field amplitude $\xi(0) \in [0,5]$ $(\ell = 1)$ in the Kerr medium as characterized by the second-order preferred value of the phase, $\kappa t \in [0, \pi]$.

the nonclassical behaviour of these pure states. That is why for $\xi(0)$ large a shorter period arises, $\frac{\pi}{k\kappa}$ at t. This period is absent for moderate values of $\xi(0)$.

The second-order preferred value of phase in figure 5.21 has been plotted using the phase unwrapping technique. It is evident that the near vacuum states exhibit a phase shift of the second order, but there is no phase shift of the first order in the near vacuum states. The explication of the form of the whole graph based on its silhouette at $\kappa t = \pi$ is still useful, but we must remember that the possible values of the second-order preferred phase are negative multiples of π, generally those of $\frac{2\pi}{k}$. It is interesting that also the phase shift of the second order exhibits a proportionality to the mean photon number $|\xi(0)|^2$. When combined what was said of the "predetermined" values, a parabolic dependence on $\xi(0)$ can transpire only after a rounding off to the multiples of $\frac{2\pi}{k}$. With the splitting of the plot of the dispersion we may also associate the presence of k very pictorial staircases. This feature of the graph can also be derived on the basis of the continuity principle from the periodicity of the second kind, which for $\xi(0)$ large is almost present with a period of $\frac{\pi}{k\kappa}$.

For the ℓ-photon coherent state with $\ell \geq 2$, the phase dispersion of the first order is unity (maximum) for all times, because of (5.4.58). As noted above, this is accompanied by the absence of the first-order preferred value of phase. In particular,

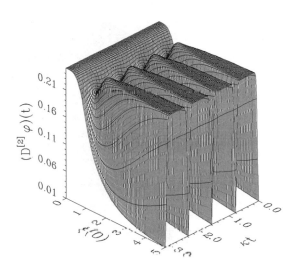

Figure 5.22: The Kerr evolution of the second-order dispersion of the phase versus the complex field amplitude $\xi(0) \in [0, 5]$ $(\ell = 2)$, $\kappa t \in [0, \pi]$.

for $\ell = 2$, we may study the phase characteristics of the second order. The three-dimensional plot in figure 5.22 exhibits a splitting into four parts and the minimum of the phase dispersion is attained three times within the period. It can be found numerically that at the times where the minimum is attained, the evolved state is a rotated initial ℓ-photon coherent state.

The preferred value of phase of the same order in figure 5.23 exhibits a smoothed phase shift in the clockwise direction (towards more negative values of the preferred phase) at the times, where the dispersion of phase is minimum. The phase shift in the anticlockwise direction occurs "covertly" at the times when the phase dispersion of this order is maximum as if it serves the purpose of achieving a multiple of π at the end of the period $\frac{\pi}{\kappa}$. In more detail, the anticlockwise phase shift occurs for $|\xi(0)|^2$ slightly smaller than the jump value, whereas for $|\xi(0)|^2$ slightly greater than the jump value the phase shift realizes as a swing in the usual clockwise direction. It is very interesting that the second-order preferred phase also increases proportionally to $|\xi(0)|^2$, which still resembles the traditional properties of the Kerr medium.

Similarly as in figure 5.21, the phase shift as detected by a higher-order preferred value of the phase in figure 5.24 already undergoes a shift for near vacuum states. Piecewise constant behaviour in dependence on the complex field amplitude $\xi(0)$ at the end of the period seems to be typical of this anharmonic oscillator model, but

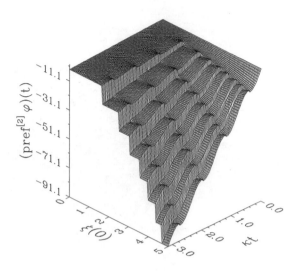

Figure 5.23: The effect of the initial complex field amplitude $\xi(0) \in [0,5]$ ($\ell = 2$) on the Kerr evolution of the second-order preferred value of the phase, $\kappa t \in [0, \pi]$.

the choice of constants from among the negative multiples of $\frac{\pi}{2}$ is not very obvious except the fact that it is affected by the increased ℓ. The fourth-order dispersion is to detect an improvement at least three times ($(k-1)$ times) during the period, but with respect to $\ell = 2$ it exhibits the improvement 7 times ($(2k-1)$ times).

For three-photon coherent state not only the first-order phase dispersion, but also the second-order one is maximum. The third-order characteristics of the phase take into account the trefoil form of a phase-space quasidistribution. The third-order dispersion of the phase in figure 5.25 is expected to drop to the minimum at least two times during period according to the theory. Since $\ell = 3$, the minimum value occurs eight times (generally $(k\ell - 1)$ times). It is quite interesting that similarly as in figure 5.18, the dependence on $\xi(0)$ exhibits a minimum which is quite uniform with respect to the time. This occurs for those values of complex field amplitude which are neither near to vacuum nor too large. It seems that the period is $\frac{\pi}{9\kappa}$, i. e., $\frac{\pi}{k\ell\kappa}$, when we allow the values of preferred phase to approximately repeat. The appropriate third-order preferred value of the phase in figure 5.26 behaves similarly to that in figure 5.23.

As expounded in section 3.10, in the statistical operator of a superposition state

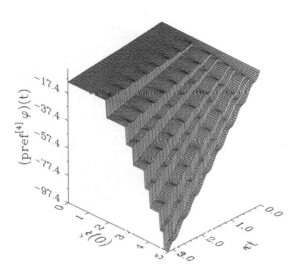

Figure 5.24: The behaviour of the fourth-order preferred value of the phase versus the initial complex field amplitude $\xi(0) \in [0,5]$ ($\ell = 2$) in the Kerr medium, $\kappa t \in [0, \pi]$.

$\hat{\rho}\left(\frac{L\pi}{N\kappa}\right) = |\psi\left(\frac{L\pi}{N\kappa}\right)\rangle\langle\psi\left(\frac{L\pi}{N\kappa}\right)|$, the "mix" and "interference" parts are distinguished,

$$\hat{\rho}\left(\frac{L\pi}{N\kappa}\right) = \hat{\rho}_{\mathrm{mix}}\left(\frac{L\pi}{N\kappa}\right) + \hat{\rho}_{\mathrm{int}}\left(\frac{L\pi}{N\kappa}\right), \qquad (5.4.67)$$

where

$$\hat{\rho}_{\mathrm{mix}}\left(\frac{L\pi}{N\kappa}\right) = Z^2 \sum_{j=0}^{2K-1} |\tilde{v}_j|^2 \left|\exp\left(-i\frac{\pi j}{K}\right)\xi(0)\right\rangle\left\langle\exp\left(-i\frac{\pi j}{K}\right)\xi(0)\right|, \qquad (5.4.68)$$

and the interference part

$$\hat{\rho}_{\mathrm{int}}\left(\frac{L\pi}{N\kappa}\right) = Z^2 \sum_{\substack{j,j'=1 \\ j\neq j'}}^{2K-1} \tilde{v}_j\tilde{v}_{j'}^* \left|\exp\left(-i\frac{\pi j}{K}\right)\xi(0)\right\rangle\left\langle\exp\left(-i\frac{\pi j'}{K}\right)\xi(0)\right|. \qquad (5.4.69)$$

The canonical phase distribution exhibits the similar decomposition

$$P\left(\varphi, \frac{L\pi}{N\kappa}\right) = P_{\mathrm{mix}}\left(\varphi, \frac{L\pi}{N\kappa}\right) + P_{\mathrm{int}}\left(\varphi, \frac{L\pi}{N\kappa}\right), \qquad (5.4.70)$$

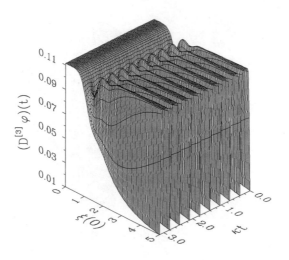

Figure 5.25: The Kerr evolution of the third-order dispersion of the phase versus the initial complex field amplitude $\xi(0) \in [0, 5]$ ($\ell = 3$), $\kappa t \in [0, \pi]$.

where

$$P_{\text{mix}}\left(\varphi, \frac{L\pi}{N\kappa}\right) = Z^2 \sum_{j=0}^{2K-1} |\tilde{v}_j|^2 \left|\psi_j\left(\varphi, \frac{L\pi}{N\kappa}\right)\right|^2, \qquad (5.4.71)$$

$$P_{\text{int}}\left(\varphi, \frac{L\pi}{N\kappa}\right) = Z^2 \sum_{\substack{j,j'=1 \\ j \neq j'}}^{2K-1} \tilde{v}_j \tilde{v}_{j'}^* \psi_j\left(\varphi, \frac{L\pi}{N\kappa}\right) \psi_{j'}^*\left(\varphi, \frac{L\pi}{N\kappa}\right), \qquad (5.4.72)$$

with

$$\psi_j\left(\varphi, \frac{L\pi}{N\kappa}\right) = \frac{1}{\sqrt{2\pi}} \exp\left[-\frac{1}{2}|\xi(0)|^2\right] \sum_{n=0}^{\infty} \frac{[\xi(0)]^n}{\sqrt{n!}} \exp\left[-in\left(\varphi + \frac{\pi j}{K}\right)\right]. \qquad (5.4.73)$$

The phase properties of the Yurke–Stoler superposition states are completely described by the phase distribution (5.4.70). For $t = \frac{\pi}{2\kappa}$, the component $P_{\text{mix}}(\varphi, t)$ appropriately normalized leads to the maximum of the phase dispersion and to the loss of the preferred phase. But the phase distribution $P(\varphi, t)$ has the peaks not precisely at $\varphi = \pm\frac{\pi}{2}$, which means that the phase dispersion is slightly lowered and there does exist the preferred phase. It is zero (mod 2π) independent of $\xi(0)$. The

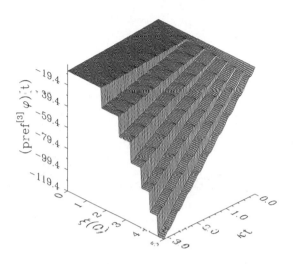

Figure 5.26. The effect of the initial complex field amplitude $\xi(0) \in [0,5]$ ($\ell = 3$) on the Kerr evolution of the third-order preferred value of the phase, $\kappa t \in [0, \pi]$.

quantum average takes the form

$$\langle \widehat{\exp}(i\varphi) \rangle(t) = \int_{-\pi}^{\pi} \exp(i\varphi) P(\varphi, t)\, d\varphi. \qquad (5.4.74)$$

Since

$$\int_{-\pi}^{\pi} \exp(i\varphi) P_{\text{mix}}(\varphi, t)\, d\varphi = 0, \qquad (5.4.75)$$

it holds also that

$$\langle \widehat{\exp}(i\varphi) \rangle(t) = \int_{-\pi}^{\pi} \exp(i\varphi) P_{\text{int}}(\varphi, t)\, d\varphi \qquad (5.4.76)$$

or the interference term saves the existence of the first-order preferred phase. Because of a lack of orthogonality in the coherent states, the concept of a superposition state is not quite clear in the limit of small $\xi(0)$. For example, at $t = \frac{\pi}{4\kappa}$, the initial coherent state with the complex field amplitude $\xi(0) = 1$ evolves into a superposition state which exhibits only three peaks, the fourth peak is not realized, and the existence of the preferred phase locked, perhaps, to the biggest peak is evident.

The phase distributions $P(\varphi, t)$ and their components $P_{\text{int}}(\varphi, t)$ at $t = \frac{\pi}{2\kappa}$ for various complex field amplitudes $\xi(0)$ are plotted in figures 5.27 and 5.28, respectively. Although evolved, the two-photon coherent state does not exhibit any Yurke–Stoler character, which is obvious, because the evolution reduces to harmonic rotation as

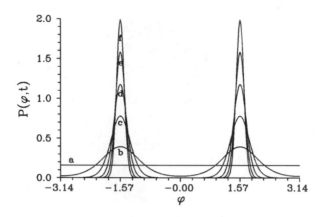

Figure 5.27: The phase distributions $P(\varphi, t)$ at $t = \frac{\pi}{2\kappa}$ for the initial two-photon coherent state and the complex field amplitudes $\xi(0) = 0, 1, 2, 3, 4, 5$ (curves a,b,...,f, respectively).

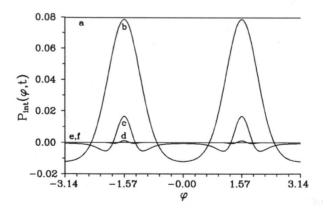

Figure 5.28: The interference components $P_{\text{int}}(\varphi, t)$ at $t = \frac{\pi}{2\kappa}$ for the initial two-photon coherent state and the complex field amplitudes $\xi(0) = 0, 1, 2, 3, 4, 5$ (curves a,b,...,f, respectively).

has been mentioned with respect to figures 5.22 and 5.23. With increasing $\xi(0)$ two Dirac delta peaks are developing at $\varphi \equiv \pm\frac{\pi}{2} \pmod{\pi}$ in figure 5.27. It is worth noting that the constant densities of the normalized and unnormalized phase distributions correspond to the vacuum state. In figure 5.28 there exists a constant interference term corresponding to the limit case of the "superposition of two vacuum states". With increasing $\xi(0)$ the interference term can also assume negative values and this property is more pronounced with increasing "orthogonality" of the coherent states in superposition. But the strong-field limit of $\zeta(0)$ also causes the decay of the interference term.

5.5   Particle interferometry

Wave–particle duality is the well-established phenomenon which can serve a basis for demonstrations of interference with particles having non-zero mass. After photons great attention was concentrated on electrons as particles of quantum electrodynamics. Extended reviews of the wave–particle duality of electrons were presented by Komrska (1971) and Matteucci (1990). Nice demonstrations of wave–particle duality were performed with neutrons as discussed by Badurek, Rauch and Tuppinger (1986), Badurek, Rauch and Summhammer (1988), Gähler and Zeilinger (1991), and Rauch (1995). With neutron beams interesting effects of quantum optics can be suggested and observed, such as spectral interference [Rauch (1993), Jacobson, Werner and Rauch (1994)], squeezing of vacuum fluctuations [Jacobson, Werner and Rauch (1994)], multiphoton transitions between a neutron beam and an oscillating magnetic field [Summhammer, Hamacher, Kaiser, Weinfurter, Jacobson and Werner (1995)], and topological phenomena in neutron interferometry [Peshkin and Lipkin (1995)]. In the last few years, great attention has been devoted to atom and ion interferometers giving new frontiers in quantum optics and physics. Extremely small wavelengths associated with atoms, which are of about 10^4 times shorter than those of visible light, make it possible to increase essentially the accuracy of measurements. Interferometric effects for neutral atoms were observed for the first time in diffraction from standing waves by Gould, Ruff and Pritchard (1986). The first two experiments performed by Carnal and Mlynek (1991) and Keith, Ekstrom, Turchette and Pritchard (1991) adopted nanofabricated structures to realize Young's two-slit experiment with helium atoms and to build an atom interferometer employing sodium atoms. These interferometers use beam splitting and recombining mechanisms. The temperature of the source of metastable helium atoms was 295 K corresponding to de Broglie wavelength of $\lambda = 0.56$ Å or 83 K corresponding to $\lambda = 1.03$ Å. Even if not very regular interference pattern was observed, this was a clear demonstration of the interference phenomenon of atoms. In other experiments, atomic paths were separated using gratings [Keith, Ekstrom, Turchette and Pritchard (1991)] or splitting due to a coherent transition between two internal states [Kasevich and Chu (1991), Riehle, Kisters, Wille, Helmcke and Borde (1991), Miniatura, Perales, Vassilev, Reinhardt,

Robert and Baudon (1991)]. Laser cooling of atoms can be used to increase the
de Broglie wavelength of the atoms, which increases the phase sensitivity to per-
turbations [Shimizu, Shimizu and Takuma (1992)]. Eichmann, Bergquist, Bollinger,
Gilligan, Itano, Wineland and Raizen (1993) observed the interference effects in light
scattered from two trapped atoms. The visibility of interference fringes was explained
in the framework of the Bragg scattering and on the basis of which-path information
for scattered photons. The authors demonstrated the increase of fringe spacing with
decreasing ion separation varied from 5.4 μm to 3.7 μm. The highest fringe visibility
was found for forward scattering directions and it decreased with increasing inclination angle. In the case that the initial and final states of ions after scattering event are
the same, one cannot distinguish from which ion a given photon came to the screen of
observation and, consequently, the interference pattern is observable. On the other
hand, if the final state differs from the initial one, one can, in principle, distinguish
the scattering event on the corresponding ion and one can obtain the which-path
information. Then there is no interference in scattered light from both the ions. In
this sense, the existence of interference fringes indicates wave-like behaviour, whereas
the absence of fringes, which is consistent with a single-photon trajectory beginning
at the source intersecting one of the ions and going to the detector, indicates particle-
like behaviour. These two ideally complementary pictures are contained in quantum
probabilistic interpretation related to operator representations of observables or, in
terms of the Feynman path amplitudes, the presence or absence of the interference
pattern depends on whether or not there are more than one possible paths from the
initial to the final states.

Atomic interferometers use particles of a greater mass than electron or neutron
interferometers, which provides increased sensitivity to changes in gravitational potential. Also internal degrees of freedom of atoms can be employed in interaction
with light for various positional measurements and for demonstrations of the relation
between interference and indistinguishability of particle paths. Also quantum nondemolition measurements of the atomic inversion can be realized. A phase–position
measurement of the field can provide atomic position information.

The interaction of a single atom with a radiation field can be well described on
the basis of the Jaynes–Cummings model (for some reviews, see [Zaheer and Zubairy
(1991), Meystre and Sargent III (1991), Walls and Milburn (1994), Vogel and Welsch
(1994), Mandel and Wolf (1995)]) with the Hamiltonian

$$\hat{H} = \hat{H}_0 + \hat{H}_{int}, \tag{5.5.1}$$

where the free Hamiltonian

$$\hat{H}_0 = \hbar\omega\hat{a}^\dagger\hat{a} + \frac{1}{2}\hbar\Omega\hat{\sigma}_z \tag{5.5.2}$$

represents the free energy of the radiation mode (renormalized zero-point energy $\hbar\omega/2$
of the radiation mode plays no role in this problem) and that of a two-level atom with

the transition frequency Ω, $\hat{\sigma}_z$ being one of the Pauli matrices and the interaction Hamiltonian is

$$\hat{H}_{\text{int}} = -e\hat{\mathbf{r}}\cdot\hat{\mathbf{E}} = \hbar g(\hat{a}^\dagger\hat{\sigma}_- + \hat{a}\hat{\sigma}_+), \qquad (5.5.3)$$

where $\hat{\sigma}_-$ and $\hat{\sigma}_+$ are the Pauli lowering and raising atomic operators, $\hat{\mathbf{E}}$ is the electric strength operator, $e\hat{\mathbf{r}}$ is the atomic dipole momentum operator, $g = \mathcal{P}\mathcal{E}\sin(\mathbf{k}\cdot\mathbf{r})/2\hbar$ is the coupling constant between the electromagnetic field and the atom, $\mathcal{P}$ is the electric-dipole matrix element, $\mathcal{E}$ is the electric field envelope, and the atom is assumed to be situated at the point $\mathbf{r}$. We have also used the rotating-wave approximation neglecting rotating terms, such as $\hat{a}^\dagger\hat{\sigma}_+$ and $\hat{a}\hat{\sigma}_-$, which are virtual and do not conserve the energy. Nevertheless, such terms can be a source of a chaotic behaviour in the atom–field system [Milonni, Shih and Ackerhalt (1987)]. This description correctly includes photon absorption and emission and the corresponding closed solutions provide various dynamical phenomena really observed, such as collapses and revivals of atomic population inversion, when the two-level atoms interact with an electromagnetic field, for instance, initially in a coherent state [Eberly, Narozhny and Sanchez-Mondragon (1980)]. Interesting applications of this model can be realized in micromasers (for a review, see [Walther (1993), Peřina, Hradil and Jurčo (1994), section 8.5]). The energy eigenvalues of the Hamiltonian (5.5.1) can be written in the form

$$E_{1n} = \hbar\left(n + \frac{1}{2}\right)\omega + \frac{\hbar}{2}\Omega_n - \hbar\left[-\frac{\Omega}{2} + (n+1)\omega + \frac{1}{2}(\Omega_n + \delta)\right],$$

$$E_{2n} = \hbar\left(n + \frac{1}{2}\right)\omega - \frac{\hbar}{2}\Omega_n = \hbar\left[\frac{\Omega}{2} + n\omega - \frac{1}{2}(\Omega_n + \delta)\right], \qquad (5.5.4)$$

where $\delta = \Omega - \omega$ is the atom–field frequency detuning and the n-photon Rabi flopping frequency is

$$\Omega_n = \sqrt{\delta^2 + 4g^2(n+1)}. \qquad (5.5.5)$$

The corresponding eigenvectors can be written in the form

$$|1, n\rangle = \sin(\theta_n)|a, n\rangle + \cos(\theta_n)|b, n+1\rangle,$$

$$|2, n\rangle = \cos(\theta_n)|a, n\rangle - \sin(\theta_n)|b, n+1\rangle, \qquad (5.5.6)$$

where $|a, n\rangle = |a\rangle|n\rangle$, $|b, n\rangle = |b\rangle|n\rangle$, $|a\rangle$ and $|b\rangle$ are the upper and lower atomic states, respectively, and $|n\rangle$ is the Fock state of the field. The angle θ_n is defined by means of the relations

$$\cos(2\theta_n) = -\frac{\delta}{\Omega_n}, \quad \sin(2\theta_n) = \frac{2g\sqrt{n+1}}{\Omega_n} \qquad (5.5.7)$$

and, in particular, we have for $n = 0$ and on frequency resonance $\omega = \Omega$ that levels $|1, 0\rangle$ and $|2, 0\rangle$ are separated by the frequency $\Omega_0 = 2g$ giving the vacuum Rabi splitting.

For the frequency resonance $\omega = \Omega$ and a two-level atom in its upper state $|a\rangle$ and the field in the Fock state $|n\rangle$ initially, the probability $p_{an}(t)$ that the atom will be in the upper state at time t is given by

$$p_{an}(t) = \cos^2(gt\sqrt{n+1})\tag{5.5.8}$$

and the transition probability on the condition that the atom is in its lower state $|b\rangle$ initially reads

$$p_{bn}(t) = \sin^2(gt\sqrt{n}).\tag{5.5.9}$$

This shows that the atomic population oscillates periodically at the Rabi frequency, similarly as in the case of classical fields. The difference is that, for the quantum field, the Rabi frequency depends on whether the atom starts in the upper level or in the lower level, because the creation operator $\hat{a}^\dagger$ and the annihilation operator $\hat{a}$ do not commute, which reflects the presence of the spontaneous emission. The initial field is mostly in another state than in the Fock state and the resulting probabilities are expressed as

$$\begin{aligned}
p_a(t) &= \sum_{n=0}^{\infty} \rho_{nn} \cos^2(gt\sqrt{n+1}), \\
p_b(t) &= \sum_{n=0}^{\infty} \rho_{nn} \sin^2(gt\sqrt{n}).
\end{aligned}\tag{5.5.10}$$

If the initial field is in the coherent state $|\alpha\rangle$ specified by the diagonal statistical operator elements in the form of the Poisson distribution, $\rho_{nn} = |\alpha|^{2n} \exp(-|\alpha|^2)/n!$, we arrive at

$$p_a(t) \approx \frac{1}{2} + \frac{1}{2}\cos(2gt|\alpha|)\exp(-g^2 t^2),\tag{5.5.11}$$

provided that a strong field is applied, $|\alpha|^2 \gg 1$, and $gt|\alpha| \ll 1$ [Meystre, Quattropani and Baltes (1974)]. Collapses and revivals of the probability $p_a(t)$ can occur as shown in figure 5.29. The revivals are of quantal origin and they spread progressively so that they overlap for later times and provide a quasi-random time evolution. These quantum effects were observed by Rempe, Walther and Klein (1987), Brecha, Orozco, Raizen, Xiao and Kimble (1995), Meekhof, Monroe, King, Itano and Wineland (1996), and Brune, Schmidt-Kaler, Maali, Dreyer, Hagley, Raimond and Haroche (1996). If a continuous spectrum of vacuum fluctuations occurs, the resulting lower-level probability amplitudes interfere with each other giving the well-known macroscopic exponential decay. For initial chaotic field, the collapses and revivals are much less pronounced. The collapses arise as a result of destructive interference of quantum Rabi floppings at different frequencies and may also occur under the influence of a classical field possessing intensity fluctuations, whereas revivals have no classical analogue [Knight and Radmore (1982), Eberly, Narozhny and Sanchez-Mondragon (1980), Barnett, Filipowicz, Javanainen, Knight and Meystre (1986)]. The collapses and revivals can also be interpreted in terms of interference in phase space using the

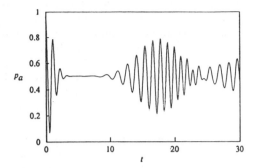

Figure 5.29: Collapses and revivals in interaction of quantum light field with an atom.

quasidistribution $\Phi_A(\alpha)$ [Eiselt and Risken (1991)] as a result of splitting a single-peaked quasidistribution.

Putting $n = 0$ in (5.5.8) and (5.5.9), we describe the spontaneous effect in the atom being initially in the upper state by means of $p_{a0}(t) = \cos^2(gt)$ and $p_{b0}(t) = 0$, respectively. This quantum effect is related to the quantum expectation value of the electric strength operator $\langle \hat{E}^2 \rangle \propto \langle 0|(\hat{a} + \hat{a}^\dagger)^2|0\rangle = 1$. This means that the vacuum fluctuations stimulate the excited atom to emit a spontaneous photon. Of course, no spontaneous emission is possible if the atom is in the ground state. If the atom is situated inside a high-quality cavity, the spontaneous emission rate can be enhanced or inhibited compared to its free-space value [Haroche and Kleppner (1989), Haroche (1992), Meystre (1992)]. Experiments with a single atom in interaction with radiation can be made in the microwave region, where high-quality factor of single-mode cavities can be reached, using highly excited Rydberg atoms, for which the dipole matrix element g between neighbouring levels scales as n^2, n being the principal quantum number. Then for sufficiently high n, stimulated effects can overcome spontaneous emission for small photon numbers. In the visible region, the dipole matrix elements are small and linear size of cavities available in laboratories is much larger than a transition wavelength. Therefore, cooperative effects are to be used to obtain great effective coupling constants and to realize cavity quantum optics experiments in the visible region.

If the atom in free space interacts with a continuum of the electromagnetic modes, the system is described by the multimode Hamiltonians

$$\hat{H}_0 = \hbar \sum_\lambda \omega_\lambda \hat{a}_\lambda^\dagger \hat{a}_\lambda + \frac{1}{2}\hbar\Omega\hat{\sigma}_z,$$

$$\hat{H}_{\text{int}} = \hbar \sum_\lambda \left(g_\lambda \hat{a}_\lambda^\dagger \hat{\sigma}_- + \text{H. c.}\right), \tag{5.5.12}$$

where ω_λ is the frequency of the λth radiation mode, $\hat{a}_\lambda$ and $\hat{a}_\lambda^\dagger$ are the corresponding annihilation and creation operators, respectively, and g_λ are the dipole coupling constants between the atom and the mode λ of the field. For moving atom, we have to add the operator $\hat{p}^2/2m$ to the free Hamiltonian, where $\hat{p}$ is the momentum operator of the atom and m is its mass. Then the standard Wigner–Weisskopf analysis [Louisell (1973), Meystre and Sargent III (1991), Peřina (1991)] leads to an irreversible exponential decay of the upper population state without exhibition of any collapses and revivals or any periodic exchange of energy between the atom and the field. This is a result of destructive interference of the probability amplitudes for single events when summing over the continuum of modes.

It is interesting to consider Young's two-slit interference experiment with an atom as outlined in figure 5.30 including path detectors. Denoting the wave function of the atom coming from the first slit to the point $\mathbf{x}$ of the screen by $\psi_1(\mathbf{x})$ and, similarly, the wave function of an atom coming from the second slit to the point $\mathbf{x}$ by $\psi_2(\mathbf{x})$, we obtain the probability density that the atom can be found at the position $\mathbf{x}$,

$$P(\mathbf{x}) = |\psi_1(\mathbf{x})|^2 + |\psi_2(\mathbf{x})|^2 + 2\text{Re}\{\psi_1^*(\mathbf{x})\psi_2(\mathbf{x})\langle d_1|d_2\rangle\}, \qquad (5.5.13)$$

where $|d_1\rangle$ and $|d_2\rangle$ are the final path detector states. Thus, the occurrence of interference is dependent on the additional factor $\langle d_1|d_2\rangle$, which may reduce the visibility of observed interference fringes. If the detector states $|d_1\rangle$ and $|d_2\rangle$ are orthogonal, as the Fock states are, it becomes possible to distinguish them and the interference pattern disappears. If they overlap, as the coherent states do, and cannot be distinguished, the ideal interference fringes can be observed or, in partial overlapping, reduced interference fringes are available. General features are that if the separation of the slits is less than the wavelength, the visibility of interference fringes is close to unity, since the atomic position cannot be localized using a light probe with the accuracy better than about a wavelength. If the distance of slits increases, the localization of the atom is possible and the fringe visibility is reduced because of a transferred momentum to the atom from the probing photon. Finally, fringes are smoothed out.

An interesting double-slit interference experiment was performed by Noel and Stroud Jr. (1995) using Rydberg wave packets within an atom. Two phase-coherent laser pulses were used to excite a single electron into a superposition of a pair of Rydberg wave packets which are initially on opposite sides of the atom orbit. The two wave packets propagate and spread until they completely overlap. Then a third phase-coherent laser pulse probes the resulting interference pattern. The relative phase of the two-wave packets was varied so that the interference produced a single localized electron-wave packet on one or the other side of the orbit. Coherent superpositions of macroscopically distinct states can be realized in this way [Noel and Stroud Jr. (1996)].

Interesting atomic interference experiments can be performed using diffraction gratings of light for atom scattering [Walls and Milburn (1994), section 17.2]. The Hamiltonian describing such an interaction is a slight modification of the Hamiltonian

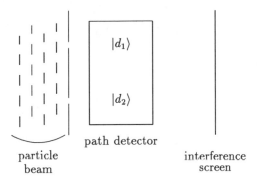

path detector

particle
beam

interference
screen

Figure 5.30: Particle two-slit experiment using path detectors.

(5.5.1),

$$\hat{H} = \hbar\omega_a\hat{\sigma}_z + \frac{\hat{p}^2}{2m} + \hbar\Omega[\hat{\sigma}_- \exp(-i\omega t) + \hat{\sigma}_+ \exp(i\omega t)]\cos(kx), \qquad (5.5.14)$$

where $\dot{p}$ is the centre-of-mass–momentum operator of the atom along the transversal x direction, m is the mass of atom, ω_a and ω are atomic and field frequencies, $k = \omega/c$ is the wave number of the standing wave, and Ω is the Rabi frequency. Usually, the so-called Raman–Nath regime is considered, in which the kinetic energy term $\hat{p}^2/2m$ is neglected. The scattered wave function can be obtained as a superposition of Gaussian modulated plane waves with momenta $p = 2n\hbar k$. The momentum transferred from the field to the atom is, therefore, an even multiple of the elementary momentum $\hbar k$ corresponding to the absorption of a photon from the $+k$-component followed by induced emission into $-k$-component of the standing wave. This is for the case of large atomic detunings, where spontaneous emission can be neglected. For smaller atomic detunings, spontaneous emission becomes important, because the recoil imparted to the atom by a spontaneously emitted photon occurs in random directions, exchanges of momenta are not integral multiples of $\hbar k$ and diffractive peaks are smoothed out. If the radiation is tuned to the atomic resonance, the diffractive regime is changed to a diffusive regime, which was experimentally demonstrated by Gould, Martin, Ruff, Stoner, Picque and Pritchard (1991) and explained theoretically by Tan and Walls (1992). An experiment demonstrating atomic wave interference using diffraction gratings of light was performed by Rasel, Oberthaler, Batelaan, Schmiedmayer and Zeilinger (1995). They have developed a novel interferometer for atomic de Broglie waves, where amplitudes division and recombination were achieved by diffraction at standing light waves acting as phase gratings (figure 5.31). It is similar to standard optical interferometers, in which the roles of atoms in mirrors

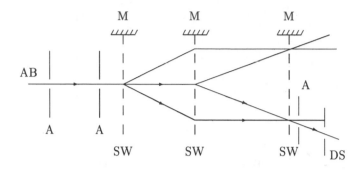

Figure 5.31: Scheme of atomic interferometer using standing waves; AB is atomic beam, SW are standing waves, A apertures, M mirrors, and DS is the detection slit. (Adapted from Rasel, Oberthaler, Batelaan, Schmiedmayer and Zeilinger (1995).)

and photons are interchanged. It can be used for demonstrations of coherence of atomic waves diffracted by standing light waves, because any manipulation of the phase, intensity, and polarization of the standing light wave permits investigation of atomic coherence properties. The authors also demonstrated the complementarity of the interference patterns on two output ports of the interferometer, which is of the Mach–Zehnder type.

This type of interferometer is also suitable for realization of molecular interferometry [Chapman, Ekstrom, Hammond, Rubenstein, Schmiedmayer, Wehinger and Pritchard (1995), Bordé, Courtier, du Burck, Goncharov and Gorlicki (1994)]. It permits to perform interference experiments with vibrational and rotational states providing large variety of superpositions between different states. Bordé, Courtier, du Burck, Goncharov and Gorlicki (1994) employed de Broglie waves of I_2 molecule in an interferometer using four laser waves as molecular beamsplitters. They have demonstrated clear interference pattern of de Broglie waves of the molecule in good agreement with the theory developed on the basis of the Schrödinger equation.

A modification of the Hamiltonian (5.5.14) [Walls and Milburn (1994), section 17.4, Matsko, Vyatchanin, Mabuchi and Kimble (1994)] can also serve a basis for nondemolition measurements of the photon number by atomic beam deflection. The momentum distribution depends on the photon-number distribution of the cavity field, which can be used to perform the quantum nondemolition measurement of the photon number by measuring the momentum distribution of the scattered atoms. Each atomic position measurement gives some information about the field statistics and reduces the statistical operator of the field, so that repeated measurements can go to complete determination of the photon statistics corresponding to a number state, provided that dissipations can be neglected.

A modification of the Hamiltonian (5.5.14) can also be employed for determination of the position of an atom passing through a standing light wave [Storey, Collett and Walls (1992)]. Particle interferometers are important tools for investigations of complementarity of wave and particle behaviour or its violation [Rangwala and Roy (1994)] including the case that particle and wave properties are not fully developed. Complementarity and quantum erasure with dispersive atom–field interactions were considered by Gerry (1996). Deflection of atoms by a quantum field was discussed by Akulin, Fam Le Kien and Schleich (1991). Quantum reflection of an atom by evanescent waves were examined by Henkel, Westbrook and Aspect (1996). A detailed analysis of atomic mirrors based on light-induced forces was developed by Tan and Walls (1994) and bichromatic atomic beamsplitters were also analyzed [Tan and Walls (1995)]. The van Cittert–Zernike theorem of optics can be transferred to matter-wave optics [Taylor, Schernthanner, Lenz and Meystre (1994)]. One can construct an atomic interferometer based on adiabatic population transfer between two ground states, which is suitable for precise measurements of the photon recoil energy [Weitz, Young and Chu (1994)]. Atomic states can be used for teleportation [Cirac and Parkins (1994)], atomic beams can be a tool for tomography [Janicke and Wilkens (1995)] and for tests of complementarity principle and realization of quantum eraser devices [Bogár and Bergou (1996)]. The atomic interferometer employing collective radiation and photon echo from atoms to measure atomic phase and to determine atomic coherence was demonstrated by Schnurr, Savard, Wang and Thomas (1995). The relationship of mutual coherence and interference in resonance fluorescence light from two atoms coupled to an optical cavity mode was discussed by Kochan, Carmichael, Morrow and Raizen (1995), which relates a which-path information to loss of coherence. Nonlinear atom optics was discussed by Lenz, Meystre and Wright (1994) and oscillation energy exchange in a coupled atom–cavity system was observed by Brecha, Orozco, Raizen, Min Xiao and Kimble (1995). Homodyne measurements of a cavity field interacting with atoms can be a source of information about a localization of an atom [Herkommer, Carmichael and Schleich (1996)]. Interaction-free preparation of well localized atom beams was suggested by Krenn, Summhammer and Svozil (1996) and quantum coherence in a cavity was discussed using two-atom correlation measurements by Davidovich, Brune, Raimond and Haroche (1996). A special issue of Quantum and Semiclassical Optics [Arimondo and Bachor (1996)] has been devoted to optics and interferometry with atoms. Quantum interference in atoms is analyzed by Brieger and Schuessler (1996), atomic wave diffraction and interference using temporal slits is experimentally realized by Szriftgiser, Guéry-Odelin, Arndt and Dalibard (1996), and ion interferometry with respect to nonclassical states is discussed by Poyatos, Cirac, Blatt and Zoller (1996). A review of atom interferometry has also been prepared by Adams, Carnal and Mlynek (1994). Experiments with correlated atom–photon states can provide a way to realize higher-order quantum correlations [Pfau, Kurtsiefer, Ekstrom and Mlynek (1996)]. Photon scattering on atoms in an atom interferometer from the viewpoint of lost and restored coherence was experimentally investigated by Pritchard, Chapman, Hammond, Lenef, Rubenstein, Schmiedmayer and Smith (1996). Multiple-beam atomic interferometer was realized by Weitz, Heupel and Hänsch (1996). Interesting decoherence effects using a Schrödinger-cat atomic states interacting with a mesoscopic superposition of quantum states involving radiation fields with classically distinct phases were observed by Brune, Hagley, Dreyer, Maître, Maali, Wunderlich, Raimond and Haroche (1996). Interference effects can also be obtained using the Bose–Einstein condensates of atoms [Naraschewski, Wallis, Schenzle, Cirac and Zoller (1996), Cirac, Gardiner, Naraschewski and Zoller (1996), Wong, Collett and Walls (1996)]. Atomic waves in crystals can interact with steady-state light waves and they can exhibit interesting interference and diffraction standing phenomena [Oberthaler, Abfalterer, Bernet, Schmiedmayer and Zeilinger (1996), Bernet, Oberthaler, Abfalterer, Schmiedmayer and Zeilinger (1996)]. Coherence properties of atom beamsplitter were examined by Choi, Wiseman, Tan and Walls (1997). Pulses of coherent atoms, which can be regarded as a pulsed atom laser, were produced by Mewes, Andrews, Kurn, Durfee, Townsend and Ketterle (1997). Atom interferometer represents a very sensitive tool for rotation sensing [Lenef, Hammond, Smith, Chapman, Rubenstein and Pritchard (1997), Gustavson, Bouyer and Kasevich (1997)]. An excellent review of atom and molecule interferometry from various points of view involving classical and nonclassical effects, quantum theory of measurement, effects of pulses and some technological problems can be found in the book of invited chapters edited by Berman (1997). Quantum statistics of a coherently driven Bose–Einstein condensate on the simplified condition of strong classical laser beam were discussed by Hu Huang and Shi-qun Li (1997) and collapses and revivals of the fringe pattern in a multiple-beam atomic interferometer were observed by Weitz, Heupel and Hänsch (1997). Interference effects of independent Bose–Einstein condensates were discussed by Röhrl, Naraschewski, Schenzle and Wallis (1997) including collapses and revivals in the interference [Wright, Wong, Collett, Tan and Walls (1997)]. Time-domain atom interferometry was realized by Cahn, Kumarakrishnan, Shim, Sleator, Berman and Dubetsky (1997). Also experimental determination of the motional quantum state of a trapped atom using homodyne tomography for the Wigner function (section 4.13) can be realized and the corresponding statistical operator can be determined [Leibfried, Meekhof, King, Monroe, Itano and Wineland (1996)]. Diffraction and interference effects of atomic matter waves using a Fresnel bi-prism device [Brouri, de Tomasi, Baudon, Reinhardt, Lorent, Robert and Gorlicki (1997)] can have an interesting analogy with a similar experiment with electrons [Komrska (1971)].

Chapter 6

Conclusions

In this book we have described the phase in optics and related branches of science from the classical and quantum points of view taking into account the most recent results obtained in quantum optics. We have discussed the role of phase in classical linear and nonlinear optics concentrating on interference and diffraction of light, nonlinear process of phase conjugation, and on phase retrieval methods and have treated quantum optical methods for phase detection, such as homodyne and heterodyne detections and homodyne tomography. Adopting phase-space description, which uses various operator orderings and yields quantum optical description based on operator generating functions, quantum characteristic functions, resolutions of the identity, and quasidistributions, we have presented various concepts of quantum phase, such as traditional Dirac non-Hermitian concept, Susskind–Glogower cosine and sine operators, well-known Pegg–Barnett Hermitian operator concept, concept adopting the antinormal ordering of exponential phase operators, concept of phase distributions derived from quasidistributions, operational approach to quantum phase, concepts of ideal and feasible phases, and have provided some information about geometrical Berry–Pancharatnam phase. The quantum phase can be involved in quantum dynamics of physical systems in the framework of the Heisenberg picture (Heisenberg–Langevin equations, quantum characteristic functions, quasidistributions, generating functions, photocount statistics) or the Schrödinger picture (generalized Fokker–Planck equations for quasidistributions, generating functions, photocount statistics). As particular applications we have treated quantum interferometry, which is able to provide new information about wave–particle behaviour of single photons and about its nonlocal properties involving coincidence measurements in quantum interferometers and quantum nondemolition measurements related to various quantum noise manipulations. Nonlinear optical processes, such as optical parametric processes, four-wave mixing, phase conjugation, Raman scattering, etc., sometimes combined in optical nonlinear couplers involving the coupling of nonlinear waveguides through evanescent waves, provide good ways of phase measurements or exhibit various phase-dependent effects. A number of these effects can also be observed with particles with nonzero

411

mass, such as electrons, neutrons, atoms, and ions.

Recently, interesting applications of quantum laws and, especially, of quantum interference started to be used in optical communications, quantum cryptography, optical processing and computing, and high-precision measurements. We hope that this book will be useful to stimulate further research in this field and in its applications with respect to the crucial role of the phase.

We have compared the concepts of ideal and feasible phases, the ideal phase being able to be expressed in the case of a quantum mechanical oscillator or, equivalently, in the case of a single radiation mode directly by a resolution of the identity, while the feasible phase having to be derived from an expression for the resolution of the identity that belongs to the complex amplitude of the mode.

We have also mentioned the history of the quantum phase problem, of course, which involves early knowledge that the action–angle variables, which were of great importance in classical mechanics, would be dethroned, if they exist at all, in quantum theory. An appropriate proof of the non-existence of an Hermitian phase operator has been based on the impossibility of a unitary phase operator which should assign a state with negative photon number to the vacuum state.

Mathematically, the solution of the quantum phase problem consists in finding an Hermitian phase operator in an enlarged Hilbert space, which can be made concrete for both the ideal and feasible phases. Whereas we obtain an interesting connection with a plane rotator for the ideal phase, it is for the feasible phase that we get useful relationships with the heterodyne and homodyne detections.

A definition of unitary phase operator is possible not only in an enlarged Hilbert space, but also in a Hilbert space restricted to a finite subset of Fock states. The ideal phase can be regained using a limiting procedure in the latter case as D. T. Pegg and S. M. Barnett have instructed us. Since one can get acquainted with it in three-fold way, a suitable term for the ideal phase is a canonical phase.

Both ideal and feasible phases are given by positive operator-valued measures (resolutions of the identity), especially, the feasible phase is related to the antinormal ordering of the field operators. The antinormal ordering of the exponential phase operators, which are obtained through an operator polar decomposition of the field operators, can be seen in the ideal phase. The symmetrical ordering of the field operators does not lead to a positive operator-valued measure, but it seems to express the statistical operator and therefore—in despite of all—also the quantum phase more directly. The goal of the "solution of the quantum phase problem" is to find a well-behaved Hermitian phase operator. But the resolution of the identity from which the solution departs (and similar resolutions of the identity) enable to construct "less well-behaved" Hermitian phase operators. The literature abounds in interesting (but also deterrent-like) analyses of these operators.

The study of various special states suggests that at least in the attractive case of a single mode the ideal phase does not afford closed formulae for real field states, while there exist states which can be described by closed formulae, because these states have been defined respecting the phase. The feasible phase and, more generally, the phase based on the orderings of the field operators allows to determine closed formulae for the real field states. We have approached this contrast also by a general consideration of ordering of exponential phase operators.

From the perspective of numerical computation, the contrast between ideal and feasible phases need not always be interesting, selected phase statistics can always be calculated from the statistical operator, nevertheless, we observe that the statistical operator need not always be defined in a theory, it can be estimated by experimental measurements and using a reconstruction technique.

References

Abbas, G. L., W. V. S. Chan and T. K. Yee, 1983, *Opt. Lett.* **8**, 419.

Abe, S., 1995, *Phys. Lett. A* **200**, 239.

Abe, S., 1996, *Phys. Lett. A* **213**, 112.

Ablekov, V. K., P. I. Zubkov and A. V. Frolov, 1976, Optical and Opto-Electronical Data Processing, Mashinostroyeniye, Moscow, in Russian.

Abramowitz, M. and I. A. Stegun, eds., 1964, Handbook of Mathematical Functions, Dover, New York.

Adam, P., J. Janszky and An. V. Vinogradov, 1991, *Phys. Lett. A* **160**, 506.

Adam, P., S. Szabo and J. Janszky, 1996, *Phys. Lett. A* **215**, 229.

Adams, C. S., O. Carnal and J. Mlynek, 1994, Advances in Atomic, Molecular and Optical Physics, Vol. 34, eds. B. Bederson and H. Walther, Academic Press, San Diego, CA, p. 1.

Agarwal, G. S., 1973, Progress in Optics, Vol. 9, ed. E. Wolf, North-Holland, Amsterdam, p. 1.

Agarwal, G. S., 1981, *Phys. Rev. A* **24**, 2889.

Agarwal, G. S., 1986, *Phys. Rev. Lett.* **57**, 827.

Agarwal, G. S., 1988, *J. Opt. Soc. Am. B* **5**, 1940.

Agarwal, G. S., 1990, *Quant. Opt.* **2**, 1.

Agarwal, G. S., 1993, *Opt. Commun.* **100**, 479.

Agarwal, G. S., 1995, *Found. Physics* **25**, 219.

Agarwal, G. S. and P. Adam, 1989, *Phys. Rev. A* **39**, 6259.

Agarwal, G. S., S. Chaturvedi, K. Tara and V. Srinivasan, 1992, *Phys. Rev. A* **45**, 4904.

Agarwal, G. S. and P. Hariharan, 1993, *Opt. Commun.* **103**, 111.

Agarwal, G. S. and R. R. Puri, 1989, *Phys. Rev. A* **40**, 5179.

Agarwal, G. S. and R. R. Puri, 1990, Coherence and Quantum Optics VI, eds. J. H. Eberly, L. Mandel and E. Wolf, Plenum Press, New York, p. 15.

Agarwal, G. S., M. O. Scully and H. Walther, 1993, *Phys. Scr.* **T48**, 128.

Agarwal, G. S. and R. Simon, 1990, *Phys. Rev. A* **42**, 6924.

Agarwal, G. S. and R. P. Singh, 1996, *Phys. Lett. A* **217**, 215.

Agarwal, G. S. and K. Tara, 1991, *Phys. Rev. A* **43**, 492.

Agarwal, G. S. and E. Wolf, 1970a, *Phys. Rev. D* **2**, 2161.

Agarwal, G. S. and E. Wolf, 1970b, *Phys. Rev. D* **2**, 2187.

Agarwal, G. S. and E. Wolf, 1970c, *Phys. Rev. D* **2**, 2206.

Aharonov, Y. and J. Anandan, 1987, *Phys. Rev. Lett.* **58**, 1593.

Akhiezer, A. I. and V. B. Berestetsky, 1965, Quantum Electrodynamics, Interscience, New York.

Akulin, V. M., Fam Le Kien and W. P. Schleich, 1991, *Phys. Rev. A* **44**, R1462.

Albeverio, S., J. E. Fenstad, R. Høegh-Krohn and T. Lindstrøm, 1986, Nonstandard Methods in Stochastic Analysis and Mathematical Physics, Academic Press, New York.

Albrecht, A., 1992, *Phys. Rev. D* **46**, 5504.

Alter, O. and Y. Yamamoto, 1995, *Phys. Rev. Lett.* **74**, 4106.

Anandan, J., J. Christian and K. Wanelik, 1997, *Am. J. Phys.* **65**, 180.

Andĕl, J., 1978, Mathematical Statistics, Technical Literature Publishers, Prague, in Czech.

Ansari, N. A. and V. I. Man'ko, 1994, *Phys. Rev. A* **50**, 1942.

Appelt, S., G. Wäckerle and M. Mehring, 1995, *Phys. Lett. A* **204**, 210.

Aragone, C., E. Chalbaud and S. Salamó, 1976, *J. Math. Phys.* **17**, 1963.

Aragone, C., G. Guerri, S. Salamó and J. L. Tani, 1974, *J. Phys. A: Math. Nucl. Gen.* **7**, L149.

Arecchi, F. T., E. Courtens, R. Gilmore and H. Thomas, 1972, *Phys. Rev. A* **6**, 2211.

Arik, M. and D. D. Coon, 1976, *J. Math. Phys.* **17**, 4.

Arimondo, E. and H.-A. Bachor, eds., 1996, Quantum and Semiclassical Optics, Vol. **8**, No. 3, IOP Publishing, Bristol.

Arnoldus, H. F., 1994, *J. Mod. Opt.* **41**, 503.

Arthurs, E. and M. S. Goodman, 1988, *Phys. Rev. Lett.* **60**, 2447.

Arthurs, E. and J. L. Kelly Jr., 1965, *Bell. Syst. Tech. J.* **44**, 725.

Aspect, A. and P. Grangier, 1991, International Trends in Optics, ed. J. W. Goodman, Academic Press, Boston, p. 247.

Aspect, A., P. Grangier and G. Roger, 1989, *J. Optics* **20**, 119.

Atakishiyev, N. M., S. M. Chumakov, A. L. Rivera and K. B. Wolf, 1996, *Phys. Lett. A* **215**, 128.

Badurek, G., H. Rauch and J. Summhammer, 1988, *Physica B* **151**, 82.

Badurek, G., H. Rauch and D. Tuppinger, 1986, *Phys. Rev. A* **34**, 2600.

Bajer, J. and J. Peřina, 1985, *Czech. J. Phys. B* **35**, 1146.

Bajer, J. and J. Peřina, 1991, *Opt. Commun.* **85**, 261.

Baltes, H. P., ed., 1978, Inverse Source Problems in Optics, Springer-Verlag, Berlin.

Baltes, H. P., ed., 1980, Inverse Scattering Problems in Optics, Springer-Verlag, Berlin.

Ban, M., 1991a, *Phys. Lett. A* **152**, 223.

Ban, M., 1991b, *Phys. Lett. A* **155**, 397.

Ban, M., 1991c, *J. Math. Phys.* **32**, 3077.

Ban, M., 1991d, *Physica A* **179**, 103.

Ban, M., 1992, *J. Opt. Soc. Am. B* **9**, 1189.

Ban, M., 1993, *Phys. Lett. A* **176**, 47.

Ban, M., 1994, *Phys. Rev. A* **50**, 2785.

Ban, M., 1995a, *Phys. Lett. A* **199**, 275.

Ban, M., 1995b, *Phys. Rev. A* **51**, 2469.

Ban, M., 1997, *J. Mod. Opt.* **44**, 1175.

Banaszek, K. and P. L. Knight, 1997, *Phys. Rev. A* **55**, 2368.

Banaszek, K. and K. Wódkiewicz, 1996, *Phys. Rev. Lett.* **76**, 4344.

Banaszek, K. and K. Wódkiewicz, 1997, *Phys. Rev. A* **55**, 3117.

Bandilla, A., 1993, *Phys. Scr.* **T48**, 49.

Bandilla, A., G. Drobný and I. Jex, 1995, *Phys. Rev. Lett.* **75**, 4019.

Bandilla, A. and H. Paul, 1969, *Ann. Phys.* (Leipzig) **23**, 323.

Bandilla, A. and H. Paul, 1970, *Ann. Phys.* (Leipzig) **24**, 119.

Bandilla, A., H. Paul and H.-H. Ritze, 1991, *Quant. Opt.* **3**, 267.

Bandilla, A. and H.-H. Ritze, 1993, *Quant. Opt.* **5**, 213.

Bandilla, A. and H.-H. Ritze, 1994, *Phys. Rev. A* **49**, 4912.

Bandyopadhyay, A. and J. Rai, 1995, *Phys. Rev. A* **51**, 1597.

Barbosa, G. A., 1996, *Phys. Rev. A* **54**, 4473.

Bardroff, P. J., E. Mayr and W. P. Schleich, 1995, *Phys. Rev. A* **51**, 4963.

Bargmann, V., 1947, *Ann. Math.* **48**, 568.

Barnett, S. M., P. Filipowicz, J. Javanainen, P. L. Knight and P. Meystre, 1986, Frontiers in Quantum Optics, eds. E. R. Pike and S. Sarkar, A. Hilger, Bristol, p. 485.

Barnett, S. M. and D. T. Pegg, 1986, *J. Phys. A: Math. Gen.* **19**, 3849.

Barnett, S. M. and D. T. Pegg, 1989, *J. Mod. Opt.* **36**, 7.

Barnett, S. M. and D. T. Pegg, 1990a, *Phys. Rev. A* **41**, 3427.

Barnett, S. M. and D. T. Pegg, 1990b, *Phys. Rev. A* **42**, 6713.

Barnett, S. M. and D. T. Pegg, 1992, *J. Mod. Opt.* **39**, 2121.

Barnett, S. M. and D. T. Pegg, 1996, *Phys. Rev. Lett.* **76**, 4148.

Barut, A. O., 1957, *Phys. Rev.* **108**, 565.

Barut, A. O. and L. Girardello, 1971, *Commun. Math. Phys.* **21**, 41.

Baseia, B., M. H. Y. Moussa and V. S. Bagnato, 1997, *Phys. Lett. A* **231**, 331.

Baseia, B., R. Vyas, C. M. A. Dantas and V. S. Bagnato, 1994, *Phys. Lett. A* **194**, 153.

Bates, R. H. T. and W. R. Fright, 1983, *J. Opt. Soc. Am.* **73**, 358.

Bates, R. H. T. and M. J. McDonnell, 1986, Image Restoration and Reconstruction, Clarendon Press, Oxford.

Beck, M., D. T. Smithey, J. Cooper and M. G. Raymer, 1993, *Opt. Lett.* **18**, 1259.

Beck, M., D. T. Smithey and M. G. Raymer, 1993, *Phys. Rev. A* **48**, R890.

Belavkin, V. P., 1990, *Lett. Math. Phys.* **20**, 85.

Belavkin, V. P. and C. Bendjaballah, 1994, *Quant. Opt.* **6**, 169.

Belavkin, V. P. and P. Staszewski, 1989, *Phys. Lett.* **140A**, 359.

Beltrametti, E. and G. Cassinelli, 1981, The Logic of Quantum Mechanics, Addison Wesley, Reading, MA.

Bencheikh, K., J. A. Levenson, Ph. Grangier and O. Lopez, 1995, *Phys. Rev. Lett.* **75**, 3422.

Benedek, C. and M. G. Benedict, 1997, *Europhys. Lett.* **39**, 347.

Beran, M. and G. B. Parrent, 1964, Theory of Partial Coherence, Prentice-Hall, Englewood Cliffs, New Jersey.

Bergou, J. and B.-G. Englert, 1991, *Ann. Phys.* (N. Y.) **209**, 479.

Berman, P. R., ed., 1997, Atom Interferometry, Academic Press, San Diego, CA.

Bernet, S., M. K. Oberthaler, R. Abfalterer, J. Schmiedmayer and A. Zeilinger, 1996, *Phys. Rev. Lett.* **77**, 5160.

Berry, M. V., 1984, *Proc. Roy. Soc. A* **392**, 45.

Berry, M. V., 1985, *J. Phys. A* **18**, 15.

Berry, M. V., 1986, Fundamental Aspects of Quantum Theory, NATO ASI series Vol. 144, ed. V. Gorini and A. Frigerio, Plenum Press, New York, p. 267.

Berry, M. V., 1987, *J. Mod. Opt.* **34**, 1401.

Berry, M. V. and S. Klein, 1996, *J. Mod. Opt.* **43**, 165.

Bertero, M. and C. De Mol, 1996, Progress in Optics, Vol. 36, ed. E. Wolf, North-Holland, Amsterdam, p. 129.

Białynicka-Birula, Z., 1968, *Phys. Rev.* **173**, 1207.

Białynicka-Birula, Z. and I. Białynicki-Birula, 1994, *J. Mod. Opt.* **41**, 2203.

Białynicki-Birula, I., 1977, *Acta Phys. Austr. Suppl.* **XVIII**, 111.

Białynicki-Birula, I., 1980, Quantum Electrodynamics, ed. A. O. Barut, Plenum Press, New York, p. 119.

Białynicki-Birula, I. and Z. Białynicka-Birula, 1976, *Phys. Rev. A* **14**, 1101.

Białynicki-Birula, I., M. Freyberger and W. Schleich, 1993, *Phys. Scr.* **T48**, 113.

Biederharn, L. C., 1989, *J. Phys. A* **22**, L873.

Bizarro, J. P., 1994, *Phys. Rev. A* **49**, 3255.

Bhandari, R., 1988, *Phys. Lett. A* **133**, 1.

Bhandari, R. and J. Samuel, 1988, *Phys. Rev. Lett.* **60**, 1211.

Blow, K. J., R. Loudon and S. J. D. Phoenix, 1993, *J. Mod. Opt.* **40**, 2515.

Blow, K. J., R. Loudon and S. J. D. Phoenix, 1994, *Opt. Commun.* **110**, 239.

Bluhm, R., V. A. Kostelecký and B. Tudose, 1995, *Phys. Rev. A* **52**, 2234.

Bogár, P. and A. Bergou, 1996, *Phys. Rev. A* **53**, 49.

Böhm, A., 1978, The Rigged Hilbert Space and Quantum Mechanics (Lecture Notes in Physics Vol. 78), Springer-Verlag, Berlin.

Bohr, N., 1949, Albert Einstein: Philosopher-Scientist, ed. P. A. Schilpp, Library of Living Philosophers, Inc., Evanston, Ill.

Bondurant, R. S. and J. H. Shapiro, 1984, *Phys. Rev. D* **30**, 2548.

Bordé, Ch. J., N. Courtier, F. du Burck, A. N. Goncharov and M. Gorlicki, 1994, *Phys. Lett. A* **188**, 187.

Born, M. and E. Wolf, 1959, Principles of Optics, Pergamon Press, Oxford.

Boschi, D., F. De Martini and G. Di Giuseppe, 1997, *Phys. Lett. A* **228**, 208.

Bouchal, Z., 1993, *J. Mod. Opt.* **40**, 1325.

Bouchal, Z. and M. Olivík, 1995, *J. Mod. Opt.* **42**, 1555.

Bouchal, Z. and J. Peřina, 1992, *J. Mod. Opt.* **39**, 1365.

Braginsky, V. B. and F. Y. Khalili, 1992, Quantum Measurements, ed. K. S. Thorne, Cambridge University Press, Cambridge.

Braginsky, V. B. and Y. I. Vorontsov, 1974, *Usp. Fiz. Nauk* **114**, 41.

Braginsky, V. B., Y. I. Vorontsov and K. S. Thorne, 1980, *Science* **209**, 547.

Brandt, R. A. and O. W. Greenberg, 1969, *J. Math. Phys.* **10**, 1168.

Braunstein, S. L., 1990, *Phys. Rev. A* **42**, 474.

Braunstein, S. L., 1992a, *Phys. Rev. A* **45**, 6803.

Braunstein, S. L., 1992b, *Phys. Rev. Lett.* **69**, 3598.

Braunstein, S. L., 1992c, *J. Phys. A* **25**, 3813.

Braunstein, S. L., 1994, *Phys. Rev. A* **49**, 69.

Braunstein, S. L. and C. M. Caves, 1990, *Phys. Rev. A* **42**, 4115.

Braunstein, S. L., C. M. Caves and G. J. Milburn, 1991, *Phys. Rev. A* **43**, 1153.

Braunstein, S. L., A. S. Lane and C. M. Caves, 1992, *Phys. Rev. Lett.* **69**, 2153.

Braunstein, S. L. and A. Mann, 1995, *Phys. Rev. A* **51**, R1727.

Braunstein, S. L. and R. I. McLachlan, 1987, *Phys. Rev. A* **35**, 1659.

Brecha, R. J., L. A. Orozco, M. G. Raizen, M. Xiao and H. J. Kimble, 1995, *J. Opt. Soc. Am.* **12**, 2329.

Brendel, J., W. Dultz and W. Martienssen, 1995, *Phys. Rev. A* **52**, 2551.

Brendel, J., E. Mohler and W. Martienssen, 1991, *Phys. Rev. Lett.* **66**, 1142.

Brendel, J., E. Mohler and W. Martienssen, 1992, *Europhys. Lett.* **20**, 575.

Brieger, M. and H. A. Schuessler, 1996, *Europhys. Lett.* **35**, 1.

Brif, C., 1995, *Quant. Semiclass. Opt.* **7**, 803.

Brif, C. and Y. Ben-Aryeh, 1994a, *Phys. Rev. A* **50**, 2727.

Brif, C. and Y. Ben-Aryeh, 1994b, *Phys. Rev. A* **50**, 3505.

Brif, C. and Y. Ben-Aryeh, 1994c, *J. Phys. A: Math. Gen.* **27**, 8185.

Brif, C. and Y. Ben-Aryeh, 1996, *Quant. Semiclass. Opt.* **8**, 1.

Brif, C. and A. Mann, 1996a, *Phys. Rev. A* **54**, 4505.

Brif, C. and A. Mann, 1996b, *Phys. Lett. A* **219**, 257.

Brouri, R., F. de Tomasi, J. Baudon, J. Reinhardt, V. Lorent, J. Robert and M. Gorlicki, 1997, *Opt. Commun.* **141**, 329.

Bruckmeier, R., K. Schneider, S. Schiller and J. Mlynek, 1997, *Phys. Rev. Lett.* **78**, 1243.

Brune, M., E. Hagley, J. Dreyer, X. Maître, A. Maali, C. Wunderlich, J. M. Raimond and S. Haroche, 1996, *Phys. Rev. Lett.* **77**, 4887.

Brune, M., S. Haroche, V. Lefevre, J. M. Raimond and N. Zagury, 1990, *Phys. Rev. Lett.* **65**, 976.

Brune, M., S. Haroche and J. M. Raimond, 1992, *Phys. Rev. A* **45**, 5193.

Brune, M., F. Schmidt-Kaler, A. Maali, J. Dreyer, E. Hagley, J. M. Raimond and S. Haroche, 1996, *Phys. Rev. Lett.* **76**, 1800.

Brunet, H., 1964, *Phys. Lett.* **10**, 172.

Burak, D. and K. Wódkiewicz, 1992, *Phys. Rev. A* **46**, 2744.

Burge, R. E., M. A. Fiddy, A. H. Greenaway and G. Ross, 1974, *J. Phys. D: Appl. Phys.* **7**, L65.

Burge, R. E., M. A. Fiddy, A. H. Greenaway and G. Ross, 1976, *Proc. Roy. Soc. A* **350**, 191.

Bush, P., P. J. Lahti and P. Mittelstaedt, 1991, The Quantum Theory of Measurement, Springer-Verlag, Berlin.

Bužek, V., 1989a, *Phys. Rev. A* **39**, 3196

Bužek, V., 1989b, *Phys. Rev. A* **39**, 5432.

Bužek, V., 1991, *J. Mod. Opt.* **38**, 801.

Bužek, V. and G. Adam, 1995, *Acta Phys. Slov.* **45**, 425.

Bužek, V., G. Adam and G. Drobný, 1996a, *Phys. Rev. A* **54**, 804.

Bužek, V., G. Adam and G. Drobný, 1996b, *Ann. Phys.* (N. Y.) **245**, 37.

Bužek, V., Ts. Gantsog and M. S. Kim, 1993, *Phys. Scr.* **T48**, 131.

Bužek, V. and M. Hillery, 1996, *J. Mod. Opt.* **43**, 1633.

Bužek, V. and I. Jex, 1989, *Acta Phys. Slov.* **39**, 351.

Bužek, V., I. Jex and Tran Quang, 1990, *J. Mod. Opt.* **37**, 159.

Bužek, V., C. H. Keitel and P. L. Knight, 1995a, *Phys. Rev. A* **51**, 2575.

Bužek, V., C. H. Keitel and P. L. Knight, 1995b, *Phys. Rev. A* **51**, 2594.

Bužek, V. and P. L. Knight, 1995, Progress in Optics, Vol. 34, ed. E. Wolf, North-Holland, Amsterdam, p. 1.

Bužek, V., P. L. Knight and I. K. Kudryavtsev, 1991, *Phys. Rev. A* **44**, 1931.

Bužek, V., P. L. Knight and A. Vidiella-Barranco, 1992, Workshop on Squeezed States and Uncertainty Relations, eds. D. Han., Y. S. Kim, and W. W. Zachary, NASA, Washington, D. C., p. 181.

Bužek, V., A. Vidiella-Barranco and P. L. Knight, 1992, *Phys. Rev. A* **45**, 6570.

Bužek, V., A. D. Wilson-Gordon, P. L. Knight and W. K. Lai, 1992, *Phys. Rev. A* **45**, 8079.

Cahill, K. E. and R. J. Glauber, 1969a, *Phys. Rev.* **177**, 1857.

Cahill, K. E. and R. J. Glauber, 1969b, *Phys. Rev.* **177**, 1882.

Cahn, S. B., A. Kumarakrishnan, U. Shim, T. Sleator, P. R. Berman and B. Dubetsky, 1997, *Phys. Rev. Lett.* **79**, 784.

Campos, R. A., 1994, *Phys. Lett. A* **184**, 173.

Campos, R. A., B. E. A. Saleh and M. C. Teich, 1989, *Phys. Rev. A* **40**, 1371.

Campos, R. A., B. E. A. Saleh and M. C. Teich, 1990, *Phys. Rev. A* **42**, 4127.

Carmichael, H. J., 1993, An Open Systems Approach to Quantum Optics (Lecture Notes in Physics Vol. 18), Springer-Verlag, Berlin.

Carmichael, H., ed., 1996, Quantum and Semiclassical Optics, Vol. **8**, No. 1, IOP Publishing, Bristol.

Carnal, O. and J. Mlynek, 1991, *Phys. Rev. Lett.* **66**, 2689.

Carrasco, M. L. A. and H. Moya-Cessa, 1997, *Quant. Semiclass. Opt.* **9**, L1.

Carruthers, P. and M. M. Nieto, 1965, *Phys. Rev. Lett.* **14**, 387.

Carruthers, P. and M. M. Nieto, 1968, *Rev. Mod. Phys.* **40**, 411.

Casado, A., A. Fernández-Rueda, T. Marshall, R. Risco-Delgado and E. Santos, 1997, *Phys. Rev. A* **55**, 3879.

Casado, A., T. W. Marshall and E. Santos, 1997, *J. Opt. Soc. Am. B* **14**, 494.

Caves, C. M., 1981, *Phys. Rev. D* **23**, 1693.

Caves, C. M., 1982, *Phys. Rev. D* **26**, 1817.

Caves, C. M. and B. L. Schumaker, 1985, *Phys. Rev. A* **31**, 3068.

Caves, C. M., K. S. Thorne, R. V. P. Drever, V. D. Sandberg and M. Zimmermann, 1980, *Rev. Mod. Phys.* **52**, 341.

Caves, C. M., Ch. Zhu, G. J. Milburn and W. Schleich, 1991, *Phys. Rev. A* **43**, 3854.

Cerveró, J. M. and J. D. Lejarreta, 1997, *Quant. Semiclass. Opt.* **9**, L5.

Chai, C.-L., 1992, *Phys. Rev. A* **46**, 7187.

Chaichian, M. and D. Ellinas, 1990, *J. Phys. A* **23**, L291.

Chaichian, M. and P. Kulish, 1990, *Phys. Lett.* **234B**, 72.

Chapman, M., C. Ekstrom, T. Hammond, R. Rubenstein, J. Schmiedmayer, S. Wehinger and D. Pritchard, 1995, *Phys. Rev. Lett.* **74**, 4783.

Chiao, R. Y., A. Antaramian, K. M. Ganga, H. Jiao, S. R. Wilkinson and H. Nathel, 1988, *Phys. Rev. Lett.* **60**, 1214.

Chiao, R. Y. and T. F. Jordan, 1988, *Phys. Lett. A* **132**, 77.

Chiao, R. Y., P. G. Kwiat and A. M. Steinberg, 1992, Workshop on Squeezed States and Uncertainty Relations, eds. J. D. Han, Y. S. Kim, and W. W. Zachary, NASA, Washington, D. C., p. 61.

Chiao, R. Y., P. G. Kwiat and A. M. Steinberg, 1994, Advances in Atomic, Molecular and Optical Physics, Vol. 34, eds. B. Bederson and H. Walther, Academic Press, San Diego, CA, p. 35.

Chiao, R. Y. and Y.-S. Wu, 1986, *Phys. Rev. Lett.* **57**, 933.

Chiu, S.-H., R. W. Gray and C. A. Nelson, 1992, *Phys. Lett. A* **164**, 237.

Chizhov, A. V., J. W. Haus and K. C. Yeong, 1995, *Phys. Rev. A* **52**, 1698.

Choi, S., H. M. Wiseman, S. M. Tan and D. F. Walls, 1997, *Phys. Rev. A* **55**, 527.

Christanell, R., H. Weinfurter and A. Zeilinger, 1993, Tech. Digest EQEC'93, eds. P. De Natale, R. Meucci and S. Pelli, Vol. 2, p. 827.

Chyba, T. H., L. J. Wang, L. Mandel and R. Simon, 1988, *Opt. Lett.* **13**, 562.

Cibils, M. B., Y. Cuche, V. Marvulle and W. F. Wreszinski, 1991, *Phys. Rev. A* **43**, 4044.

Cirac, J. I., C. W. Gardiner, M. Naraschewski and P. Zoller, 1996, *Phys. Rev. A* **54**, R3714.

Cirac, J. I. and A. S. Parkins, 1994, *Phys. Rev. A* **50**, R4441.

Cohendet, O., Ph. Combe, M. Sirugue and M. Sirugue-Collin, 1988, *J. Phys. A* **21**, 2875.

Collett, M. J., 1993a, *Phys. Scr.* **T48**, 124.

Collett, M. J., 1993b, *Phys. Rev. Lett.* **70**, 3400.

Cook, R. J., 1982, *Phys. Rev. A* **25**, 2164.

Cramér, H., 1946, Mathematical Methods of Statistics, Princeton University Press, Princeton, NJ.

Crosignani, B., P. Di Porto and M. Bertolotti, 1975, Statistical Properties of Scattered Light, Academic Press, New York.

Czirják, A. and M. G. Benedict, 1996, *Quant. Semiclass. Opt.* **8**, 975.

Daeubler, B., Ch. Miller, H. Risken and L. Schoendorff, 1993, *Phys. Scr.* **T48**, 119.

Dainty, J. C., ed., 1984, Laser Speckle and Related Phenomena, Springer-Verlag, Berlin.

Dakna, M., L. Knöll and D.-G. Welsch, 1997a, *Quant. Semiclass. Opt.* **9**, 331.

Dakna, M., L. Knöll and D.-G. Welsch, 1997b, *Phys. Rev. A* **55**, 2360.

Dalibard, J., Y. Castin and K. Mølmer, 1992, *Phys. Rev. Lett.* **68**, 580.

Dalton, B. J. and P. L. Knight, 1990, *Phys. Rev. A* **42**, 3034.

Damaskinsky, E. V. and V. S. Yarunin, 1978, *The Academic Research Journal (Tomsk University): Physics* **6**, 59.

Daniel, D. J. and G. J. Milburn, 1989, *Phys. Rev. A* **39**, 4628.

D'Ariano, G. M., 1992, *Int. J. Mod. Phys. B* **6**, 1291.

D'Ariano, G. M., 1995, *Quant. Semiclass. Opt.* **7**, 693.

D'Ariano, G. M., 1997a, Quantum Optics and the Spectroscopy of Solids, eds. T. Hakioğlu and A. S. Shumovsky, Kluwer, Dordrecht, p. 139.

D'Ariano, G. M., 1997b, Quantum Optics and the Spectroscopy of Solids, eds. T. Hakioğlu and A. S. Shumovsky, Kluwer, Dordrecht, p. 175.

D'Ariano, G. M., U. Leonhardt and H. Paul, 1995, *Phys. Rev. A* **52**, R1801.

D'Ariano, G. M., C. Machiavello and M. G. A. Paris, 1994a, *Phys. Rev. A* **50**, 4298.

D'Ariano, G. M., C. Machiavello and M. G. A. Paris, 1994b, *Phys. Lett. A* **195**, 31.

D'Ariano, G. M., C. Macchiavello and M. G. A. Paris, 1995, *Phys. Lett. A* **198**, 286.

D'Ariano, G. M., C. Macchiavello, N. Sterpi and H. P. Yuen, 1996, *Phys. Rev. A* **54**, 4712.

D'Ariano, G. M., S. Mancini, V. I. Man'ko and P. Tombesi, 1996, *Quant. Semiclass. Opt.* **8**, 1017.

D'Ariano, G. M., S. Mancini and P. Tombesi, 1997, *Opt. Commun.* **133**, 165.

D'Ariano, G. M. and M. G. A. Paris, 1993, *Phys. Rev. A* **48**, 4039.

D'Ariano, G. M. and M. G. A. Paris, 1994, *Phys. Rev. A* **49**, 3022.

D'Ariano, G. M. and M. G. A. Paris, 1997, *Phys. Rev. A* **55**, 2267.

D'Ariano, G. M., M. G. A. Paris and R. Seno, 1996, *Phys. Rev. A* **54**, 4495.

D'Ariano, G., M. Rasetti and M. Vadacchino, 1985, *Phys. Rev. D* **32**, 1034.

D'Ariano, G. M. and M. F. Sacchi, 1995, *Phys. Rev. A* **52**, R4309.

D'Ariano, G. M. and M. F. Sacchi, 1997, *Phys. Lett. A* **231**, 325.

Davidović, D. M., D. Lalović and A. R. Tančić, 1994, *J. Phys. A: Math. Gen.* **27**, 8247.

Davidovich, L., M. Brune, J. M. Raimond and S. Haroche, 1996, *Phys. Rev. A* **53**, 1295.

Davies, E. B., 1976, Quantum Theory of Open Systems, Academic Press, London.

De Moura, F. A. B. F. and M. L. Lyra, 1996, *J. Mod. Opt.* **43**, 1671.

de Oliveira, F. A. M., 1992, *Phys. Rev. A* **45**, 3113.

de Oliveira, F. A. M., M. S. Kim, P. L. Knight and V. Bužek, 1990, *Phys. Rev. A* **41**, 2645.

Derka, R., V. Bužek, G. Adam and P. L. Knight, 1996, *Fine Mechanics and Optics* **41**, 341.

De Velis, B. and G. O. Reynolds, 1967, Theory and Applications of Holography, Addison–Wesley, Reading, MA.

De Vito, E. and A. Levrero, 1994, *J. Mod. Opt.* **41**, 2233.

Di Giuseppe, G., F. De Martini and D. Boschi, 1997, *Phys. Rev. A* **56**, 176.

Di Giuseppe, G., L. Haiberger, F. De Martini and A. V. Sergienko, 1997, *Phys. Rev. A* **56**, R21.

Dirac, P. A. M., 1925, *Proc. Roy. Soc.* **109**, 642.

Dirac, P. A. M., 1926a, *Proc. Roy. Soc.* **110**, 561.

Dirac, P. A. M., 1926b, *Proc. Roy. Soc.* **111**, 281.

Dirac, P. A. M., 1926c, *Proc. Roy. Soc.* **111**, 405.

Dirac, P. A. M., 1927, *Proc. Roy. Soc. A* **114**, 243.

Dirac, P. A. M., 1958, The Principles of Quantum Mechanics, 4th ed., Clarendon Press, Oxford.

Dodonov, V. V. and V. I. Man'ko, 1997, *Phys. Lett. A* **229**, 335.

Dodonov, V. V., V. I. Man'ko and D. E. Nikov, 1995, *Phys. Rev. A* **51**, 3328.

Domokos, P., P. Adam and J. Janszky, 1994, *Phys. Rev. A* **50**, 4293.

Domokos, P. and J. Janszky, 1994, *Phys. Lett. A* **186**, 289.

Domokos, P., J. Janszky and P. Adam, 1994, *Phys. Rev. A* **50**, 3340.

Dong, B., J. Zhuang and O. K. Ersoy, 1994, *J. Mod. Opt.* **41**, 1575.

Drobný, G. and I. Jex, 1992, *Phys. Lett. A* **169**, 273.

Drummond, P. D., J. Breslin and R. M. Shelby, 1994, *Phys. Rev. Lett.* **73**, 2837.

Drummond, P. D. and C. W. Gardiner, 1980, *J. Phys. A* **13**, 2353.

Dubin, D. A., M. A. Hennings and T. B. Smith, 1994, *Publ. RIMS Kyoto Univ.* **30**, 479.

Dubin, D. A., M. A. Hennings and T. B. Smith, 1995, *Int. J. Mod. Phys. B* **9**, 2597.

Dung, H. T. and A. S. Shumovsky, 1992, *Phys. Lett. A* **169**, 379.

Dupertuis, M.-A., 1988, *Phys. Rev. A* **37**, 4752.

Dupertuis, M.-A., S. M. Barnett and S. Stenholm, 1987a, *J. Opt. Soc. Am. B* **4**, 1102.

Dupertuis, M.-A., S. M. Barnett and S. Stenholm, 1987b, *J. Opt. Soc. Am. B* **4**, 1124.

Dupertuis, M.-A. and S. Stenholm, 1987, *J. Opt. Soc. Am. B* **4**, 1124.

Duran, P. L., 1970, Theory of H^p Spaces, Academic Press, New York.

Durnin, J., 1987, *J. Opt. Soc. Am. A* **4**, 651.

Dušek, M., 1997, *J. Mod. Opt.* **44**, 593.

Dutra, S. M., P. L. Knight and H. Moya-Cessa, 1993, *Phys. Rev. A* **48**, 3168.

Eberly, J. H., N. B. Narozhny and J. J. Sanchez-Mondragon, 1980, *Phys. Rev. Lett.* **44**, 1323.

Edwards, S. F. and G. B. Parrent, 1959, *Opt. Acta* **6**, 367.

Eichmann, U., J. C. Bergquist, J. J. Bollinger, J. M. Gilligan, W. M. Itano, D. J. Wineland and M. G. Raizen, 1993, *Phys. Rev. Lett.* **70**, 2359.

Einstein, A., B. Podolsky and N. Rosen, 1935, *Phys. Rev. A* **47**, 777.

Eiselt, J. and H. Risken, 1991, *Phys. Rev. A* **43**, 346.

Ekert, A. K. and P. L. Knight, 1990, *Phys. Rev. A* **42**, 487.

Ekert, A. K., J. G. Rarity, P. R. Tapster and G. M. Palma, 1992, *Phys. Rev. Lett.* **69**, 1293.

Ellinas, D., 1991a, *J. Math. Phys.* **32**, 1.

Ellinas, D., 1991b, *J. Mod. Opt.* **38**, 2393.

Ellinas, D., 1992, *Phys. Rev. A* **45**, 3358.

Englert, B.-G. and K. Wódkiewicz, 1995, *Phys. Rev. A* **51**, R2661.

Englert, B.-G., K. Wódkiewicz and P. Riegler, 1995, *Phys. Rev. A* **52**, 1704.

Enzer, D. and G. Gabrielse, 1997, *Phys. Rev. Lett.* **78**, 1211.

Fain, V. M., 1967, *Soviet Physics JETP* **52**, 1544.

Fan, H.-Y. and M. Xiao, 1996, *Phys. Rev. A* **54**, 5295.

Fan, H.-Y. and M. Xiao, 1997, *Phys. Lett. A* **227**, 192.

Fan, H.-Y. and H. R. Zaidi, 1988, *Opt. Commun.* **68**, 143.

Feller, W., 1966, An Introduction to Probability Theory and Its Applications, Vol. I., J. Wiley, New York.

Ferwerda, H. A., 1978, Inverse Source Problems, ed. H. P. Baltes, Springer-Verlag, Berlin, p. 13.

Ferwerda, H. A. and B. J. Hoenders, 1975a, *Opt. Acta* **22**, 25.

Ferwerda, H. A. and B. J. Hoenders, 1975b, *Opt. Acta* **22**, 35.

Fienup, J. R., 1978, *Opt. Lett.* **3**, 27.

Fienup, J. R., 1982, *Appl. Opt.* **21**, 2758.

Fienup, J. R., 1984, Optics in Modern Science and Technology, Proc. ICO 13, ed. H. Ohzu, Sapporo, p. 606.

Fienup, J. R. and C. C. Wakerman, 1986, *J. Opt. Soc. Am. A* **3**, 1897.

Figurny, P., A. Orłowski and K. Wódkiewicz, 1993, *Phys. Rev. A* **47**, 5151.

Fittinghoff, D. N., J. L. Bowie, J. N. Sweetser, R. T. Jennings, M. A. Krumbügel, K. W. DeLong and R. Trebino, 1996, *Opt. Lett.* **21**, 884.

Fougères, A., C. H. Monken and L. Mandel, 1994, *Opt. Lett.* **19**, 1771.

Fougères, A., J. W. Noh, T. P. Grayson and L. Mandel, 1994, *Phys. Rev. A* **49**, 530.

Franson, J. D., 1989, *Phys. Rev. Lett.* **62**, 2205.

Franson, J. D., 1991a, *Phys. Rev. Lett.* **67**, 290.

Franson, J. D., 1991b, *Phys. Rev. A* **44**, 4552.

Franson, J. D., 1994, *Phys. Rev. A* **49**, 3221.

Freyberger, M., 1995, *Phys. Rev. A* **51**, 3347.

Freyberger, M., M. Heni and W. P. Schleich, 1995, *Quant. Semiclass. Opt.* **7**, 187.

Freyberger, M. and A. M. Herkommer, 1994, *Phys. Rev. Lett.* **72**, 1952.

Freyberger, M. and W. Schleich, 1993, *Phys. Rev. A* **47**, R30.

Freyberger, M. and W. Schleich, 1994, *Phys. Rev. A* **49**, 5056.

Freyberger, M., K. Vogel and W. Schleich, 1993a, *Quant. Opt.* **5**, 65.

Freyberger, M., K. Vogel and W. P. Schleich, 1993b, *Phys. Lett. A* **176**, 41.

Friberg, A. T. and E. Wolf, 1995, *Opt. Lett.* **20**, 623.

Frieden, B. R., 1983, Probability, Statistical Optics, and Data Testing, Springer-Verlag, Berlin.

Frieden, B. R., 1991, Probability, Statistical Optics, and Data Testing, 2nd ed., Springer-Verlag, Berlin.

Frins, E. M. and W. Dultz, 1997, *Opt. Commun.* **136**, 354.
Fujikawa, K., 1995, *Phys. Rev. A* **52**, 3299.
Gabor, D., 1948, *Nature* **161**, 777.
Gabor, D., 1949, *Proc. Roy. Soc. A* **197**, 454.
Gabor, D., 1951, *Proc. Phys. Soc. B* **64**, 449.
Gaeta, A. L. and R. W. Boyd, 1988, *Phys. Rev. Lett.* **60**, 2618.
Gagen, M. J., 1995, *Phys. Rev. A* **51**, 2715.
Gähler, R. and A. Zeilinger, 1991, *Am. J. Phys.* **59**, 316.
Galetti, D. and A. F. R. De Toledo Piza, 1988, *Physica* **140A**, 267.
Galindo, A., 1984a, *Lett. Math. Phys.* **8**, 495.
Galindo, A., 1984b, *Anales De Fisica* **81**, 191.
Gamo, H., 1963, Electromagnetic Theory and Antennas, Part 2, ed. E. C. Jordan, Macmillan, New York, p. 801.
Gangopadhyay, G., 1994, *J. Mod. Opt.* **41**, 525.
Gantsog, Ts., 1992, *Phys. Lett. A* **170**, 249.
Gantsog, Ts., A. Joshi and R. Tanaś, 1994, *Quant. Opt.* **6**, 517.
Gantsog, Ts., A. Miranowicz and R. Tanaś, 1992, *Phys Rev. A* **46**, 2870.
Gantsog, Ts. and R. Tanaś, 1991a, *Opt. Commun.* **82**, 145.
Gantsog, Ts. and R. Tanaś, 1991b, *Quant. Opt.* **3**, 33.
Gantsog, Ts. and R. Tanaś, 1991c, *J. Mod. Opt.* **38**, 1021.
Gantsog, Ts. and R. Tanaś, 1991d, *Phys. Rev. A* **44**, 2086.
Gantsog, Ts. and R. Tanaś, 1991e, *J. Mod. Opt.* **38**, 1537.
Gantsog, Ts. and R. Tanaś, 1996, *Phys. Rev. A* **53**, 562.
Gantsog, Ts., R. Tanaś and R. Zawodny, 1991a, *Phys. Lett. A* **155**, 1.
Gantsog, Ts., R. Tanaś and R. Zawodny, 1991b, *Opt. Commun.* **82**, 345.
Gantsog, Ts., R. Tanaś and R. Zawodny, 1993, *Acta Phys. Slov.* **43**, 74.
Gardiner, C. W., 1986, *Phys. Rev. Lett.* **56**, 1917.
Gardiner, C. W., 1991, Quantum Noise, Springer-Verlag, Berlin.
Garraway, B. M. and P. L. Knight, 1992, *Phys. Rev. A* **46**, 5346.
Garraway, B. M. and P. L. Knight, 1993, *Phys. Scr.* **T48**, 66.
Garraway, B. M., B. Sherman, H. Moya-Cessa, P. L. Knight and G. Kurizki, 1994, *Phys. Rev. A* **49**, 535.
Garrison, J. C. and J. Wong, 1970, *J. Math. Phys.* **11**, 2242.
Garuccio, A., 1995, *Phys. Rev. A* **52**, 2535.
Gea-Banacloche, J. and G. Leuchs, 1987, *J. Opt. Soc. Am. B* **4**, 1667.
Gelfand, I. M. and G. E. Shilov, 1964, Generalized Functions, Vol. I, Academic Press, New York.
Gerhardt, H., U. Büchler and G. Liftin, 1974, *Phys. Lett.* **49A**, 119.
Gerhardt, H., H. Welling and D. Frölich, 1973, *Appl. Phys.* **2**, 91.
Gerry, C. C., 1987, *Opt. Commun.* **63**, 278.
Gerry, C. C., 1988, *Phys. Rev. A* **38**, 1734.
Gerry, C. C., 1990, *Opt. Commun.* **75**, 168.
Gerry, C. C., 1996, *Phys. Rev. A* **53**, 1179.
Gerry, C. C., 1997, *Phys. Rev. A* **55**, 2478.
Gerry, C. C. and R. Grobe, 1995a, *Phys. Rev. A* **51**, 1698.
Gerry, C. C. and R. Grobe, 1995b, *Phys. Rev. A* **51**, 4123.
Gerry, C. C. and K. E. Urbanski, 1990, *Phys. Rev. A* **42**, 662.
Gevorkyan, S. T. and W. H. Maloyan, 1997, *J. Mod. Opt.* **44**, 1443.
Gilchrist, A., C. W. Gardiner and P. D. Drummond, 1997, *Phys. Rev. A* **55**, 3014.

Gilson, C. R., S. M. Barnett and S. Stenholm, 1987, *J. Mod. Opt.* **34**, 949.

Glauber, R. J., 1963a, *Phys. Rev.* **131**, 2766.

Glauber, R. J., 1963b, *Phys. Rev. Lett.* **10**, 84.

Glauber, R. J. and M. Lewenstein, 1991, *Phys. Rev. A* **43**, 467.

Goldberger, M. L., H. W. Lewis and K. M. Watson, 1963, *Phys. Rev.* **132**, 2764.

Goldenberg, L. and L. Vaidman, 1995, *Phys. Rev. Lett.* **75**, 1239.

Goldhirsch, I., 1980, *J. Phys. A: Math. Gen.* **13**, 3479.

Gould, P. L., P. J. Martin, G. A. Ruff, R. E. Stoner, J. L. Picque and D. E. Pritchard, 1991, *Phys. Rev. A* **43**, 585.

Gould, P. L., G. A. Ruff and D. E. Pritchard, 1986, *Phys. Rev. Lett.* **56**, 827.

Grangier, P., G. Roger and A. Aspect, 1986, *Europhys. Lett.* **1**, 173.

Grayson, T. P., J. R. Torgerson and G. A. Barbosa, 1994, *Phys. Rev. A* **49**, 626.

Grayson, T. P., X. Y. Zou, D. Branning, J. R. Torgerson and L. Mandel, 1993, *Phys. Rev. A* **48**, 4793.

Greenberger, D. M., M. A. Horne, A. Shimony and A. Zeilinger, 1990, *Am. J. Phys.* **58**, 1131.

Grice, W. P. and I. A. Walmsley, 1997, *Phys. Rev. A* **56**, 1627.

Grønbech-Jensen, N., P. L. Christiansen and P. S. Ramanujam, 1989, *J. Opt. Soc. Am. B* **6**, 2423.

Gudder, S., 1979, Stochastic Methods in Quantum Mechanics, North-Holland, Amsterdam.

Gureyev, T. E. and K. A. Nugent, 1996, *J. Opt. Soc. Am. A* **13**, 1670.

Gureyev, T. E., A. Roberts and K. A. Nugent, 1995a, *J. Opt. Soc. Am. A* **12**, 1932.

Gureyev, T. E., A. Roberts and K. A. Nugent, 1995b, *J. Opt. Soc. Am. A* **12**, 1942.

Gustavson, T. L., P. Bouyer and M. A. Kasevich, 1997, *Phys. Rev. Lett.* **78**, 2046.

Haake, F., 1973, Springer Tracts in Modern Physics, Vol. 66, ed. G. Höhler, Springer-Verlag, Berlin, p. 98.

Hach III, E. E. and C. C. Gerry, 1993, *Quant. Opt.* **5**, 327.

Haken, H., 1970, Handbuch der Physik, Vol. 25/2c, ed. S. Flügge, Springer-Verlag, Berlin.

Hakioğlu, T., A. S. Shumovsky and O. Aytür, 1994, *Phys. Lett. A* **194**, 304.

Hall, M. J. W., 1991, *Quant. Opt.* **3**, 7.

Hall, M. J. W., 1993, *J. Mod. Opt.* **40**, 809.

Hall, M. J. W., 1994, *Phys. Rev. A* **50**, 3295.

Hall, M. J. W., 1995, *Phys. Rev. Lett.* **74**, 3307.

Hall, M. J. W., 1997, *Phys. Rev. A* **55**, 100.

Hall, M. J. W. and I. G. Fuss, 1991, *Quant. Opt.* **3**, 147.

Hannay, J. H., 1985, *J. Phys. A* **18**, 221.

Hardy, L., 1994, *Phys. Rev. Lett.* **73**, 2279.

Harel, G., G. Kurizki, J. K. McIver and E. Coutsias, 1996, *Phys. Rev. A* **53**, 4543.

Hariharan, P., N. Brown and B. C. Sanders, 1993a, *J. Mod. Opt.* **40**, 113.

Hariharan, P., N. Brown and B. C. Sanders, 1993b, *J. Mod. Opt.* **40**, 1573.

Hariharan, P. and P. E. Ciddor, 1994, *Opt. Commun.* **110**, 13.

Hariharan, P., I. Fujima, N. Brown and B. C. Sanders, 1993, *J. Mod. Opt.* **40**, 1477.

Hariharan, P., K. G. Larkin and M. Roy, 1994, *J. Mod. Opt.* **41**, 663.

Hariharan P., H. Ramachandran, K. A. Suresh and J. Samuel, 1997, *J. Mod. Opt.* **44**, 707.

Hariharan, P., M. Roy, P. A. Robinson and J. W. O'Byrne, 1993, *J. Mod. Opt.* **40**, 871.

Hariharan, P. and B. C. Sanders, 1996, Progress in Optics, Vol. 36, ed. E. Wolf, North-Holland, Amsterdam, p. 49.

Harms, J. and J. Lorigny, 1964, *Phys. Lett.* **10**, 173.

Haroche, S., 1992, Fundamental Systems in Quantum Optics, eds. J. Dalibard, J. M. Raimond and J. Zinn-Justin, North-Holland, Amsterdam, p. 767.

Haroche, S. and D. Kleppner, 1989, *Phys. Today* **42**, No. 1, 24.

Harrison, F. E. and D. F. Walls, 1996, *Opt. Commun.* **123**, 331.

Hasegawa, Y., M. Zawiski, H. Rauch and A. I. Joffe, 1996, *Phys. Rev. A* **53**, 2486.

Heisenberg, W., 1925, *Z. Phys.* **33**, 879.

Heitler, W., 1954, The Quantum Theory of Radiation, 3rd ed., Clarendon Press, Oxford.

Helstrom, C. W., 1976, Quantum Detection and Estimation Theory, Academic Press, New York.

Helstrom, C. W., M. Charbit and C. Bendjaballah, 1987, *Opt. Commun.* **64**, 253.

Henkel, C., C. I. Westbrook and A. Aspect, 1996, *J. Opt. Soc. Am. B* **13**, 233.

Hennings, M. A., T. B. Smith and D. A. Dubin, 1995a, *J. Phys. A: Math. Gen.* **28**, 6779.

Hennings, M. A., T. B. Smith and D. A. Dubin, 1995b, *J. Phys. A: Math. Gen.* **28**, 6809.

Herkommer, A. M., H. J. Carmichael and W. P. Schleich, 1996, *Quant. Semiclass. Opt.* **8**, 189.

Herzog, T. J., P. G. Kwiat, H. Weinfurter and A. Zeilinger, 1995, *Phys. Rev. Lett.* **75**, 3034.

Herzog, U. and H. Paul, 1993, *Opt. Commun.* **103**, 519.

Herzog, U., H. Paul and Th. Richter, 1993, *Phys. Scr.* **T48**, 61.

Herzog, T. J., J. G. Rarity, H. Weinfurter and A. Zeilinger, 1994, *Phys. Rev. Lett.* **72**, 629.

Herzog, U. and Th. Richter, 1994, *J. Mod. Opt.* **41**, 553.

Hilgevoord, J., 1996, *Am. J. Phys.* **64**, 1451.

Hillery, M., 1987a, *Opt. Commun.* **62**, 135.

Hillery, M., 1987b, *Phys. Rev. A* **36**, 3796.

Hillery, M., 1989, *Phys. Rev. A* **40**, 3147.

Hillery, M., M. Freyberger and W. Schleich, 1995, *Phys. Rev. A* **51**, 1792.

Hillery, M. and L. Mlodinow, 1993, *Phys. Rev. A* **48**, 1548.

Hillery, M., D. Yu and J. Bergou, 1992, Workshop on Squeezed States and Uncertainty Relations, eds. D. Han, Y. S. Kim, and W. W. Zachary, NASA, Washington, D. C., p. 125.

Hillery, M., M. Zou and V. Buzek, 1996, *Quant. Semiclass. Opt.* **8**, 1041

Hlubina, P., 1993, *J. Mod. Opt.* **40**, 1893.

Hlubina, P., 1995, *J. Mod. Opt.* **42**, 1407.

Hlubina, P., 1997, *J. Mod. Opt.* **44**, 163.

Hoenders, B. J., 1978, Inverse Source Problems, ed. H. P. Baltes, Springer-Verlag, p. 41.

Holevo, A. S., 1982, Probabilistic and Statistical Aspects of Quantum Theory, North-Holland, Amsterdam.

Holland, M. J. and K. Burnett, 1993, *Phys. Rev. Lett.* **71**, 1355.

Holland, M. J., D. F. Walls and P. Zoller, 1991, *Phys. Rev. Lett.* **67**, 1716.

Hollenhorst, J. N., 1979, *Phys. Rev. D* **19**, 1669.

Home, D. and G. S. Agarwal, 1995, *Phys. Lett. A* **209**, 1.

Hong, C. K. and L. Mandel, 1985a, *Phys. Rev. Lett.* **54**, 323.

Hong, C. K. and L. Mandel, 1985b, *Phys. Rev. A* **32**, 974.

Hong, C. K., Z. Y. Ou and L. Mandel, 1987, *Phys. Rev. Lett.* **59**, 2044.

Hradil, Z., 1990a, *Phys. Rev. A* **41**, 400.

Hradil, Z., 1990b, *Phys. Lett. A* **146**, 1.

Hradil, Z., 1991, *Phys. Rev. A* **44**, 792.

Hradil, Z., 1992a, *Quant. Opt.* **4**, 93.

Hradil, Z., 1992b, *Phys. Rev. A* **46**, R2217.

Hradil, Z., 1993a, *Phys. Rev. A* **47**, 2376.

Hradil, Z., 1993b, *Phys. Rev. A* **47**, 4532.

Hradil, Z., 1995, *Phys. Rev. A* **51**, 1870.

Hradil, Z., 1997, *Phys. Rev. A* **55**, R1561.

Hradil, Z. and J. Bajer, 1993, *Phys. Rev. A* **48**, 1717.

Hradil, Z., R. Myška, T. Opatrný and J. Bajer, 1996, *Phys. Rev. A* **53**, 3738.

Hradil, Z., R. Myška, J. Peřina, M. Zawisky, Y. Hasegawa and H. Rauch, 1996, *Phys. Rev. Lett.* **76**, 4295.

Hradil, Z. and J. H. Shapiro, 1992, *Quant. Opt.* **4**, 31.

Hu, X. M., 1995, *J. Mod. Opt.* **42**, 1505.

Huang, Hu and Shi-qun Li, 1997, *Opt. Commun.* **137**, 51.

Huttner, B., N. Imoto, N. Gisin and T. Mor, 1995, *Phys. Rev. A* **51**, 1863.

Ifantis, E. K., 1971, *Lett. Nuovo Cimento* **2**, 1096.

Imoto, N., H. A. Haus and Y. Yamamoto, 1985, *Phys Rev. A* **32**, 2287.

Imoto, N., J. R. Jeffers and R. Loudon, 1992, Quantum Measurements in Optics, eds. P. Tombesi and D. F. Walls, Plenum Press, New York.

Imoto, N., M. Ueda and T. Ogawa, 1990, *Phys. Rev. A* **41**, 4127.

Inönü, E. and E. P. Wigner, 1953, *Proc. Natl. Acad. Sci. USA* **39**, 510.

Jackiw, R., 1968, *J. Math. Phys.* **9**, 339.

Jacobson, D. L., S. A. Werner and H. Rauch, 1994, *Phys. Rev. A* **49**, 3196.

Jaeger, G., M. A. Horne and A. Shimony, 1993, *Phys. Rev. A* **48**, 1023.

Jaeger, G., A. Shimony and L. Vaidman, 1995, *Phys. Rev. A* **51**, 54.

Janicke, U. and M. Wilkens, 1995, *J. Mod. Opt.* **42**, 2183.

Janszky, J., P. Domokos and P. Adam, 1992, *Phys. Rev. A* **48**, 213.

Janszky, J., P. Domokos, S. Szabo and P. Adam, 1995, *Phys. Rev. A* **51**, 4191.

Jarzynski, C., 1995, *Phys. Rev. Lett.* **74**, 1264.

Jaynes, E. T., 1957a, *Phys. Rev.* **106**, 620.

Jaynes, E. T., 1957b, *Phys. Rev.* **108**, 171.

Jeffers, J. R., N. Imoto and R. Loudon, 1993, *Phys. Rev. A* **47**, 3346.

Jex, I., G. Drobný and M. Matsuoka, 1992, *Opt. Commun.* **94**, 619.

Jex, I., M. Matsuoka and M. Koashi, 1993, *Quant. Opt.* **5**, 275.

Ji, J.-Y. and J. K. Kim, 1995, *Phys. Lett. A* **208**, 25.

Ji, J.-Y., J. K. Kim, S. P. Kim and K.-S. Soh, 1995, *Phys. Rev. A* **52**, 3352.

Jones, K. R. W., 1993, *Phys. Scr.* **T48**, 100.

Joobeur, A., B. E. A. Saleh, T. S. Larchuk and M. C. Teich, 1996, *Phys. Rev. A* **53**, 4360.

Joobeur, A., B. E. A. Saleh and M. C. Teich, 1994, *Phys. Rev. A* **50**, 3349.

Jordan, P., 1935, *Z. Phys.* **94**, 531.

Joshi, A. and S. V. Lawande, 1992, *Phys. Rev. A* **46**, 5906.

Joshi, A., A. K. Pati and A. Banerjee, 1994, *Phys. Rev. A* **49**, 5131.

Judge, D., 1963, *Phys. Lett.* **5**, 189.

Judge, D., 1964, *Nuovo Cimento* **31**, 332.

Kano, Y. and E. Wolf, 1962, *Proc. Phys. Soc.* **80**, 1273.

Kar, T. K. and D. Bhaumik, 1995, *Phys. Lett. A* **207**, 243.

Kärtner, F. X. and H. A. Haus, 1993, *Phys. Rev. A* **47**, 4585.

Kasevich, M. and S. Chu, 1991, *Phys. Rev. Lett.* **67**, 181.

Keith, D. W., C. R. Ekstrom, Q. A. Turchette and D. E. Pritchard, 1991, *Phys. Rev. Lett.* **66**, 2693.

Keller, T. E. and M. H. Rubin, 1997, *Phys. Rev. A* **56**, 1534.

Kennedy, R. S., 1973, *MIT Res. Lab. Electron. Quart. Prog. Rep.* **108**, 219.

Kennedy, T. A. B. and P. D. Drummond, 1988, *Phys. Rev. A* **38**, 1319.

Khalfin, L. A., 1960, *Proc. Acad. Sci. USSR* **132**, 1051.

Kim, M. S., V. Bužek and M. G. Kim, 1994, *Phys. Lett. A* **186**, 283.

Kim, M. S., F. A. M. de Oliveira and P. L. Knight, 1989a, *Phys. Rev. A* **40**, 2494.

Kim, M. S., F. A. M. de Oliveira and P. L. Knight, 1989b, *Opt. Commun.* **72**, 99.

Kim, M. S. and N. Imoto, 1995, *Phys. Rev. A* **52**, 2401.

Kimler IV, W. C. and C. A. Nelson, 1996, *SUNY BING* **2**, 19.

Kiss, T., U. Herzog and U. Leonhardt, 1995, *Phys. Rev. A* **52**, 2433.

Kitagawa, M., N. Imoto and Y. Yamamoto, 1987, *Phys. Rev. A* **35**, 5270.

Kitagawa, M. and Y. Yamamoto, 1986, *Phys. Rev. A* **34**, 3974.

Klauder, J. R. and B.-S. Skagerstam, eds., 1985, Coherent States, World Scientific, Singapore.

Klauder, J. R. and E. C. G. Sudarshan, 1968, Fundamentals of Quantum Optics, W. A. Benjamin, New York.

Klyshko, D. N., 1989, *Phys. Lett. A* **140**, 19.

Klyshko, D. N., 1996, *Sov. Phys.-Usp.* **39**, 573.

Knight, P. L. and P. M. Radmore, 1982, *Phys. Lett. A* **90**, 342.

Kochan, P., H. J. Carmichael, P. R. Morrow and M. G. Raizen, 1995, *Phys. Rev. Lett.* **75**, 45.

Koďousek, J. and J. Peřina, 1984, *Opt. Commun.* **50**, 411.

Kohler, D. and L. Mandel, 1970, *J. Opt. Soc. Am.* **60**, 280.

Kohler, D. and L. Mandel, 1973, *J. Opt. Soc. Am.* **63**, 126.

Komrska, J., 1971, Advances in Electronics and Electron Physics, Vol. 30, ed. L. Marton, Academic Press, New York, p. 139.

Korolkova, N. and J. Peřina, 1996, *Opt. Commun.* **136**, 135.

Korolkova, N. and J. Peřina, 1997, *Opt. Commun.* **137**, 263.

Kottler, F., 1965, Progress in Optics, Vol. 4, ed. E. Wolf, North-Holland, Amsterdam, p. 281.

Kowarz, M. W. and G. S. Agarwal, 1995, *J. Opt. Soc. Am. A* **12**, 1324.

Krakovsky, A. and J. L. Birman, 1995, *Phys. Rev. A* **51**, 50.

Král, P., 1990a, *J. Mod. Opt.* **37**, 889.

Král, P., 1990b, *Phys. Rev. A* **42**, 4177.

Krenn, G., J. Summhammer and K. Svozil, 1996, *Phys. Rev. A* **53**, 1228.

Krenn, G. and A. Zeilinger, 1996, *Phys. Rev. A* **54**, 1793.

Kuang, L.-M. and X. Chen, 1994a, *Phys. Lett. A* **186**, 8.

Kuang, L.-M. and X. Chen, 1994b, *Phys. Rev. A* **50**, 4228.

Kuang, L.-M., F.-B. Wang and Y.-G. Zhou, 1993, *Phys. Lett. A* **183**, 1.

Kuang, L.-M., F.-B. Wang and Y.-G. Zhou, 1994, *J. Mod. Opt.* **41**, 1307.

Kühn, H., D.-G. Welsch and W. Vogel, 1994, *J. Mod. Opt.* **41**, 1607.

Kühn, H., D.-G. Welsch and W. Vogel, 1995, *Phys. Rev. A* **51**, 4240.

Kulish, P. and E. V. Damaskinsky, 1990, *J. Phys. A* **23**, L415.

Kumar, A. and P. S. Gupta, 1997, *Opt. Commun.* **136**, 441.

Kurtsiefer, C., O. Dross, D. Voigt, C. R. Ekstrom, T. Pfau and J. Mlynek, 1997, *Phys. Rev. A* **55**, R2539.

Kwiat, P. G. and R. Y. Chiao, 1991, *Phys. Rev. Lett.* **66**, 588.

Kwiat, P. G., K. Mattle, H. Weinfurter and A. Zeilinger, 1995, *Phys. Rev. Lett.* **75**, 4337.

Kwiat, P. G., A. M. Steinberg and R. Y. Chiao, 1993a, *Phys. Rev. A* **47**, R2472.

Kwiat, P. G., A. M. Steinberg and R. Y. Chiao, 1993b, *Phys. Rev. A* **48**, 3999.

Kwiat, P. G., A. M. Steinberg and R. Y. Chiao, 1994, *Phys. Rev. A* **49**, 61.

Kwiat, P. G., W. A. Vareka, C. K. Hong, H. Nathel and R. Y. Chiao, 1990, *Phys. Rev. A* **41**, 2910.

Lai, Y. and H. A. Haus, 1989, *Quant. Opt.* **1**, 199.

Landau, L. D. and E. M. Lifshitz, 1964, Statistical Physics, Nauka, Moscow, in Russian.

Landau, L. D. and E. M. Lifshitz, 1977, Quantum Mechanics, 3rd ed., Pergamon, Oxford.

Lane, A. S., S. L. Braunstein and C. M. Caves, 1993, *Phys. Rev. A* **47**, 1667.

Lanzerotti, M. Y. and A. L. Gaeta, 1995, *Phys. Rev. A* **51**, 4057.

Lanzerotti, M. Y., A. L. Gaeta and R. W. Boyd, 1995, *Phys. Rev. A* **51**, 3182.

Lanzerotti, M. Y., R. W. Schirmer, A. L. Gaeta and G. S. Agarwal, 1996, *Phys. Rev. Lett.* **77**, 2202.

Larchuk, T. S., R. A. Campos, J. G. Rarity, P. R. Tapster, E. Jakeman, B. E. A. Saleh and M. C. Teich, 1993, *Phys. Rev. Lett.* **70**, 1603.

Larchuk, T. S., M. C. Teich and B. E. A. Saleh, 1995, *Phys. Rev. A* **52**, 4145.

Lax, M., 1968, Fluctuations and Coherence Phenomena in Classical and Quantum Physics, Gordon and Breach, New York.

Lee, C. T., 1991, *Phys. Rev. A* **44**, R2775.

Leibfried, D., D. M. Meekhof, B. E. King, C. Monroe, W. M. Itano and D. J. Wineland, 1996, *Phys. Rev. Lett.* **77**, 4281.

Lenef, A., T. D. Hammond, E. T. Smith, M. S. Chapman, R. A. Rubenstein and D. E. Pritchard, 1997, *Phys. Rev. Lett.* **78**, 760.

Leng, B., S.-D. Du and C.-D. Gong, 1995, *J. Mod. Opt.* **42**, 435.

Lenz, G., P. Meystre and E. M. Wright, 1994, *Phys. Rev. A* **50**, 1681.

Leong, K. W. and J. H. Shapiro, 1986, *Opt. Commun.* **58**, 73.

Leonhardt, U., 1993, *Phys. Rev. A* **48**, 3265.

Leonhardt, U., 1995, *Phys. Rev. Lett.* **74**, 4101.

Leonhardt, U., 1996, *Phys. Rev. A* **53**, 2998.

Leonhardt, U., 1997, Measuring the Quantum State of Light, Cambridge University Press, Cambridge.

Leonhardt, U., B. Böhmer and H. Paul, 1995, *Opt. Commun.* **119**, 296.

Leonhardt, U. and I. Jex, 1994, *Phys. Rev. A* **49**, R1555.

Leonhardt, U. and M. Munroe, 1996, *Phys. Rev. A* **54**, 3682.

Leonhardt, U., M. Munroe, T. Kiss, Th. Richter and M. G. Raymer, 1996, *Opt. Commun.* **127**, 144.

Leonhardt, U. and H. Paul, 1993a, *Phys. Rev. A* **47**, R2460.

Leonhardt, U. and H. Paul, 1993b, *J. Mod. Opt.* **40**, 1745.

Leonhardt, U. and H. Paul, 1993c, *Phys. Scr.* **T48**, 45.

Leonhardt, U. and H. Paul, 1993d, *Phys. Rev. A* **48**, 4598.

Leonhardt, U. and H. Paul, 1994a, *Phys. Rev. Lett.* **72**, 4086.

Leonhardt, U. and H. Paul, 1994b, *J. Mod. Opt.* **41**, 1427.

Leonhardt, U., H. Paul and G. M. D'Ariano, 1995, *Phys. Rev. A* **52**, 4899.

Leonhardt, U., J. A. Vaccaro, B. Böhmer and H. Paul, 1995, *Phys. Rev. A* **51**, 84.

Lerner, E. C., 1968, *Nuovo Cimento B* **56**, 183.

Lerner, E. C., H. W. Huang and G. E. Walters, 1970, *J. Math. Phys.* **11**, 1679.

Lévy-Leblond, J.-M., 1973, *Rev. Mex. Fis.* **22**, 15.

Lévy-Leblond, J.-M., 1976, *Ann. Phys.* (N. Y.) **101**, 319.

Lévy-Leblond, J.-M., 1977, *Phys. Lett.* **64A**, 159.

Lévy-Leblond, J.-M., 1986, *Am. J. Phys.* **54**, 135.

Lewis, H. R., W. E. Lawrence and J. D. Harris, 1996, *Phys. Rev. Lett.* **77**, 5157.

Lewis Jr., H. R. and W. B. Riesenfeld, 1969, *J. Math. Phys.* **10**, 1458.

Li, Y., M. Baba and M. Matsuoka, 1997, *Phys. Rev. A* **55**, 3177.

Lindblad, G., 1976, *Commun. Math. Phys.* **48**, 199.

Lindner, A., D. Reiss, G. Wassiliadis and H. Freese, 1996, *Phys. Lett. A* **218**, 1.

Liu, S. and Y. Chen, 1995, *J. Opt. Soc. Am. B* **12**, 829.

Lohmann, A. W., J. Ojeda-Castaneda and N. Streibl, 1983, *Opt. Acta* **30**, 1259.

London, F., 1926, *Z. Phys.* **37**, 915.

London, F., 1927, *Z. Phys.* **40**, 193.

Loss, D. and K. Mullen, 1991, *Phys. Rev. A* **43**, 2129.

Loudon, R., 1973, The Quantum Theory of Light, 1st ed., Clarendon Press, Oxford.

Loudon, R., 1989, Coherence and Quantum Optics VI, eds. J. H. Eberly, L. Mandel and E. Wolf, Plenum Press, New York, p. 703.

Loudon, R. and P. L. Knight, 1987, *J. Mod. Opt.* **34**, 709.

Louisell, W. H., 1963, *Phys. Lett.* **7**, 60.

Louisell, W. H., 1964, Radiation and Noise in Quantum Electronics, McGraw-Hill, New York.

Louisell, W. H., 1973, Quantum Statistical Properties of Radiation, J. Wiley, New York.

Louisell, W. H., A. Yariv and A. E. Siegman, 1961, *Phys. Rev.* **124**, 1646.

Luis, A. and J. Peřina, 1996a, *Phys. Rev. A* **53**, 1886.

Luis, A. and J. Peřina, 1996b, *Phys. Rev. A* **54**, 4564.

Luis, A. and J. Peřina, 1996c, *Phys. Rev. Lett.* **76**, 4340.

Luis, A. and J. Peřina, 1996d, *Quant. Semiclass. Opt.* **8**, 873.

Luis, A. and J. Peřina, 1996e, *Quant. Semiclass. Opt.* **8**, 887.

Luis, A. and J. Peřina, 1998, *J. Phys. A: Math. Gen.* **31**, 1423.

Luis, A. and L. L. Sánchez-Soto, 1993a, *Phys. Rev. A* **47**, 1492.

Luis, A. and L. L. Sánchez-Soto, 1993b, *Phys. Rev. A* **48**, 4702.

Luis, A. and L. L. Sánchez-Soto, 1993c, *Quant. Opt.* **5**, 33.

Luis, A. and L. L. Sánchez-Soto, 1995, *Phys. Rev. A* **51**, 861.

Luis, A. and L. L. Sánchez-Soto, 1996, *Phys. Rev. A* **53**, 495.

Luis, A. and L. L. Sánchez-Soto, 1997a, *Opt. Commun.* **133**, 159.

Luis, A. and L. L. Sánchez-Soto, 1997b, *Phys. Rev. A* **56**, 994.

Luis, A., L. L. Sánchez-Soto and R. Tanaś, 1995, *Phys. Rev. A* **51**, 1634.

Lukš, A. and V. Peřinová, 1987, *Czech. J. Phys. B* **37**, 1224.

Lukš, A. and V. Peřinová, 1990, *Phys. Rev. A* **42**, 5805.

Lukš, A. and V. Peřinová, 1991, *Czech. J. Phys.* **41**, 1205.

Lukš, A. and V. Peřinová, 1992, *Phys. Rev. A* **45**, 6710.

Lukš, A. and V. Peřinová, 1993a, *Quant. Opt.* **5**, 287.

Lukš, A. and V. Peřinová, 1993b, *Phys. Scr.* **T48**, 94.

Lukš, A. and V. Peřinová, 1994, *Quant. Opt.* **6**, 125.

Lukš, A. and V. Peřinová, 1996, *J. Phys. A: Math. Gen.* **29**, 4665.

Lukš, A. and V. Peřinová, 1997a, *Phys. Lett. A* **229**, 8.

Lukš, A. and V. Peřinová, 1997b, *Fine Mechanics and Optics* **42**, 25.

Lukš, A. and V. Peřinová, 1997c, *Acta Phys. Slov.* **47**, 315.

Lukš, A. and V. Peřinová, 1997d, Wave and Quantum Aspects of Contemporary Optics, ed. M. Pluta, SPIE, Bellingham, WA, in print.

Lukš, A., V. Peřinová and Z. Hradil, 1988, *Acta Phys. Pol.* **74**, 713.

Lukš, A., V. Peřinová and J. Křepelka, 1992a, *Czech. J. Phys.* **42**, 59.

Lukš, A., V. Peřinová and J. Křepelka, 1992b, *Phys. Rev. A* **46**, 489.

Lukš, A., V. Peřinová and J. Křepelka, 1994a, *Phys. Rev. A* **50**, 818.

Lukš, A., V. Peřinová and J. Křepelka, 1994b, *J. Mod. Opt.* **41**, 2325.

Lukš, A., V. Peřinová and J. Křepelka, 1997, unpublished.

Lukš, A., V. Peřinová and J. Peřina, 1988, *Opt. Commun.* **67**, 149.

Lütkenhaus, N. and S. M. Barnett, 1995, *Phys. Rev. A* **51**, 3340.

Lutterbach, L. G. and L. Davidovich, 1997, *Phys. Rev. Lett.* **78**, 2547.

Lynch, R., 1986, *J. Opt. Soc. Am. B* **3**, 1106.

Lynch, R., 1987, *J. Opt. Soc. Am. B* **4**, 1723.

Lynch, R., 1988, *Opt. Commun.* **67**, 67.

Lynch, R., 1990, *Phys. Rev. A* **41**, 2841.

Lynch, R., 1993, *Phys. Rev. A* **47**, 1576.

Lynch, R., 1995, *Phys. Rep.* **256**, 367.

Ma, X. and W. Rhodes, 1991, *Phys. Rev. A* **43**, 2576.

MacFarlane, A. J., 1989, *J. Phys. A* **22**, 4581.

Malkin, I. A. and V. I. Man'ko, 1979, Dynamical Symmetries and Coherent States of Quantum Systems, Nauka, Moscow, in Russian.

Mancini, S., P. Tombesi and V. I. Man'ko, 1997, *Europhys. Lett.* **37**, 79.

Mandel, L., 1963, Progress in Optics, Vol. 2, ed. E. Wolf, North-Holland, Amsterdam, p. 181.

Mandel, L., 1966, *Phys. Rev.* **152**, 438.

Mandel, L., 1976, Progress in Optics, Vol. 13, ed. E. Wolf, North-Holland, Amsterdam, p. 27.

Mandel, L., 1982a, *Phys. Rev. Lett.* **49**, 136.

Mandel, L., 1982b, *Opt. Commun.* **42**, 437.

Mandel, L., 1983, *Phys. Rev. A* **28**, 929.

Mandel, L., 1991, *Opt. Lett.* **16**, 1882.

Mandel, L. and E. Wolf, 1995, Optical Coherence and Quantum Optics, Cambridge University Press, Cambridge.

Martinez, J. C., 1995, *Phys. Lett. A* **204**, 205.

Massar, S. and S. Popescu, 1995, *Phys. Rev. Lett.* **74**, 1259.

Mathai, A. M., 1993, A Handbook of Generalized Special Functions for Statistical and Physical Sciences, Clarendon Press, Oxford.

Matsko, A. B., S. P. Vyatchanin, H. Mabuchi and H. J. Kimble, 1994, *Phys. Lett. A* **192**, 175.

Matteucci, G., 1990, *Am. J. Phys.* **58**, 1143.

Matthys, D. R. and E. T. Jaynes, 1980, *J. Opt. Soc. Am.* **70**, 263.

McAlister, D. F., M. Beck, L. Clarke, A. Mayer, M. G. Raymer, 1995, *Opt. Lett.* **20**, 1181.

McNeil, K. J., P. D. Drummond and D. F. Walls, 1978, *Opt. Commun.* **27**, 292.

McNeil, K. J. and D. F. Walls, 1974, *J. Phys. A* **7**, 617.

Mead, C. A. and D. G. Truhlar, 1979, *J. Chem. Phys.* **70**, 2384.

Meekhof, D. M., C. Monroe, B. E. King, W. M. Itano and D. J. Wineland, 1996, *Phys. Rev. Lett.* **76**, 1796.

Mehta, C. L., 1965, *Nuovo Cimento* **36**, 202.

Mehta, C. L., 1968, *J. Opt. Soc. Am.* **58**, 1233.

Mehta, C. L., A. K. Roy and G. M. Saxena, 1992, *Phys. Rev. A* **46**, 1565.

Mehta, M. L., 1987, *J. Math. Phys.* **28**, 781.

Mendaš, I., 1997, *Phys. Rev. A* **55**, 1514.

Mendaš, I. and D. B. Popović, 1994, *Phys. Rev. A* **50**, 947.

Mendaš, I. and D. B. Popović, 1995, *Phys. Rev. A* **52**, 4356.

Messiah, A., 1961, Quantum Mechanics, Vol. I, J. Wiley, New York.

Messiah, A., 1962, Quantum Mechanics, Vol. II, J. Wiley, New York.

Mewes, M.-O., M. R. Andrews, D. M. Kurn, D. S. Durfee, C. G. Townsend and W. Ketterle, 1997, *Phys. Rev. Lett.* **78**, 582.

Meystre, P., 1992, Progress in Optics, Vol. 30, ed. E. Wolf, North-Holland, Amsterdam, p. 261.

Meystre, P., A. Quattropani and H. P. Baltes, 1974, *Phys. Lett. A* **49**, 85.

Meystre, P. and M. Sargent III, 1991, Elements of Quantum Optics, Springer-Verlag, Berlin.

Michler, M., K. Mattle, H. Weinfurter and A. Zeilinger, 1996, *Phys. Rev. A* **53**, R1209.

Milburn, G. J., 1984, *J. Phys. A* **17**, 737.

Milburn, G. J., 1986, *Phys. Rev. A* **33**, 674.

Milburn, G. J., 1989, Squeezed and Nonclassical Light, eds. P. Tombesi and E. R. Pike, Plenum Press, New York, p. 151.

Milburn, G. J., W.-Y. Chen and K. R. Jones, 1994, *Phys. Rev. A* **50**, 801.

Milburn, G. J. and C. A. Holmes, 1986, *Phys. Rev. Lett.* **56**, 2237.

Milburn, G. J. and D. F. Walls, 1983a, *Phys. Rev. A* **28**, 2065.

Milburn, G. J. and D. F. Walls, 1983b, *Phys. Rev. A* **28**, 2646.

Milburn, G. J. and D. F. Walls, 1988, *Phys. Rev. A* **38**, 1087.

Millane, R. P., 1990, *J. Opt. Soc. Am. A* **7**, 394.

Milonni, P. W., M. L. Shih and J. R. Ackerhalt, eds., 1987, Chaos in Laser–Matter Interactions, World Scientific, Singapore.

Miniatura, Ch., F. Perales, G. Vassilev, J. Reinhardt, J. Robert and J. Baudon, 1991, *J. Physique* **1**, 425.

Miranowicz, A., J. Bajer, A. Ekert and W. Leoński, 1997, *Acta Phys. Slov.* **47**, 319.

Miranowicz, A., T. Opatrný and J. Bajer, 1997, Quantum Optics and the Spectroscopy of Solids, eds. T. Hakioğlu and A. S. Shumovsky, Kluwer, Dordrecht, p. 225.

Miranowicz, A., R. Tanaś and S. Kielich, 1990, *Quant. Opt.* **2**, 253.

Misell, D. L., 1973a, *J. Phys. D: Appl. Phys.* **6**, 2200.

Misell, D. L., 1973b, *J. Phys. D: Appl. Phys.* **6**, 2217.

Misell, D. L., R. E. Burge and A. H. Greenaway, 1974, *Nature* **247**, 401.

Misell, D. L. and A. H. Greenaway, 1974a, *J. Phys. D* **7**, 832.

Misell, D. L. and A. H. Greenaway, 1974b, *J. Phys. D* **7**, 1660.

Mišta, L. and J. Peřina, 1977a, *Acta Phys. Pol. A* **52**, 425.

Mišta, L. and J. Peřina, 1977b, *Czech. J. Phys. B* **27**, 831.

Mišta, L. and J. Peřina, 1978, *Czech. J. Phys. B* **28**, 392.

Mišta, L., V. Peřinová, J. Peřina and Z. Braunerová, 1977, *Acta Phys. Pol. A* **51**, 739.

Miura, N. and N. Baba, 1996, *Opt. Lett.* **13**, 979.

Mogilevtsev, D. and S. Ya. Kilin, 1996, *Opt. Commun.* **132**, 452.

Monken, C. H., D. Branning and L. Mandel, 1996, Coherence and Quantum Optics VII, eds. J. H. Eberly, L. Mandel and E. Wolf, Plenum Press, New York, p. 701.

Monken, C. H., A. Garuccio, D. Branning, J. R. Torgerson, F. Narducci and L. Mandel, 1996, *Phys. Rev. A* **53**, 1782.

Mostafazadeh, A., 1997a, *Phys. Rev. A* **55**, 1653.

Mostafazadeh, A., 1997b, *Phys. Lett. A* **232**, 395.

Moyal, J. E., 1949, *Proc. Cambridge Phil. Soc.* **45**, 99.

Mukunda, N., 1979, *Am. J. Phys.* **47**, 182.

Munro, W. J. and M. D. Reid, 1995, *Phys. Rev. A* **52**, 2388.

Munroe, M., D. Boggavarapu, M. E. Anderson and M. G. Raymer, 1995, *Phys. Rev. A* **52**, R924.

Myška, R., 1996, *Acta Phys. Slov.* **46**, 457.

Nakajima, N. and T. Asakura, 1982, *Optik* **60**, 289.

Naraschewski, M., H. Wallis, A. Schenzle, J. I. Cirac and P. Zoller, 1996, *Phys. Rev. A* **54**, 2185.

Nath, R. and P. Kumar, 1990, *Opt. Commun.* **76**, 51.

Nath, R. and P. Kumar, 1991, *J. Mod. Opt.* **38**, 263.

Nath, R. and P. Kumar, 1996, *J. Mod. Opt.* **43**, 7.

Natterer, F., 1986, The Mathematics of Computerized Tomography, J. Wiley, New York.

Negrete-Regagnon, P., 1996, *Opt. Lett.* **21**, 275.

Nelson, C. A., 1993, Symmetries in Science VI, ed. B. Gruber, Plenum Press, New York, p. 523.

Nelson, C. A. and M. H. Fields, 1995, *Phys. Rev. A* **51**, 2410.

Newton, R. G., 1980, *Ann. Phys.* (N. Y.) **124**, 327.

Nibbering, E. T. J., M. A. Franco, B. S. Prade, G. Grillon, J.-P. Chambaret, A. Mysyrowicz, 1996, *J. Opt. Soc. Am. B* **13**, 317.

Nienhuis, G. and S. J. van Enk, 1993, *Phys. Scr.* **T48**, 87.

Nieto, M. M., 1977, *Phys. Lett.* **60A**, 401.

Nieto, M. M., 1993, *Phys. Scr.* **T48**, 5.

Nieto-Vesperinas, M., 1991, Scattering and Diffraction in Physical Optics, J. Wiley, New York.

Nieto-Vesperinas, M. and O. Hignette, 1979, *Opt. Pura y Apl.* **12**, 175.

Noel, M. W. and C. R. Stroud Jr., 1995, *Phys. Rev. Lett.* **75**, 1252.

Noel, M. W. and C. R. Stroud Jr., 1996, *Phys. Rev. Lett.* **77**, 1913.

Noh, J. W., A. Fougères and L. Mandel, 1991, *Phys. Rev. Lett.* **67**, 1426.

Noh, J. W., A. Fougères and L. Mandel, 1992a, *Phys. Rev. A* **45**, 424.

Noh, J. W., A. Fougères and L. Mandel, 1992b, *Phys. Rev. A* **46**, 2840.

Noh, J. W., A. Fougères and L. Mandel, 1993a, *Phys. Rev. A* **47**, 4535.

Noh, J. W., A. Fougères and L. Mandel, 1993b, *Phys. Rev. A* **47**, 4541.

Noh, J. W., A. Fougères and L. Mandel, 1993c, *Phys. Rev. Lett.* **71**, 2579.

Noh, J. W., A. Fougères and L. Mandel, 1993d, *Phys. Scr.* **T48**, 29.

Noh, J. W., A. Fougères and L. Mandel, 1993e, *Phys. Rev. A* **48**, 1719.

Nussenzveig, H. M., 1972, Causality and Dispersion Relations, Academic Press, London.

Nussenzveig, H. M., 1973, Introduction to Quantum Optics, Gordon and Breach, London.

Obada, A.-S. F., O. M. Yassin and S. M. Barnett, 1997, *J. Mod. Opt.* **44**, 149.

Oberthaler, M. K., R. Abfalterer, S. Bernet, J. Schmiedmayer and A. Zeilinger, 1996, *Phys. Rev. Lett.* **77**, 4980.

Opatrný, T., 1994, *J. Phys. A: Math. Gen.* **27**, 7201.

Opatrný, T., 1995, *J. Phys. A: Math. Gen.* **28**, 6961.

Opatrný, T., V. Bužek, J. Bajer and G. Drobný, 1995, *Phys. Rev. A* **52**, 2419.

Opatrný, T., A. Miranowicz and J. Bajer, 1996, *J. Mod. Opt.* **43**, 417.

Opatrný, T. and D.-G. Welsch, 1997, *Phys. Rev. A* **55**, 1462.

Opatrný, T., D.-G. Welsch and W. Vogel, 1997a, *Optics Commun.* **134**, 112.

Opatrný, T., D.-G. Welsch and W. Vogel, 1997b, *Phys. Rev. A* **55**, 1416.

Oppenheim, A. V. and R. W. Schafer, 1975, Digital Signal Processing, Prentice-Hall, Englewood Cliffs, NJ.

Orłowski, A. and H. Paul, 1994, *Phys. Rev. A* **50**, R921.

Orłowski, A., H. Paul and B. Böhmer, 1997, *Opt. Commun.* **138**, 311.

Ou, Z. Y., 1996, *Phys. Rev. Lett.* **77**, 2352.

Ou, Z. Y., 1997a, *Quant. Semiclass. Opt.* **9**, 599.

Ou, Z. Y., 1997b, *Phys. Rev. A* **55**, 2598.

Ou, Z. Y. and L. Mandel, 1988a, *Phys. Rev. Lett.* **59**, 2046.

Ou, Z. Y. and L. Mandel, 1988b, *Phys. Rev. Lett.* **61**, 54.

Ou, Z. Y., L. J. Wang and L. Mandel, 1989, *Phys. Rev. A* **40**, 1428.

Ou, Z. Y., L. J. Wang and L. Mandel, 1990, *J. Opt. Soc. Am. B* **7**, 211.

Ou, Z. Y., L. J. Wang, X. Y. Zou and L. Mandel, 1990, *Phys. Rev. A* **41**, 566.

Ou, Z. Y., X. Y. Zou, L. J. Wang and L. Mandel, 1990a, *Phys. Rev. A* **42**, 2957.

Ou, Z. Y., X. Y. Zou, L. J. Wang and L. Mandel, 1990b, *Phys. Rev. Lett.* **65**, 321.

Pancharatnam, S., 1956, *Proc. Ind. Acad. Sci. A* **44**, 247.

Pancharatnam, S., 1975, Collected Works of S. Pancharatnam, University Press, Oxford.

Paprzycka, M. and R. Tanaś, 1992, *Quant. Opt.* **4**, 331.

Paris, M. G. A., 1997, *Phys. Lett. A* **225**, 23.

Parmenter, R. W. and R. H. Valentine, 1996, *Phys. Lett. A* **219**, 7.

Pati, A. K. and S. V. Lawande, 1995, *Phys. Rev. A* **51**, 5012.

Pati, A. K. and S. V. Lawande, 1996, *Phys. Lett. A* **223**, 233.

Paul, H., 1974, *Fortschr. Phys.* **22**, 657.

Paul, H., 1986, *Rev. Mod. Phys.* **58**, 209.

Paul, H., 1991, *Quant. Opt.* **3**, 169.

Paul, H. and W. Brunner, 1981, *Ann. Physik* **38**, 89.

Paul, H., U. Leonhardt and G. M. D'Ariano, 1995, *Acta Phys. Slov.* **45**, 261.

Paul, H., P. Törmä, T. Kiss and I. Jex, 1996, *Phys. Rev. Lett.* **76**, 2464.

Pauli, W., 1980, General Principles of Quantum Mechanics, Springer-Verlag, Berlin.

Pegg, D. T., 1991, *J. Phys. A: Math. Gen.* **24**, 3031.

Pegg, D. T. and S. M. Barnett, 1988, *Europhys. Lett.* **6**, 483.

Pegg, D. T. and S. M. Barnett, 1989, *Phys. Rev. A* **39**, 1665.

Pegg, D. T. and S. M. Barnett, 1991, *Phys. Rev. A* **43**, 2579.

Pegg, D. T. and S. M. Barnett, 1997, *J. Mod. Opt.* **44**, 225.

Pegg, D. T. and J. A. Vaccaro, 1995, *Phys. Rev. A* **51**, 859.

Pegg, D. T., J. A. Vaccaro and S. M. Barnett, 1990, *J. Mod. Opt.* **37**, 1703.

Peiponen, K.-E., E. M. Vartiainen and T. Asakura, 1997, Progress in Optics, Vol. 37, ed. E. Wolf, North-Holland, Amsterdam.

Pereira, S. F., Z. Y. Ou and H. J. Kimble, 1994, *Phys. Rev. Lett.* **72**, 214.

Perelomov, A. M., 1972, *Commun. Math. Phys.* **26**, 222.

Perelomov, A. M., 1986, Generalized Coherent States and Their Applications, Springer-Verlag, Berlin.

Perez-Ilzarbe, M. J., M. Nieto-Vesperinas and R. Navarro, 1990, *J. Opt. Soc. Am. A* **7**, 434.

Peřina, J., 1963, *Opt. Acta* **10**, 333.

Peřina, J., 1972, Coherence of Light, Van Nostrand, London.

Peřina, J., 1981a, *Opt. Acta* **28**, 325.

Peřina, J., 1981b, *Opt. Acta* **28**, 1529.

Peřina, J., 1985, Coherence of Light, 2nd ed., D. Reidel, Dordrecht.

Peřina, J., 1991, Quantum Statistics of Linear and Nonlinear Optical Phenomena, 2nd ed., Kluwer, Dordrecht.

Peřina, J. and J. Bajer, 1990, *Phys. Rev. A* **41**, 516.

Peřina, J., J. Bajer, J. Křepelka and Z. Hradil, 1987, *J. Mod. Opt.* **34**, 965.

Peřina, J., R. Horák, Z. Hradil, C. Sibilia and M. Bertolotti, 1989, *J. Mod. Opt.* **36**, 571.

Peřina, J., Z. Hradil and B. Jurčo, 1994, Quantum Optics and Fundamentals of Physics, Kluwer, Dordrecht.

Peřina, J. and J. Peřina Jr., 1995, *Quant. Semiclass. Opt.* **7**, 863.

Peřina, J. and J. Peřina Jr., 1996, *J. Mod. Opt.* **43**, 1951.

Peřina, J. and V. Peřinová, 1976, *Czech. J. Phys. B* **26**, 489.

Peřina, J., V. Peřinová and R. Horák, 1973, *Czech. J. Phys. B* **23**, 993.

Peřina, J., V. Peřinová and J. Koďousek, 1984, *Opt. Commun.* **49**, 210.

Peřina, J., V. Peřinová, C. Sibilia and M. Bertolotti, 1984, *Opt. Commun.* **49**, 285.

Peřina, J., B. E. A. Saleh and M. C. Teich, 1983, *Opt. Commun.* **48**, 212.

Peřina, J. and J. Tillich, 1966, *Acta Univ. Palack. Olom.* **21**, 153.

Peřina Jr., J., 1993, *Czech. J. Phys.* **43**, 615.

Peřinová, V., 1981, *Opt. Acta* **28**, 747.

Peřinová, V. and J. Křepelka, 1993, *Phys. Rev. A* **48**, 3881.

Peřinová, V., J. Křepelka and J. Peřina, 1986, *Opt. Acta* **33**, 1263.

Peřinová, V. and A. Lukš, 1990, *Phys. Rev. A* **41**, 414.

Peřinová, V. and A. Lukš, 1994, Progress in Optics, Vol. 33, ed. E. Wolf, North-Holland, Amsterdam, p. 129.

Peřinová, V. and A. Lukš, 1995, *Acta Phys. Slov.* **45**, 395.

Peřinová, V. and A. Lukš, 1996, *Acta Phys. Slov.* **46**, 481.

Peřinová, V., A. Lukš and M. Kárská, 1990, *J. Mod. Opt.* **37**, 1055.

Peřinová, V., A. Lukš and J. Křepelka, 1994, *Laser Physics* **4**, 717.

Peřinová, V., A. Lukš and J. Křepelka, 1996a, *Phys. Rev. A* **53**, 525.

Peřinová, V., A. Lukš and J. Křepelka, 1996b, *Phys. Rev. A* **54**, 821.

Peřinová, V., A. Lukš and J. Křepelka, 1997a, *Quant. Semiclass. Opt.* **9**, 465.

Peřinová, V., A. Lukš and J. Křepelka, 1997b, *Quant. Semiclass. Opt.* **9**, 995.

Peřinová, V., A. Lukš, J. Křepelka, C. Sibilia and M. Bertolotti, 1991, *J. Mod. Opt.* **38**, 2429.

Peřinová, V., A. Lukš, J. Křepelka, C. Sibilia and M. Bertolotti, 1996, *J. Mod. Opt.* **43**, 13.

Peřinová, V., A. Lukš and P. Szlachetka, 1989, *J. Mod. Opt.* **36**, 1435.

Peřinová, V. and J. Peřina, 1978a, *Czech. J. Phys. B* **28**, 1183.

Peřinová, V. and J. Peřina, 1978b, *Czech. J. Phys. B* **28**, 1196.

Peřinová, V. and J. Peřina, 1978c, *Czech. J. Phys. B* **28**, 306.

Peřinová, V. and J. Peřina, 1981, *Opt. Acta* **28**, 769.

Peřinová, V., J. Peřina, P. Szlachetka and S. Kielich, 1979a, *Acta Phys. Pol. A* **56**, 267.

Peřinová, V., J. Peřina, P. Szlachetka and S. Kielich, 1979b, *Acta Phys. Pol. A* **56**, 275.

Peřinová, V., V. Vrana and A. Lukš, 1995, *Phys. Rev. A* **51**, 2499.

Peshkin, M. and H. J. Lipkin, 1995, *Phys. Rev. Lett.* **74**, 2847.

Pfau, T., Ch. Kurtsiefer, C. R. Ekstrom and J. Mlynek, 1996, Coherence and Quantum Optics VII, eds. J. H. Eberly, L. Mandel and E. Wolf, Plenum Press, New York, p. 123.

Phoenix, S. J. D., 1990, *Phys. Rev. A* **41**, 5132.

Pieczonková, A., 1982a, *Czech. J. Phys. B* **32**, 831.

Pieczonková, A., 1982b, *Opt. Acta* **29**, 1509.

Pieczonková, A. and J. Peřina, 1981, *Czech. J. Phys. B* **31**, 837.

Pittman, T. B., 1995, *Phys. Lett. A* **204**, 193.

Pittman, T. B., Y. H. Shih, A. V. Sergienko and M. H. Rubin, 1995, *Phys. Rev. A* **51**, 3495.

Pittman, T. B., Y. H. Shih, D. V. Strekalov and A. V. Sergienko, 1995, *Phys. Rev. A* **52**, R3429.

Pittman, T. B., Y. H. Shih, D. V. Strekalov, A. V. Sergienko and M. H. Rubin, 1996, Fourth International Conference on Squeezed States and Uncertainty Relations, eds. D. Han, K. Peng, Y. S. Kim and V. I. Man'ko, NASA, Greenbelt, MD, p. 139.

Pittman, T. B., D. V. Strekalov, D. N. Klyshko, M. H. Rubin, A. V. Sergienko and Y. H. Shih, 1996, *Phys. Rev. A* **53**, 2804.

Pittman, T. B., D. V. Strekalov, A. Migdall, M. H. Rubin, A. V. Sergienko and Y. H. Shih, 1996, *Phys. Rev. Lett.* **77**, 1917.

Popov, V. N. and V. S. Yarunin, 1973, *The Leningrad University Journal: Physics* **22**, 7.

Popov, V. N. and V. S. Yarunin, 1992, *J. Mod. Opt.* **39**, 1525.

Poyatos, J. F., J. I. Cirac, R. Blatt and P. Zoller, 1996, *Phys. Rev. A* **54**, 1532.

Pritchard, D. E., M. S. Chapman, T. D. Hammond, A. Lenef, R. A. Rubenstein, J. Schmiedmayer and E. T. Smith, 1996, Coherence and Quantum Optics VII, eds. J. H. Eberly, L. Mandel and E. Wolf, Plenum Press, New York, p. 133.

Pták, P. and S. Pulmannová, 1991, Orthomodular Structures as Quantum Logics, Kluwer, Dordrecht.

Radcliffe, J. M., 1971, *J. Phys. A* **4**, 313.

Rangwala, S. and S. M. Roy, 1994, *Phys. Lett. A* **190**, 1.

Rao, R. C., 1973, Linear Statistical Inference and Its Applications, J. Wiley, New York.

Rarity, J. G., J. Burnett, P. R. Tapster and R. Paschotta, 1993, *Europhys. Lett.* **22**, 95.

Rarity, J. G. and P. R. Tapster, 1989, *J. Opt. Soc. Am. B* **6**, 1221.

Rarity, J. G. and P. R. Tapster, 1990a, *Phys. Rev. A* **41**, 5139.

Rarity, J. G. and P. R. Tapster, 1990b, *Phys. Rev. Lett.* **64**, 2495.

Rarity, J. G. and P. R. Tapster, 1992, *Phys. Rev. A* **45**, 2052.

Rarity, J. G., P. R. Tapster, E. Jakeman, T. Larchuk, R. A. Campos, M. C. Teich and B. E. A. Saleh, 1990, *Phys. Rev. Lett.* **65**, 1348.

Rasel, E. M., M. K. Oberthaler, H. Batelaan, J. Schmiedmayer and A. Zeilinger, 1995, *Phys. Rev. Lett.* **75**, 2633.

Rathjen, C., 1995, *J. Opt. Soc. Am. A* **12**, 1997.

Rauch, H., 1993, *Phys. Lett. A* **173**, 240.

Rauch, H., 1995, *Nuovo Cimento* **110B**, 557.

Raymer, M. G., J. Cooper and M. Beck, 1993, *Phys. Rev. A* **48**, 4617.

Raymer, M. G., D. F. McAlister and U. Leonhardt, 1996, *Phys. Rev. A* **54**, 2397.

Řeháček, J. and J. Peřina, 1996a, *Opt. Commun.* **125**, 82.

Řeháček, J. and J. Peřina, 1996b, *Opt. Commun.* **132**, 549.

Rempe, G., H. Walther and N. Klein, 1987, *Phys. Rev. Lett.* **58**, 353.

Reynaud, S., A. Heidmann, E. Giacobino and C. Fabre, 1992, Progress in Optics, Vol. 30, ed. E. Wolf, North-Holland, Amsterdam, p. 1.

Ribeiro, P. H. S. and G. A. Barbosa, 1996, *Phys. Rev. A* **54**, 3489.

Ribeiro, P. H. S. and G. A. Barbosa, 1997, *Opt. Commun.* **139**, 139.

Ribeiro, P. H. S., S. Pádua, J. C. Machado da Silva and G. A. Barbosa, 1994, *Phys. Rev. A* **49**, 4176.

Ribeiro, P. H. S., S. Pádua, J. C. Machado da Silva and G. A. Barbosa, 1995, *Phys. Rev. A* **51**, 1631.

Richter, Th., 1996a, *Phys. Rev. A* **54**, 2499.

Richter, Th., 1996b, *Phys. Rev. A* **53**, 1197.

Richter, Th., 1996c, *Phys. Lett. A* **211**, 327.

Riegler, P. and K. Wódkiewicz, 1994, *Phys. Rev. A* **49**, 1387.

Riehle, F., Th. Kisters, A. Wille, J. Helmcke and Ch. J. Borde, 1991, *Phys. Rev. Lett.* **67**, 177.

Risken, H., 1984, The Fokker–Planck Equation, Springer-Verlag, Berlin.

Rocca, F. and M. Sirugue, 1973, *Commun. Math. Phys.* **34**, 111.

Roch, J.-F., K. Vigneron, Ph. Grelu, A. Sinatra, J.-Ph. Poizat and Ph. Grangier, 1997, *Phys. Rev. Lett.* **78**, 634.

Rühl, A., M. Naraschewski, A. Schenzle and H. Wallis, 1997, *Phys. Rev. Lett.* **78**, 4143.

Roman, P. and A. S. Marathay, 1963, *Nuovo Cimento* **30**, 1452.

Rosenblum, M., 1962, *Proc. Am. Math. Soc.* **13**, 590.

Ross, G., M. A. Fiddy and M. Nieto-Vesperinas, 1980, Inverse Scattering Problems, ed. H. P. Baltes, Springer-Verlag, Berlin, p. 15.

Roy, A. K. and C. L. Mehta, 1995, *Quant. Semiclass. Opt.* **7**, 877.

Royer, A., 1996, *Phys. Rev. A* **53**, 70.

Rubin, M. H., D. N. Klyshko, Y. H. Shih and A. V. Sergienko, 1994, *Phys. Rev. A* **50**, 5122.

Rudin, W., 1970, Real and Complex Analysis, McGraw-Hill, London.

Ryff, L. C. B., 1995, *Phys. Rev. A* **51**, 79.

Saks, S. and A. Zygmund, 1971, Analytical Functions, Elsevier Science, Amsterdam.

Saleh, B. E. A. and M. C. Teich, 1991, Fundamentals of Photonics, J. Wiley, New York.

Saletan, E. J. and A. H. Cromer, 1971, Theoretical Mechanics, J. Wiley, New York.

Sánchez-Ruiz, J., 1994, *J. Phys. A: Math. Gen.* **27**, L843.

Sánchez-Soto, L. L. and A. Luis, 1994, *Opt. Commun.* **105**, 84.

Sanders, B. C., 1992, *Phys. Rev. A* **45**, 7746.

Sanders, B. C., S. M. Barnett and P. L. Knight, 1986, *Opt. Commun.* **58**, 290.

Sanders, B. C. and G. J. Milburn, 1989, *Phys. Rev. A* **39**, 694.

Sanders, B. C. and G. J. Milburn, 1995, *Phys. Rev. Lett.* **75**, 2944.

Sanders, B. C., G. J. Milburn and Z. Zhang, 1997, *J. Mod. Opt.* **44**, 1309.

Santhanam, T. S., 1976, *Phys. Lett.* **56A**, 345.

Santhanam, T. S., 1977a, *Found. Phys.* **7**, 121.

Santhanam, T. S., 1977b, *Lett. Nuovo Cimento* **20**, 13.

Saxton, W. O., 1978, Computer Techniques for Image Processing in Electron Microscopy, Academic Press, New York.

Schaufler, S., M. Freyberger and W. P. Schleich, 1994, *J. Mod. Opt* **41**, 1765.

Schiller, S., G. Breitenbach, S. F. Pereira, T. Müller and J. Mlynek, 1996, *Phys. Rev. Lett.* **77**, 2933.

Schiller, S., S. F. Pereira, G. Breitenbach, T. Müller, A. G. White and J. Mlynek, 1996, Coherence and Quantum Optics VII, eds. J. H. Eberly, L. Mandel and E. Wolf, Plenum Press, New York, p. 475.

Schleich, W. P., 1988, Interference in Phase Space, Habilitation Thesis (Max-Planck-Institute für Quantenoptik, Garching).

Schleich, W., 1989, Squeezed and Nonclassical Light, eds. P. Tombesi and E. R. Pike, Plenum Press, New York, p. 129.

Schleich, W., A. Bandilla and H. Paul, 1992, *Phys. Rev. A* **45**, 6652.

Schleich, W. P., J. P. Dowling and R. J. Horowicz, 1991, *Phys. Rev. A* **44**, 3365.

Schleich, W. P., J. P. Dowling, R. J. Horowicz and S. Varro, 1990, New Frontiers in Quantum Electrodynamics and Quantum Optics, ed. A. O. Barut, Plenum Press, New York, p. 31.

Schleich, W., R. J. Horowicz and S. Varro, 1989a, Quantum Optics V, eds. D. F. Walls and J. Harvey, Springer-Verlag, Heidelberg, p. 133.

Schleich, W., R. J. Horowicz and S. Varro, 1989b, *Phys. Rev. A* **40**, 7405.

Schleich, W. P., M. Pernigo and Fam Le Kien, 1991, *Phys. Rev. A* **44**, 2172.

Schleich, W. P. and M. G. Raymer, eds., 1997, Journal of Modern Optics, Vol. **44**, No. 11/12, Taylor and Francis, London.

Schleich, W., D. F. Walls and J. A. Wheeler, 1988, *Phys. Rev. A* **38**, 1177.

Schleich, W. P., H. Walther and J. A. Wheeler, 1988, *Found. Physics* **18**, 953.

Schleich, W. P. and J. A. Wheeler, 1987a, *Nature* **326**, 574.

Schleich, W. P. and J. A. Wheeler, 1987b, *J. Opt. Soc. Am. B* **4**, 1715.

Schleich, W. P. and J. A. Wheeler, 1987c, Proceedings of First International Conference of Physics in Phase Space, ed. W. W. Zachary, Springer-Verlag, New York, p. 200.

Schnurr, C., T. Savard, L. J. Wang and J. E. Thomas, 1995, *Opt. Lett.* **20**, 413.

Schröder, M. R., 1990, Number Theory in Science and Communication, Springer-Verlag, Berlin.

Schrödinger, E., 1935, *Naturwissenschaften* **23**, 807.

Schubert, M. and B. Wilhelmi, 1986, Nonlinear Optics and Quantum Electronics, J. Wiley, New York.

Schumaker, B. L., 1984, *Opt. Lett.* **9**, 189.

Schumaker, B. L. and C. M. Caves, 1985, *Phys. Rev. A* **31**, 3093.

Schwinger, J., 1965, Quantum Theory of Angular Momenta, eds. L. C. Biederharn and H. van Dam, Academic Press, New York.

Scully, M. O. and L. Cohen, 1987, Physics of Phase Space, eds. Y. S. Kim, W. W. Zachary, Springer-Verlag, Berlin, p. 253.

Scully, M. O. and K. G. Whitney, 1972, Progress in Optics, Vol. 10, ed. E. Wolf, North-Holland, Amsterdam, p. 89.

Selvadoray, M. and M. S. Kumar, 1997, *Opt. Commun.* **136**, 125.

Selvadoray, M., M. S. Kumar and R. Simon, 1994, *Phys. Rev. A* **49**, 4957.

Sergienko, A. V., Y. H. Shih, T. B. Pittman and M. H. Rubin, 1996, Fourth International Conference on Squeezed States and Uncertainty Relations, eds. D. Han, K. Peng, Y. S. Kim, and V. I. Man'ko, NASA, Greenbelt, MD, p. 159.

Sergienko, A. V., Y. H. Shih and M. H. Rubin, 1993, *J. Mod. Opt.* **40**, 1425.

Sergienko, A. V., Y. H. Shih and M. H. Rubin, 1995, *J. Opt. Soc. Am. B* **12**, 859.

Seshadri, S., S. Lakshmibala and V. Balakrishnan, 1997, *Phys. Rev. A* **55**, 869.

Shanta, P., S. Chaturvedi, V. Srinivasan, G. S. Agarwal and C. L. Mehta, 1994, *Phys. Rev. Lett.* **72**, 1447.

Shapiro, J. H., 1980, *IEEE Trans. Inf. Theory* **26**, 490.

Shapiro, J. H., 1985, *IEEE J. Quant. Electron.* **QE-21**, 237.

Shapiro, J. H., 1992, Workshop on Squeezed States and Uncertainty Relations, eds. D. Han, Y. S. Kim, and W. W. Zachary, NASA, Washington, D. C., p. 107.

Shapiro, J. H., 1993, *Phys. Scr.* **T48**, 105.

Shapiro, J. H., 1995, *Opt. Lett.* **20**, 1059.

Shapiro, J. H., 1996, *J. Opt. Soc. Am. B* **13**, 751.

Shapiro, J. H. and S. R. Shepard, 1991, *Phys. Rev. A* **43**, 3795.

Shapiro, J. H., S. R. Shepard and N. C. Wong, 1989, *Phys. Rev. Lett.* **62**, 2377.

Shapiro, J. H., S. R. Shepard and N. C. Wong, 1990, Coherence and Quantum Optics VI, eds. J. H. Eberly, L. Mandel, and E. Wolf, Plenum Press, New York, p. 1071.

Shapiro, J. H. and S. S. Wagner, 1984, *IEEE J. Quant. Electron.* **QE-20**, 803.

Shapiro, J. H., H. P. Yuen and J. A. Machado Mata, 1979, *IEEE Trans. Inf. Theory* **25**, 179.

Shepard, S. R., 1995, Fundamental Problems in Quantum Theory, ed. D. M. Greenberger, A. Zeilinger, New York Academy of Sciences, New York, p. 812.

Sherman, B. and G. Kurizki, 1992, *Phys. Rev. A* **45**, R7674.

Sherman, B., G. Kurizki and A. Kadyshevitch, 1992, *Phys. Rev. Lett.* **69**, 1927.

Shih, Y. H. and M. H. Rubin, 1993, *Phys. Lett. A* **182**, 16.

Shih, Y. H. and A. V. Sergienko, 1994, *Phys. Lett. A* **186**, 29.

Shih, Y. H., A. V. Sergienko, T. B. Pittman, D. V. Strekalov and D. N. Klyshko, 1996, Fourth International Conference on Squeezed States and Uncertainty Relations, eds. D. Han, K. Peng, Y. S. Kim, and V. I. Man'ko, NASA, Greenbelt, MD, p. 169.

Shih, Y., A. V. Sergienko and M. H. Rubin, 1993a, *Phys. Rev. A* **47**, 1288.

Shih, Y., A. V. Sergienko and M. H. Rubin, 1993b, *Phys. Rev. A* **48**, 4001.

Shimizu, F., K. Shimizu and H. Takuma, 1992, *Phys. Rev. A* **46**, 17.

Shumovsky, A. S., 1997, *Opt. Commun.* **136**, 219.

Simon, R., H. J. Kimble and E. C. G. Sudarshan, 1988, *Phys. Rev. Lett.* **61**, 19.

Smith, T. B., D. A. Dubin and M. A. Hennings, 1992, *J. Mod. Opt.* **39**, 1603.

Smithey, D. T., M. Beck, J. Cooper and M. G. Raymer, 1993, *Phys. Rev. A* **48**, 3159.

Smithey, D. T., M. Beck, J. Cooper, M. G. Raymer and A. Faridani, 1993, *Phys. Scr.* **T48**, 35.

Smithey, D. T., M. Beck, M. G. Raymer and A. Faridani, 1993, *Phys. Rev. Lett.* **70**, 1244.

Solimeno, S., B. Crosignani and P. Di Porto, 1986, Guiding, Diffraction, and Confinement of Optical Radiation, Academic Press, Orlando, FLA.

Soroko, L. M., 1971, Fundamentals of Holography and Coherent Optics, Nauka, Moscow, in Russian.

Soroko, L. M., 1978, Holography and Coherent Optics, Plenum Press, New York.

Spence, J. C. H., 1974, *Opt. Acta* **21**, 835.

Srinivas, M. D. and E. B. Davies, 1981, *Opt. Acta* **28**, 981.

Srinivas, M. D. and E. B. Davies, 1982, *Opt. Acta* **29**, 235.

Steinberg, A. M., P. G. Kwiat and R. Y. Chiao, 1992a, *Phys. Rev. Lett.* **68**, 2421.

Steinberg, A. M., P. G. Kwiat and R. Y. Chiao, 1992b, *Phys. Rev. A* **45**, 6659.

Stenholm, S., 1992, *Ann. Phys.* (N. Y.) **218**, 233.

Stenholm, S., 1993, *Phys. Scr.* **T48**, 77.

Stoler, D., 1970, *Phys. Rev. D* **1**, 3217.

Stoler, D., 1971a, *Phys. Rev. D* **4**, 1925.

Stoler, D., 1971b, *Phys. Rev. D* **4**, 2309.

Stoler, D., 1974, *Phys. Rev. Lett.* **33**, 1397.

Stoler, D., 1975, *Phys. Rev.* **11**, 3033.

Stoler, D., B. E. A. Saleh and M. C. Teich, 1985, *Opt. Acta* **32**, 345.

Storey, P., M. Collett and D. Walls, 1992, *Phys. Rev. Lett.* **68**, 472.

Stratonovich, R. L., 1956, *Sov. Phys.-JETP* **31**, 1012.

Strekalov, D. V., T. B. Pittman, A. V. Sergienko, Y. H. Shih and P. G. Kwiat, 1996, *Phys. Rev. A* **54**, R1.

Strekalov, D. V., A. V. Sergienko, D. N. Klyshko and Y. H. Shih, 1995, *Phys. Rev. Lett.* **74**, 3600.

Sudarshan, E. C. G., 1963, *Phys. Rev. Lett.* **10**, 277.

Sukumar, C. V., 1989a, *J. Mod. Opt.* **36**, 1591.

Sukumar, C. V., 1989b, *Phys. Rev. A* **40**, 5426.

Sukumar, C. V., 1993, *Phys. Rev. A* **47**, 1554.

Summhammer, J., K. A. Hamacher, H. Kaiser, H. Weinfurter, D. L. Jacobson and S. A. Werner, 1995, *Phys. Rev. Lett.* **75**, 3206.

Summy, G. S. and D. T. Pegg, 1990, *Opt. Commun.* **77**, 75.

Sun, P. C. and H. C. Fu, 1989, *J. Phys. A* **22**, L983.

Sun, P. C., Y. Mazurenko and Y. Fainman, 1995, *Opt. Lett.* **20**, 1062.

Susskind, L. and J. Glogower, 1964, *Physics* **1**, 49.

Szabo, S., P. Adam, J. Janszky and P. Domokos, 1996, *Phys. Rev. A* **53**, 2698.

Szłachetka, P., S. Kielich, J. Peřina and V. Peřinová, 1979, *J. Phys. A* **12**, 1921.

Szłachetka, P., S. Kielich, J. Peřina and V. Peřinová, 1980, *Opt. Acta* **27**, 1609.

Szriftgiser, P., D. Guéry-Odelin, M. Arndt and J. Dalibard, 1996, *Phys. Rev. Lett.* **77**, 4.

Tabor, M., 1989, Chaos and Integrability in Nonlinear Dynamics, J. Wiley, New York.

Tan, S. M. and D. F. Walls, 1992, *Appl. Phys. B* **54**, 434.

Tan, S. M. and D. F. Walls, 1994, *Phys. Rev. A* **50**, 1561.

Tan, S. M. and D. F. Walls, 1995, *Opt. Commun.* **118**, 412.

Tanaś, R., 1991, *Opt. & Spektrosk.* **70**, 637.

Tanaś, R. and Ts. Gantsog, 1991, *J. Opt. Soc. Am. B* **8**, 2505.

Tanaś, R. and Ts. Gantsog, 1992a, *Quant. Opt.* **4**, 245.

Tanaś, R. and Ts. Gantsog, 1992b, *Phys. Rev. A* **45**, 5031.

Tanaś, R. and Ts. Gantsog, 1992c, *J. Mod. Opt.* **39**, 749.

Tanaś, R. and Ts. Gantsog, 1992d, *Opt. Commun.* **87**, 369.

Tanaś, R., Ts. Gantsog, A. Miranowicz and S. Kielich, 1991, *J. Opt. Soc. Am. B* **8**, 1576.

Tanaś, R., Ts. Gantsog and R. Zawodny, 1991a, *Quant. Opt.* **3**, 221.

Tanaś, R., Ts. Gantsog and R. Zawodny, 1991b, *Opt. Commun.* **83**, 278.

Tanaś, R. and S. Kielich, 1983, *Opt. Commun.* **45**, 351.

Tanaś, R. and S. Kielich, 1984, *Opt. Acta* **31**, 81.

Tanaś, R. and S. Kielich, 1990, *J. Mod. Opt.* **37**, 1935.

Tanaś, R., A. Miranowicz and Ts. Gantsog, 1993, *Phys. Scr.* **T48**, 53.

Tanaś, R., A. Miranowicz and Ts. Gantsog, 1996, Progress in Optics, Vol. 35, ed. E. Wolf, North-Holland, Amsterdam, p. 355.

Tang, Z., 1995, *Phys. Rev. A* **52**, 3448.

Tänzler, W. and F. J. Schütte, 1981, *Ann. Physik* **38**, 73.

Tapster, P. R., J. G. Rarity and P. C. M. Owens, 1994, *Phys. Rev. Lett.* **73**, 1923.

Taylor, B., K. J. Schernthanner, G. Lenz and P. Meystre, 1994, *Opt. Commun.* **110**, 569.

Teich, M. C., 1968, *Proc. IEEE* **56**, 37.

Teich, M. C. and B. E. A. Saleh, 1988, Progress in Optics, Vol. 26, ed. E. Wolf, North-Holland, Amsterdam, p. 1.

Teich, M. C. and B. E. A. Saleh, 1989, *Quant. Optics* **1**, 153.

Teich, M. C., B. E. A. Saleh and J. Peřina, 1984, *J. Opt. Soc. Am. B* **1**, 366.

Tewari, S. P. and P. Hariharan, 1997, *J. Mod. Opt.* **44**, 543.

Titulaer, U. M. and R. J. Glauber, 1966, *Phys. Rev.* **145**, 1041.

Tiwari, S. C., 1992, *J. Mod. Opt.* **39**, 1097.

Tiwari, S. C., 1997, *Phys. Rev. A* **56**, 157.

Toll, J., 1956, *Phys. Rev.* **104**, 1760.

Tombesi, P., 1997, Quantum Optics and the Spectroscopy of Solids, eds. T. Hakioğlu and A. S. Shumovsky, Kluwer, Dordrecht, p. 203.

Tombesi, P. and A. Mecozzi, 1987, *J. Opt. Soc. Am. B* **4**, 1700.

Tomita, A. and R. Y. Chiao, 1986, *Phys. Rev. Lett.* **57**, 937.

Torgerson, J. R. and L. Mandel, 1996, *Phys. Rev. Lett.* **76**, 3939.

Torgerson, J. R. and L. Mandel, 1997, *Opt. Commun.* **133**, 153.

Törmä, P., 1996, *J. Mod. Opt.* **43**, 2437.

Torres-Vega, Go., A. Zúñiga-Segundo and J. D. Moralez-Guzmán, 1996, *Phys. Rev. A* **53**, 3792.

Tsui, Y. K. and M. F. Reid, 1992, *Phys. Rev. A* **46**, 549.

Tu, H.-T. and C.-D. Gong, 1993, *J. Mod. Opt.* **40**, 57.

Turski, A. L., 1972, *Physica* **57**, 432.

Turunen, J., A. Vasara, A. T. Friberg, 1991, *J. Opt. Soc. Am. A* **8**, 282.

Ueda, M., 1989, *Quant. Opt.* **1**, 131.

Ueda, M., 1990, *Phys. Rev. A* **41**, 3875.

Ueda, M., N. Imoto, H. Nagaoka and T. Ogawa, 1992, *Phys. Rev. A* **46**, 2859.

Ueda, M., N. Imoto and T. Ogawa, 1990a, *Phys. Rev. A* **41**, 3891.

Ueda, M., N. Imoto and T. Ogawa, 1990b, *Phys. Rev. A* **41**, 6331.

Ueda, M. and M. Kitagawa, 1992, *Phys. Rev. Lett.* **68**, 3424.

Unruh, W. G., 1978, *Phys. Rev. D* **18**, 1764.

Unruh, W. G., 1979, *Phys. Rev. D* **19**, 2888.

Vaccaro, J. A., 1995, *Phys. Rev. A* **51**, 3309.

Vaccaro, J. and S. M. Barnett, 1995, *J. Mod. Opt.* **42**, 2165.

Vaccaro, J. A., S. M. Barnett and D. T. Pegg, 1992, *J. Mod. Opt.* **39**, 603.

Vaccaro, J. A. and Y. Ben-Aryeh, 1995, *Opt. Commun.* **113**, 427.

Vaccaro, J. A. and R. F. Bonner, 1995, *Phys. Lett. A* **198**, 167.

Vaccaro, J. A. and A. Orłowski, 1995, *Phys. Rev. A* **51**, 4172.

Vaccaro, J. A. and D. T. Pegg, 1989, *Opt. Commun.* **70**, 529.

Vaccaro, J. A. and D. T. Pegg, 1990a, *J. Mod. Opt.* **37**, 17.

Vaccaro, J. A. and D. T. Pegg, 1990b, *Phys. Rev. A* **41**, 5156.

Vaccaro, J. A. and D. T. Pegg, 1993, *Phys. Scr.* **T48**, 22.

Vaccaro, J. A. and D. T. Pegg, 1994a, *Opt. Commun.* **105**, 335.

Vaccaro, J. A. and D. T. Pegg, 1994b, *Phys. Rev. A* **49**, 4985.

Vaglica, A. and G. Vetri, 1984, *Opt. Commun.* **51**, 239.

Vanden Bergh, G. and H. DeMeyer, 1978, *J. Phys. A* **11**, 1569.

van Kampen, N. G., 1953, *Phys. Rev.* **89**, 1072.

van Kampen, N. G., 1981, Stochastic Processes in Physics and Chemistry, North-Holland, Amsterdam.

Varadarajan, V. S., 1985, Geometry of Quantum Theory, Springer-Verlag, Berlin.

Várilly, J. C. and J. M. Gracia-Bondía, 1989, *Ann. Phys.* (N.Y.) **190**, 107.

Vartiainen, E. M., K.-E. Peiponen, H. Kishida and T. Koda, 1996, *J. Opt. Soc. Am. B* **13**, 2106.

Vidiella-Barranco, A. and J. A. Roversi, 1994, *Phys. Rev. A* **50**, 5233.

Vilenkin, N. Ya., 1968, Special Functions and the Theory of Group Representations, American Mathematical Society, Providence, RI.

Voevodin, V. V. and E. E. Tyrtyshnikov, 1987, Numerical Procedures with the Toeplitz Matrices, Nauka, Moscow, in Russian.

Vogel, K., V. M. Akulin and W. P. Schleich, 1993, *Phys. Rev. Lett.* **71**, 1816.

Vogel, K. and H. Risken, 1989, *Phys. Rev. A* **40**, 2847.

Vogel, K. and W. Schleich, 1992, Fundamental Systems in Quantum Optics, eds. J. Dalibard, J. M. Raimond and J. Zinn-Justin, Elsevier Science, Amsterdam, p. 715.

Vogel, W. and J. Grabow, 1993, *Phys. Rev. A* **47**, 4227.

Vogel, W. and W. Schleich, 1991, *Phys. Rev. A* **44**, 7642.

Vogel, W. and D.-G. Welsch, 1994, Lectures on Quantum Optics, Akademie-Verlag, Berlin.

Vogel, W. and D.-G. Welsch, 1995, *Acta Phys. Slov.* **45**, 313.

Voitsekhovich, V. V., 1995, *J. Opt. Soc. Am. A* **12**, 2194.

von Mises, R., 1931, Wahrscheinlichkeitsrechnung und ihre Anwendung in der Statistik und theoretischen Physik, Deuticke, Leipzig.

von Neumann, J., 1932, Mathematische Grundlagen der Quantenmechanik, Springer-Verlag, Berlin (Engl. transl. 1955, Princeton University Press, Princeton, NJ).

Vourdas, A., 1990, *Phys. Rev. A* **41**, 1653.

Vourdas, A., 1992, *Opt. Commun.* **91**, 236.

Vourdas, A., 1993, *Phys. Scr.* **T48**, 84.

Vourdas, A. and R. F. Bishop, 1989, *Phys. Rev. A* **39**, 214.

Vourdas, A. and R. H. Wiener, 1987, *Phys. Rev. A* **36**, 5866.

Wagle A. G., V. C. Rakhecha, J. Summhammer, G. Badurek, H. Weinfurter, B. E. Allman, H. Kaiser, K. Hamacher, D. L. Jacobson and S. A. Werner, 1997, *Phys. Rev. Lett.* **78**, 755.

Walker, N. G., 1987, *J. Mod. Opt.* **34**, 15.

Walker, N. G. and J. E. Carroll, 1984, *Electron. Lett.* **20**, 981.

Walker, N. G. and J. E. Carroll, 1986, *Opt. Quant. Electron.* **18**, 355.

Wallentowitz, S. and W. Vogel, 1996, *Phys. Rev. A* **53**, 4528.

Walls, D. F., 1970, *Z. Phys.* **237**, 224.

Walls, D. F., 1983, *Nature* **306**, 141.

Walls, D. F. and R. Barakat, 1970, *Phys. Rev. A* **1**, 446.

Walls, D. F. and G. J. Milburn, 1985, *Phys. Rev. A* **31**, 2403.

Walls, D. F. and G. J. Milburn, 1994, Quantum Optics, Springer-Verlag, Berlin.

Walther, A., 1963, *Opt. Acta* **10**, 41.

Walther, H., 1993, Physics and Probability Essays in Honor of Edwin T. Jaynes, eds. W. T. Grandy Jr. and P. W. Milonni, Cambridge University Press, Cambridge.

Wang, L. J., X. Y. Zou and L. Mandel, 1991a, *Phys. Rev. A* **44**, 4614.

Wang, L. J., X. Y. Zou and L. Mandel, 1991b, *J. Opt. Soc. Am. B* **8**, 978.

Watanabe, K. and Y. Yamamoto, 1988, *Phys. Rev. A* **38**, 3556.

Wedberg, T. C. and J. J. Stamnes, 1995, *Pure Appl. Opt.* **4**, 39.

Weitz, M., T. Heupel and T. W. Hänsch, 1996, *Phys. Rev. Lett.* **77**, 2356.

Weitz, M., T. Heupel and T. W. Hänsch, 1997, *Europhys. Lett.* **37**, 517.

Weitz, M., B. C. Young and S. Chu, 1994, *Phys. Rev. Lett.* **73**, 2563.

Weyl, H., 1927, *Z. Phys.* **46**, 1.

Weyl, H., 1932, Theory of Groups and Quantum Mechanics, E. P. Dutton Co., New York (reprinted 1950, Dover, New York).

Wigner, E. P., 1931, Gruppentheorie und ihre Anwendung auf die Quantenmechanik der Atomspektren, F. Vieweg und Sohn, Braunschweig.

Wigner, E. P., 1932, *Phys. Rev.* **40**, 749.

Wigner, E. P., 1959, Group Theory and Its Applications to the Quantum Mechanics of Atomic Spectra, Academic Press, New York.

Wiseman, H. M. and G. J. Milburn, 1993, *Phys. Rev. A* **47**, 642.

Wódkiewicz, K., 1984, *Phys. Rev. Lett.* **52**, 1064.

Wódkiewicz, K., 1986, *Phys. Lett. A* **115**, 304.

Wódkiewicz, K., 1988, *Phys. Lett. A* **129**, 1.

Wódkiewicz, K. and J. H. Eberly, 1985, *J. Opt. Soc. Am. B* **2**, 458.

Wolf, E., 1962, *Proc. Phys. Soc.* **80**, 1269.

Wolf, E., 1970, *J. Opt. Soc. Am.* **60**, 18.

Wolf, E., 1996, Trends in Optics, ed. A. Consortini, Academic Press, San Diego, CA, p. 83.

Wong, T., M. J. Collett and D. F. Walls, 1996, *Phys. Rev. A* **54**, R3718.

Wong, T., S. M. Tan, M. J. Collett and D. F. Walls, 1997, *Phys. Rev. A* **55**, 1288.

Wootters, W. K., 1987, *Ann. Phys.* (N. Y.) **176**, 1.

Wright, E. M., T. Wong, M. J. Collett, S. M. Tan and D. F. Walls, 1997, *Phys. Rev. A* **56**, 591.

Wünsche, A., 1996a, *Phys. Rev. A* **54**, 5291.

Wünsche, A., 1996b, *Quant. Semiclass. Opt.* **8**, 343.

Wünsche, A. and V. Bužek, 1997, *Quant. Semiclass. Opt.* **9**, 631.

Xin, Z. Z., Q. Zhao, M. Hirayama and K. Matumoto, 1994, *Phys. Rev. A* **50**, 4419.

Yamamoto, Y., N. Imoto and S. Machida, 1986, *Phys. Rev. A* **33**, 3243.

Yao, D.-M., 1987, *Phys. Lett. A* **122**, 77.

Yu, Z., 1993, *Phys. Lett. A* **175**, 391.

Yu, T., 1996, *Phys. Lett. A* **223**, 9.

Yu, S., 1997, *Phys. Rev. Lett.* **79**, 780.

Yuen, H. P., 1976, *Phys. Rev. A* **13**, 2226.

Yuen, H. P., 1986, *Phys. Rev. Lett.* **56**, 2176.

Yuen, H. P. and W. V. S. Chan, 1983, *Opt. Lett.* **8**, 177.

Yuen, H. P. and J. H. Shapiro, 1978, *IEEE Trans. Inf. Theory* **24**, 657.

Yuen, H. P. and J. H. Shapiro, 1980, *IEEE Trans. Inf. Theory* **26**, 78.

Yurke, B., 1986, *Phys. Rev. Lett.* **56**, 1515.

Yurke, B., S. L. McCall and J. R. Klauder, 1986, *Phys. Rev. A* **33**, 4033.

Yurke, B. and D. Stoler, 1986, *Phys. Rev. Lett.* **57**, 13.

Zaheer, K. and M. S. Zubairy, 1991, Advances in Atomic, Molecular, and Optical Physics, Vol. 28, eds. D. Bates and B. Bederson, Academic Press, New York, p. 143.

Zeilinger, A., T. Herzog, M. A. Horne, P. G. Kwiat, K. Mattle and H. Weinfurter, 1996, Coherence and Quantum Optics VII, eds. J. H. Eberly, L. Mandel and E. Wolf, Plenum Press, New York, p. 305.

Zeilinger, A., M. A. Horne, H. Weinfurter and M. Zukowski, 1997, *Phys. Rev. Lett.* **78**, 3031.

Zeng, J. Y. and Y. A. Lei, 1995, *Phys. Rev. A* **51**, 4415.

Zhang, W. M., D. H. Feng and R. Gilmore, 1990, *Rev. Mod. Phys.* **62**, 867.

Zheng, S.-B. and G.-C. Guo, 1997a, *Opt. Commun.* **137**, 308.

Zheng, S.-B. and G.-C. Guo, 1997b, *J. Mod. Opt.* **44**, 963.

Zhu, K., H. Tang and D. Huang, 1996, *Opt. Commun.* **124**, 266.

Zou, X. Y., T. Grayson, G. A. Barbosa and L. Mandel, 1993, *Phys. Rev. A* **47**, 2293.

Zou, X. Y., T. P. Grayson and L. Mandel, 1992, *Phys. Rev. Lett.* **69**, 3041.

Zou, X. Y., L. J. Wang and L. Mandel, 1991, *Phys. Rev. Lett.* **67**, 318.

Zucchetti, A., W. Vogel, M. Tasche and D.-G. Welsch, 1996, *Phys. Rev. A* **54**, 1678.

Zucchetti, A., W. Vogel and D.-G. Welsch, 1996, *Phys. Rev. A* **54**, 856.

Zukowski, M., A. Zeilinger and M. A. Horne, 1997, *Phys. Rev. A* **55**, 2564.

Index